Modern Electrical Communications

Theory And Systems

MODERN ELECTRICAL COMMUNICATIONS

Theory And Systems

Henry Stark, *D. Eng. Sc.*

Associate Professor
of Electrical and Systems Engineering
Rensselaer Polytechnic Institute

Franz B. Tuteur, *Ph. D.*

Professor of
Engineering and Applied Science
Yale University

PRENTICE-HALL, INC., ENGLEWOOD CLIFFS, NEW JERSEY 07632

Library of Congress Cataloging in Publication Data

STARK, HENRY, (date)
    Modern electrical communications.

    Includes bibliographical references and index.
    1. Telecommunication.   I. Tuteur, Franz B.,
joint author.   II. Title.
TK5101.S67   621.38   78-8635
ISBN   0-13-593202-5

© 1979 by Prentice-Hall, Inc.
Englewood Cliffs, N.J.   07632

Printed in the United States of America

10 9 8 7 6 5 4 3 2 1

PRENTICE-HALL INTERNATIONAL, INC., *London*
PRENTICE-HALL OF AUSTRALIA PTY. LIMITED, *Sydney*
PRENTICE-HALL OF CANADA, LTD., *Toronto*
PRENTICE-HALL OF INDIA PRIVATE LIMITED, *New Delhi*
PRENTICE-HALL OF JAPAN, INC., *Tokyo*
PRENTICE-HALL OF SOUTHEAST ASIA PTE. LTD., *Singapore*
WHITEHALL BOOKS LIMITED, *Wellington, New Zealand*

To Anna Stark

Contents

Preface

This book grew out of a set of notes prepared for a two-term senior level course in electrical communication systems and a one-term graduate course in statistical communications and signal processing. In the five years or so that we have taught the undergraduate communication sequence at Yale, it has been our common experience that because of the simultaneous proliferation of newer disciplines and student interests, we could not count on the students' having had the traditional background courses required to tackle a fairly mature course devoted exclusively to communication engineering. For this reason we had to develop a course that was relatively complete and self-contained.

We had to develop fundamental topics such as Fourier methods and linear-systems theory, but we also felt that the students deserved to be informed about advanced topics, such as statistical communication theory, signal processing, television, radar, and sonar, in as much depth as possible. We felt that although theory had to be properly covered, it was necessary to show application of the theory in at least a few concrete modern devices. In short we attempted to make our course a broad and integrated study of the entire field of electrical communication engineering.

We believe that our experience with our own students is in no sense unique. The field of electrical engineering has changed so vastly in the last few decades and there are so many topics that our students must know today, that what used to be the "traditional" electrical engineering curriculum has ceased to exist in many places. Even schools with more traditional electrical engineering curricula have developed a greater range of options and have given their students additional freedom in crossing the boundaries between various disciplines. Faced with this greater loosening of the traditional course-of-study hierarchy in electrical engineering, we felt that a text that presented the essential background to communications theory in a few brief introductory chapters and then pushed on to more advanced topics would fill a general need. This is why we have written this book.

In teaching first-year graduate students we found that, if anything, the variations in students' preparedness, sophistication, and interests are even greater than they are for undergraduates. Some graduate students had not even been electrical engineering majors but came from such allied fields as computer science or physics. Others came from undergraduate schools that offered relatively unstructured programs that often left gaps in the students' appreciation of fundamental notions in electrical engineering. In many cases, the students were inadequately prepared in linear systems, random processes, and basic communication systems. In teaching students with such varied backgrounds, we found it convenient to have available a book that not only discussed signal processing and the central ideas in statistical communication theory, but also contained the necessary background material to enable the student to "catch up" on his or her own time.

In line with the experience that we have had in our course, we have tried to keep the prerequisite knowledge required from our readers to a minimum; we expect basically college-level calculus and a first course in linear-system theory that includes some exposure to the Laplace transform. Beyond this base level we have tried to make our explanations and derivations as complete and as self-contained as possible. Much use is made of mathematical arguments, and we have included very few results, curves, or tables that the reader could not generate by himself on the basis of the information furnished in the text. In some sections we have used elementary concepts of complex-variable theory, but there is little loss in either continuity or substance for the reader not familiar with this theory.

As might have been expected, the book contains more material than we normally covered in our undergraduate course. For instance in the introductory chapter that deals with Fourier methods, we included material on convergence, vector representation of signals, and concepts such as inner product and projection. We did this partly for completeness and partly because these concepts were needed in later chapters. The book contains a survey of linear systems and active filters going considerably beyond the treatment that we were able to give in our course. We also felt that with so much signal processing being done digitally today we would be amiss if we did not discuss digital systems and filters in some depth. We therefore included a chapter on digital filters that contains a discussion of the discrete Fourier transform (DFT) and the fast Fourier transform (FFT). Digital methods were also integrated into several of the other chapters, notably the one on signal processing.

If the book is used in an undergraduate course, the first half, that is Chapters 1-7, could form the basis for a relatively demanding 16-week semester in which classes normally meet for three class hours per week. In contrast to what seems to be the prevailing practice we have separated probabilistic concepts from modulation, filtering, and detection practice. Although this means that comparison of various modulation schemes based on signal-to-noise analysis is delayed to the second half of the book, we found that our approach had the practical advantage of permitting us to get through all of AM, FM, and pulse modulation in one term.

If all of Chapters 1 through 7 prove too taxing for a single semester's study, the more advanced sections of Chapter 2, most of the discussion of active filters and frequency transformations of Chapter 3, Chapter 5 (on digital systems) and the section on FM interference in Chapter 7 can be omitted without serious loss of continuity.

Chapters 8–13 normally comprise the second half of the course sequence. This part makes greater demands on the mathematical background of the students than does the first part, and it may be difficult to cover everything if this background is missing. If some of these chapters are to be omitted, we suggest the following alternatives: For students with no previous course in probability theory a careful study of Chapters 8, 9, and 10, followed by portions of 11 and either 12 or 13; for students with a good background in probability theory, we recommend omitting Chapters 8 and 9 and proceeding directly to Chapter 10 followed by 11, 12, and 13.

The book can be used for a single-semester undergraduate course to be given to students with a background in Fourier methods, linear systems, and some probability theory. Such a course would be based on Chapters 1, 4, 6, 7, and portions of Chapters 5, 10, and 11.

The book can also be used as a basis for a single-semester first year graduate course in which the student is held responsible for knowing the fundamentals of Fourier theory, linear systems, and probability theory. The course would then be based on a careful study of Chapters 10, 11, 12, and supplementary material on coding and information theory. Chapter 13 is an option for students interested in two-dimensional systems and image processing.

The authors are extremely grateful to Dr. Reed Even who critically read the entire manuscript and made numerous excellent suggestions. Thanks are due to Helen Brady for doing the typing of most of the first draft, to Dan Tuteur who helped with two of the chapters, and to Joy Breslauer and Michelle Gall who exhibited patience and skill in typing the manuscript to completion. Thanks are due also to the administrations of the Department of Engineering and Applied Science at Yale and the School of Engineering at Rensselaer Polytechnic Institute who permitted us to take some of the large amount of time needed to complete this project from our normal duties.

Finally thanks are due to our wives Alice and Ruth without whose unfailing patience and encouragement the work could not have been completed.

Henry Stark / Franz B. Tuteur

Troy, N. Y.   *New Haven, CT.*

Modern Electrical Communications

Theory And Systems

Introduction

1.1 Definitions

There is a humorous expression that "pressing a suit" does not mean the same thing to a lawyer that it does to a tailor [1-1]. It's somewhat like that with the word *communication*. *Webster's New Collegiate Dictionary* offers, among others, these definitions: " . . . a process by which information is exchanged between individuals through a common system of symbols. . . ," and also ". . . a technique for expressing ideas effectively. . ." (one thinks of a public-speaking improvement course), and finally ". . . a system of routes for moving troops, supplies, and vehicles."

From our point of view, the first definition is obviously closest to the subject matter of this book. The word *electrical* in electrical communication means that our concern is less with original symbols such as written or spoken speech than with the transmission and reception of the electrical symbols into which the original symbols have been translated. Not all electrical communication is concerned with communication between persons; in the modern world, computers also talk to each other.

1.2 Problems in Communication Theory

In a classic work by Claude Shannon and Warren Weaver [1-2] it is argued that problems in communication theory fall into one of three levels. Calling these levels A, B, and C, the authors describe them as follows:

Level A. With what precision can the symbols in communication be transmitted?

Level B. How precisely do the transmitted symbols convey the desired meaning?

Level C. How effectively does the received meaning affect conduct in the desired way?

Problems in level A are basically technical and of primary concern to electrical and communication engineers. Problems in level B are semantic in nature and are related to how well the meaning of the symbols is interpreted at the receiver. Linguists, and others concerned with the structure and theory of language, are on comfortable ground here. Effectiveness problems (level C) are perhaps harder to define as they involve esthetic and psychological factors. The following question, however, arises: Does level A encompass the most superficial problem in communication? Shannon and Weaver write that this is decidedly not so. Initially one might be led to feel that levels B and C, dealing with the philosophical foundations of communication and behavior, are somehow "deeper" than level A whose primary concern is, ostensibly, good design. Shannon and Weaver, however, argue that any limitations discovered in level A must, assuredly, affect levels B and C. More basically still, the mathematical theory of communication discloses that level A influences the other levels far more fundamentally than one would initially suspect.

We shall not pursue these points further. They were made to suggest that the technological aspects of electrical communication transcend engineering considerations of economy, efficiency, and accuracy (all-important as these are in the real world). Indeed, fundamental results in communication technology have a profound influence on the broader aspects of communication. These thought-provoking ideas are discussed in Reference [1-2], and the reader is urged to consult this milestone work.

1.3   A General Communication System

Figure 1.3-1 is a symbolic representation of an electrical communication system. As it stands, it is a more general representation than is required for electrical communications. For example, if used to model a speaker-listener "system," the source might be the speaker's brain, the message is the thought to be expressed, the transmitter is the vocal system that encodes the message in a form suitable for transmission through air, the carrier signal is the variations in sound pressure, the channel is the air, the receiver is the listener's ear (and associated nerves), and the destination is the listener's brain. However, if Fig. 1.3-1 is used to model an electrical communication system, the message $s(t)$ will be taken to be an electrical signal, it being assumed that the original message, i.e., speech, written text, pictures, etc., has already been converted into $s(t)$. We shall take it for granted that in

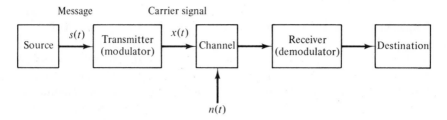

Figure 1.3-1. Block diagram of a communication system.

electrical communications means exist to convert the original message into an electrical signal. In addition to the term *message*, $s(t)$ is also sometimes called intelligence, baseband signal, audio signal, envelope, information,† etc. Examples of message signals are the signals out of a microphone, the induced voltage in a magnetic phonograph cartridge, the induced current in the winding of a tape recording head, the binary voltage levels in a computer, the video current out of a TV camera, and the current in a photo-conductor exposed to light. Obviously there are many more.

The unprocessed messages are generally not suited for transmission through the channel. This stems from the fact that $s(t)$ is generally a low-frequency waveform which implies long wavelengths. The basic laws of elec-tromagnetic radiation require that for efficient radiation the radiating element, i.e., the antenna, be a significant fraction of the wavelength of the signal. A 1000-Hz‡ signal has a wavelength of 3×10^5 meters in air; a quarter-wave antenna therefore would be 75 kilometers long! This is much too impractical for ordinary broadcasting.§ On the other hand, a 1.0-MHz‖ signal has a wavelength of 300 meters for which a quarter-wave antenna would be only 75 meters long.

In addition to there being, so to speak, a mismatch between the medium and the message, there are other problems with the transmission of low-frequency signal components as well. In radar and microwave communica-tion, it is frequently necessary to generate a narrow beam in order for the power intensity at the receiver to be sufficiently great to overcome electrical noise. However, waves undergo the phenomenon of *diffraction*, which results in a spreading of the beam. The diffraction angle is proportional to λ/D, where λ is the wavelength and D is the diameter of the radiator. The spread-

---

†As we shall see in later chapters, some of these terms have rather special technical mean-ing and should be used only in the proper context.

‡1 cycle per second = 1 hertz, abbreviated Hz. The unit is named in honor of the great experimentalist Heinrich Hertz (1857–1894), who verified Maxwell's theory.

§Although very large antennas may be in use for low-frequency communications in the military.

‖1 MHz = 1 megahertz = 10^6 Hz. A gigahertz is 10^9 Hz and is written 1 GHz. A kilohertz is 10^3 Hz and is written 1 kHz.

ing therefore is seen to be *inversely* proportional to frequency, and at low frequencies the beam intensity may drop off too sharply for satisfactory communication.

There are several other important reasons why $s(t)$ must be processed before transmission. We shall mention only one more here. Consider ordinary commercial radio broadcasting. This is perhaps the electrical communication system that we know best. Most programs consist of speech or music in which the signals cover a band from tens of hertz to perhaps 15 kHz. How could we ever have more than one listening choice in one locality if we could not separate the signals? By modulating the various messages $\{s(t)\}$ onto different carriers, overlap and interference between them are avoided and the listener is given a large selection of programs to choose from. The same holds true for TV and telephony. In the latter, it is desired to send many messages between two specific points on the same communication link. To prevent interference, the signals are *multiplexed. Frequency multiplexing* separates the various messages by modulating them onto different carriers and assigning them nonoverlapping frequency bands. In point of fact, the telephone company sends hundreds and even thousands of signals simultaneously over some of its links.

Henceforth we shall refer to the preprocessing of $s(t)$ for transmission simply as *modulation*. A precise meaning of this term will be given in Chapter 6. A significant portion of this book (Chapters 4, 5, 6, and 7 and portions of Chapter 11) is devoted to modulation, its effect on the signal, and its ability to overcome noise. In the next section we shall furnish a brief survey of some common modulation schemes.

Returning now to Fig. 1.3-1, we designate the medium between transmitter and receiver as the channel. In most electrical communication systems the channel is characterized by a long transmission path. This means that signals become attenuated and, therefore, that contamination by random noise and interference is more likely. The longest channels are found in deep-space probes. In the Mars/Viking mission, the channel consists of 225 million miles of space, and it takes the signal roughly 18 minutes to reach earth. When possible, relay stations are added to break up the overall channel length into more manageable sections. In microwave systems, relay stations are placed every 25–30 miles and reamplify the microwave beams. The use of relays in long-range communication channels is widespread, and even the Viking system uses a relay station called the Orbiter.

As already stated, the carrier frequency must be matched to the channel medium. Equivalently, for a fixed carrier, the channel medium must be matched to the carrier. Ordinary copper wire is adequate for transmitting low-frequency signals but introduces excessive attenuation and radiation losses for signals in the microwave region. Parallel-wire transmission lines are better when the wavelength is less than a kilometer, but even here shielding and radiation losses become serious problems at high frequencies. Coaxial cables are still better because the fields are practically perfectly shielded inside the line and confined between the inner and outer conductors. For wavelengths shorter than a meter, metallic waveguides are the preferred medium for guiding electromagnetic waves. At optical frequencies where wavelengths are measured in microns (1 millionth of a meter) dielectric waveguides called fibers are useful for guiding energy. A comparison of the losses associated with various wires, cables, and waveguides is furnished in [1-3], p. 615. A useful table of preferred transmission media versus frequency is furnished in [1-4], p. 10. The underlying physics which governs the guiding and transmission of electrical energy is covered in many places, e.g., [1-5].

Returning once more to Fig. 1.3-1, we have somewhat unfairly attributed all forms of noise to the channel. Actually noise is introduced at the receiver and transmitter also. The term noise is used very broadly here and includes interference from other stations, nonlinear effects in the system itself, effects due to signal fading, multipath propagation, other problems, and finally random "natural" noise. The last reflects the two facts that we live in a warm universe (thermal noise) and that electromagnetic radiation as well as electrons are quantized and exhibit particle-type behavior (shot noise). Shot noise is less a problem for the more common communication systems than thermal noise. Although interference and nonlinear distortions can conceivably, at impractical costs, be reduced below any predetermined threshold, such is not the case with natural noise. The latter represents a "hard" constraint put in our way by (an inflexible) nature. Because thermal noise is pervasive in electrical communication systems, we shall refer to it simply as *noise* and use more specific terms to describe other distorting effects.

The ultimate quality of a communication channel is measured by its capacity C, in bits per second. The term bit here has a rather special meaning and is defined as the information gained when the outcome of a binary experiment (e.g., an experiment involving a yes-no answer) with equiprobable outcomes is disclosed. A very basic theorem,† first proved by Shannon [1-2], says that given a channel with capacity C and a source with information rate $R < C$ (C, R both in bits per second), then, by appropriate coding, the output of the source can be transmitted through the channel with *arbitrarily small probability of error even though random noise is present.* Another important result, also attributed to Shannon, says that the capacity of the bandlimited *Gaussian* channel is given by

$$C = W \log_2 \left(1 + \frac{P_x}{P_n}\right),$$

where W is the channel bandwidth, P_x is the average power in the transmitted signal, and P_n is the average power in the noise. The latter is "white" thermal

†This theorem is frequently called the *fundamental theorem of information.*

noise with uniform power density over the band W and fluctuations subject to the Gaussian probability law.

The study of Shannon's fundamental theorem, associated source coding schemes, and the determination of C for different channels is a branch of communication theory called *information theory*. Although, with rare exception, information theory has had little influence on the design of electrical communication schemes, it has aroused great interest among such diverse groups as electrical engineers, social scientists, and linguists. It is not uncommon in some professional circles to lump all electrical communication theory dealing with noise and other random processes under the heading of information theory.

Returning one final time to Fig. 1.3-1, we see that the attempted recovery of the message $s(t)$ is done with the receiver. We add the term *attempted* because the received carrier is invariably contaminated to some degree by noise. For high ratios of signal power to noise power, a very good replica of $s(t)$ can be recovered. The receiver strips away $s(t)$ from the carrier signal and basically does the inverse operation of the transmitter. If the transmitter is fundamentally a modulator, the receiver is fundamentally a demodulator. *Optimum* receivers try to recover the signal while minimizing noise effects. "Optimum" implies a criterion of sorts, and there is no single optimum receiver for all criteria. Chapter 11 is devoted to optimal signal-processing schemes and associated receivers.

Last, the demodulated message is absorbed at the destination. The latter is frequently a person but could be a machine such as the Viking lander on Mars or a digital computer.

1.4 Modulation

We saw in Sec. 1.3 that modulation holds a key position in electrical communication systems. There are many modulation schemes, and they can be grouped, somewhat broadly, into the following categories: continuous-wave (CW), pulse, analog, and digital.

CW Modulation

In CW modulation, the carrier is most frequently a fixed-frequency sinusoid upon which the signal $s(t)$ is impressed. Common examples are double-sideband modulation (DSB), amplitude modulation (AM), single-sideband modulation (SSB), vestigial sideband modulation (VSB), frequency modulation (FM), and phase modulation (PM). In DSB,† the signal is direct-

---

†Double-sideband modulation is also abbreviated DSBSC, meaning "double-sideband suppressed carrier." We shall use both DSB and DSBSC. Still another term in use is AM-SC, meaning "AM with suppressed carrier."

ly impressed on the carrier amplitude. Conceptually the simplest type of modulation, DSB is not so easy to demodulate. AM is very similar to DSB except that the signal also contains a strong unmodulated carrier. The result is a wave that is extremely easy to demodulate—a fact that accounts, in part, for the widespread use of AM. SSB and VSB are bandwidth-conserving schemes that are generally not so easy to demodulate as AM. Basically, in SSB and VSB, only a portion of the modulated signal band is transmitted. The redundant part is omitted. The result is that more channels can be allocated in the same spectral band. VSB is widely used in TV systems. In all these schemes, the information is basically stored in the low-frequency amplitude fluctuations of the carrier envelope. FM and PM are examples of _angle_ modulation. In FM the instantaneous frequency of the carrier is proportional to the signal. Unlike AM, DSB, VSB, and SSB, there is no information in the carrier envelope; the information is stored in the zero crossings of the carrier wave. PM is very much like FM except that it is the _instantaneous phase_ that is proportional to the signal. FM is widely used in high-quality broadcasting and also is the method by which the TV sound signal is modulated.

Pulse Modulation

Pulse modulation differs from CW modulation in that the carrier signal exists only at certain intervals of time. The carrier in this case consists most often of a periodic sequence of pulses that are on only during a portion of the cycle. The signal $s(t)$ is used to vary some parameter of the pulse such as its amplitude (which leads to pulse amplitude modulation or PAM for short), its duration (pulse duration modulation—PDM), or its displacement from a reference point on the time scale (pulse position modulation—PPM). In PAM, which is the pulse analog of AM, the signal is impressed on the _amplitude_ of the pulse. In PDM, the duration of the pulse is made proportional to the signal amplitude during the instant when the pulse is on. In PPM, the _delay_ of the pulse is proportional to the signal amplitude. PPM is easily derived from PDM by differentiating and clipping. Unlike PAM, there is no information stored in the amplitude variations of the pulses in PDM and PPM.

The theoretical foundation for pulse modulation is the famous _sampling theorem_, to be discussed in Chapter 4. It might have occurred to the reader that a carrier consisting only of pulses that are not on at all times cannot transmit the message $s(t)$ continuously. Are we then throwing away valuable information? The answer, surprisingly, is _no_ if the message is band limited and the pulse-repetition rate is high enough. The sampling theorem is not only important in pulse modulation; it also is invaluable for explaining numerous phenomena in communication theory and forms the basis for the discrete Fourier transform (DFT), discussed in Chapter 5.

Analog Modulation

The CW pulse modulation schemes described above all share the property that some parameter of the carrier wave is *proportional* to the instantaneous values of the message and that this parameter varies *continuously* (at least within a predetermined range). Such is not the case in digital modulation. We shall consider an example below.

Digital Modulation—PCM

PCM, which stands for pulse code modulation, starts out essentially as PAM with an important difference: The amplitudes of the pulses are altered so that, in the end, each of them acquires only one of N preselected levels. Here information is really thrown away. However, because even the finest receivers, including the human ear, are not sensitive to fine variations below some threshold, the throwing away of information need not be bothersome in a practical sense. The preselected levels are assigned numerical values, and these, properly coded, are transmitted over the channel. Thus to generate PCM, three key steps are required: sampling, quantizing, and coding. Sampling is done as in PAM by using a periodic sequence of pulses for a carrier. Quantizing, however, generates a finite set of values from a possible infinite set of values and corresponds mathematically to mapping from an infinite-dimensional space [the set of all values $s(t)$ can take on] to a finite-dimensional space (the set of all quantized levels). The final step in PCM is to assign to each level a code word, preferably made up of simple binary symbols. For example, a quantized level of 7 volts might be coded into the code word 0111, which uses only the two symbols 0 and 1. Because *two* symbols are in use, this code is called binary. The conversion of these symbols to signals for transmission can be done in several ways. In amplitude-shift keying (ASK), a constant-frequency carrier is switched between two levels, say 0 volts and V volts. The 0-volt level might signify a 0 in which case the V-volt level would signify a 1. Frequency-shift keying (FSK) uses two frequencies to denote the 0, 1 symbols, while phase-shift keying (PSK) switches a constant-frequency carrier between two phases, say 0 radians to signify a 0 and π radians to signify a 1. In Chapter 4, these schemes will be discussed in greater detail.

1.5 Historical Review

One might begin a historical review of the development of communication with Volta's discovery of the voltaic battery. Volta's discovery was more than a useful device: It laid to rest Galvani's theory of "animal electricity" which proposed that the convulsions observed in severed frogs' legs were due to a mysterious vital force hidden in the tissues of the leg. In 1801, when

Volta demonstrated his battery to an important audience which included Napoleon, he took great care to make his battery appear like an electric eel, probably to mock Galvani's theory of animal electricity.

At about the same time, Michael Faraday was an errand boy in a bookshop near London. Faraday's apprenticeship in the bookshop led to his becoming a bookbinder, a job he apparently despised. Faraday's interest in science ultimately led to a job with Sir Humphrey Davy. On August 29, 1831, Faraday produced an induced current by moving a magnet near a conductor. Conceivably one of the greatest discoveries of all time, electromagnetic induction did not strike Faraday the same way (at least immediately). He wrote to a friend, ". . . It may be a weed instead of a fish that, after all my labor, I may have at last pulled up."

In 1834, Gauss and Weber designed one of the first telegraphs to be operated over any significant distance. Gauss, of course, is better known as one of the finest mathematicians who ever lived. In addition to mathematics and telegraphy, Gauss made fundamental contributions in astronomy, electromagnetic theory, and actuarial science. The Gauss-Weber telegraph receiver used a free-swinging magnetic needle inside the coil carrying the signal current. The direction in which the magnet swung depended on the direction of the current in the coil. The code that Gauss-Weber used is shown in Table 1.5-1.

TABLE 1.5-1   TELEGRAPH CODE OF GAUSS AND WEBER

$l = left$ $r = right$

A	*r*	M	*lrl*	0	*rlrl*
B	*ll*	N	*rll*	1	*rllr*
C, K	*rrr*	O	*rl*	2	*lrrl*
D	*rrl*	P	*rrrr*	3	*lrlr*
E	*l*	R	*rrrl*	4	*llrr*
F, V	*rlr*	S	*rrlr*	5	*lllr*
G	*lrr*	T	*rlrr*	6	*llrl*
H	*lll*	U	*lr*	7	*lrll*
I, J	*rr*	W	*lrrr*	8	*rlll*
L	*llr*	Z	*rrll*	9	*llll*

SOURCE: W. R. Bennett and J. R. Davey, *Data Transmission*, McGraw-Hill Book Company, New York, 1965. Used with permission of McGraw-Hill Book Company.

The telegraph system of Cooke and Wheatstone, first demonstrated in 1837 and enthusiastically acclaimed in 1845 when it was used to capture a murderer who was subsequently hanged, led to the formation of the English Electric Telegraph Company in 1846. By 1852, the company had laid 4000 miles of telegraph lines in England.

In the United States, it was left to Samuel Morse, assisted by some very able people, to devise a revolutionary new telegraph system using the now-famous "dot-dash" Morse code. With Congress assisting via a $30,000 appropriation, the first telegraph line went into operation on May 24, 1844. It linked Washington to Baltimore, a distance of about 40 miles.

The first transatlantic telegraph cable was the result of a partnership between Cyrus Field, John Brett, and Charles Bright. Field, an American, and his two English partners faced incredible hardships in their attempts at laying a transatlantic cable. Finally on July 27, 1866, a successful and permanent telegraph link between Europe and the United States was established.

Although the first practical telephone system is attributed to Alexander Graham Bell, a German schoolteacher by the name of Philipp Reis is actually credited with developing the first telephone (1860). In 1877, Bell established the Bell Telephone Company, and the first telephone exchange was opened in New Haven, Connecticut in 1878. It wasn't, however, until 1915 that, using the then-new concept of electronic amplification, Bell held the first transcontinental telephone conversation with Thomas Watson (New York to San Francisco). And it wasn't until an almost unbelievably recent year (1953) that the first transatlantic underwater telephone line was completed.

Many people played important roles in the early development of so-called wireless communications, but none played greater parts than James Maxwell, Heinrich Hertz, Oliver Lodge, Marchesi Marconi, and A. S. Popov.

From all the electromagnetic phenomena discovered by Oersted, Faraday, and others, Maxwell synthesized a general theory which, to this day, forms the basis of radio communication. Although Maxwell published his results in 1864, it wasn't until 1887, when Hertz experimentally verified key predictions† in Maxwell's theory, that universal acceptance of the theory was achieved. Hertz died in 1894, at the age of 37. Impetus to the development of radio was furnished by the invention of the *coherer* by Oliver Lodge in 1877. This sensitive device could detect radio signals that were far fainter than any signals that could be picked up with other devices. Lodge demonstrated wireless signaling over a distance of 150 yards in 1894 at Oxford, England.

Marconi and Popov, working independently and at around the same time, put the finishing touches on the first practical wireless systems. In 1895, Marconi transmitted radio signals through a distance of over 2 kilometers. By 1898, after having obtained a patent and having founded the Wireless Telegraph and Signal Company, Marconi could communicate using radio signals over 60-mile links. The future of the "wireless" was assured. On December 12, 1901, at 12:30 in the afternoon, a prearranged signal of three

---

†Which included a demonstration that radio waves and light waves were fundamentally the same entity.

faint clicks was heard on Signal Hill in Newfoundland. The signals originated 1700 miles away in Cornwall, England.

The discovery of the vacuum diode by Fleming in 1904 and the triode by Lee De Forest in 1906 signaled the dawn of wireless voice communication. The triode, a device of tremendous importance, was, at least until the development of the transistor, the main device for electronic amplification.

At around this time developments came very rapidly. By 1907, speech was being transmitted over 200-mile channels in the eastern United States. By 1920, station KBKA in Pittsburgh began broadcasting on a scheduled basis. In 1923, over 500 transmitters were operating in the United States. They all used nearly the same wavelength. Shortly thereafter (1927), the Federal Radio Commission was formed to put things in order.

Lee De Forest, who was not exactly a modest man (his autobiography was titled *Father of Radio*), decried the intellectual mismatch between the high-level intelligence that produced radio and the low-level thinking that characterized the broadcasting industry. He called their programs a "... stench in the nostrils of the gods of the ionosphere."

During World War I, a young electrical engineer named Edwin Armstrong designed a greatly improved version of the broadcast receiver which became known as the superheterodyne. Almost all modern receivers are of this type. If the invention of the superheterodyne was all that Armstrong could claim, his name would still be honored in the history of communications. However, in 1933, Armstrong demonstrated a revolutionary new system of communication which he called *frequency modulation* (FM). Bitter squabbles with De Forest,† industry disinterest, and other factors delayed the wide acceptance of FM. By 1949, however, there were 600 FM stations operating in the United States.

In 1929, a Russian emigre by the name of V. K. Zworykin demonstrated the first television system using the now-obsolescent iconoscope. By 1939, the British Broadcasting Corp. (BBC) was broadcasting on a commercial basis, and over 20,000 TV sets had been sold in London. The iconoscope has been replaced by other cameras such as the image orthicon and vidicon. However, the underlying principle, i.e., scanning, remains the same. Color TV, a very complicated extension of black and white TV, began in the United States in 1954. Through some very ingenious engineering, color TV has been made *compatible* with black/white TV. This means that no increase in channel bandwidth is necessary and that black/white receivers can receive signals broadcast in color as black/white pictures and vice versa.

The developments of solid-state devices and integrated circuits have their roots in the invention of the transistor by Brattain, Bardeen, and Shockley

---

†These apparently went on for a good portion of his life and filled him with a great sense of failure and depression.

in 1948. The ultimate impact of large-scale integration (LSI) of circuits is still to be felt. Desk-size computers now give more computing capacity than house-size computers of 20 years ago.

The possibility of almost unlimited bandwidth capacity was suggested by the observation of laser action by Maiman in the United States in 1960. The initial proposals for a practical laser came from Schawlow and Townes of the United States and Basov and Prokhorov in the U.S.S.R. in 1958. They were awarded the 1964 Nobel Prize for physics for their contributions to the development of the laser.

A new way to produce images, made practically possible by the invention of the laser, is holography. Originally invented by D. Gabor in 1947, holography is a way of generating three-dimensional imagery by reconstructing the original wavefronts of light scattered from the object. Very significant improvements in holography were obtained in the early 1960's by E. Leith and J. Upatnicks, who invented *off-axis holography*. Conceivably the ultimate in image reproduction, holography has encountered many difficulties in its practical incorporation in TV and movies. Gabor received the 1971 Nobel prize in physics for his discovery of holography.

How about future trends? We can look forward to further miniaturization, more computer communication networks, sophisticated deep-space probes with pictures relayed from distant parts of the solar system, optical communication via fibers and lasers, widespread use of cable TV, further improvements in modern radio telescopes, new techniques of image processing, and the increased use of communications nets to bring us closer together still. How about the *distant* future? Will tachyons† replace electromagnetic waves and enable rapid communication to the most distant corners of the universe? Does the extrasensory perception channel exist, and can it be used for communication? As the old expression goes, only time will tell.

A brief table of significant events in communications is given below.

DATES OF SIGNIFICANT EVENTS IN COMMUNICATION

Year	Event
1600–1750:	Initial studies of electric and magnetic phenomena started.
1780–1790:	Coulomb experiments on the force between electric charges.
1790–1800:	Voltaic cell discovered.
1820:	Oersted shows that electric currents produce magnetic fields.
1830–1840:	Faraday and others show that electric fields are produced by changing magnetic fields.
1834:	Gauss and Weber build electromagnetic telegraph.
1837:	Cooke and Wheatstone telegraph built.
1842:	Morse's telegraph installed between Baltimore and Washington.

†After the Greek word *tachys*, meaning "swift." Tachyons are particles postulated to travel faster than light. To date, they have not been observed.

Year   Event

1860:	First telephone built by Reis.
1864:	Maxwell's theory of electromagnetism published.
1866:	Permanent transatlantic telegraph cable installed.
1876:	Telephone developed by Bell.
1887:	Hertz experiments on electromagnetic radiation.
1894:	Lodge demonstrates electromagnetic wireless communication over a distance of 150 yards.
1898:	Marconi and Jackson transmit a signal over a distance of 60 miles.
1901:	Transatlantic wireless communication developed by Marconi.
1904:	Fleming develops vacuum-tube diode.
1906:	Lee De Forest invents vacuum-tube triode.
1907:	Fessenden broadcasts speech over a distance of 200 miles.
1908:	Campbell-Swinton publishes basic ideas on television broadcasting.
1920:	First scheduled broadcast by station KBKA, Pittsburgh.
1929:	Zworykin demonstrates his television system.
1933:	Armstrong develops frequency modulation.
1936:	Commercial television broadcasting started by BBC.
1948:	Transistor amplifier built by Brattain, Bardeen, and Shockley.
1950–1960:	Microwave communication links developed.
1953:	First transatlantic telephone cable.
1954:	Color television broadcasting started in the United States.
1954:	First maser built.
1960:	First laser built.
1960:	First communication satellite launched (Echo I).
1960:	First TV pictures from space (TIROSI weather satellite).
1962:	Era of practical satellite communication begins with Telstar I.
1962:	Era of practical holography begins with off-axis holography.
1963:	Perfection of solid-state oscillators.
1965:	Mariner IV deep-space probe sends pictures from Mars to Earth.
1969:	First voice messages and TV from the Moon (Apollo XI).
1960–1970:	Cable TV, laser communication systems, "side-looking" radar, computer networks and large-scale growth in digital technology, adaptive filters, electro-optical processing for radar, radio astronomy.
1976:	First Viking spacecraft lands on Mars, sends back very high-quality imagery of Mars, and carries out biological tests to determine whether life exists on Mars.

A great deal of the material in this section was obtained from a delightful book by P. Davidovits called *Communication* [1-6]. Written in an informal and humorous manner, the book gives the history of communication and surveys how communication systems work in a qualitative manner.

1.6 The Book

Our quantitative study of electrical communication systems begins with Chapter 2 in which we discuss the basic mathematical tools required for the rest of the book. The most important mathematical theory required for the understanding of communication systems is *Fourier theory*. This is the heart of Chapter 2. After completing the discussion of Fourier theory, we continue with a quantitative description of *linear systems*. This is the main subject of Chapter 3. Here we shall also discuss the theory of filters, including *active filters*. In Chapter 4 we discuss the sampling theorem and its implications and continue with an analysis of *pulse modulation*. With the notion of sampling under our belt, we continue our study with *digital systems* and fast Fourier transforms, which are the subject matter of Chapter 5. Digital filters—a very rapidly growing technology—is discussed here. In Chapters 6 and 7 we deal with amplitude (AM) and angle modulation, respectively. In Chapter 6, which is conceptually straightforward, we discuss AM and the systems derived from it. Also the theory and practice of *television*, both black/white and color, are treated here. Because TV involves so much of the high technology of communication and because of its great impact on our civilization, it represents an almost perfect case study to illustrate the concepts dealt with in detail in this chapter. In Chapter 7, also conceptually simple, we use somewhat more mathematics because we discuss frequency (FM) and phase (PM) modulation. Chapter 7 concludes the first phase of the book.

The second phase of the book, Chapters 8–13, is much more concerned with *statistical communications*. The tools of statistical communications are *probability*, *random variables*, and *random processes*. Chapter 8 is a very brief summary of the central ideas in probability theory. In Chapter 9, the focus is shifted to random variables. Basically of a mathematical nature, these chapters are needed for understanding Chapters 10 and 11. Chapter 10, in which we deal with random processes, power spectra, signal-to-noise calculations for AM and FM systems, etc., is conceptually much harder than earlier portions of the book. The same is true of Chapter 11, in which we deal with signal processing and optimal systems. Chapter 11 is conceivably the most advanced chapter in the book, and in it we touch upon existing and active research areas. Included here are discussions of Wiener filters, recursive filters, the matched filter, the Bayes receiver, and optimum signaling schemes.

Chapter 12 is more similar to Chapters 6 and 7 in the sense that we deal with existing systems. The subject matter here is radar and, to some extent, sonar.

Last, we come to Chapter 13, where we deal with a subject which we feel has long been overlooked: two-dimensional systems. In an era when TV and facsimile systems are commonplace, when imagery is the primary input to our senses, the inclusion of two-dimensional systems in a book such as this is overdue.

Items marked with an asterisk are either more advanced, more specific, or more peripheral than is necessary for a first reading. They can be omitted without serious loss the first time around.

REFERENCES

[1-1]   G. RAISBECK, *Information Theory*, M.I.T. Press, Cambridge, Mass., 1963.

[1-2]   C. E. SHANNON and W. WEAVER, *The Mathematical Theory of Communication*, University of Illinois Press, Urbana, Ill., 1964.

[1-3]   *Reference Data for Radio Engineers*, 4th ed., International Telephone and Telegraph Corporation, New York, 1956.

[1-4]   A. B. CARLSON, *Communication Systems*, 2nd ed., McGraw-Hill, New York, 1975.

[1-5]   S. A. SCHELKUNOFF, *Electromagnetic Fields*, Ginn/Blaisdell, Waltham, Mass., 1963.

[1-6]   P. DAVIDOVITS, *Communication*, Holt, Rinehart and Winston, New York, 1972.

Fourier Methods

2.1 Introduction

The interrelation between time and frequency plays a central role in communication theory. Signals can be characterized by such parameters as duration, period, voltage level, and time-history. The use of such parameters leads to the *time-domain* characterization of signals. But signals can also be characterized by bandwidth, spectral content, phase, and frequency. The use of these parameters leads to a *frequency-domain* characterization of signals. How are time-domain parameters related to frequency-domain parameters? This is the main concern of time-frequency analysis.

The analytical relations between time and frequency are established by the Fourier series and the Fourier integral, studied in this chapter. We start with a brief discussion of linear time-invariant systems since Fourier analysis is particularly useful in input-output calculations for these systems. We consider periodic signals and their expansion into Fourier series and eventually extend the analysis to nonperiodic signals and the Fourier integral, whose properties we deal with in some detail. The discussion of the Fourier series is itself extended to incorporate general orthogonal-function expansions of which the Fourier series is only one example. This leads us to a brief examination of the minimum mean-square-error property of such expansions as well as concepts such as completeness, L_2 spaces, vector-space representation of signals, orthogonal projections, and the Gram-Schmidt process for generating a set of orthonormal functions.

2.2 The Superposition Principle

Finding the response of a system to an arbitrary input signal is a central problem encountered in communication theory. It is a problem that arises in the description of small systems, such as amplifiers and filter circuits, as well as in large systems, such as telephone networks, television, or radio. For nonlinear or time-varying systems this problem is generally quite difficult, and in most cases there are no analytic solutions. Fortunately we are often able to deal with linear, time-invariant systems for which the input-output problem has well-known solutions.

Practical systems are, of course, never exactly linear or time-invariant. Quite aside from the fact that certain circuits such as the rectifiers, modulators, and demodulators considered in later chapters are intrinsically nonlinear, time-varying, or both, even a circuit that is nominally linear will generally exhibit some nonlinear behavior. For instance, a "linear" amplifier is linear only for signal amplitudes up to a certain maximum level and will saturate if one tries to go beyond this. Saturation is a nonlinear effect. Similarly, heating of the components of an electric circuit will cause parameter values to change and thereby destroy strict time invariance. However, in practice it is often possible to disregard these effects, to deal with nominally linear systems as though they were exactly linear, and to treat departures from linearity and time invariance as distortions that can be examined separately.

Consider a linear system represented as shown in Fig. 2.2-1. The input is an arbitrary signal $x(t)$; the output is $y(t)$. This system is sometimes referred

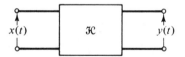

Figure 2.2-1. Linear system.

to as a *two-port* because it has an input port and an output port. Systems may have several input ports or several output ports or both, in which case they are called multiport systems; we shall not deal with these here.

Without inquiring into exactly what is inside the box shown in Fig. 2.2-1, we can represent the operation of the system by†

$$y(t) = \mathcal{K}[x(t)]. \qquad (2.2\text{-}1)$$

†The symbolic relation in Eq. (2.2-1) should not be read "y at time t depends on x at time t through the operator $\mathcal{K}$." Generally y at time t depends on the values of x during an interval of time which may be of finite or infinite duration.

The symbol $\mathfrak{IC}$ stands simply for an arbitary transformation, or operator. The operator $\mathfrak{IC}$ is *real* if real input x results in real output y.

Let $x_1(t)$ and $x_2(t)$ be two arbitrary inputs and $y_1(t)$ and $y_2(t)$ be the corresponding outputs. Let a_1, a_2 be two arbitrary constants. Then $\mathfrak{IC}$ is a *linear operator* if and only if

$$\mathfrak{IC}[a_1 x_1(t) + a_2 x_2(t)] = a_1 \mathfrak{IC}[x_1(t)] + a_2 \mathfrak{IC}[x_2(t)]$$
$$= a_1 y_1(t) + a_2 y_2(t). \tag{2.2-2}$$

This is the *superposition principle*. It implies, for instance, that doubling the input doubles the output. By letting $x_1(t) = x_2(t)$ and $a_1 = -a_2$, we also observe that linearity (or superposition) implies that zero input gives zero output. Thus linearity means more than just straight line—the straight line must pass through the origin as well.

If there are three inputs and three constants, we can group any two of them and add the third:

$$\mathfrak{IC}[a_1 x_1(t) + a_2 x_2(t) + a_3 x_3(t)] = \mathfrak{IC}\{[a_1 x_1(t) + a_2 x_2(t)] + a_3 x_3(t)\}$$
$$= a_1 \mathfrak{IC}[x_1(t)] + a_2 \mathfrak{IC}[x_2(t)]$$
$$+ a_3 \mathfrak{IC}[x_3(t)]. \tag{2.2-3}$$

This process can obviously be repeated, and therefore the basic definition of superposition easily extends to any finite number of inputs. The definition is, in fact, usually extended to encompass an infinite number of inputs, even though this does not follow quite so obviously from Eq. (2.2-3). Thus, if

$$y_i(t) = \mathfrak{IC}[x_i(t)],$$

Figure 2.2-2. Illustration of time invariance.

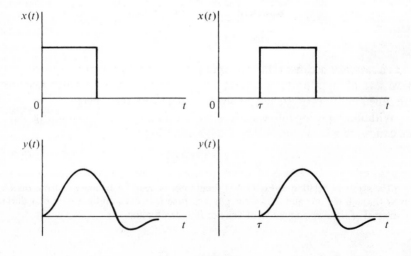

linearity implies that

$$\mathfrak{IC}\left[\sum_{i=1}^{\infty} a_i x_i(t)\right] = \sum_{i=1}^{\infty} a_i y_i(t). \tag{2.2-4}$$

In addition to being linear, a system or operator may or may not be time-invariant. For a time-invariant operator

$$\mathfrak{IC}[x(t - \tau)] = y(t - \tau) \tag{2.2-5}$$

for all τ; in other words the response at any particular time t depends on the difference between this time and some arbitrary reference; it does not depend on absolute time. This is illustrated in Fig. 2.2-2.

2.3 Linear System Response

To find out something about a linear system one may apply certain standard test signals $z_n(t)$ to the input and observe the output. Typical of these test signals are impulses, steps, or sine waves. The response of the system to such a test signal is a characteristic description of the system's more general input-output behavior. In fact, vis-à-vis input-output behavior, the test-signal response may be a better characteristic than a complete circuit diagram of the insides of the box because it tells us more readily what the system actually does.†

If we want to find the response of the system to an arbitary input $x(t)$, we can try to represent $x(t)$ in terms of a sum of elementary test functions and then use superposition to find the output. Typically such a representation will have the form

$$x(t) = \sum_{n=1}^{\infty} a_n z_n(t). \tag{2.3-1}$$

Then, if the response to the elementary signal $z_n(t)$ is $h_n(t)$, the superposition principle gives

$$y(t) = \sum_{n=1}^{\infty} a_n h_n(t). \tag{2.3-2}$$

We should point out that Eq. (2.3-1) is generally not exact, and therefore Eq. (2.3-2) is also only an approximation. We hope, however, that it will be a good enough approximation so that we don't leave out anything essential.

In the following sections we consider a particular signal representation, the Fourier series. Some of the details of the superposition procedure are dealt with in the following chapter. Signal representation is one of the central

---

†The detailed internal structure and inner working of the system, furnished by the circuit diagram, is not generally conveyed by the test-signal response.

issues in communications theory, and it will come up again in various guises throughout this book.

2.4 Periodic Signals

The Fourier series is a representation for periodic signals. A real function of time $x(t)$ is periodic if there is some time interval τ for which

$$x(t + \tau) = x(t) \qquad (2.4\text{-}1)$$

for all t in $(-\infty, \infty)$. The smallest value of τ for which Eq. (2.4-1) is true is called the period T. The fundamental frequency is defined by

$$f_0 = \frac{1}{T} \qquad (2.4\text{-}2a)$$

in cycles per second or hertz (abbreviated Hz) or

$$\omega_0 = \frac{2\pi}{T} \qquad (2.4\text{-}2b)$$

in radians per second.

The Fourier series representation is given by

$$x(t) = \sum_{n=0}^{\infty} (a_n \cos n\omega_0 t + b_n \sin n\omega_0 t). \qquad (2.4\text{-}3)$$

Thus the test functions $z_n(t)$ referred to earlier are sines and cosines. If the input to a linear time-invariant system is a sinusoid with frequency ω, the output is a sinusoid of the same frequency, but the amplitude is multiplied by an amplitude factor $A(\omega)$ and the phase is shifted by a phase angle $\theta(\omega)$. Therefore if the input is represented by a Fourier series as in Eq. (2.4-3), the output will be

$$y(t) = \sum_{n=0}^{\infty} A(n\omega_0)\{a_n \cos [n\omega_0 t + \theta(n\omega_0)] + b_n \sin [n\omega_0 t + \theta(n\omega_0)]\}. \qquad (2.4\text{-}4)$$

This relation is the raison d'être for the use of the Fourier series in the analysis of linear systems. The complex number

$$H(\omega) = A(\omega) \exp [j\theta(\omega)]$$

is called the *transfer function* or *frequency response* and is probably the most commonly used characteristic of linear time-invariant systems. The Fourier series permits the transfer function to be used in studying the behavior of the system with arbitrary, i.e., nonsinusoidal periodic input signals. The restriction to periodic signals can be removed, as shown in Sec. 2.8. The transfer function is considered in more detail in Chapter 3.

An equivalent and somewhat more convenient form of the Fourier series is

$$x(t) = \sum_{n=-\infty}^{\infty} c_n e^{jn\omega_0 t}. \qquad (2.4\text{-}5)$$

The equivalence between Eqs. (2.4-3) and (2.4-5) is easily demonstrated by using the exponential representation for the sine and cosine functions:

$$\sin\theta = \frac{-j}{2}(e^{j\theta} - e^{-j\theta}),\qquad(2.4\text{-}6)$$

$$\cos\theta = \frac{1}{2}(e^{j\theta} + e^{-j\theta}).\qquad(2.4\text{-}7)$$

Substituting this into Eq. (2.4-3) results in

$$x(t) = \sum_{n=0}^{\infty}\frac{a_n}{2}(e^{jn\omega_0 t} + e^{-jn\omega_0 t}) - \frac{jb_n}{2}(e^{jn\omega_0 t} - e^{-jn\omega_0 t})$$

$$= \sum_{n=0}^{\infty}\left(\frac{a_n - jb_n}{2}\right)e^{jn\omega_0 t} + \left(\frac{a_n + jb_n}{2}\right)e^{-jn\omega_0 t}.\qquad(2.4\text{-}8)$$

We now define
$$\left[= \sum_{n=0}^{\infty} c_n e^{jn\omega_0 t} + c^*_n e^{-jn\omega_0 t}\right]$$

$$c_n = \frac{a_n - jb_n}{2}$$
$$\qquad\qquad (n \neq 0)$$
$$c^*_n = \frac{a_n + jb_n}{2}\qquad(2.4\text{-}9)$$

$$c_0 = a_0,$$

where the asterisk means complex conjugation. Because $x(t)$ is assumed real, $x(t) = x^*(t)$. It then follows that $c^*_n = c_{-n}$, because

$$x^*(t) = \sum_{n=-\infty}^{\infty} c^*_n e^{-jn\omega_0 t}$$

$$x(t) = \sum_{n=-\infty}^{\infty} c_n e^{j\omega_0 t}$$

$$= \sum_{n=-\infty}^{\infty} c^*_{-n} e^{jn\omega_0 t}\qquad \text{(if } n \text{ is replaced by } -n)$$

$$= \sum_{n=-\infty}^{\infty} c_n e^{jn\omega_0 t} = x(t).$$

$$c_n = c^*_{-n}$$
$$c_{-n} = c^*_n$$

Equation (2.4-8) can be written in the form

$$x(t) = \sum_{n=1}^{\infty} c_n e^{jn\omega_0 t} + \sum_{n=1}^{\infty} c^*_n e^{-jn\omega_0 t} + c_0,$$

which, when use is made of $c^*_n = c_{-n}$, can be written as

$$x(t) = \sum_{n=1}^{\infty} c_n e^{jn\omega_0 t} + \sum_{n=-1}^{-\infty} c_n e^{jn\omega_0 t} + c_0$$

$$= \sum_{n=-\infty}^{\infty} c_n e^{jn\omega_0 t}.\qquad(2.4\text{-}10)$$

Some important results that follow from the fact that $c^*_n = c_{-n}$ are that

$$|c_n| = |c_{-n}|$$

and

$$\angle c_n = -\angle c_{-n};$$

i.e., the magnitude has even symmetry and the phase has odd symmetry.

Simply writing down the series as we have done in Eq. (2.4-3) or (2.4-5) is, of course, no guarantee that it converges to the desired periodic function $x(t)$, or that it converges at all. We shall investigate the convergence problem in the next section. However, these expressions suggest a simple way to calculate the coefficients. One reason for preferring the exponential form of the Fourier series is that the procedure for doing this is slightly simpler. We multiply Eq. (2.4-10) by $e^{-jm\omega_0 t}$, giving

$$x(t)e^{-jm\omega_0 t} = \sum_{n=-\infty}^{\infty} c_n e^{j(n-m)\omega_0 t}. \tag{2.4-11}$$

Assuming convergence of the series, we can integrate term by term over one period with the result

$$\int_{-T/2}^{T/2} x(t)e^{-jm\omega_0 t}\, dt = \sum_{n=-\infty}^{\infty} c_n \int_{-T/2}^{T/2} e^{j(n-m)\omega_0 t}\, dt$$

$$= \sum_{n=-\infty}^{\infty} c_n \left[\frac{e^{j(n-m)\omega_0 T/2} - e^{-j(n-m)\omega_0 T/2}}{j(n-m)\omega_0} \right]$$

$$= T \sum_{n=-\infty}^{\infty} c_n \frac{\sin(n-m)\omega_0 T/2}{(n-m)\omega_0 T/2}. \tag{2.4-12}$$

By Eq. (2.4-2) $\omega_0 T/2 = \pi$, and therefore Eq. (2.4-12) can be written in the form

$$\int_{-T/2}^{T/2} x(t)e^{-jm\omega_0 t}\, dt = T \sum_{n=-\infty}^{\infty} c_n \operatorname{sinc}(n-m), \tag{2.4-13}$$

where the function sinc $(\cdot)$ is defined by

$$\operatorname{sinc}(z) \equiv \frac{\sin \pi z}{\pi z}. \tag{2.4-14}$$

The sinc function is an important and useful function which will reappear many more times in this book. It is plotted in Fig. 2.4-1. For the purposes of this section the important property to note is that it is zero for all nonzero

Figure 2.4-1. Function sinc (z).

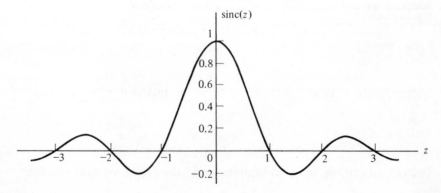

integer values of z and that its value for $z = 0$ is 1. Hence in Eq. (2.4-13) all the elements of the sum vanish except the one for which $n = m$. This then results in the formula for the Fourier coefficients:

$$c_m = \frac{1}{T} \int_{-T/2}^{T/2} x(t) e^{-jm\omega_0 t} \, dt. \tag{2.4.-15}$$

The formulas for the a_n and b_n used in Eq. (2.4-4) follow directly from Eqs. (2.4-15) and (2.4-9). Thus, for $n \neq 0$,

$$a_n = c_n + c_{-n} = \frac{1}{T} \int_{-T/2}^{T/2} x(t)(e^{-jn\omega_0 t} + e^{jn\omega_0 t}) \, dt$$

$$= \frac{2}{T} \int_{-T/2}^{T/2} x(t) \cos n\omega_0 t \, dt, \tag{2.4-16}$$

$$a_0 = c_0 = \frac{1}{T} \int_{-T/2}^{T/2} x(t) \, dt, \tag{2.4-17}$$

$$b_n = \frac{1}{j}(c_{-n} - c_n) = \frac{1}{T} \int_{-T/2}^{T/2} x(t) \left(\frac{e^{jn\omega_0 t} - e^{-jn\omega_0 t}}{j} \right) dt$$

$$= \frac{2}{T} \int_{-T/2}^{T/2} x(t) \sin n\omega_0 t \, dt. \tag{2.4-18}$$

Another form of the Fourier series, which is often convenient, is the cosine series:

$$x(t) = c_0 + \sum_{n=1}^{\infty} x_n \cos(\omega_n t + \phi_n), \tag{2.4-19}$$

where $x_n = 2|c_n| = \sqrt{a_n^2 + b_n^2}$,
$\phi_n = \angle c_n = -\tan^{-1}(b_n/a_n)$,
$\omega_n = n\omega_0$.

The phase angle† ϕ_n is zero for even functions, i.e., for functions $x(t)$ such that $x(-t) = x(t)$. Odd functions, for which $x(t) = -x(-t)$, can be represented by a sine series $\sum_{n=1}^{\infty} x_n \sin \omega_n t$.

EXAMPLES

1. *Square pulse train* (Fig. 2.4-2). This function has the value 1 in the interval $-\tau/2, \tau/2$; hence

$$c_n = \frac{1}{T} \int_{-\tau/2}^{\tau/2} e^{-j2\pi n t/T} \, dt = \frac{1}{-j2\pi n} (e^{-j2\pi n t/T}) \Big|_{-\tau/2}^{\tau/2}$$

$$= \frac{\tau}{T} \operatorname{sinc} n \frac{\tau}{T}$$

$$= d \operatorname{sinc} nd,$$

†The phase or angle of c_n is sometimes called its argument, written arg c_n.

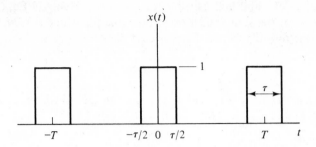

Figure 2.4-2. Square pulse train.

where $d \equiv \tau/T$ is the duty ratio. The magnitude and phase of the first few c's are shown in Fig. 2.4-3(a) for $d = \frac{1}{4}$. This is called a *line spectrum* since only discrete frequencies appear in it. Line spectra are characteristic of periodic functions. In this particular example the c's are all real. This means that the trigonometric form of the series [Eq. (2.4-3)] has only cosine terms; i.e., the b's of Eq. (2.4-18) are all zero. This is so because we chose the time origin to make $x(t)$ an even function. The effect of a shift in the time origin is considered in another example which we shall consider shortly.

We have already seen that for real $x(t)$ the magnitude of c_n has even symmetry and the phase of c_n has odd symmetry. The odd phase symmetry is shown in Figs. 2.4-3(b) and (c).

Observe that $c_0 = d$ and that the smaller the value of d, the larger the numbers n for which sinc nd has significantly large values. In fact we can define the significant width of the spectrum in terms of those values of n around $n = 0$ for which sinc nd exceeds zero.† These are given by $-1 \leq nd \leq 1$ or $-1/d \leq n \leq 1/d$.

The corresponding bandwidth is f_0/d, where f_0 is the fundamental frequency in hertz. In Fig. 2.4-3 the significant values of n go from -4 to 4, and therefore the bandwidth is $4f_0$. If d were smaller than $\frac{1}{4}$, the bandwidth would be larger. The reciprocal relation between duration and bandwidth is one of the most important properties of the Fourier expansion.

2. *Shifted square pulse train* (Fig. 2.4-4). Here

$$c_n = \frac{1}{T} \int_{a-\tau/2}^{a+\tau/2} e^{-j2\pi nt/T} \, dt = \frac{1}{-j2\pi n}(e^{-j2\pi n(a+\tau/2)/T} - e^{-j2\pi n(a-\tau/2)/T})$$

$$= e^{-j2\pi na/T} \frac{\tau}{T} \text{ sinc } n\frac{\tau}{T}. \tag{2.4-20}$$

The magnitude spectrum (and therefore the bandwidth) is not changed by the time shift, but there is now an additional phase shift of $-2\pi na/T$ radians in the c_n's. This is shown in Fig. 2.4-3(c).

---

†This is, obviously, an arbitrary definition. The lines around $n = 0$ for which $|n| \leq d^{-1}$ are sometimes called the *main lobe* of the sinc function.

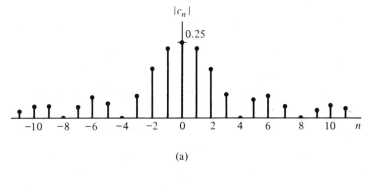

(a)

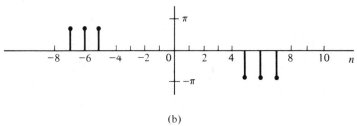

(b)

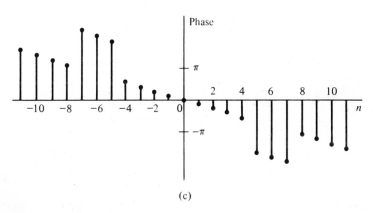

(c)

Figure 2.4-3. Line spectrum for the square pulse train with duty ratio $d = \frac{1}{4}$: (a) magnitude; (b) phase for even function; (c) phase for shifted pulse train.

3. *Cosine pulse train* (Fig. 2.4-5). The pulse centered about the origin is described by

$$x(t) = \begin{cases} \cos 2\pi f_1 t, & -\frac{\tau}{2} < t < \frac{\tau}{2}, \\ 0, \text{ for the remainder of the period.} \end{cases}$$

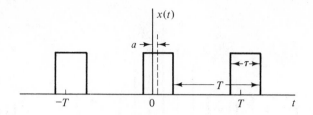

Figure 2.4-4. Shifted square pulse train.

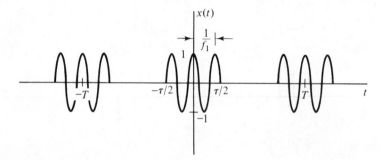

Figure 2.4-5. Cosine pulse train.

The cosine pulse train is given by $\sum_{n=-\infty}^{\infty} x(t - nT)$. Here c_n is computed from

$$c_n = \frac{1}{T} \int_{-\tau/2}^{\tau/2} e^{-j2\pi nt/T} \cos 2\pi f_1 t \, dt. \tag{2.4-21}$$

This is most conveniently evaluated by expressing the cosine itself in terms of exponentials:

$$c_n = \frac{1}{T} \int_{-\tau/2}^{\tau/2} e^{-j2\pi nt/T} \frac{e^{j2\pi f_1 t} + e^{-j2\pi f_1 t}}{2} \, dt$$

$$= \frac{1}{2T} \int_{-\tau/2}^{\tau/2} (e^{-j2\pi(n/T+f_1)t} + e^{-j2\pi(n/T-f_1)t}) \, dt$$

$$= \frac{\tau}{2T} \left[\operatorname{sinc}\left(\frac{n}{T} + f_1\right)\tau + \operatorname{sinc}\left(\frac{n}{T} - f_1\right)\tau \right]. \tag{2.4-22}$$

The line spectrum (i.e., values of $|c_n|$) for $d \equiv \tau/T = \frac{1}{4}, f_1 = 4/T$ is shown in Fig. 2.4-6.

We now have two spectral lobes centered at $\pm f_1$. The significant width of these spectral lobes is still given by $n = \pm T/\tau$; i.e., it depends on the duty cycle and not on the details of the pulse shape. For the particular values of d and f_1 used in Fig. 2.4-6, c_0 is zero. This reflects the fact that the cosine pulse, with these values of d and f_1, has no dc value. The pulse shown in Fig. 2.4-5 *does* have a dc value (since the area above the t axis is greater than that below); hence for that cosine pulse c_0 would be positive.

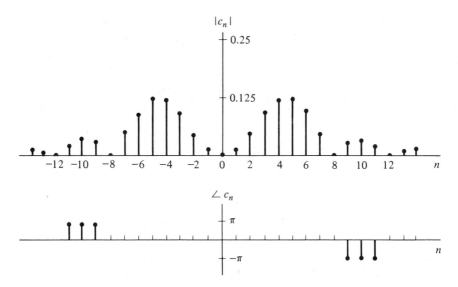

Figure 2.4-6. Line spectrum corresponding to a cosine pulse train. Duty cycle is $\frac{1}{4}$, $f_1 = 4/T$.

2.5   Convergence

The formal procedure of the previous section for calculating the coefficients was based on the assumption that Eqs. (2.4-3) and (2.4-5) are truly equalities, i.e., that an arbitrary periodic function can be represented in terms of a sum of sines and cosines. There is a priori no reason to think that this is always possible. However, if $x(t)$ is integrable, the coefficients as defined in Eqs. (2.4-15)–(2.4-18) will exist. We can therefore start with the function $x(t)$, find the coefficients, and then examine the properties of the series formed with these coefficients. We shall see that the series does converge to $x(t)$ after a fashion, provided that $x(t)$ is not too complicated. We should not expect, however, that the series will converge to $x(t)$ at every t, for arbitrary $x(t)$.

To see this, consider a function $y(t)$ which is equal to $x(t)$ except at a single point, where it differs by a finite amount. Since this finite difference does not affect the calculation of c_n, the Fourier coefficients for $x(t)$ and $y(t)$ will be identical; hence the series cannot represent both functions at the point in question. More generally, functions that are equal "almost everywhere"— that is, everywhere except at a finite number of isolated points—will have the same Fourier series representation, and the series cannot represent all of them at every point. Convergence of the Fourier series is therefore, in general, *not pointwise*.

To see how the Fourier series converges to the desired time function, consider the truncated series

$$x_N(t) = \sum_{n=-N}^{N} c_n e^{jn\omega_0 t},$$ (2.5-1)

and observe how this series behaves as N becomes very large. Substituting from Eq. (2.4-15) and exchanging the order of summation and integration, we can write

$$x_N(t) = \sum_{n=-N}^{N} \left[\frac{1}{T} \int_{-T/2}^{T/2} x(\tau) e^{-jn\omega_0 \tau} \, d\tau \right] e^{jn\omega_0 t}$$

$$= \frac{1}{T} \int_{-T/2}^{T/2} x(\tau) \sum_{n=-N}^{N} e^{jn\omega_0(t-\tau)} \, d\tau.$$ (2.5-2)

The summation is a geometric series which is easily put into a closed form (Prob. 2-10):

$$S_N(t-\tau) \equiv \sum_{n=-N}^{N} e^{jn\omega_0(t-\tau)} = \frac{\sin\left(N + \frac{1}{2}\right)\omega_0(t-\tau)}{\sin(\omega_0/2)(t-\tau)}.$$ (2.5-3)

Then Eq. (2.5-2) becomes

$$x_N(t) = \frac{1}{T} \int_{-T/2}^{T/2} x(\tau) S_N(t-\tau) \, d\tau.$$ (2.5-4)

A plot of the function $S_N(t-\tau)$ for fixed t and $N = 5$ is shown in Fig. 2.5-1. In performing the integration, t is treated as a constant. Since the integrand is periodic, we can always arrange the interval $[-T/2, T/2]$ so that the peak of $S_N(t-\tau)$ at $\tau = t$ is well within the range of integration. We see that $x_N(t)$ is a weighted average of $x(\tau)$, with the function $S_N(t-\tau)$ placing most of the weight at $\tau = t$. As N becomes larger the peak and the number

Figure 2.5-1. Plot of the function $S_N(t-\tau)$ for $N = 5$.

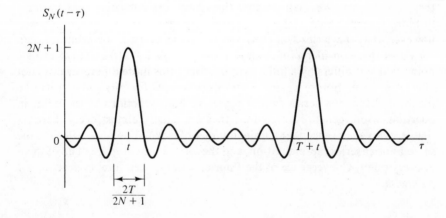

of oscillations in one period of $S_N(t - \tau)$ increase so that the weighting of $x(\tau)$ becomes more and more concentrated in the neighborhood of $\tau = t$. Formally it is easily shown (Prob. 2-11) that for τ near t and for large N

$$S_N(t - \tau) \approx 2N \text{ sinc } 2Nf_0(t - \tau), \qquad (2.5\text{-}5)$$

where $f_0 = \omega_0/2\pi$. This function has a large and narrow peak at $\tau = t$ and negligible magnitude for $|\tau - t| \gg 1/Nf_0$. Hence if $x(\tau)$ is "smooth" near $\tau = t$ (this means that it is continuous and has a derivative), we can approximate Eq. (2.5-4) by

$$x_N(t) \approx \frac{1}{T} x(t) \int_{-T/2}^{T/2} S_N(t - \tau) \, d\tau$$

$$\approx x(t) \int_{-\infty}^{\infty} 2Nf_0 \text{ sinc } 2Nf_0(t - \tau) \, d\tau, \qquad (2.5\text{-}6)$$

where we have replaced $1/T$ by f_0 and extended the limits of integration to $\pm\infty$ because $S_N(t - \tau)$ is a narrowly peaked function. An elementary property of the sinc function is

$$\int_{-\infty}^{\infty} \text{sinc } z \, dz = 1 \qquad (2.5\text{-}7)$$

(see, for instance, formula 499 on p. 68 of Peirce and Foster's table of integrals [2-1]). Hence, by making the change of variable $2Nf_0(t - \tau) = z$ we find that if $x(t)$ is a smooth function,

$$\lim_{N \to \infty} x_N(t) = x(t).$$

Thus we see that the finite Fourier series does indeed converge to the function $x(t)$, provided that $x(t)$ is sufficiently well behaved.

Convergence at a Discontinuity

Functions with isolated discontinuities, such as the square pulses considered in the examples, occur quite frequently. The Fourier series cannot represent such functions at the points of discontinuity, and the question therefore arises as to what the series does there.

Accordingly, consider a periodic function $x(t)$, discontinuous at $t = t_0$ but possessing a derivative on each side of the point t_0. A piece of the function might look like Fig. 2.5-2. In the vicinity of $t = t_0$, we can regard $x(t)$ as the sum of a continuous function and a step; i.e.,

$$x(t) = x_c(t) + [x(t_+) - x(t_-)]u(t - t_0), \qquad (2.5\text{-}8)$$

where the function $u(\cdot)$ is defined by

$$u(t - t_0) = \begin{cases} 1, & t > t_0, \\ 0, & t \le t_0. \end{cases} \qquad (2.5\text{-}9)$$

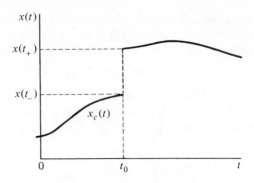

Figure 2.5-2. Discontinuous function.

We do not have to be concerned about points at some distance from t_0 because the weighting function $S_N(t - \tau)$ of Eq. (2.5-3) will not give them any appreciable weight. Substituting Eq. (2.5-8) into Eq. (2.5-4) results in

$$x_N(t) = \frac{1}{T} \int_{-T/2}^{T/2} \{x_c(\tau) + [x(t_+) - x(t_-)]u(\tau - t_0)\}S_N(t - \tau)\,d\tau.$$

We assume that the N-term Fourier series is a good approximation of the continuous part of $x(t)$, and therefore the integration of the component $x_c(\tau)S_N(t - \tau)$ results approximately in $x_c(t)$. This leaves

$$x_N(t) \approx x_c(t) + [x(t_+) - x(t_-)]\frac{1}{T} \int_{-T/2}^{T/2} u(\tau - t_0)S_N(t - \tau)\,d\tau$$

$$= x_c(t) + [x(t_+) - x(t_-)]\frac{1}{T} \int_{t_0}^{T/2} S_N(t - \tau)\,d\tau.$$

We define

$$u_N(t - t_0) = \frac{1}{T} \int_{t_0}^{T/2} S_N(t - \tau)\,d\tau \qquad (2.5\text{-}10)$$

and obtain, finally,

$$x_N(t) = x_c(t) + [x(t_+) - x(t_-)]u_N(t - t_0). \qquad (2.5\text{-}11)$$

We now investigate the function $u_N(t - t_0)$.

Without materially affecting the discussion we can assume that t_0 is not near the end points $\pm T/2$. (If it is, one can simply redefine the starting point of the period.) For large N, $S_N(t - \tau)$ is sharply peaked near t and essentially negligible elsewhere. Hence it is permissible to replace the upper limit of integration in the definition of $u_N(t - t_0)$ by ∞. Also, instead of $S_N(t - \tau)$ we use $2N$ sinc $2Nf_0(t - \tau)$ as in Eq. (2.5-5). These changes result in

$$u_N(t - t_0) = \frac{2N}{T} \int_{t_0}^{\infty} \text{sinc } 2Nf_0(t - \tau)\,d\tau. \qquad (2.5\text{-}12)$$

We now make the change of variable $\lambda = 2\pi N f_0(t - \tau) = N\omega_0(t - \tau)$, and this results in

$$u_N(t - t_0) = \frac{1}{\pi} \int_{-\infty}^{N\omega_0(t - t_0)} \frac{\sin \lambda}{\lambda} \, d\lambda = \frac{1}{2} + \frac{1}{\pi} \operatorname{Si} [N\omega_0(t - t_0)], \quad (2.5\text{-}13)$$

where

$$\operatorname{Si}(x) = \int_0^x \frac{\sin \lambda}{\lambda} \, d\lambda \qquad (2.5\text{-}14)$$

is the sine-integral function. This is a tabulated function (see, for instance, Jahnke and Emde [2-2], pp. 1–9). A plot of this function and of $u_N(t)$ is shown in Fig. 2.5-3. We see that the step at $t = t_0$ is replaced by the oscillating function $u_N(t - t_0)$, scaled by the magnitude of the step. The N-term Fourier series representation of the discontinuous function $x(t)$ shown in Fig. 2.5-2 will therefore appear as in Fig. 2.5-4. At the point $t = t_0$, $u_N(t - t_0) = \frac{1}{2}$;

Figure 2.5-3. (a) Function Si (x). (b) Function $u_N(t)$.

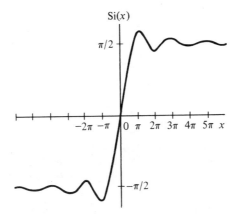

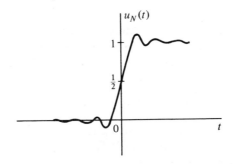

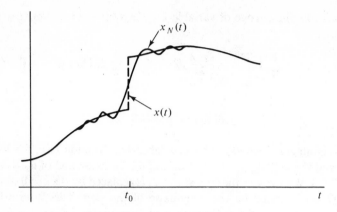

Figure 2.5-4. N-Term representation of a discontinuous function near the discontinuity.

therefore, since $x(t_-) = x_c(t_0)$,

$$x_N(t_0) = x(t_-) + \tfrac{1}{2}[x(t_+) - x(t_-)]$$

$$= \frac{x(t_+) + x(t_-)}{2}. \tag{2.5-15}$$

This is independent of N and will hold for $N \rightarrow \infty$. Thus we find that at points of discontinuity the Fourier series gives the *average* between the two extreme points.

At points near t_0 we see that the series representation oscillates around the true value. The period of the oscillation is equal to $\pi/N\omega_0$, and it decreases as N becomes larger. However, the amplitude of the oscillation is independent of N. In particular, the size of the first overshoot is 18% of the step size, and this does not decrease. This effect is called the *Gibbs phenomenon*. Although usually of little consequence, it shows that the Fourier series does not converge uniformly in the vicinity of a discontinuity. In other words, the finite series $x_N(t)$ does not converge to a definite limit with increasing N near the discontinuity point. There are modifications of the Fourier series, achieved by the inclusion of weighting functions called the Fejér polynomials, that will eliminate this phenomenon ([2-3] and [2-4], p. 46). However, besides resulting in a more complicated expression for the coefficients, the Fejér modification also destroys the minimum mean-square-error property of the Fourier transform, about which more will be said in Sec. 2.6.

2.6 Orthogonal Function Expansion

The Fourier expansion is only a particular example of expanding a given function in terms of a set of orthogonal functions.

A set of complex-valued functions $\{\phi_n(t)\}$, $n = 1, 2, \ldots$, is said to be *orthogonal* over the interval $(0, T)$ if for all integers m, n

$$\int_0^T \phi_n(t)\phi_m^*(t)\, dt = \begin{cases} C & \text{if } m = n \quad (C \text{ a constant}), \\ 0 & \text{if } m \neq n. \end{cases} \tag{2.6-1}$$

The set is *orthonormal* if

$$\int_0^T \phi_n(t)\phi_m^*(t)\, dt = \delta_{mn} = \begin{cases} 1 & \text{if } m = n, \\ 0 & \text{if } m \neq n. \end{cases} \tag{2.6-2}$$

The number δ_{mn} is called the *Kronecker delta* and appears frequently in analyses. An orthogonal set of functions can obviously be made orthonormal by dividing each function by $\sqrt{C}$. Thus, the functions $e^{j2\pi nt/T}$, $n = 0, \pm 1, \pm 2, \ldots$, are orthogonal over $[0, T]$, and the corresponding orthonormal functions are $\{(1/\sqrt{T})e^{j2\pi nt/T}\}$.

An orthogonal function expansion has the form

$$x(t) = \sum_{n=1}^{\infty} c_n\phi_n(t). \tag{2.6-3}$$

This is sometimes referred to as a *generalized Fourier series*. It is identical with the Fourier expansion given in Eq. (2.4-5) if we make the correspondence $\phi_1(t) = 1$, $\phi_2(t) = e^{j\omega_0 t}$, $\phi_3(t) = e^{-j\omega_0 t}$, $\phi_4(t) = e^{j2\omega_0 t}$, $\phi_5(t) = e^{-j2\omega_0 t}$, ... The fact that the index in one sum runs from $-\infty$ to $+\infty$ and the other from 1 to ∞ is seen to make no real difference. The set of c_n's is called the *spectrum* of the function $x(t)$.

As pointed out earlier, writing down the series is no guarantee that it converges. However, Eq. (2.6-3) suggests that we can find the coefficients c_n by multiplying by $\phi_m^*(t)$ and integrating term by term. If we take the $\phi_n(t)$ to be orthonormal, this gives the formula

$$\int_0^T x(t)\phi_m^*(t)\, dt = \sum_{n=1}^{\infty} c_n \int_0^T \phi_n(t)\phi_m^*(t)\, dt$$

$$= c_m. \tag{2.6-4}$$

Note the similarity to the procedure for finding the Fourier coefficients. We can define the partial sum

$$x_N(t) = \sum_{n=1}^{N} c_n\phi_n(t), \tag{2.6-5}$$

where the c_n are given by Eq. (2.6-4). This will exist if the c_n exist, i.e., if $x(t)$ is integrable. The mean-square error between $x(t)$ and $x_N(t)$ is defined by

$$\langle \epsilon^2 \rangle = \int_0^T |x(t) - x_N(t)|^2\, dt$$

$$= \int_0^T |x(t) - \sum_{n=1}^{N} c_n\phi_n(t)|^2\, dt. \tag{2.6-6}$$

(We use the absolute-value bars so that complex signals can be included in the definition.)

The set of functions $\phi_n(t)$, $n = 1 \ldots$, is often referred to as the *basis* set and the functions themselves as the *basis*. The choice of a suitable basis is generally dictated by the type of problem being solved. For instance, if one deals with linear, time-invariant systems, the exponential or trigonometric functions are convenient because they form the general solution of the differential equations describing such systems. (A somewhat more pompous way of saying the same thing is that the exponential functions are eigenfunctions of time-invariant linear operators.) In other applications, other basis sets such as the Bessel or Laguerre functions may be more appropriate. In digital applications, functions such as the Walsh and Haar functions [2-5] which can take on only two values (e.g., zero and one) are sometimes convenient.

An important requirement of a basis is that it should be *complete*. A set of orthonormal functions $\{\phi_n(t)\}$ is said to be *complete* if, by choosing N large enough, the mean-square error can be made arbitrarily small, i.e., if

$$\lim_{N \to \infty} \int_0^T |x(t) - \sum_{n=1}^N c_n \phi_n(t)|^2 \, dt = 0. \tag{2.6-7}$$

If the $\phi_n(t)$ form a complete set, the generalized Fourier series is said to converge *in the mean* to the function $x(t)$. Note that this does not imply pointwise convergence unless the series converges uniformly. It can be shown [2-6] that the exponential functions

$$\frac{1}{\sqrt{T}}, \frac{e^{j2\pi t/T}}{\sqrt{T}}, \frac{e^{-j2\pi t/T}}{\sqrt{T}}, \cdots \quad \text{as well as} \quad \frac{1}{\sqrt{T}}, \sqrt{\frac{2}{T}} \cos \frac{2\pi t}{T},$$

$$\sqrt{\frac{2}{T}} \sin \frac{2\pi t}{T}, \sqrt{\frac{2}{T}} \cos \frac{4\pi t}{T}, \sqrt{\frac{2}{T}} \sin \frac{4\pi t}{T} \cdots$$

form complete orthonormal sets. Other complete sets include the Bessel functions, Legendre polynomials, etc. The task of demonstrating completeness of a given set of orthonormal functions is not trivial, and we shall not pursue it here. We assume completeness of the set of orthogonal functions $\{\phi_n(t)\}$ in the following discussions.

By expanding the square and integrating term by term, we can write Eq. (2.6-7) as

$$\lim_{N \to \infty} \int_0^T |x(t)|^2 \, dt - 2 \operatorname{Re} \left[\sum_{n=1}^N c_n^* \int_0^T x(t) \phi_n(t) \, dt \right] + \sum_{n=1}^N \sum_{m=1}^N c_n c_m^* \int_0^T \phi_n(t) \phi_m^*(t) \, dt$$

$$= \lim_{N \to \infty} \int_0^T |x(t)|^2 \, dt - 2 \sum_{n=1}^N |c_n|^2 + \sum_{n=1}^N |c_n|^2$$

$$= \int_0^T |x(t)|^2 \, dt - \lim_{N \to \infty} \sum_{n=1}^N |c_n|^2 = 0$$

or

$$\int_0^T |x(t)|^2 \, dt = \sum_{n=1}^\infty |c_n|^2. \tag{2.6-8}$$

Equation (2.6-8) is a form of *Parseval's theorem*. If $x(t)$ is regarded as the current in a 1-ohm resistor, then $|x(t)|^2$ is the instantaneous power in watts. Therefore the expression on the left is the *energy* per period. One can regard $|c_n|^2$ as the energy associated with the orthonormal function $\phi_n(t)$. For instance, consider the trigonometric Fourier series for an even function $x(t)$. This series has only cosine terms, and if we want it to be in terms of *orthonormal* functions, it will have the form

$$x(t) = \frac{c_0}{\sqrt{T}} + \sum_{n=1}^{\infty} \sqrt{\frac{2}{T}}\, c_n \cos n\omega_0 t.$$

The energy, E_n, in the nth term is given by

$$E_n = \frac{2}{T}|c_n|^2 \int_0^T \cos^2 n\omega_0 t \, dt = |c_n|^2.$$

Thus $|c_n|^2$ is the energy per period of the nth frequency component. Since the c_n's constitute the spectrum of the function $x(t)$, Parseval's theorem says that the energy in the spectrum is identical to the energy in the time function.

Signals having finite energy in the time interval T are referred to as being in the space $L_2(T)$; that is, for $x \in L_2(T)$

$$\int_0^T |x(t)|^2 \, dt < \infty.$$

Every signal in $L_2(T)$ can be represented by a Fourier series of the form of Eq. (2.6-3).

Optimum Property of the Fourier Series

As with any infinite series, the usefulness of the (generalized) Fourier expansion depends on a small number of terms furnishing an adequate approximation. This raises the question of whether the Fourier series using a given basis is necessarily the best possible way to represent a periodic function by a few terms.†

This question is easily answered by supposing that the finite sum $\sum_{n=1}^{N} \gamma_n \phi_n(t)$ is a better approximation to $x(t)$ than the sum using the c_n's defined in Eq. (2.6-4). If we use the mean-square error as our criterion, we get

$$\langle \epsilon^2 \rangle = \int_0^T |x(t) - \sum_{n=1}^{N} \gamma_n \phi_n(t)|^2 \, dt,$$

which by adding and subtracting $c_n \phi_n(t)$ can be written as

$$\langle \epsilon^2 \rangle = \int_0^T |x(t) - \sum_{n=1}^{N} (\gamma_n - c_n)\phi_n(t) - \sum_{n=1}^{N} c_n \phi_n(t)|^2 \, dt. \qquad (2.6-9)$$

---

†It is clear that some bases will fit a specified function better than others. However, since the choice of basis is generally determined by factors external to the problem of approximating a specific function, we assume the basis to be given.

If we expand the square, there will be cross-product terms of the form

$$\int_0^T x^*(t) \sum_{n=1}^N c_n \phi_n(t)\, dt = \sum_{n=1}^N c_n \int_0^T x^*(t)\phi_n(t)\, dt = \sum_{n=1}^N |c_n|^2 \qquad (2.6\text{-}10)$$

as well as terms like

$$\int_0^T \sum_{n=1}^N \sum_{m=1}^N c_n c_m^* \phi_n(t)\phi_m^*(t)\, dt = \sum_{n=1}^N |c_n|^2. \qquad (2.6\text{-}11)$$

The first of these follows from the definition of the c_n's and the second from the orthonormality of the ϕ_n's. We leave the details to the reader, but after all reductions of this sort are made one is left with

$$\langle \epsilon^2 \rangle = \int_0^T |x(t)|^2\, dt + \sum_{n=1}^N |\gamma_n - c_n|^2 - \sum_{n=1}^N |c_n|^2, \qquad (2.6\text{-}12)$$

and this is clearly minimized if $\gamma_n = c_n$. Thus the partial Fourier series is indeed an optimum approximation, at least in the mean-square-error sense.

The fact that the summation index goes from 1 to N doesn't necessarily mean that one gets the best approximation by using the first N harmonic terms. We see from Eq. (2.6-8) that if the $\phi_n(t)$ form a complete set, then

$$\int_0^T |x(t)|^2\, dt = \sum_{n=1}^\infty |c_n|^2. \qquad (2.6\text{-}13)$$

Therefore the minimum mean-square error becomes

$$\langle \epsilon^2 \rangle_{\min} = \sum_{n=1}^\infty |c_n|^2 - \sum_{n=1}^N |c_n|^2 = \sum_{n=N+1}^\infty |c_n|^2. \qquad (2.6\text{-}14)$$

Thus the mean-square error is equal to the sum of the squared magnitude of all the coefficients "left out" of the expansion. This suggests that the best approximation will be obtained by including in the finite sum the terms with the largest coefficients. It also shows that the mean-square error of the approximation becomes negligibly small if the magnitude of the left-out coefficients is small relative to the ones retained in the sum.

*2.7   Representation of Signals by Vectors†

It is sometimes convenient to think of the orthonormal functions $\phi_n(t)$ as unit vectors in an n-dimensional vector space. As an example consider the function

$$x_1(t) = 0.5\sqrt{\frac{2}{T}} \sin \frac{2\pi t}{T} + 0.3\sqrt{\frac{2}{T}} \sin \frac{6\pi t}{T}, \qquad (2.7\text{-}1)$$

which we may regard as a Fourier series consisting of two terms. The functions $\phi_1(t) = \sqrt{2/T} \sin(2\pi t/T)$ and $\phi_2(t) = \sqrt{2/T} \sin(6\pi t/T)$ are orthonormal over the period T. We can represent this function in a two-dimensional vector space as shown in Fig. 2.7-1(a). Another signal, $x_2(t) =$

†This material is used in Chapter 11.

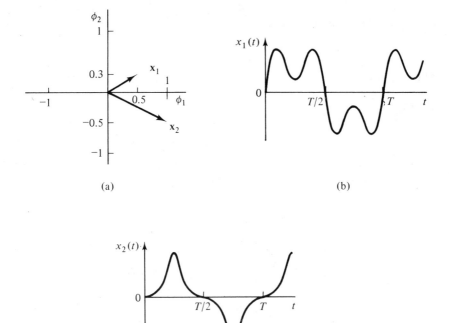

Figure 2.7-1. (a) Vector-space representation of two signals. (b) $x_1(t)$ as a time function. (c) $x_2(t)$ as a time function.

$\sqrt{2/T} \sin (2\pi t/T) - 0.5\sqrt{2/T} \sin (6\pi t/T)$ is also shown. The corresponding waveforms are shown in Figs. 2.7-1(b) and (c).

The characterization of signals by finite-dimensional vectors permits us to use such geometrical and vectorial concepts as distance, length, and dot product. Of particular importance in some of the later chapters is the concept of closeness: "How close is $x_1(t)$ to $x_2(t)$?" We shall find that the vector characterization of signals facilitates our visualization of this concept.

Consider all the signals $x(t)$ in $L_2(T)$ that can be *exactly* represented by the finite sum

$$x(t) = \sum_{n=1}^{N} x_n \phi_n(t), \qquad (2.7\text{-}2)$$

where the $\phi_n(t)$, $n = 1, \ldots, N$, constitute a particular set of orthonormal functions. Each of these signals can be represented by a single point or vector in an *N-dimensional vector subspace* M_N. This vector space is defined by the orthonormal functions $\phi_n(t)$ which constitute its unit vectors. The functions $\phi_n(t)$ are the *basis* of the vector space M_N, and one says that the space M_N

is *spanned* by the basis set $\{\phi_n\}$. Because Eq. (2.7-2) is an exact relation, we know that the components are

$$x_n = \int_0^T x(t)\phi_n^*(t) \, dt. \tag{2.7-3}$$

Thus every signal $x(t)$ in M_N is completely determined by the vector $\mathbf{x} = (x_1, x_2, \ldots, x_N)$; in fact, is equivalent to it.

The energy of $x(t) \in M_N$ is given by

$$E = \int_0^T |x(t)|^2 \, dt = \sum_{n=1}^N |x_n|^2 \equiv \|\mathbf{x}\|^2, \tag{2.7-4}$$

where $\|\mathbf{x}\|$ is called the *norm* (more precisely, *Euclidean norm*) of the vector $\mathbf{x}$. The energy E is essentially the Euclidean length.

The integral or inner product of two functions in M_N is denoted (x, y) and is defined by

$$(x, y) \equiv \int_0^T x(t)y^*(t) \, dt \tag{2.7-5}$$

$$= \sum_{n=1}^N \sum_{n=1}^N x_n y_m^* \int_0^T \phi_n(t)\phi_m^*(t) \, dt$$

$$= \sum_{n=1}^N x_n y_n^*. \tag{2.7-6}$$

The quantity $\sum_{n=1}^N x_n y_n^*$ is recognized as the dot or inner product of the two N-vectors $\mathbf{x} = (x_1, \ldots, x_N)$ and $\mathbf{y} = (y_1, \ldots, y_N)$. Hence the inner product of two functions is the dot product of their vector representations in an orthogonal space. The dot product is frequently symbolized by $\mathbf{x} \cdot \mathbf{y}$. Hence Eq. (2.7-6) is conveniently written $(x, y) = \mathbf{x} \cdot \mathbf{y}$. If we define ϕ_i $(i = 1, \ldots, N)$ as a unit vector with a one in the ith position and zeros in all other positions, then the vector $\mathbf{x}$ can be represented by

$$\mathbf{x} = \sum_{i=1}^N x_i \phi_i. \tag{2.7-7}$$

In this form $\mathbf{x}$ is expressed as the vectorial sum of its components along the orthogonal axes.

It is useful to talk about distances between vectors. It seems reasonable to expect a distance function d to satisfy

1. $d(\mathbf{x}, \mathbf{y}) = 0 \Rightarrow \mathbf{x} = \mathbf{y}$.
2. $d(\mathbf{x}, \mathbf{y}) \geq 0$ (negative distance is a meaningless concept).
3. $d(\mathbf{x}, \mathbf{y}) = d(\mathbf{y}, \mathbf{x})$ (the distance from $\mathbf{x}$ to $\mathbf{y}$ should be the same as the distance from $\mathbf{y}$ to $\mathbf{x}$).
4. $d(\mathbf{x}, \mathbf{y}) \leq d(\mathbf{x}, \mathbf{z}) + d(\mathbf{y}, \mathbf{z})$ (triangle inequality).

A useful distance is the generalized Euclidean distance, given by

$$d(\mathbf{x}, \mathbf{y}) = \left[\sum_{n=1}^N |x_n - y_n|^2 \right]^{1/2}. \tag{2.7-8}$$

With $\mathbf{z} = \mathbf{x} - \mathbf{y}$ and $z_n = x_n - y_n$, we obtain

$$\sum_{n=1}^{N} |x_n - y_n|^2 = \sum_{n=1}^{N} |z_n|^2 = \|\mathbf{z}\|^2. \tag{2.7-9}$$

It follows therefore from Eq. (2.7-4) that

$$\int_0^T |x(t) - y(t)|^2 \, dt = \|\mathbf{x} - \mathbf{y}\|^2. \tag{2.7-10}$$

The relation between norm, distance, and dot products is summarized by

$$\|\mathbf{x} - \mathbf{y}\| = d(\mathbf{x}, \mathbf{y}),$$

$$(\mathbf{x} - \mathbf{y}) \cdot (\mathbf{x} - \mathbf{y}) = d^2(\mathbf{x}, \mathbf{y}).$$

Projection

In Sec. 2.6 we considered the problem of how to best approximate a given finite energy signal $x(t)$ by a finite sum $x_N(t)$. In terms of the vector-space point of view we regard $x(t)$ as being in the infinite-dimensional vector space $L_2(T)$, and what we are doing when we approximate it is to map it into the finite subspace M_N which contains $x_N(t)$. We have already seen that the best approximation in the minimum mean-square-error sense is given by

$$x_N(t) = \sum_{n=1}^{N} x_n \phi_n(t), \tag{2.7.11}$$

where

$$x_n = (x, \phi_n) = \int_0^T x(t)\phi_n^*(t) \, dt. \tag{2.7-12}$$

We regard $x(t)$ as being represented by the infinite-dimensional vector $\mathbf{x} = \sum_{n=1}^{\infty} x_n \phi_n$ and $x_N(t)$ by the N-dimensional vector $\mathbf{x}_N$. The error in the approximation corresponds to the vector $\boldsymbol{\epsilon} = \mathbf{x} - \mathbf{x}_N$. The dot product of $\mathbf{x}_N$ and $\boldsymbol{\epsilon}$ is

$$\mathbf{x}_N \cdot \boldsymbol{\epsilon} = \mathbf{x}_N \cdot \mathbf{x} - \mathbf{x}_N \cdot \mathbf{x}_N = 0 \tag{2.7-13}$$

because

$$\mathbf{x}_N \cdot \mathbf{x} = \int_0^T x(t) x_N^*(t) \, dt$$

$$= \int_0^T x(t) \sum_{n=1}^{N} x_n^* \phi_n^*(t) \, dt$$

$$= \sum_{n=1}^{N} |x_n|^2 = \mathbf{x}_N \cdot \mathbf{x}_N.$$

Thus we see that $\mathbf{x}_N$ and $\boldsymbol{\epsilon}$ are orthogonal, i.e., perpendicular.

A useful geometrical interpretation is obtained by considering a three-dimensional vector $\mathbf{x}_3$ which is optimally approximated by a two-dimensional vector $\mathbf{x}_2$. This is shown in Fig. 2.7-2. Observe that the best approximation is given by $\mathbf{x}_2$ such that $\mathbf{x}_2$ and $\boldsymbol{\epsilon}$ are perpendicular. We therefore refer to $\mathbf{x}_2$

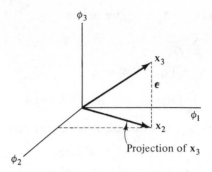

Figure 2.7-2. Projection of a three-dimensional vector onto two dimensions.

as the *orthogonal projection*, or simply the *projection*, of $\mathbf{x}_3$ into the two-dimensional vector subspace spanned by ϕ_1 and ϕ_2. Similarly, we can think of

$$x_N(t) = \sum_{n=1}^{N} x_n \phi_n(t)$$

as the *projection* of $x(t)$ into the N-dimensional subspace M_N if $x_n = (x, \phi_n)$. These ideas are formalized in the *projection theorem*, one form of which is the following: Let x be an element of the vector space M spanned by the orthogonal basis set $\{\phi_n\}$. Let x_N be an element in a lower-dimensional subspace $M_N \subset M$.† Then the norm of the error $\|x - x_N\|$ is a minimum if and only if x_N is the projection of x into M_N. Furthermore, the minimum error and the projection are orthogonal.

In the statement of the projection theorem we have omitted the boldface notation for vectors because the elements $\{x\}$ can be signals $\{x(t)\}$ in $L_2(T)$ function space or N-tuples $\{\mathbf{x}\}$ in the space R^N. It is not difficult to show that the set of real or complex functions in $L_2(T)$ is a linear vector space and, therefore, that treating $x(t)$ as a vector is justified. The norm of $x(t)$ in function space is defined by

$$\|x\| = \left[\int_0^T |x(t)|^2 \, dt \right]^{1/2}.$$

Two vectors $x(t), y(t) \in L_2(T)$ are orthogonal if their inner product is zero, i.e.,

$$\int_0^T x(t) y^*(t) \, dt = 0.$$

The close relationship between $x_N(t) \in M_N$ and $\mathbf{x}_N \in R^N$ is called an *isomorphism*. Specifically, the isomorphism is a one-to-one and onto map, T, from

---

†The set inclusion symbol $\subset$ implies that M_N is spanned by a subset of the same basis as M.

M_N to R^N described by

$$T[x_N(t)] = [(x_N, \phi_1), (x_N, \phi_2), \ldots, (x_N, \phi_N)].$$

The construction of an orthonormal basis for an arbitrary N-dimensional subspace is possible through use of the Gram-Schmidt procedure. It is discussed below.

Gram-Schmidt Procedure for Generating Orthonormal Functions

We have considered orthonormal function expansions such as the Fourier series that use "naturally" orthogonal functions such as the exponential or the trigonometric functions. However, it is possible to take any *arbitrary* set of functions $s_1(t), s_2(t), \ldots, s_M(t)$ and convert it into a set of orthonormal functions. We shall find use for this concept later in this book, especially in Chapter 11.

A simple method for generating a set of orthonormal functions is the Gram-Schmidt procedure, explained below. For simplicity we assume that the functions $s_i(t)$ are real; then the procedure is as follows:

1. *Construction of $\phi_1(t)$.* Any one of the functions, say $s_1(t)$, can be used to start the process. Thus

$$\phi_1(t) = \frac{s_1(t)}{\sqrt{E_1}}, \tag{2.7-14}$$

where

$$E_1 = \int_0^T s_1^2(t)\, dt. \tag{2.7-15}$$

It is clear that ϕ_1 is properly normalized since

$$\int_0^T \phi_1^2(t)\, dt = \frac{1}{E_1} \int_0^T s_1^2(t)\, dt = 1. \tag{2.7-16}$$

2. *Construction of $\phi_2(t)$.* First define an auxiliary function $\theta_2(t)$ by

$$\theta_2(t) = s_2(t) - (s_2, \phi_1)\phi_1(t), \tag{2.7-17}$$

where the notation

$$(s_2, \phi_1) = \int_0^T s_2(t)\phi_1(t)\, dt$$

signifies the inner product as in Eq. (2.7-5). This subtracts out any component of $s_2(t)$ in the direction of $\phi_1(t)$ and leaves a residual θ_2 which is orthogonal to ϕ_1. We again normalize by defining

$$E_2 = \int_0^T \theta_2^2(t)\, dt. \tag{2.7-18}$$

Then

$$\phi_2(t) = \frac{\theta_2(t)}{\sqrt{E_2}}. \tag{2.7-19}$$

From Eqs. (2.7-17) and (2.7-19) we obtain

$$\sqrt{E_2}\,\phi_2(t) = s_2(t) - (s_2, \phi_1)\phi_1(t). \tag{2.7-20}$$

Taking the inner product with ϕ_2 and noting that $(\phi_2, \phi_2) = 1$ and $(\phi_2, \phi_1) = 0$, we get

$$\sqrt{E_2} = (s_2, \phi_2). \tag{2.7-21}$$

Hence

$$s_2(t) = (s_2, \phi_1)\phi_1(t) + (s_2, \phi_2)\phi_2(t). \tag{2.7-22}$$

3. *General steps.* Assume that $\phi_1(t), \ldots, \phi_{l-1}(t)$ have been computed. To compute $\phi_l(t)$, first form the residual

$$\theta_l(t) = s_l(t) - \sum_{j=1}^{l-1} (s_l, \phi_j)\phi_j(t) \tag{2.7-23}$$

and then normalize by

$$\phi_l(t) = \frac{\theta_l(t)}{\sqrt{E_l}}. \tag{2.7-24}$$

It is easily shown that the set $\{\phi_l\}$ is an orthogonal set. For example, for $l \neq i$,

$$
\begin{aligned}
\int_0^T \phi_l(t)\phi_i(t)\,dt &= \frac{1}{\sqrt{E_l}} \int_0^T \theta_l(t)\phi_i(t)\,dt \\
&= \frac{1}{\sqrt{E_l}} \left[\int_0^T s_l(t)\phi_i(t)\,dt - \sum_{j=1}^{l-1} (s_l, \phi_j) \int_0^T \phi_j(t)\phi_i(t)\,dt \right] \\
&= \frac{1}{\sqrt{E_l}} [(s_l, \phi_i) - (s_l, \phi_i)] = 0,
\end{aligned}
\tag{2.7-25}
$$

where the last step follows from the orthonormality of the ϕ_i which cause all terms in the sum except the ith to vanish.

We see that each new function introduces at most one new orthonormal function. If $s_l(t)$ is not independent, it can be expressed as a linear combination of the earlier orthonormal functions. Then $\theta_l(t) = 0$, and no new function is created. Hence the number N of orthonormal functions satisfies $N \leq M$, where M is the number of distinct functions $s_k(t)$.

To illustrate the procedure, consider the three waveforms $s_1(t)$, $s_2(t)$, and $s_3(t)$ shown in Fig. 2.7-3. We leave the details of the computation to the reader (Prob. 2-18), but it is easily demonstrated by following steps 1, 2, and

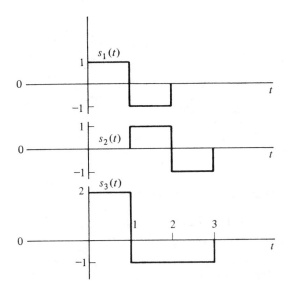

Figure 2.7-3. Set of three waveforms from which to construct an orthonormal set.

3 above that there are only two orthonormal functions:

$$\phi_1(t) = \frac{1}{\sqrt{2}} s_1(t), \tag{2.7-26}$$

$$\phi_2(t) = \frac{1}{\sqrt{6}} [s_1(t) + 2s_2(t)]. \tag{2.7-27}$$

The three signals are seen to be linearly dependent; in fact

$$s_1(t) = \sqrt{2}\,\phi_1(t), \tag{2.7-28a}$$

$$s_2(t) = -\frac{1}{\sqrt{2}}\phi_1(t) + \frac{3}{\sqrt{6}}\phi_2(t), \tag{2.7-28b}$$

$$s_3(t) = \frac{3}{\sqrt{2}}\phi_1(t) + \frac{3}{\sqrt{6}}\phi_2(t). \tag{2.7-28c}$$

A two-dimensional space is sufficient to represent the three signals, and the three vectors representing $s_1(t)$, $s_2(t)$, and $s_3(t)$ in the ϕ_1, ϕ_2 coordinate system are

$$s_1 = (\sqrt{2}, 0)$$

$$s_2 = \left(-\frac{1}{\sqrt{2}}, \frac{3}{\sqrt{6}}\right) \tag{2.7-29}$$

$$s_3 = \left(\frac{3}{\sqrt{2}}, \frac{3}{\sqrt{6}}\right).$$

These are shown in Fig. 2.7-4.

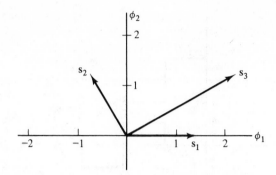

Figure 2.7-4. Vector representations of $s_i(t)$, $i = 1, 2, 3$.

The reader is invited to sketch the functions ϕ_1 and ϕ_2 to convince himself that Eqs. (2.7-28) give the original signal set.

2.8 The Fourier Integral

Fourier analysis can be extended to nonperiodic functions by regarding a nonperiodic function as a limit of a periodic signal with an infinitely long period. We first rewrite the Fourier series in the form

$$x(t) = \frac{1}{2\pi} \sum_{n=-\infty}^{\infty} X(\omega_n) e^{j\omega_n t} \, \Delta\omega, \qquad (2.8\text{-}1)$$

where

$$X(\omega_n) = \int_{-T/2}^{T/2} x(t) e^{-j\omega_n t} \, dt. \qquad (2.8\text{-}2)$$

Equation (2.8-2) is basically the same as Eq. (2.4-15) except that we have chosen to replace c_n by $X(\omega_n)/T$. Also, Eq. (2.8-1) is the same as Eq. (2.4-5), but we use the fact that $1/T = f_0 = \Delta\omega/2\pi$; i.e., the fundamental frequency is regarded as the frequency increment between successive harmonics.

As the period T increases, the frequency increment $\Delta\omega$ shrinks, and the discrete frequencies ω_n come closer together. Equation (2.8-1) can be approximated by an integral; in fact, this process is similar to the standard Riemann definition of the integral. Thus,

$$x(t) = \frac{1}{2\pi} \sum_{n=-\infty}^{\infty} X(\omega_n) e^{j\omega_n t} \, \Delta\omega \xrightarrow[T \to \infty]{} \frac{1}{2\pi} \int_{-\infty}^{\infty} X(\omega) e^{j\omega t} \, d\omega, \qquad (2.8\text{-}3)$$

or†

$$x(t) = \int_{-\infty}^{\infty} X(f) e^{j2\pi f t} \, df, \qquad \text{where } f = \frac{\omega}{2\pi}. \qquad (2.8\text{-}4)$$

†Observe that $X(\omega) = X(2\pi f) = \tilde{X}(f)$, where the $\tilde{X}$ indicates a different function. However, because $\tilde{X}(\cdot)$ differs from $X(\cdot)$ only by an argument factor of 2π, it is standard to use the same notation for both.

The expression given in Eq. (2.8-4) is sometimes preferred since it eliminates the factor 2π and results in a more symmetrical formulation.

The same limiting operation can formally be applied to Eq. (2.8-2) by simply extending the limits of integration to $\pm\infty$ and by replacing ω_n by ω. When this is done, however, we must make sure that the resulting infinite integral exists. The conditions under which this is true are known as the *Dirichlet conditions*. Essentially they require the function $x(t)$ to have only a finite number of discontinuities for $-\infty < t < \infty$ and to satisfy

$$\int_{-\infty}^{\infty} |x(t)|\, dt < \infty.†$$

(2.8-5)

Then we have the Fourier integral pair:

$$x(t) = \int_{-\infty}^{\infty} X(f)e^{j2\pi ft}\, df,$$

(2.8-6)

$$X(f) = \int_{-\infty}^{\infty} x(t)e^{-j2\pi ft}\, dt.$$

(2.8-7)

We refer to the second of these two equations as the Fourier transform and the first as the inverse Fourier transform. Instead of writing the integral expressions we frequently use the symbolism

$$X(f) = \mathcal{F}[x(t)]$$
$$x(t) = \mathcal{F}^{-1}[X(f)].$$

(2.8-8)

Also we generally use lowercase symbols to denote time functions and uppercase for Fourier transforms.

Observe that Eq. (2.8-5) is not satisfied by periodic time functions or functions having finite power at all times. However, by introducing the concept of the Dirac delta function (see Sec. 2.11) the Fourier integral definition can be extended to such functions as well.

Although a completely rigorous proof of the Fourier integral relations is beyond the scope of this text, we can present a simple demonstration of their validity. This is quite similar to the one given for the Fourier series in Sec. 2.5. We assume that $x(t)$ is absolutely integrable so that $X(f)$ exists. Then we write the inverse Fourier transform in the form

$$x(t) = \lim_{A\to\infty} \int_{-A}^{A} X(f)e^{j2\pi ft}\, df$$

(2.8-9)

and investigate the behavior of this limit. Substituting Eq. (2.8-7) for $X(f)$ results in

$$x(t) = \lim_{A\to\infty} \int_{-A}^{A} df\, e^{j2\pi ft} \int_{-\infty}^{\infty} x(\tau)e^{-j2\pi f\tau}\, d\tau.$$

(2.8-10)

†The precise form of the conditions is slightly more complicated. See, for instance, LePage [2-7].

Exchanging the orders of integration is generally permissible† and results in

$$x(t) = \lim_{A \to \infty} \int_{-\infty}^{\infty} x(\tau)\, d\tau \int_{-A}^{A} e^{j2\pi f(t-\tau)}\, df$$

$$= \lim_{A \to \infty} \int_{-\infty}^{\infty} x(\tau) \cdot 2A \operatorname{sinc}\left[2A(t - \tau)\right]\, d\tau. \qquad (2.8\text{-}11)$$

This expression is very similar to Eq. (2.5-4). As before, we can regard the integral as a weighted average of $x(\tau)$ with the weighting function $2A \operatorname{sinc} 2A(t - \tau)$. This function is shown in Fig. 2.8-1 for relatively large

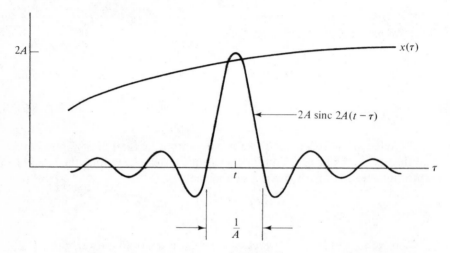

Figure 2.8-1. Weighting function $2A \operatorname{sinc} 2A(t - \tau)$ together with the smooth function $x(\tau)$.

A. It has a large narrow peak near $\tau = t$ and is otherwise oscillatory. As $A \to \infty$ the peak becomes infinitely large and infinitely narrow; hence the weighting is more and more concentrated near $\tau = t$. Then if $x(\tau)$ is sufficiently smooth (continuous and possessing a derivative), it can be replaced by $x(t)$ and taken outside of the integral as a constant. The remaining integration yields unity by Eq. (2.5-7), and Eq. (2.8-11) is shown to be an identity.

If $x(\tau)$ has an isolated discontinuity at $\tau = t$, the effect is exactly the same as in the Fourier series. At the discontinuity point the inverse Fourier transform has the value $\frac{1}{2}[x(t_+) + x(t_-)]$. For finite A in Eq. (2.8-11) the step in $x(\tau)$ becomes an oscillatory function whose frequency increases with A but whose amplitude is independent of A. This is the Gibbs phenomenon already referred to.

---

†A sufficient condition is that $x(\tau)$ be piecewise continuous and that the integral converge uniformly with respect to its upper limit [2-7].

2.9 Elementary Properties of the Fourier Transform

Parseval's Theorem

This theorem has already been discussed in connection with the Fourier series. If $X(f) = \mathcal{F}[x(t)]$, Parseval's theorem takes the form

$$\int_{-\infty}^{\infty} |x(t)|^2 \, dt = \int_{-\infty}^{\infty} |X(f)|^2 \, df. \tag{2.9-1}$$

The absolute-value signs are used since both $x(t)$ and $X(f)$ may be complex. As noted earlier, $|x(t)|^2$ can be identified with the instantaneous power in the signal $x(t)$, and therefore the expression on the left is the signal energy. This is assumed to be finite; i.e., the signal is assumed to be in class L_2. The expression on the right must then also be the signal energy, and the integrand $|X(f)|^2$ is called the *energy density*, expressed in joules per hertz. If $|X(f)|$ is larger in a certain range of frequencies $[f_1, f_2]$ than in the range $[f_3, f_4]$, we say that $x(t)$ has more energy in the band $[f_1, f_2]$ than in the band $[f_3, f_4]$. A useful generalization of Parseval's theorem is

$$\int_{-\infty}^{\infty} x(t)y^*(t) \, dt = \int_{-\infty}^{\infty} X(f)Y^*(f) \, df. \tag{2.9-2}$$

The derivation of both Eqs. (2.9-1) and (2.9-2) is left as an exercise.

Linearity

If $X(f) = \mathcal{F}[x(t)]$, $Y(f) = \mathcal{F}[y(t)]$, then

$$\mathcal{F}[ax(t) + by(t)] = aX(f) + bY(f). \tag{2.9-3}$$

This follows directly from the definition and the linearity of the integral operation.

Time Shift

If $X(f) = \mathcal{F}[x(t)]$, then

$$\mathcal{F}[x(t - a)] = X(f)e^{-j2\pi fa}. \tag{2.9-4}$$

Proof:

$$\mathcal{F}[x(t - a)] = \int_{-\infty}^{\infty} x(t - a)e^{-j2\pi ft} \, dt$$

$$= \int_{-\infty}^{\infty} x(\tau)e^{-j2\pi f(\tau + a)} \, d\tau$$

$$= e^{-j2\pi fa} \int_{-\infty}^{\infty} x(\tau)e^{-j2\pi f\tau} \, d\tau = X(f)e^{-j2\pi fa} \tag{2.9-5}$$

by a simple change of variable in line 2. Similarly, if $x(t)$ and $X(f)$ form a Fourier pair, then the inverse transform of $X(f - f_0)$ is $x(t)e^{j2\pi f_0 t}$. This result is called the *frequency-shift* property of the Fourier transform.

Scale Change

If $X(f) = \mathfrak{F}[x(t)]$,

$$\mathfrak{F}[x(at)] = \frac{1}{|a|} X\left(\frac{f}{a}\right). \tag{2.9-6}$$

Proof: Let $a > 0$; then

$$\int_{-\infty}^{\infty} x(at)e^{j2\pi ft}\, dt = \int_{-\infty}^{\infty} x(\tau)e^{j2\pi f(\tau/a)}\, \frac{d\tau}{a} = \frac{1}{a} X\left(\frac{f}{a}\right).$$

If $a < 0$, the change of variable also results in a reversal of the limits of integration, which causes a further sign change; Eq. (2.9-6) results.

The scale-change property is another manifestation of the reciprocal relation between time and frequency, already mentioned. It has the important consequence that short-duration signals have wide spectral widths and that, conversely, narrow spectra correspond to long-duration signals. This is demonstrated by the Fourier transform of the square pulse:

$$x(t) = \text{rect}\,(t) = \begin{cases} 1, & |t| \le \tfrac{1}{2}, \\ 0, & |t| > \tfrac{1}{2}. \end{cases} \tag{2.9-7}$$

The Fourier transform of this function is

$$X(f) = \int_{-1/2}^{1/2} e^{-j2\pi ft}\, dt = \text{sinc}\,(f). \tag{2.9-8}$$

By the scale-change property the spectrum of rect (at) is $(1/|a|)$ sinc $(f/|a|)$. The time function and its spectrum are shown in Fig. 2.9-1 for three different values of a.

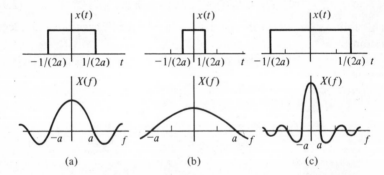

Figure 2.9-1. Reciprocal relation between time and frequency: (a) medium-length pulse, medium-width spectrum; (b) short pulse, wide spectrum; (c) long pulse, narrow spectrum.

Duality

If $X(f) = \mathfrak{F}[x(t)]$, then

$$\mathfrak{F}[X(t)] = x(-f).$$

This property, which follows directly from the defining integrals of the Fourier transformation, generates additional simple relationships. For example, we showed that

$$\mathcal{F}[\text{rect }(t)] = \text{sinc }(f). \tag{2.9-9}$$

Hence, by duality

$$\mathcal{F}[\text{sinc }(t)] = \text{rect }(f). \tag{2.9-10}$$

Modulation Property

If $X(f) = \mathcal{F}[x(t)]$, then

$$\mathcal{F}[x(t) \cos 2\pi f_0 t] = \tfrac{1}{2}[X(f - f_0) + X(f + f_0)]. \tag{2.9-11}$$

This follows by writing $\cos 2\pi f_0 t$ as $\tfrac{1}{2}(e^{j2\pi f_0 t} + e^{-j2\pi f_0 t})$ and then using the frequency-shift property. The effect is illustrated in Fig. 2.9-2.

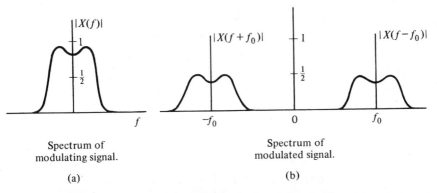

Spectrum of modulating signal.

(a)

Spectrum of modulated signal.

(b)

Figure 2.9-2. Illustration of the modulation property: (a) spectrum of modulating signal; (b) spectrum of modulated signal.

Differentiation and Integration

If $X(f) = \mathcal{F}[x(t)]$, then

$$\mathcal{F}\left[\frac{dx}{dt}\right] = j2\pi f X(f). \tag{2.9-12}$$

This follows directly by differentiating the Fourier integral under the integral sign. The process can obviously be repeated, so that if the nth derivative exists, then

$$\mathcal{F}\left[\frac{d^n x}{dt^n}\right] = (j2\pi f)^n X(f). \tag{2.9-13}$$

If the nth derivative does not exist, then $(j2\pi f)^n X(f)$ will generally not satisfy the Dirichlet conditions [Eq. (2.8-5)].

Equation (2.9-12) shows that if we set

$$x(t) = \frac{dy(t)}{dt},$$ (2.9-14)

so that $X(f) = j2\pi f Y(f)$, then

$$Y(f) = \frac{X(f)}{j2\pi f}.$$ (2.9-15)

It is tempting to conclude from Eq. (2.9-15) that

$$\mathcal{F}\left[\int_{-\infty}^{t} x(\tau)\, d\tau\right] = \frac{X(f)}{j2\pi f}.$$ (2.9-16)

This approach, although it sometimes results in the right answer, does not consider arbitrary constants of integration.

A better method to obtain the formula for the transform of the integral is to evaluate it formally using integration by parts:

$$\mathcal{F}\left[\int_{-\infty}^{t} x(\tau)\, d\tau\right] = \int_{-\infty}^{\infty} e^{-j2\pi ft} \int_{-\infty}^{t} x(\tau)\, d\tau\, dt$$ (2.9-17)

$$= -\frac{e^{-j2\pi ft}}{j2\pi f} \int_{-\infty}^{t} x(\tau)\, d\tau \Big|_{-\infty}^{\infty} + \frac{X(f)}{j2\pi f}.$$ (2.9-18)

The first term on the right will be zero if

$$X(0) = \int_{-\infty}^{\infty} x(\tau)\, d\tau = 0,$$ (2.9-19)

and if this is true, Eq. (2.9-16) holds. If it is not true, then $y(t)$ does not satisfy the Dirichlet condition, and formally, therefore, the Fourier transform of the integral does not exist. We shall find, however, that this difficulty, as well as several others in the formal Fourier theory, can be circumvented by the use of delta functions. We shall see below (Sec. 2.11) that with this addition to the theory, the Fourier transform of the integral does exist as long as $X(0)$ is finite, and its form is

$$\mathcal{F}\left[\int_{-\infty}^{t} x(\tau)\, d\tau\right] = X(f)\left[\frac{\delta(f)}{2} + \frac{1}{j2\pi f}\right].$$ (2.9-20)

The duality theorem can be applied to these results to generate two additional theorems: If $X(f) = \mathcal{F}[x(t)]$, and if the indicated Fourier transforms exist, then

$$\mathcal{F}[t^n x(t)] = (-j2\pi)^{-n} \frac{d^n X(f)}{df^n},$$ (2.9-21)

and

$$\mathcal{F}\left[\frac{x(t)}{t}\right] = -j2\pi \int_{-\infty}^{f} X(u)\, du.$$ (2.9-22)

Symmetry Properties

Although both $x(t)$ and its Fourier transform $X(f)$ can be complex functions, important simplifications are possible when these functions are either purely real or imaginary.

Thus, suppose that $x(t)$ is real. Then by inspection of the Fourier integral it follows that

$$X(-f) = X^*(f). \qquad (2.9\text{-}23)$$

The function $X(f)$ is said to have *Hermitian* symmetry. It implies that

$$|X(-f)| = |X(f)|,$$
$$\arg X(-f) = -\arg X(f). \qquad (2.9\text{-}24)$$

Thus for a real function of time we find that the amplitude of the Fourier transform is an even function, while the phase is odd.

If we set

$$X(f) = A(f) + jB(f), \qquad (2.9\text{-}25)$$

it follows from Eq. (2.9-23) that

$$A(f) = A(-f)$$
$$B(f) = -B(-f); \qquad (2.9\text{-}26)$$

i.e., the real part of $X(f)$ is even and the imaginary part odd. Analogous relations hold if $x(t)$ is pure imaginary. In this case it is easily seen that $X(f)$ is *skew Hermitian*; i.e.,

$$X(-f) = -X^*(f).$$

Multiplication and Convolution

Let $X(f)$, $Y(f)$, and $Z(f)$ be the Fourier transforms of $x(t)$, $y(t)$, and $z(t)$, respectively. Then if

$$Z(f) = X(f)Y(f),$$

it follows that

$$z(t) = \int_{-\infty}^{\infty} x(\tau)y(t - \tau)\, d\tau. \qquad (2.9\text{-}27)$$

Proof: Fourier-transforming Eq. (2.9-27) gives

$$\mathcal{F}[z(t)] = \int_{-\infty}^{\infty} \int_{-\infty}^{\infty} x(\tau)y(t - \tau)\, d\tau\, e^{-j2\pi ft}\, dt$$

$$= \int_{-\infty}^{\infty} \int_{-\infty}^{\infty} x(\tau)\, e^{-j2\pi f\tau}\, y(t - \tau)e^{-j2\pi f(t-\tau)}\, d\tau\, dt, \qquad (2.9\text{-}28)$$

where we have simply multiplied the integrand by $e^{j\pi f(t-\tau)} = 1$. Exchanging orders of integration and making the change of variable $s = t - \tau$ in the second integral, we obtain

$$\mathcal{F}[z(t)] = \int_{-\infty}^{\infty} x(\tau)e^{-j2\pi ft}\, d\tau \int_{-\infty}^{\infty} y(s)\, e^{-j2\pi fs}\, ds$$

$$= X(f)Y(f), \tag{2.9-29}$$

which is seen to be an identity.

The integral on the right in Eq. (2.9-27) is referred to as the *convolution* integral, and we say that $x(t)$ is convolved with $y(t)$. A simple notation for Eq. (2.9-27) is

$$z(t) = x(t) * y(t). \tag{2.9-30}$$

This notation symbolizes the fact that the operation of convolution is similar to that of multiplication, i.e., that it is commutative, associative, and distributive: $x * y = y * x$; $x * (y * z) = (x * y) * z$; $(x + y) * z = x * z + y * z$. These relations are easily derived; see Prob. 2-23.

Generally speaking, convolution is a kind of smoothing or averaging operation in which one of the functions is weighted by the other. Of course the nature of the smoothing depends on the functions involved, and if one of the functions is "badly behaved," e.g., the derivative of the Dirac delta of Sec. 2.11, the signal $z(t)$ may be anything but smooth. As a simple example, if in Eq. (2.9-27) we set $y(t)$ $= (1/T)$ rect (t/T),

$$z(t) = \frac{1}{T} \int_{t-T/2}^{t+T/2} x(\tau)\, d\tau. \tag{2.9-31}$$

This shows that $z(t)$ is simply a running average of $x(t)$ with averaging time T.

By use of the duality theorem it is easily shown that if

$$z(t) = x(t)y(t),$$

then

$$Z(f) = \int_{-\infty}^{\infty} X(\lambda)Y(f - \lambda)\, d\lambda$$

or

$$Z(f) = X(f) * Y(f). \tag{2.9-32}$$

The fact that convolution of two time functions is equivalent to multiplication of their Fourier transforms [Eq. (2.9-27)] is doubtlessly the most important property of the Fourier integral. It permits the complicated process of convolution to be replaced by the much simpler process of multiplication. We shall see in the next chapter that the output of a linear time-invariant system is the convolution of the input with the impulse response of the system.† The Fourier transform of the impulse response is the *transfer function* already briefly mentioned in Sec. 2.4. Because of the convolution-multiplication property of the Fourier integral, the transform of the output is the product of the transform of the input and the transfer function. This

---

†The impulse response is precisely that: the response of a system to an input that consists of an impulse.

is the extension of the important result given in Eq. (2.4-4) for periodic signals to nonperiodic signals.

The properties of the Fourier transform discussed in this chapter are summarized in Table 2.9-1.

TABLE 2.9-1   PROPERTIES OF THE FOURIER TRANSFORMATION

Name	Time Function	Fourier Spectrum
1. Parseval's Theorem	$\int_{-\infty}^{\infty} x(t)y^*(t)\,dt$	$\int_{-\infty}^{\infty} X(f)Y^*(f)\,df$
2. Linearity	$ax(t) + by(t)$	$aX(f) + bY(f)$
3. Time shift	$x(t - a)$	$X(f)e^{-j2\pi fa}$
4. Scale change	$x(at)$	$\dfrac{1}{|a|}X(f/a)$
5. Duality	$X(t)$	$x(-f)$
6. Frequency shift	$x(t)e^{j2\pi f_0 t}$	$X(f - f_0)$
7. Modulation property	$x(t)\cos 2\pi f_0 t$	$\frac{1}{2}[X(f - f_0) + X(f + f_0)]$
8. Differentiation	dx/dt	$j2\pi f X(f)$
9. Integration	$\int_{-\infty}^{t} x(\tau)\,d\tau$	$X(f)\left[\dfrac{\delta(f)}{2} + \dfrac{1}{j2\pi f}\right]$
10. Multiplication by t^n (also called frequency differentiation)	$t^n x(t)$	$(-j2\pi)^{-n}\,d^n X(f)/df^n$
11. Symmetry	Real $x(t)$	$X(-f) = X^*(f)$
	Imaginary $x(t)$	$X(-f) = -X^*(f)$
12. Convolution	$\int_{-\infty}^{\infty} x(\tau)y(t - \tau)\,d\tau$	$X(f)\cdot Y(f)$
13. Multiplication	$x(t)y(t)$	$\int_{-\infty}^{\infty} X(\lambda)Y(f - \lambda)\,d\lambda$

2.10 Some Useful Fourier Pairs

Square Pulse

This function and its Fourier transform have already been discussed; cf. Eqs. (2.9-7) and (2.9-8). Thus we have the pair

$$\text{rect}\,(t) \longleftrightarrow \text{sinc}\,(f). \tag{2.10-1}$$

The two functions are shown in Fig. 2.9-1. They are both real and even. The notation rect $(\cdot)$ and sinc $(\cdot)$ is due to P. M. Woodward [2-9].

Triangular Pulse

The triangular pulse (see Fig. 2.10-1) is given by

$$x(t) = \begin{cases} 1 - |t|, & |t| \le 1, \\ 0, & |t| > 1. \end{cases} \tag{2.10-2}$$

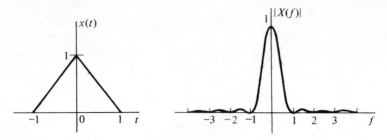

Figure 2.10-1. Triangular pulse and its Fourier spectrum.

One way to compute the Fourier transform of $x(t)$ is to think of it as the convolution of two rectangular pulses:

$$x(t) = \int_{-\infty}^{\infty} \text{rect } (\tau) \text{ rect } (t - \tau) \, d\tau. \tag{2.10-3}$$

The integrand of this expression is shown for several values of t in Fig. 2.10-2; the area of the shaded overlap region is the value of the integral.

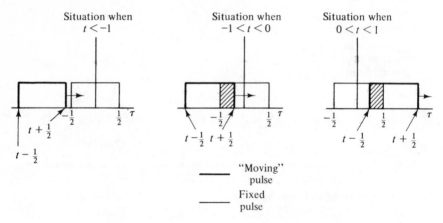

Figure 2.10-2. Convolution of two rectangular pulses.

Observe that there is no overlap for $|t| > 1$ so that the area is zero; for $|t| < 1$ the overlap is a linear function of t which is maximum when $t = 0$. Thus the triangular shape results. Because of the convolution property and Eq. (2.10-1), the corresponding spectrum is immediately written as

$$X(f) = \text{sinc}^2 (f). \tag{2.10-4}$$

The triangular pulse and its spectrum are shown in Fig. 2.10-1. The triangular pulse is "smoother" than the rectangular pulse since it is continuous, whereas the rectangular pulse is not. This greater smoothness is reflected in a reduced high-frequency content—the amplitude of $\text{sinc}^2 (f)$

is much smaller than that of $|\text{sinc}\,(f)|$ for $f > 1$. This is a general and very important observation—the smoother the time function, the smaller its high-frequency content.

Gaussian Pulse

By using the identity

$$\int_{-\infty}^{\infty} e^{-(ax^2+bx)}\, dx = \sqrt{\frac{\pi}{a}}\, e^{b^2/4a}, \tag{2.10-5}$$

which holds for complex a and b as long as $\text{Re}\,a > 0$, the following Fourier pair is easily derived:

$$e^{-\pi t^2} \longleftrightarrow e^{-\pi f^2}. \tag{2.10-6}$$

The pulse $e^{-\pi t^2}$ has a Gaussian, i.e., bell, shape and is known as the Gaussian pulse. Thus the Gaussian pulse and its Fourier transform are identical in form. By using the scale-change theorem (property 4 in Table 2.9-1) it is easily shown that for

$$x(t) = e^{-at^2},$$

we have

$$X(f) = \sqrt{\left|\frac{\pi}{a}\right|}\, e^{-\pi^2 f^2/a},$$

or

$$X(\omega) = \sqrt{\left|\frac{\pi}{a}\right|}\, e^{-\omega^2/4a}. \tag{2.10-7}$$

Exponential Pulse

In linear systems theory the asymmetrical exponential pulse occurs frequently. Its form is

$$x(t) = e^{-at}u(t), \quad \text{Re}\,a > 0, \tag{2.10-8}$$

where $u(t)$ is the unit step function:

$$u(t) = \begin{cases} 1, & t > 0, \\ 0, & t < 0. \end{cases} \tag{2.10-9}$$

The Fourier transform of this pulse is easily shown to be

$$X(f) = \frac{1}{j\omega + a} = \frac{1}{\sqrt{a^2 + \omega^2}}\, e^{-j\,\tan^{-1}(\omega/a)}. \tag{2.10-10}$$

See Fig. 2.10-3. Since

$$\frac{d^n}{d\omega^n}\left(\frac{1}{j\omega + a}\right) = \frac{n!(-j)^n}{(j\omega + a)^{n+1}},$$

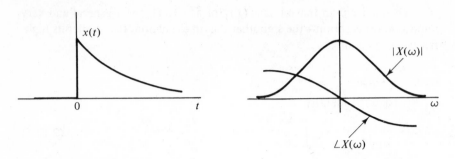

Figure 2.10-3. Asymmetrical exponential pulse $\exp(-at)\cdot u(t)$ and its Fourier transform.

it easily follows from the frequency-differentiation rule (property 10 in Table 2.9-1) that

$$\mathcal{F}\left[\frac{t^n}{n!}\,e^{-at}u(t)\right] = \frac{1}{(j\omega + a)^{n+1}}. \tag{2.10-11}$$

Also, since the symmetrical exponential pulse $e^{-a|t|}$ can be written as $x(t) + x(-t)$, where $x(t)$ is given by Eq. (2.10-8), we have the additional pair

$$\mathcal{F}[e^{-a|t|}] = \frac{1}{j\omega + a} + \frac{1}{-j\omega + a} = \frac{2a}{\omega^2 + a^2}. \tag{2.10-12}$$

Many additional Fourier pairs can be derived from the exponential pulse pair. For instance, since a in Eq. (2.10-8) can be complex, Fourier transforms for functions of the form $e^{-at}\sin(bt)\cdot u(t)$ or $e^{-at}\cos(bt)\cdot u(t)$ are straightforward extensions. (See Prob. 2-27.)

Signum Function

The signum function, written sgn (t), is defined by

$$\text{sgn}(t) = \begin{cases} 1, & t > 0, \\ 0, & t = 0, \\ -1, & t < 0. \end{cases} \tag{2.10-13}$$

This function does not satisfy the Dirichlet conditions and therefore, strictly speaking, has no Fourier transform. The usual approach is to deal with the function $e^{-a|t|}\,\text{sgn}(t)$, which for $a > 0$ does satisfy the Dirichlet condition [see Fig. 2.10-4(a)]. This function is very similar to the symmetrical exponential pulse of Eq. (2.10-12) except that the portion for $t < 0$ is negative. Therefore

$$\mathcal{F}[e^{-a|t|}\,\text{sgn}(t)] = \frac{1}{j\omega + a} - \frac{1}{-j\omega + a} = \frac{-2j\omega}{\omega^2 + a^2}. \tag{2.10-14}$$

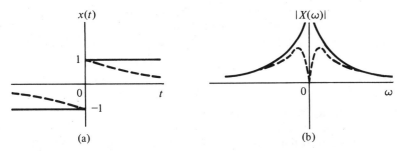

Figure 2.10-4. Signum function and its Fourier spectrum. (a) The solid line shows the signum function, and the dashed line the approximation $\exp(-a|t|) \cdot \text{sgn}(t)$, which satisfies the Dirichlet conditions. (b) The Fourier spectrum of the signum function (solid lines) and of the approximation (dashed lines). Observe that the approximation is excellent except at the frequency origin.

It is now possible to go to the limit of $a \to 0$, so that

$$\mathscr{F}[\text{sgn}\,(t)] = \lim_{a \to 0} \frac{-2j\omega}{\omega^2 + a^2} = \frac{2}{j\omega}. \qquad (2.10\text{-}15)$$

Spectra for finite a and for the limit as $a \to 0$ are shown in Fig. 2.10-4(*b*). Observe that the approximation for small a is very good except near the frequency origin.

Equation (2.10-15) implies that the inverse Fourier transform of $1/j\omega$ is $1/2\,\text{sgn}\,(t)$. The formal demonstration of this is illuminating. Thus consider

$$x(t) = \frac{1}{2\pi} \int_{-\infty}^{\infty} \frac{e^{j\omega t}}{j\omega}\,d\omega$$

$$= \frac{1}{2\pi j} \int_{-\infty}^{\infty} \frac{\cos \omega t}{\omega}\,d\omega + \frac{1}{2\pi} \int_{-\infty}^{\infty} \frac{\sin \omega t}{\omega}\,d\omega. \qquad (2.10\text{-}16)$$

The second term gives the desired $\frac{1}{2}\,\text{sgn}\,(t)$; this follows from the fact that

$$\int_{-\infty}^{\infty} \frac{\sin mx}{x}\,dx = \begin{cases} \pi, & m > 0, \\ 0, & m = 0, \\ -\pi, & m < 0. \end{cases} \qquad (2.10\text{-}17)$$

(See any standard table of integrals, e.g., Peirce and Foster [2-1], formula 499.) Thus we get the correct result if we interpret the first term as being zero. This is true if the improper integral is defined as the *Cauchy principal value* (CPV):

$$\int_{-\infty}^{\infty} \frac{\cos \omega t}{\omega}\,d\omega \equiv \lim_{\substack{\tau \to \infty \\ \epsilon \to 0}} \left(\int_{-\tau}^{-\epsilon} \frac{\cos \omega t}{\omega} + \int_{\epsilon}^{\tau} \frac{\cos \omega t}{\omega}\,d\omega \right). \qquad (2.10\text{-}18)$$

Since the integrand is an odd function of ω and the limits are symmetrical (by definition of CPV), the integral vanishes.

More generally, use of the CPV integration in evaluating inverse Fourier transforms makes it possible to obtain transforms of functions that do not satisfy the Dirichlet conditions. Therefore when the defining integrals of the Fourier transform or its inverse are improper, the CPV definition is always implied.

2.11 The Dirac Delta Function

The restriction of formal Fourier integral theory to functions satisfying the Dirichlet conditions is inconvenient in applications, since one often wants to deal with signals that have finite average power at all times and hence infinite energy. Examples of such signals are dc and periodic waveforms and certain random signals such as continuous noise, speech, music, or television signals. Also, it is desirable to combine the Fourier series and the Fourier integral into a unified theory so that the Fourier series can be regarded as just a special case of the Fourier integral. It turns out that the proper use of the Dirac delta function takes care of practically all of these objectives. We emphasize the words "proper use," because the delta function is not really a function in the normal sense, and some of its properties must be obtained by means of complicated, and not always clearly rigorous, limiting arguments. Many of the theoretical difficulties surrounding the delta function can be eliminated by using the theory of distributions which puts many intuitive arguments on a rigorous foundation ([2-4], p. 269, and [2-8]). For our purposes a fairly elementary treatment will suffice, but the interested reader is referred to the book by Friedman [2-8].

The delta function $\delta(t)$ is often "defined" as a function that is zero everywhere except for $t = 0$ where it is infinite and such that

$$\int_{-\infty}^{\infty} \delta(t) \, dt = 1. \tag{2.11-1}$$

A more satisfactory definition is to regard $\delta(t)$ as the limit of one of several pulses; for instance,

$$\delta(t) = \lim_{a \to \infty} a \operatorname{rect}(at), \tag{2.11-2}$$

or

$$\delta(t) = \lim_{a \to \infty} a \operatorname{sinc}(at), \tag{2.11-3}$$

or

$$\delta(t) = \lim_{a \to \infty} a e^{-\pi a^2 t^2}. \tag{2.11-4}$$

These are shown, for large a, in Fig. 2.11-1. The exact shape of these functions is immaterial. Their important features are (1) unit area and (2) rapid decrease to zero for $t \neq 0$.

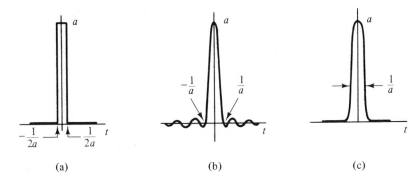

Figure 2.11-1. Pulses that approach the delta function as a limit: (a) rectangular pulse; (b) sinc pulse; (c) Gaussian pulse.

Still another definition is to call any function a delta function if for arbitrary continuous $f(t)$ it satisfies the integral equation

$$\int_{-\infty}^{\infty} f(\tau)\delta(\tau - t)\, d\tau = f(t). \tag{2.11-5}$$

This definition can, of course, be related to the previous one, since any of the pulses in Eqs. (2.11-2)–(2.11-4), when substituted for $\delta(t)$ in Eq. (2.11-5), will essentially furnish the same result when a is very large. This follows because the integrand is significantly nonzero only for $\tau \simeq t$. It can therefore be approximately evaluated by replacing $f(\tau)$ by the constant $f(t)$. Then since the delta function has unit area, the result follows. The way in which the delta function picks out a particular value of $f(t)$ is sometimes referred to as *sifting*, and Eq. (2.11-5) is therefore called a *sifting* integral. Because delta functions ordinarily appear in such sifting integrals, Eq. (2.11-5) is often regarded as a more "fundamental" definition than the ones given earlier in this paragraph.†

Parenthetically, it might be noted that integral equations similar to Eq. (2.11-5), i.e., of the form

$$\int_{a}^{b} f(\tau)g(t, \tau)\, d\tau = \lambda f(t), \tag{2.11-6}$$

play an important role in analysis [2-6]. The function $g(t, \tau)$ is referred to as the *kernel* of the integral equation, and for a given kernel such an equation generally has solutions only for particular values of λ called *eigenvalues*. The corresponding solutions are the *eigenfunctions*. According to this point of view, we find that when the kernel is a delta function any continuous function is an eigenfunction, and the eigenvalues are all equal to 1.

Observe that $\delta(t) = \delta(-t)$; therefore Eq. (2.11-5) is a convolution of

---

†It also can be shown that the earlier definitions do not uniquely specify $\delta(t)$. See, for example, [2-4], p. 270.

$f(t)$ and $\delta(t)$. Hence if the Fourier transform of $f(t)$ is $F(\omega)$, we should have, from property 12 in Table 2.9-1, that

$$F(\omega)\mathfrak{F}[\delta(t)] = F(\omega),$$

or

$$\mathfrak{F}[\delta(t)] = \int_{-\infty}^{\infty} \delta(t)e^{-j\omega t}\, dt = 1. \tag{2.11-7}$$

This can be regarded as still another definition of the delta function: A delta function is the (inverse) Fourier transform of 1. This gives rise to the further useful relation

$$\int_{-\infty}^{\infty} e^{\pm j2\pi ft}\, df = \int_{-\infty}^{\infty} \cos 2\pi ft\, df = \delta(t), \tag{2.11-8}$$

and similarly,

$$\int_{-\infty}^{\infty} e^{\pm j2\pi ft}\, dt = \int_{-\infty}^{\infty} \cos 2\pi ft\, dt = \delta(f). \tag{2.11-9}$$

Note that if the infinite integral is regarded as a limit, one has

$$\lim_{a\to\infty} \int_{-a/2}^{a/2} e^{j2\pi ft}\, df = \lim_{a\to\infty} a\,\text{sinc}\,(at) = \delta(t). \tag{2.11-10}$$

Thus Eq. (2.11-8) is consistent with Eq. (2.11-3).

There is a possible source of error in using these definitions with the variable ω rather than f. If in Eq. (2.11-8) we make the change of variable $\omega = 2\pi f$, we get

$$\int_{-\infty}^{\infty} e^{\pm j\omega t}\, d\omega = 2\pi\delta(t), \tag{2.11-11}$$

$$\int_{-\infty}^{\infty} e^{\pm j\omega t}\, dt = 2\pi\delta(\omega). \tag{2.11-12}$$

Equation (2.11-12) is exactly the same as Eq. (2.11-11) except that the roles of t and ω are interchanged. However, we see from this that the Fourier transform of 1 is $\delta(f)$ or $2\pi\delta(\omega)$; it is *not* $\delta(\omega)$.

One occasionally encounters derivatives of delta functions. By using Eq. (2.11-7) with the differentiation rule (property 8 in Table 2.9-1), we find that the Fourier transform of $\delta'(t)$ is $j\omega$. Therefore

$$\int_{-\infty}^{\infty} f(\tau)\delta'(\tau - t)\, d\tau \longleftrightarrow j\omega F(\omega) \longleftrightarrow \int_{-\infty}^{\infty} f'(\tau)\delta(\tau - t)\, d\tau, \tag{2.11-13}$$

or, in general,

$$\int_{-\infty}^{\infty} f(\tau)\frac{d^n}{d\tau^n}[\delta(\tau - t)]\, d\tau = \int_{-\infty}^{\infty} \frac{d^n f(\tau)}{d\tau^n}\,\delta(\tau - t)\, d\tau. \tag{2.11-14}$$

This result can also be obtained by using integration by parts to evaluate the left-hand side of Eq. (2.11-13).

2.12 Applications of the Delta Function

Unit Step Function

The unit step function is commonly used in circuit analysis. It has the value 1 for $t > 0$ and 0 for $t < 0$ and can be written in the form

$$u(t) = \tfrac{1}{2} + \tfrac{1}{2}\, \text{sgn}\,(t), \tag{2.12-1}$$

where sgn (t) is the signum function introduced in Sec. 2.10 [Eq. (2.10-13)]. The transform of the signum function was shown in Eq. (2.10-15) to be given by $2/j\omega$; hence

$$\mathcal{F}[u(t)] = \pi\delta(\omega) + \frac{1}{j\omega} = \frac{\delta(f)}{2} + \frac{1}{j2\pi f}. \tag{2.12-2}$$

This result can be used to show that the formula for the Fourier transform of the integral is as given in Eq. (2.9-20). Thus let

$$y(t) = \int_{-\infty}^{t} x(\tau)\, d\tau. \tag{2.12-3}$$

This can be regarded as the convolution of $x(t)$ with a unit step function; i.e.,

$$y(t) = \int_{-\infty}^{\infty} x(\tau)u(t - \tau)\, d\tau = x(t) * u(t). \tag{2.12-4}$$

Then the multiplication-convolution rule (property 12 in Table 2.9-1) can be applied to give

$$Y(\omega) = X(\omega)\left[\pi\delta(\omega) + \frac{1}{j\omega}\right] = X(f)\left[\frac{\delta(f)}{2} + \frac{1}{j2\pi f}\right], \tag{2.12-5}$$

which is the same result given in Eq. (2.9-20).

Observe the fallacy of arguing that since the unit step function is the integral of the delta function its Fourier transform should just be $1/j\omega$. This fallacy results from neglecting a constant of integration.

Ramp Function

This function is obtained directly as the integral of a square pulse or equivalently by convolution of the square pulse with the unit step function:

$$r(t) = \int_{-\infty}^{\infty} \text{rect}\,(\tau)u(t - \tau)\, d\tau. \tag{2.12-6}$$

The Fourier transform is therefore

$$R(f) = \text{sinc}\,(f)\left[\frac{\delta(f)}{2} + \frac{1}{j2\pi f}\right]$$

$$= \frac{\delta(f)}{2} + \frac{\text{sinc}\,(f)}{j2\pi f}$$

$$= \pi\delta(\omega) + \frac{\sin\,(\omega/2)}{j\omega^2/2}. \tag{2.12-7}$$

In going from the first line to the second we use the fact that for any continuous function $\phi(t)$, $\delta(t)\phi(t) = \delta(t)\phi(0)$, which is easily checked by use of the limiting relations for the delta function, i.e., Eqs. (2.11-2)–(2.11-4). In the third line we have used the fact that $\omega = 2\pi f$ and $\pi\delta(\omega) = \frac{1}{2}\delta(\omega/2\pi)$. The ramp and its Fourier transform are shown in Fig. 2.12-1.

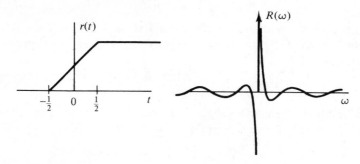

Figure 2.12-1. Ramp function and its Fourier transform.

Periodic Signals

The use of delta functions permits us to easily extend the definition of the Fourier integral to periodic functions. In particular, for

$$x(t) = \cos 2\pi f_0 t, \tag{2.12-8}$$

we have

$$X(f) = \int_{-\infty}^{\infty} \cos 2\pi f_0 t\, e^{-j2\pi ft}\, dt$$

$$= \frac{1}{2} \int_{-\infty}^{\infty} (e^{j2\pi(f_0 - f)t} + e^{-j2\pi(f_0 + f)t})\, dt$$

$$= \frac{1}{2}[\delta(f - f_0) + \delta(f + f_0)], \tag{2.12-9}$$

by Eq. (2.11-9). This is pictured in Fig. 2.12-2. The spectrum of the cosine consists of two *lines* or *spikes*. Similarly, for $x(t) = \sin \omega_0 t$, $X(f) = \frac{1}{2}[-j\delta(f - f_0) + j\delta(f + f_0)]$; thus the spectrum consists of the same two lines, but they are associated with 90° phase shifts. For a zero frequency

Figure 2.12-2. Cosine wave and its Fourier spectrum.

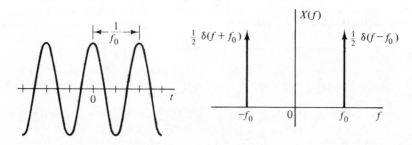

cosine wave, i.e., a dc signal, the two spikes coalesce, and therefore the spectrum of the dc signal is a single line at the origin of the frequency scale.

Observe that for a cosine wave lasting for a finite time T we have

$$x(t) = \cos 2\pi f_0 t \cdot \text{rect}\left(\frac{t}{T}\right), \qquad (2.12\text{-}10)$$

for which the spectrum, by the convolution theorem, is given by

$$X(f) = \frac{1}{2}\int_{-\infty}^{\infty} [\delta(f' - f_0) + \delta(f' + f_0)]T \text{ sinc } T(f - f')\, df'$$

$$= \frac{T}{2}[\text{sinc } T(f - f_0) + \text{sinc } T(f + f_0)]. \qquad (2.12\text{-}11)$$

This is shown in Fig. 2.12-3 for $Tf_0 \gg 1$. The two infinitely narrow lines in Fig. 2.12-2 have become spectral lobes of finite width. Equation (2.12-11) is a particular instance of a sifting integral. It illustrates the ease with which convolutions involving delta functions can be performed.

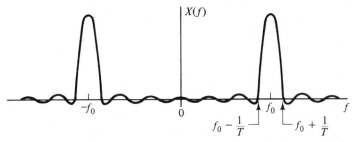

Figure 2.12-3. Spectrum of finite-length cosine wave. The frequency is f_0, and the length of the pulse is T.

The Fourier transform of an arbitrary periodic function $x_p(t)$ can be obtained by first expanding $x_p(t)$ in a Fourier series and then taking the Fourier transform. The transform of each term of the series produces a delta function so that the final result is an infinite series of delta functions at each of the harmonic frequencies.

This procedure can be generalized somewhat by first defining the periodic function $x_p(t)$ in terms of a nonperiodic function $x(t)$ by the series expansion

$$x_p(t) = \sum_{n=-\infty}^{\infty} x(t - nT). \qquad (2.12\text{-}12)$$

It is assumed that $x(t)$ is chosen so that this series converges to $x_p(t)$. It is easily seen that $x_p(t)$ as defined in Eq. (2.12-12) satisfies the basic definition of a periodic function given in Eq. (2.4-1):

$$x_p(t + T) = \sum_{n=-\infty}^{\infty} x[t - (n - 1)T] = x_p(t). \qquad (2.12\text{-}13)$$

A possible and obvious choice for $x(t)$ is a function which equals $x_p(t)$ over one period and is zero elsewhere. However, this is clearly not the only possibility; in fact there is generally an infinite number of ways of choosing $x(t)$ in Eq. (2.12-12) to yield a given $x_p(t)$. According to a convenient notation introduced by Woodward [2-9], Eq. (2.12-12) can be written as

$$x_p(t) = \text{rep}_T\, x(t). \qquad (2.12\text{-}14)$$

We can think of $x(t)$ as being the *generating function* which generates the periodic function $x_p(t)$.

We now expand $x_p(t)$ in a Fourier series:

$$x_p(t) = \sum_{m=-\infty}^{\infty} X_m e^{j2\pi mt/T}, \qquad (2.12\text{-}15)$$

where the coefficients are given by

$$
\begin{aligned}
X_m &= \frac{1}{T} \int_{-T/2}^{T/2} x_p(t) e^{-j2\pi mt/T}\, dt \\
&= \frac{1}{T} \int_{-T/2}^{T/2} \sum_{n=-\infty}^{\infty} x(t - nT) e^{-j2\pi mt/T}\, dt \\
&= \sum_{n=-\infty}^{\infty} \frac{1}{T} \int_{-T/2}^{T/2} x(t - nT) e^{-j2\pi m(t-nT)/T}\, dt \\
&= \sum_{n=-\infty}^{\infty} \frac{1}{T} \int_{-T/2-nT}^{T/2-nT} x(\tau) e^{-j2\pi m\tau/T}\, d\tau.
\end{aligned}
\qquad (2.12\text{-}16)
$$

In going from the second line to the third line we have exchanged the orders of summation and integration and multiplied by $e^{j2\pi mn} = 1$. In going from the third line to the fourth line we have made the change of variable $\tau = t - nT$. Observe now that the summands in Eq. (2.12-16) consist of integrals over adjacent segments of the τ axis; hence performing the summation is equivalent to integrating over infinite limits. Hence

$$X_m = \frac{1}{T} \int_{-\infty}^{\infty} x(\tau) e^{-j2\pi m\tau/T}\, d\tau = \frac{1}{T} X\left(\frac{m}{T}\right), \qquad (2.12\text{-}17)$$

where $X(f)$ is the Fourier transform of $x(t)$. Thus we obtain the Fourier series representation of $x_p(t)$ in terms of the Fourier transform of the generating function $x(t)$:

$$x_p(t) = \sum_{n=-\infty}^{\infty} x(t - nT) = \frac{1}{T} \sum_{m=-\infty}^{\infty} X\left(\frac{m}{T}\right) e^{j2\pi mt/T}. \qquad (2.12\text{-}18)$$

This equation is one form of *Poisson's sum formula*. The coefficients $(1/T)X(m/T)$ are identical with the coefficients c_m of the exponential form of the Fourier series [cf. Eq. (2.4-5)]; in fact, Eq. (2.12-18) is nothing more than a Fourier series. However, we now have a relation between the standard Fourier coefficients c_n and the Fourier transform of the nonperiodic generating function $x(t)$. We have found, in fact, that the Fourier coefficients are proportional to the sampled Fourier transform at points m/T, i.e., at integral multiples of the fundamental frequency. This is the principle on which much of the theory of signal sampling is based, and we shall see it again in Chapter 4.

We now proceed formally to Fourier transform equation (2.12-18) with the result

$$X_p(f) = \frac{1}{T} \int_{-\infty}^{\infty} x_p(t) e^{-j2\pi ft} \, dt = \frac{1}{T} \int_{-\infty}^{\infty} \sum_{m=-\infty}^{\infty} X\left(\frac{m}{T}\right) e^{-j2\pi[f-(m/T)]t} \, dt.$$

$$(2.12\text{-}19)$$

Exchanging orders of integration and summation and using the definition for the delta function given in Eq. (2.11-8), we obtain the Fourier pair

$$\sum_{n=-\infty}^{\infty} x(t - nT) \longleftrightarrow \frac{1}{T} \sum_{m=-\infty}^{\infty} X\left(\frac{m}{T}\right) \delta\left(f - \frac{m}{T}\right). \qquad (2.12\text{-}20)$$

This can be written in simplified notation using the rep operation defined in Eq. (2.12-14) and a *comb* operation also defined by Woodward [2-9] as follows:

$$\text{comb}_F X(f) = \sum_{m=-\infty}^{\infty} X(mF) \delta(f - mF), \qquad (2.12\text{-}21)$$

where F is the repetition frequency of the delta functions. In simplified notation Eq. (2.12-20) is

$$\text{rep}_T x(t) \longleftrightarrow \frac{1}{T} \text{comb}_{1/T} X(f). \qquad (2.12\text{-}22)$$

The comb of a sinc function is shown in Fig. 2.12-4. Note the essential similarity to Fig. 2.4-3(a). The figure does, in fact, represent the spectrum of a train of rectangular pulses with pulse length a and repetition frequency F. It can therefore also be regarded as a representation of the Fourier series coefficients of such a pulse train. Equations (2.12-20) and (2.12-21) are the desired Fourier transforms of a periodic function.

A periodic function of considerable interest in later work is the ideal sampling function given by

$$\text{rep}_T \, \delta(t) = \sum_{n=-\infty}^{\infty} \delta(t - nT) = \text{comb}_T \, (1). \qquad (2.12\text{-}23)$$

Figure 2.12-4. The function comb$_F$ sinc (af).

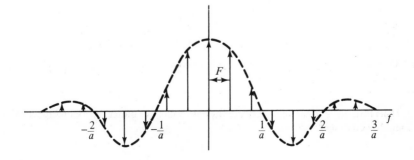

This is simply a comb of delta functions spaced T seconds apart. The Fourier transform of the delta function is 1 by Eq. (2.11-7); hence from Eq. (2.12-20) we get

$$\sum_{n=-\infty}^{\infty} \delta(t - nT) \longleftrightarrow \frac{1}{T} \sum_{m=-\infty}^{\infty} \delta\left(f - \frac{m}{T}\right), \qquad (2.12\text{-}24)$$

and we find that except for scale the Fourier transform of a comb of delta functions is another comb of delta functions†:

$$\text{comb}_T(1) \longleftrightarrow \frac{1}{T} \text{comb}_{1/T}(1). \qquad (2.12\text{-}25)$$

We observe that the derivation of Eq. (2.12-20) could have been simplified by introducing the comb operation initially and deriving Eq. (2.12-25). Then use of the multiplication-convolution rule would have given the equivalent of Eq. (2.12-20) directly:

$$x(t) * \text{comb}_T(1) \longleftrightarrow \frac{1}{T} X(f) \text{comb}_{1/T}(1) = \frac{1}{T} \text{comb}_{1/T} X(f). \qquad (2.12\text{-}26)$$

We summarize Secs. 2.10–2.12 by the Fourier pairs in Table 2.12-1. The

TABLE 2.12-1 FOURIER PAIRS

Name	$x(t)$	$X(f)$
1. Square pulse	rect (t)	sinc f
2. Triangular pulse	$1 - \lvert t \rvert$ for $\lvert t \rvert \leq 1$, 0 for $\lvert t \rvert > 1$	sinc$^2 f$
3. Gaussian pulse	$e^{-\pi t^2}$	$e^{-\pi f^2}$
4. Asymmetric exponential pulse	$e^{-at} u(t)$	$1/(j\omega + a)$
5. Symmetric exponential pulse	$e^{-a\lvert t \rvert}$	$2a/(\omega^2 + a^2)$
6. Signum function	sgn (t)	$2/j\omega$
7. Delta function	$\delta(t)$	1
8. Unit step	$u(t)$	$\dfrac{\delta(f)}{2} + \dfrac{1}{j2\pi f}$
9. Cosine	$\cos 2\pi f_0 t$	$\frac{1}{2}[\delta(f - f_0) + \delta(f + f_0)]$
10. Periodic functions	$\text{rep}_T x(t) = \sum\limits_{n=-\infty}^{\infty} x(t - nT)$	$\dfrac{1}{T}\text{comb}_{1/T} X(f) = \dfrac{1}{T} \sum\limits_{n=-\infty}^{\infty} X\left(\dfrac{n}{T}\right)\delta\left(f - \dfrac{n}{T}\right)$

†This formula is so useful in deriving various results in Fourier theory that it probably should be committed to memory.

entries in this table together with the ones in Table 2.9-1 make it possible to find the Fourier transforms of most of the functions of practical interest almost by inspection. This process has already been used in several places. For instance, we obtained the spectrum for the triangular pulse by observing that this pulse could be obtained by convolving two square pulses, and the spectrum then followed from property 12 in Table 2.9-1. We obtained the transform of the unit step function by noting that it was the sum of a constant ($\frac{1}{2}$) and $\frac{1}{2}$ sgn (t), etc. For more complicated examples, see Prob. 2-31 and 2-32.

2.13   Time Functions with One-Sided Spectra

In applications dealing with modulation, narrowband filters, or narrowband signals it is often very convenient to use time functions having one-sided spectra. We observed earlier (Sec. 2.9) that the Fourier transform $X(f)$ of a real signal $x(t)$ has Hermitian symmetry; i.e., $X(-f) = X^*(f)$. Thus, a signal having a nonzero spectrum only for positive frequencies must be a complex signal. However, since for real signals the spectrum for negative frequencies is completely determined by the spectrum for positive frequencies, there cannot be any loss of information if we simply remove the negative-frequency portion; i.e., the complex signal that results from this operation must somehow be equivalent to the original real signal.

To investigate the implications of one-sided spectra, consider the real function of time $x(t)$ having the spectrum $X(f)$ with the usual symmetry properties. To convert this to a one-sided spectrum we define the new spectrum

$$\Xi(f) = X(f) [1 + \text{sgn} (f)]; \qquad (2.13\text{-}1)$$

that is,

$$\Xi(f) = \begin{cases} 0, & f < 0, \\ 2X(f), & f > 0. \end{cases} \qquad (2.13\text{-}2)$$

This is shown in Fig. 2.13-1; in effect we remove the spectrum from the negative-frequency side and pile it on top of the spectrum on the positive-frequency side; the area under the magnitude of the spectrum remains the same.

The resulting time function is obtained by inverse-Fourier-transforming Eq. (2.13-1); the result is the complex signal

$$\xi(t) = x(t) * \left[\delta(t) - \frac{1}{j\pi t} \right], \qquad (2.13\text{-}3)$$

where we have used property 12 of Table 2.9-1 (convolution-multiplication) and pairs 6 and 7 in Table 2.12-1. The negative sign is one of the consequences of the duality property. Performing the indicated convolution results in

$$\xi(t) = x(t) + j\hat{x}(t), \qquad (2.13\text{-}4)$$

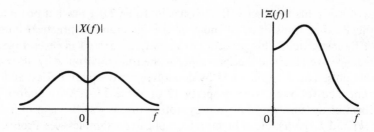

Figure 2.13-1. Symmetric spectrum and the one-sided form obtained by removing the negative-frequency portion and adding it to the positive-frequency side.

where

$$\hat{x}(t) = \frac{1}{\pi} \int_{-\infty}^{\infty} \frac{x(\tau)}{t - \tau} \, d\tau \qquad (2.13\text{-}5)$$

is the *Hilbert transform* of $x(t)$. A complex waveform whose imaginary part is the Hilbert transform of the real part is referred to as a *complex analytic waveform*. The term "analytic" is used in the sense of complex-variable theory. A complex function $f(p)$ of the complex variable $p = t + ju$ is analytic in a region R if it has no poles or other singularities in that region (more precisely if it satisfies the Cauchy-Riemann equations at every point in R). For a complex function $\xi(p)$ that is analytic over the upper half-plane (i.e., for all $p = t + ju$ such that $u > 0$) it can be shown that the real and imaginary parts of $\xi(p)$ on the t axis are related by the Hilbert transformation. A demonstration is given in LePage [2-7], Sec. 7-10.

The defining equation [Eq. (2.13-5)] for the Hilbert transform is an improper integral since the integrand generally goes to infinity for $\tau = t$. The transform is therefore defined as the Cauchy principal value:

$$\hat{x}(t) = \lim_{\substack{\epsilon \to 0 \\ A \to \infty}} \frac{1}{\pi} \int_{-A}^{-\epsilon} + \int_{\epsilon}^{A} \frac{x(\tau)}{t - \tau} \, d\tau. \qquad (2.13\text{-}6)$$

The use of the Cauchy principal value here is consistent with its use in the inverse transformation of $1/j\omega$, i.e., Eq. (2.10-18).

We see from Eq. (2.13-5) that the Hilbert transform of $x(t)$ is the convolution of $x(t)$ with $1/\pi t$:

$$\hat{x}(t) = x(t) * \frac{1}{\pi t}. \qquad (2.13\text{-}7)$$

Therefore the Fourier transform of the Hilbert transform is given by

$$\mathcal{F}[\hat{x}(t)] \equiv \hat{X}(f) = -jX(f) \, \text{sgn} \, (f). \qquad (2.13\text{-}8)$$

Thus it is theoretically possible to generate the Hilbert transform of a signal by passing it through a linear circuit that provides a 90° phase *lag* at all positive frequencies and a 90° phase *lead* at all negative frequencies. As will be seen in the next chapter, such a circuit cannot be exactly realized because of constraints imposed by causality. However, for narrowband signals (i.e., where the Fourier spectrum has significant magnitude only in a narrow range of frequencies around some frequency f_0), a circuit that provides 90° phase lag in the band around f_0 and 90° phase lead in the band around $-f_0$ can be reasonably well approximated. Such circuits are used in connection with single-sideband modulation systems and will be discussed in Chapter 6.

Some of the properties of the Hilbert transform and of complex analytic signals are best illustrated by means of examples.

EXAMPLES

1. Consider the signal

$$x(t) = \cos \omega_0 t. \tag{2.13-9}$$

The Hilbert transform is

$$\hat{x}(t) = \frac{1}{\pi} \int_{-\infty}^{\infty} \frac{\cos \omega_0 \tau}{t - \tau} \, d\tau$$

$$= \frac{1}{\pi} \int_{-\infty}^{\infty} \frac{\cos \omega_0 (\tau - t + t)}{t - \tau} \, d\tau$$

$$= \frac{\cos \omega_0 t}{\pi} \int_{-\infty}^{\infty} \frac{\cos \omega_0 (t - \tau)}{t - \tau} \, d\tau + \frac{\sin \omega_0 t}{\pi} \int_{-\infty}^{\infty} \frac{\sin \omega_0 (t - \tau)}{t - \tau} \, d\tau$$

$$= \sin \omega_0 t. \tag{2.13-10}$$

The result in Eq. (2.13-10) follows from the fact that the first CPV integral in the third line is zero, i.e., the integrand is odd, and the second CPV integral has value π for $\omega_0 > 0$. The complex analytic signal corresponding to $\cos \omega_0 t$ is

$$\xi(t) = \cos \omega_0 t + j \sin \omega_0 t = e^{j\omega_0 t}. \tag{2.13-11}$$

Thus we find that the complex analytic signal corresponding to a cosine (or sine) wave is just the complex phasor used in circuit theory. Note that the spectrum of the cosine wave consists of two delta functions at $\pm \omega_0$ having an area of $\frac{1}{2}$; the spectrum of the complex analytic signal is a single delta function at ω_0 having an area of 1. This illustrates how the two-sided spectrum of the real time function is converted to a one-sided spectrum as discussed earlier; cf. Fig. 2.13-1.

2. Communications systems often utilize signals of the form $f(t) \cos (\omega_0 t + \phi)$ where the spectrum of $f(t)$ contains only low frequencies and ω_0 is a relatively high frequency. We shall be interested in the complex analytic form for this kind of signal.

A more general version of this problem is obtained by considering the signal

$$x(t) = f(t)g(t), \tag{2.13-12}$$

where $f(t)$ is a low-frequency signal and $g(t)$ a high-frequency signal. Stated more precisely, let $F(\omega)$ and $G(\omega)$ be the Fourier transforms of $f(t)$ and $g(t)$, respectively;

then these transforms have the property

$$F(\omega) = 0, \qquad |\omega| > 2\pi W$$
$$G(\omega) = 0, \qquad |\omega| < 2\pi W. \tag{2.13-13}$$

See Fig. 2.13-2. The desired Hilbert transform of the product is obtained by an indirect method. By use of the multiplication-convolution property of the Fourier

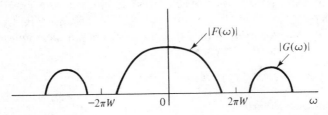

Figure 2.13-2. Nonoverlapping spectra of two functions $f(t)$ and $g(t)$.

transform and Eq. (2.13-8) expressed in terms of ω rather than f, we find that the Fourier transform of $\hat{x}(t)$ is

$$\mathcal{F}[\hat{x}(t)] = \mathcal{F}[\widehat{f(t)g(t)}] = -\frac{j}{2\pi} \operatorname{sgn}(\omega) \int_{-\infty}^{\infty} F(\omega - u)G(u)\, du. \tag{2.13-14}$$

Also,

$$\mathcal{F}[f(t)\hat{g}(t)] = -\frac{j}{2\pi} \int_{-\infty}^{\infty} \operatorname{sgn}(u)F(\omega - u)G(u)\, du. \tag{2.13-15}$$

These two results are identical. To show this, consider their difference:

$$\frac{j}{2\pi} \int_{-\infty}^{\infty} [\operatorname{sgn}(\omega) - \operatorname{sgn}(u)]F(\omega - u)G(u)\, du.$$

The product $F(\omega - u)G(u)$ is shown in Fig. 2.13-3 for $\omega > 0$. Note that if the two spectra overlap they do so for $u > 0$, but then $[\operatorname{sgn}(\omega) - \operatorname{sgn}(u)] = 0$. A similar argument can be made for $\omega < 0$. Thus as long as the spectra $F(\omega)$ and $G(\omega)$ have no region of overlap we find that

$$\hat{x}(t) = f(t)\hat{g}(t). \tag{2.13-16}$$

Figure 2.13-3. Overlap of the spectra $F(\omega - u)$ and $G(u)$ for positive ω.

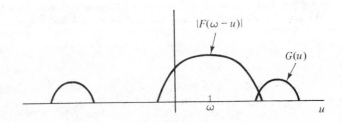

Thus we have the important result that *only the high-frequency factor in the signal $f(t)g(t)$ is Hilbert transformed* when the Hilbert transform of the product is computed.†

For the signal of the form

$$x(t) = f(t) \cos(\omega_0 t + \phi), \qquad (2.13\text{-}17)$$

where $f(t)$ is the low-frequency envelope of a high-frequency carrier, we therefore find that the complex analytic form is

$$\xi(t) = f(t)e^{j(\omega_0 t + \phi)}. \qquad (2.13\text{-}18)$$

Thus the complex analytic signal is again identical with the familiar phasor form. In some cases $e^{j\phi}$ is also a low-frequency time function. Then $\mu(t) = f(t)e^{j\phi(t)}$ is called the *complex envelope* of the signal. Its modulus $|\mu(t)| = |\xi(t)|$ is the *amplitude modulation*, and the phase $\phi(t)$ is the *phase modulation* of the signal $x(t)$. This simple representation of amplitude and phase modulation constitutes one of the main advantages of the complex signal notation. If the Fourier spectrum of the complex envelope is

$$M(f) = \int_{-\infty}^{\infty} \mu(t)e^{-j2\pi ft}\, dt, \qquad (2.13\text{-}19)$$

then

$$\Xi(f) = M(f - f_0),$$

where $f_0 = \omega_0/2\pi$ and $\Xi(f) = \mathfrak{F}[\xi(t)]$. The complex envelope function $\mu(t)$ is generally not complex analytic, and therefore $M(f)$ will not generally be zero for negative argument. However, if $\xi(t)$ is to be complex analytic, $\Xi(f)$ must be zero for negative f; this requires that $M(f)$ be zero for $f < -f_0$. (See Fig. 2.13-4.) This is approximately satisfied if $x(t)$ is *narrowband*, i.e., if the extent of $M(f)$ along the frequency axis is much smaller than the center frequency f_0. For this reason the

†Generalizations of this theorem to two- and higher-dimensional functions exist. See [2-10] and [2-11].

Figure 2.13-4. Spectrum of a narrowband complex analytic signal. For center frequency $f = f_0$ it is necessary that $M(f) = 0$ for $f < -f_0$.

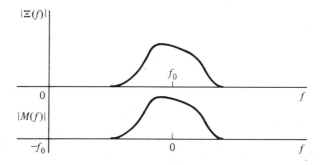

concepts of the complex analytic signal and the complex envelope are generally useful mainly for narrowband signals.

3. As a final example we consider the Hilbert transform of the square pulse:

$$x(t) = \text{rect}\left(\frac{t - T/2}{T}\right). \tag{2.13-20}$$

The computation of the transform is straightforward, although as with all convolutions involving discontinuous functions it is necessary to consider several cases in performing the integration. The result can be put in the form

$$\hat{x}(t) = \frac{1}{\pi} \ln\left|\frac{t}{T - t}\right|. \tag{2.13-21}$$

The signal and its Hilbert transform are shown in Fig. 2.13-5.

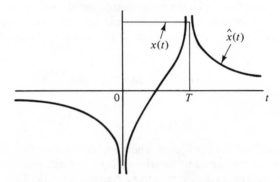

Figure 2.13-5. Square pulse and its Hilbert transform. The places where $x(t)$ goes to infinity are sometimes called "horns," and they may cause problems in communication systems in which the Hilbert transforms of signals are involved, such as in single sideband.

2.14 Summary

In this chapter we have explored the relationship between time and frequency as embodied in the Fourier series and the Fourier integral. Conversion to the frequency domain via one of these two methods is convenient because if one knows the Fourier spectrum of the input to a linear system one can find the output spectrum by simply multiplying the input spectrum by the transfer function or frequency response of the system. The use of both the Fourier series and the Fourier integral is simplified by the use of a number of standard properties and theorems which generally make it relatively easy to find the Fourier transforms of arbitrary functions. These theorems and some of the more commonly used Fourier transform pairs have been summarized in Tables 2.9-1 and 2.12-1. Although the Fourier series is generally an expansion of a function in terms of trigonometric functions (sines and cosines) or exponential functions, the idea can be generalized to arbitrary orthogonal or orthonormal functions. The implications of this generalization, and some of the more useful properties of orthogonal-function expansions, have been discussed in Secs. 2.6 and 2.7. Impulse functions are important in Fourier analysis because among other things they provide a bridge between the Fourier series and the Fourier integral. These functions and their applications have been discussed in Secs. 2.11 and 2.12. In the final section we discussed the complex analytic signal representation and the Hilbert transform, which permit the two-sided spectra of ordinary (real) signals to be replaced by single-sided spectra.

PROBLEMS

2-1. Assume that a system is characterized by an operator $\mathcal{H}$ that relates output $y(t)$ to input $x(t)$ according to

$$y(t) = \mathcal{H}[x(t)].$$

If $\mathcal{H}$ is causal, meaning that $y(t)$ doesn't depend on future values of $x(t)$, what constraints must be applied to the time interval during which values of $x(t)$ contribute to $y(t)$?

2-2. A given system, described by an operator $\mathcal{H}$, gives the response

$$y(t) = \mathcal{H}[x(t)]$$
$$= x(t^3).$$

Is the system time-invariant?

2-3. A particular zero-memory system is described by

$$y(t) = \mathcal{H}[x(t)]$$
$$= \alpha x(t) + \beta.$$

Is the system linear? Is it time-invariant?

2-4. Compute the Fourier coefficients c_n for all n in the representation

$$x(t) = \sum_{n=-\infty}^{\infty} c_n e^{jn\omega_0 t}$$

for the waveform shown in Fig. P2-4. Plot the magnitude and phase of c_n.

Figure P2-4

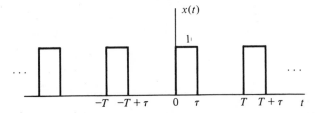

2-5. Repeat Prob. 2-4 for the waveform shown in Fig. P2-5.

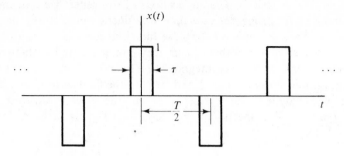

Figure P2-5

2-6. Repeat Prob. 2-4 for the waveform shown in Fig. P2-6. Use any properties of the Fourier series that can simplify the computations. *Hint*: Consider differentiating $x(t)$.

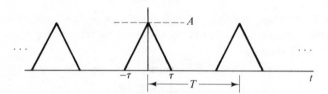

Figure P2-6

2-7. Use Eq. (2.4-22) and the appropriate Fourier series property to rapidly determine the Fourier coefficients (magnitude and phase) of the waveform shown in Fig. P2-7.

Figure P2-7

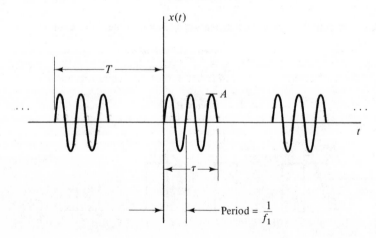

2-8. Consider the following three forms of the Fourier series,

$$x(t) = \sum_{n=-\infty}^{\infty} c_n e^{jn\omega_0 t} \qquad \text{(form 1)}$$

$$x(t) = c_0 + \sum_{n=1}^{\infty} x_n \cos(n\omega_0 t + \phi_n) \qquad \text{(form 2)}$$

$$x(t) = a_0 + \sum_{n=1}^{\infty} a_n \cos n\omega_0 t + b_n \sin n\omega_0 t \qquad \text{(form 3)},$$

and establish the relations between the coefficients. Suggest which form might be most convenient under what circumstance.

2-9. In form 3 of Prob. 2-8, what simplifications occur if
a. $x(t)$ is odd: $x(t) = -x(-t)$?
b. $x(t)$ is even: $x(t) = x(-t)$?
c. $x(t)$ has half-wave odd symmetry:
   $x(t) = -x[t + (T/2)]$?

2-10. Prove that the function

$$S_N(t) \equiv \sum_{n=-N}^{N} e^{jn\omega_0 t}$$

can be written in closed form as

$$S_N(t) = \frac{\sin(N + \tfrac{1}{2})\omega_0 t}{\sin(\omega_0/2)t}.$$

Carefully sketch the result and show the period. What happens when $N \longrightarrow \infty$?

2-11. Show that $S_N(t)$ in Prob. 2-10 is approximately given by

$$S_N(t) \simeq 2N \operatorname{sinc} 2Nf_0 t$$

when t is near the origin, i.e., $t \simeq 0$ and N is large.

2-12. In the circuit shown in Fig. P2-12 the diode is assumed ideal.

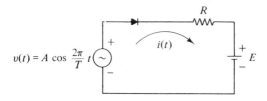

$v(t) = A \cos \frac{2\pi}{T} t$

Figure P2-12

a. With $A > E$, write an expression for the current $i(t)$.
b. Compute the Fourier coefficients $\{c_n\}$ in the Fourier series representation of $i(t)$.
c. Compute the average power dissipated in the resistor at frequencies $n/T, n = 0, 1, 2, \ldots$. How many of the first n terms account for 90% of the total power dissipated if $A = 100$ volts and $E = 50$ volts?

2-13. Obtain, by inspection, a Fourier series representation for

$$x(t) = 2 \cos \omega_0 t + 5 \cos 2\omega_0 t + 4.$$

2-14. A complete orthonormal set $\{\phi_i(t)\}$ over the interval $T = [-1, 1]$ can be obtained by applying the Gram-Schmidt procedure to the sequence $1, t, t^2, t^3, \ldots$. Show that

$$\phi_0(t) = \frac{1}{\sqrt{2}}, \quad \phi_1(t) = \left(\frac{3}{2}\right)^{1/2} t, \quad \phi_2(t) = \left(\frac{5}{2}\right)^{1/2}\left(\frac{3}{2}t^2 - \frac{1}{2}\right), \quad \text{etc.}$$

The resultant functions

$$P_n(t) \equiv \left(\frac{2}{2n+1}\right)^{1/2} \phi_n(t)$$

are known as the Legendre polynomials.

2-15. Work out the details leading up to Eq. (2.6-12), and show that the generalized Fourier coefficients are indeed optimum for minimizing the mean-square error in representing a function $x(t)$.

2-16. a. Show that for the Fourier coefficients

$$c_n = \frac{1}{T} \int_{-T/2}^{T/2} x(t) e^{-jn\omega_0 t} \, dt, \qquad n = 0, \pm 1, \ldots,$$

Parseval's theorem takes the form

$$\frac{1}{T} \int_{-T/2}^{T/2} |x(t)|^2 \, dt = \sum_{n=-\infty}^{\infty} |c_n|^2,$$

where $x(t)$ has period T.

b. What is the average power per cycle for the waveform

$$x(t) = \sum_{n=-\infty}^{\infty} \text{rect}\left(\frac{t - nT}{\tau}\right), \qquad \tau < T.$$

c. Can you suggest a way in which Parseval's theorem can be used to evaluate the series $\sum_{n=-\infty}^{\infty} |c_n|^2$?

2-17. Instead of the Euclidean distance given in Eq. (2.7-8), consider the quantity

$$q(\mathbf{x}, \mathbf{y}) = \sum_{n=1}^{N} |x_n - y_n|.$$

Show that $q(\mathbf{x}, \mathbf{y})$ satisfies the properties of a distance and hence can be considered as such.

2-18. Starting with the three waveforms shown in Fig. 2.7-3, show that

$$\phi_1(t) = \frac{1}{\sqrt{2}} s_1(t)$$

$$\phi_2(t) = \frac{1}{\sqrt{6}} [s_1(t) + 2s_2(t)]$$

and that the three signals are given by

$$s_1(t) = \sqrt{2}\phi_1(t)$$

$$s_2(t) = -\frac{1}{\sqrt{2}}\phi_1(t) + \frac{3}{\sqrt{6}}\phi_2(t)$$

$$s_3(t) = \frac{3}{\sqrt{2}}\phi_1(t) + \frac{3}{\sqrt{6}}\phi_2(t).$$

2-19. Let $x(t)$ and $X(f)$ form a Fourier transform pair. Prove Parseval's theorem,

$$\int_{-\infty}^{\infty} |x(t)|^2\, dt = \int_{-\infty}^{\infty} |X(f)|^2\, df,$$

and its generalization

$$\int_{-\infty}^{\infty} x(t)y^*(t)\, dt = \int_{-\infty}^{\infty} X(f)Y^*(f)\, df.$$

2-20. Consider the two signals

$$x_1(t) = \sqrt{2a_1}\, e^{-a_1 t}\, u(t)$$

$$x_2(t) = \sqrt{2a_2}\, e^{-a_2 t}\, u(t), \qquad a_1, a_2 > 0.$$

Assume that $a_1 \gg a_2$. How do the signals compare on the basis of energy? How do they compare on the basis of *energy density* $|X(f)|^2$? Which of the two signals would require greater bandwidth for transmission without distortion? Assume that distortion is insignificant when 90% or more of the energy is transmitted.

2-21. Consider the signal $x_T(t)$ defined by

$$x_T(t) = \begin{cases} x(t), & |t| < \dfrac{T}{2} \\ 0, & \text{otherwise} \end{cases} = x(t)\, \text{rect}\left(\frac{t}{T}\right).$$

Let $X_T(f)$ denote the Fourier transform of $x_T(t)$. Write an expression for $X_T(f)$ in terms of $X(f)$. The above is known as truncation in time. Truncation in frequency is described by

$$X_W(f) = \begin{cases} X(f), & |f| < W \\ 0, & \text{otherwise} \end{cases} = X(f)\, \text{rect}\left(\frac{f}{2W}\right).$$

If $x_W(t)$ denotes the inverse transform of $X_W(f)$, describe $x_W(t)$ in terms of $x(t)$. Justify the term "smoothing" for these operations, i.e., that $X_T(f)$ is a smoothed form of $X(f)$ and $x_W(t)$ is a smoothed form of $x(t)$.

2-22. A generalization of the notion introduced in Prob. 2-21 is the following. Describe $x_s(t)$ by

$$x_s(t) = \begin{cases} x(t)s(t), & |t| < T, \\ 0, & \text{otherwise.} \end{cases}$$

Let $s(t) = [1 - (|t|/T)]\, \text{rect}\,(t/2T)$. Describe $X_s(f)$ in terms of $X(f)$ if $X_s(f)$ is the Fourier transform of $x_s(t)$.

2-23. Prove that the convolution operation is commutative, associative, and distributive, i.e., that

$$x * y = y * x$$
$$x * (y * z) = (x * y) * z$$
$$(x + y) * z = x * z + y * z.$$

2-24. Let the nth moment M_n of $x(t)$ be defined by

$$M_n = \int_{-\infty}^{\infty} t^n x(t)\, dt, \qquad n = 0, 1, 2, \ldots.$$

Assume that all the moments exist. Show that

$$\left. \frac{d^n X(\omega)}{d\omega^n} \right|_{\omega=0} = (-j)^n M_n, \qquad n = 0, 1, 2, \ldots.$$

2-25. Consider a signal $x(t)$ that is zero for $t < 0$. Show that if its transform is written as

$$X(f) = A(f) + jB(f),$$

then $x(t)$ can be written as

$$x(t) = 2 \int_{-\infty}^{\infty} A(f) \cos 2\pi f t\, dt$$

$$= -2 \int_{-\infty}^{\infty} B(f) \sin 2\pi f t\, df.$$

What does this say about the dependency relationship between $A(f)$ and $B(f)$?

2-26. An interpolating function $p(t)$ has the property that $p(0) = 1$; $p(t) = p(t - k\tau)$, $k = 0, \pm 1, \ldots$; τ is the period; and $p(k\tau) = 0$ for $k \neq 0$. Consider the representation

$$x(t) = \sum_{k=-\infty}^{\infty} x(k\tau) p(t - k\tau),$$

where $\tau = [2W]^{-1}$, W is a parameter, and $p(t) = \text{sinc } 2Wt$. Show that this representation is adequate only if the Fourier transform of $x(t)$ is strictly zero outside $|f| > W$. *Hint*: Consider writing $x(t)$ as

$$x(t) = [x(t) \cdot \sum_{k=-\infty}^{\infty} \delta(t - k\tau)] * \text{sinc } 2Wt,$$

and consider its transform.

2-27. Compute the Fourier transforms of
a. $x(t) = e^{-at} \sin(bt) \cdot u(t), a > 0.$
b. $x(t) = e^{-at} \cos(bt) \cdot u(t), a > 0.$

2-28. Let $x(t)$ be an arbitrary function, and let $x_c(t)$ be its "causal" part; i.e.,

$$x_c(t) = \begin{cases} x(t), & t \geq 0, \\ 0, & t < 0. \end{cases}$$

a. Show that $x_c(t)$ can be written as

$$x_c(t) = \tfrac{1}{2}x(t)[1 + \operatorname{sgn} t].$$

b. Compute $X_c(f) \equiv \mathfrak{F}[x_c(t)]$ in terms of the Fourier transform of $x(t)$.

c. Let $x_c(t) = x(t)$; i.e., $x(t) = 0$ for $t < 0$. Derive a relation between the real and imaginary part of $X(f)$.

2-29. Let $X_1(f) = \operatorname{sinc} fT_1$, $X_2(f) = \operatorname{sinc} fT_2$, where $T_1 > T_2$. Compute $X_1(f) * X_2(f)$. What can you conclude with respect to smoothing one sinc function with another? *Hint*: Use the property that

$$\int_{-\infty}^{\infty} X_1(\lambda)X_2(f - \lambda)\,d\lambda = \int_{-\infty}^{\infty} x_1(t)x_2(t)e^{-j2\pi ft}\,dt.$$

2-30. Prove Eq. (2.12-24) by writing

$$g(t) \equiv \sum_{n=-\infty}^{\infty} \delta(t - nT)$$

and expanding $g(t)$ into an ordinary Fourier series; i.e.,

$$g(t) = \sum_{n=-\infty}^{\infty} c_n e^{j(2\pi/T)nt}.$$

Finally, take the Fourier transform of the Fourier series of $g(t)$ term by term.

2-31. Compute the Fourier transform of the periodic waveform shown in Fig. P2-31. There are $N = 50$ pulses altogether. Sketch the results carefully. Put the time origins where it might be most convenient and assume that exactly the same cosine pattern appears in each pulse.

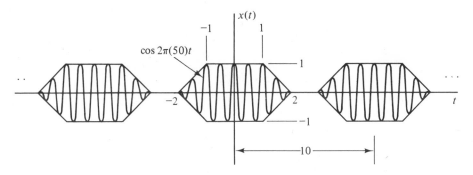

Figure P2-31

2-32. A sequence of N cosine pulses is shown in Fig. P2-32. The frequency of the cosine is f_0, the length of the pulse is a, and the pulse period is T. The string of pulses can be written in the form

$$x(t) = \operatorname{rect}\left(\frac{t}{NT}\right)\left\{\operatorname{rep}_T\left[\operatorname{rect}\left(\frac{t}{a}\right)\cos 2\pi f_0\, t\right]\right\}.$$

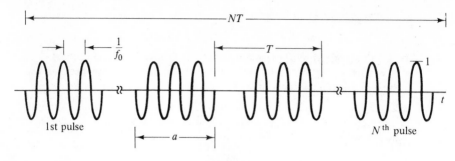

Figure P2-32

Find the Fourier transform, using the appropriate rules and Fourier pairs. Sketch the result if N is assumed to be fairly large and if $T \gg a \gg 1/f_0$.

REFERENCES

[2-1]   B. O. PEIRCE and R. M. FOSTER, *A Short Table of Integrals*, 4th ed., Ginn, Boston, 1956.

[2-2]   E. JAHNKE and F. EMDE, *Table of Functions*, Dover, New York, 1945.

[2-3]   E. A. GUILLEMIN, *The Mathematics of Circuit Analysis*, Wiley, New York, 1959, p. 496.

[2-4]   A. PAPOULIS, *The Fourier Integral and Its Applications*, McGraw-Hill, New York, 1962.

[2-5]   J. L. HAMMOND and R. S. JOHNSON, "Review of Orthogonal Square-Wave Functions and Their Application to Linear Networks," *J. Franklin Inst.*, **273**, pp. 211–225, March 1962.

[2-6]   R. COURANT and D. HILBERT, *Methods of Mathematical Physics*, Vol. 1, Wiley-Interscience, New York, 1953.

[2-7]   W. R. LePAGE, *Complex Variables and the Laplace Transform for Engineers*, McGraw-Hill, New York, 1952, Chap. 9.

[2-8]   B. FRIEDMAN, *Principles of Applied Mathematics*, Wiley, New York, 1956.

[2-9]   P. M. WOODWARD, *Probability and Information Theory with Application to Radar*, McGraw-Hill, New York, 1952, Chap. 2.

[2-10]   H. STARK, "An Extension of the Hilbert Transform Product Theorem," *Proc. IEEE*, **59**, No. 9, pp. 1359–1360, Sept. 1971.

[2-11]   P. BEDROSIAN and H. STARK, "An Extension of the Hilbert Transform Product Theorem," *Proc. IEEE*, **60**, No. 2, pp. 228–229, Feb. 1972.

Linear Circuits and Filters

3.1 Introduction

A filter is a device that passes certain frequencies of the input signal and rejects or attenuates others. In this chapter we deal with linear, lumped-parameter, time-invariant filters constructed from electric circuit elements such as resistors, capacitors, inductors, or operational amplifiers (op-amps). These devices are all assumed to be substantially linear and time-invariant, and we shall ignore effects of distributed parameters, e.g., transmission-line effects in the lines connecting the components.

We start by reviewing some simple concepts from elementary linear circuit theory. We then consider the various filter types, such as low-pass, high-pass, band-pass, etc., and the theoretical limits placed on their performance by physical constraints such as realizability and causality. We then briefly describe a few of the standard ways in which ideal filters are approximated by practical circuits.

3.2 Input-Output Using Superposition

In Chapter 2 we defined a linear system as one obeying the superposition principle. We also mentioned that we can use this principle to obtain the output for any arbitrary input if we know the response of the system to certain test functions. To do this we express the arbitrary input in terms of these test functions and then use the superposition principle to express the output in terms of the elementary responses. In Chapter 2 we dealt with the expansion of the input signal in terms of sinusoids, in which case the output is determined from the frequency response of the system. This is the *frequency-domain* point of view.

Another, and in some ways a more general, expansion of signals is in terms of damped sinusoids, or exponentials of the form e^{-st}, where $s = \sigma + j\omega$ (σ, ω real) is called a *complex frequency*. Such an expansion results in the *Laplace transform*. For functions $x(t)$ that are zero for $t < 0$ this has the one-sided form given by

$$X(s) = \int_0^\infty x(t)e^{-st}\, dt \qquad (3.2\text{-}1)$$

and is seen to be a direct generalization of the Fourier transform considered in the previous chapter.

The Laplace transform is useful in the solution of differential equations and transient problems, while the Fourier transform is more appropriate for dealing with the kind of signals encountered in communication systems. This is the reason for emphasizing Fourier transform methods in this book. We assume that the reader is generally familiar with the Laplace transform and with the notion of complex frequency. We shall find the latter useful in Sec. 3.4.

In this section we consider a representation of the signal in terms of impulses, so that the output is obtained from the superposition of impulse responses of the system. This is referred to as the *time-domain* approach.

In general the response of a linear system to an impulse depends on the time at which the impulse was applied and the time elapsed. Thus for a linear system characterized by the transformation $\mathcal{3C}$ the impulse response is

$$h(t, \tau) = \mathcal{3C}[\delta(t - \tau)].$$

This is the response at time t to an impulse applied at time τ. If the system is *time-invariant*, the response depends only on the time elapsed since the application of the pulse, and therefore for such a system

$$\mathcal{3C}[\delta(t - \tau)] = h(t - \tau), \qquad (3.2\text{-}2)$$

or since the time origin is immaterial,

$$h(t) = \mathcal{3C}[\delta(t)]. \qquad (3.2\text{-}3)$$

A causal system is one which cannot respond to a signal prior to its application. Hence, for causal systems

$$h(t, \tau) = 0 \qquad \text{for } t < \tau. \qquad (3.2\text{-}4)$$

If an input $x(t)$ is applied to a linear system with impulse response $h(t, \tau)$, the output is given by the *superposition integral*

$$y(t) = \int_{-\infty}^\infty x(\tau)h(t, \tau)\, d\tau. \qquad (3.2\text{-}5)$$

This important relation is easily derived by considering the sifting-integral expression

$$x(t) = \int_{-\infty}^{\infty} x(\tau)\delta(t - \tau) \, d\tau. \tag{3.2-6}$$

This is one of the ways of defining the delta function [cf. Eq. (2.11-5)] and holds at any point t where $x(t)$ is continuous. It may, however, also be regarded as a representation of $x(t)$ in terms of a continuum of impulse functions.

Several definitions for linearity were presented in the previous chapter. One of these [Eq. (2.2-4)] states that if $y_i(t) = \mathcal{K}[x_i(t)]$, $i = 1, \ldots, n$, and if there is a set of arbitrary constants a_i, then

$$\mathcal{K}\left[\sum_{i=1}^{n} a_i x_i(t)\right] = \sum_{i=1}^{n} a_i y_i(t).$$

This definition is generally extended to infinite sums and also to integrals [3-1], so that if $y(t, \tau) = \mathcal{K}[x(t, \tau)]$

$$\mathcal{K}\left[\int_{-\infty}^{\infty} a(\tau)x(t, \tau) \, d\tau\right] = \int_{-\infty}^{\infty} a(\tau)y(t, \tau) \, d\tau; \tag{3.2-7}$$

i.e., the operations of linear transformation and integration can be interchanged. Note that the variable of integration τ corresponds to the index i in the summation and need not have dimensions of time, although in fact it frequently does.

If we apply this definition of linearity to Eq. (3.2-6), we obtain immediately

$$y(t) = \mathcal{K}[x(t)] = \mathcal{K}\left[\int_{-\infty}^{\infty} x(\tau)\delta(t - \tau) \, d\tau\right]$$

$$= \int_{-\infty}^{\infty} x(\tau)h(t, \tau) \, d\tau. \tag{3.2-8}$$

In Eq. (3.2-6) the integrand $x(\tau)\delta(t - \tau)$ can be regarded as an impulse occurring at time τ with area proportional to $x(\tau)$. Then Eq. (3.2-8) can be regarded as the summation or superposition of the responses to all of these impulses—hence the name *superposition integral.*

Strictly speaking, neither Eq. (3.2-6) nor (3.2-8) is valid at points where $x(t)$ is discontinuous. However, both equations hold on either side of any discontinuity. The failure of the representation at points of discontinuity is therefore, in general, unimportant. In particular, it often happens that $y(t)$ is continuous even though $x(t)$ is not. In this case Eq. (3.2-8) is valid even at points of discontinuity of $x(t)$.

If the linear system $\mathcal{K}$ is time-invariant, we can use Eq. (3.2-2) in Eq. (3.2-5) to write

$$y(t) = \int_{-\infty}^{\infty} x(\tau)h(t - \tau) \, d\tau. \tag{3.2-9}$$

Thus, in this case the output is the *convolution* of the input and the impulse response, and we can use the shorthand notation

$$y(t) = x(t) * h(t). \tag{3.2-10}$$

If the linear system is causal, the integrand in Eq. (3.2-8) is zero for all $\tau > t$. Hence the upper limit of integration is effectively t, giving the result

$$y(t) = \int_{-\infty}^{t} x(\tau)h(t, \tau) \, d\tau,$$

or, for time-invariant, causal systems,

$$y(t) = \int_{-\infty}^{t} x(\tau)h(t - \tau) \, d\tau. \tag{3.2-11}$$

Finally it is often assumed that the input is zero prior to some time, say, $t = 0$ so that $x(\tau) = 0$ for $\tau < 0$. This leads to

$$y(t) = \int_{0}^{t} x(\tau)h(t, \tau) \, d\tau, \tag{3.2-12}$$

or

$$y(t) = \int_{0}^{t} x(\tau)h(t - \tau) \, d\tau \tag{3.2-13}$$

for time-varying and time-invariant systems, respectively.

EXAMPLE

See Fig. 3.2-1. Suppose that

$$x(t) = a \operatorname{rect}\left(\frac{t - T/2}{T}\right), \qquad T < 1$$

$$h(t) = \begin{cases} 0, & t \leq 0, \\ 1 - t, & 0 < t \leq 1, \\ 0, & t > 1. \end{cases}$$

The factors $x(\tau)$ and $h(t - \tau)$ in the integrand of Eq. (3.2-13) are shown in Fig. 3.2-1(c) for $0 < t < T$.

There are five distinct regimes:

for $t < 0$: $\qquad\qquad\qquad y(t) = 0;$

for $0 \leq t \leq T$: $\qquad\quad y(t) = a \int_{0}^{t} (1 - t + \tau) \, d\tau$

$$= a\left(t - \frac{t^2}{2}\right);$$

for $T \leq t \leq 1$: $\qquad\quad y(t) = a \int_{0}^{T} (1 - t + \tau) \, d\tau$

$$= a\left(T - tT + \frac{T^2}{2}\right);$$

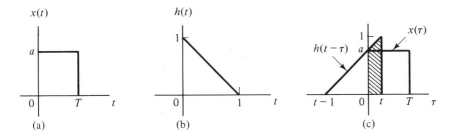

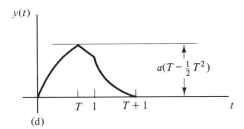

Figure 3.2-1. Example of an input $x(t)$ applied to a time-invariant linear system with impulse response $h(t)$: (a) the input; (b) the impulse response; (c) the two factors $x(\tau)$ and $h(t - \tau)$ involved in the convolution integral for $0 < t < T$ (the output at time t is proportional to the shaded area); (d) the output.

$$\text{for } 1 \leq t \leq T + 1: \quad y(t) = a \int_{t-1}^{T} (1 - t + \tau) \, d\tau$$

$$= (a/2)[t - (T + 1)]^2$$

$$\text{for } T + 1 \leq t: \quad y(t) = 0.$$

The result is shown in Fig. 3.2-1(d).

3.3 Differential Equations for Linear Time-Invariant Systems

The representation of linear systems in terms of the impulse response is not the only possibility, and for many purposes a representation using differential equations is more convenient.

A lumped-parameter, linear, time-invariant system is described by an ordinary linear differential equation with constant coefficients. By contrast, a distributed-parameter system gives rise to a description in terms of partial differential equations, and a time-varying system is represented by a differential equation with nonconstant, i.e., time-dependent, coefficients.

In practice a very wide class of electric circuits can be modeled as consisting of essentially fixed resistors, capacitors, inductors, and operational amplifiers that are linear, time-invariant, lumped-parameter elements. Although we expect the reader to be generally familiar with the analysis of such circuits, we present a brief review below. This will serve to introduce our notation and also permit us to discuss some of the properties of the resulting differential equations.

A simple circuit consisting of a resistor, inductor, and capacitor in series is shown in Fig. 3.3-1. We assume that the circuit is driven by a source of

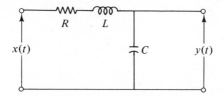

Figure 3.3-1. Series RLC circuit.

voltage $x(t)$. The source has zero impedance; i.e., its voltage is not affected by any current drawn by the network. We are interested in the open-circuit voltage across the capacitor C.

The integrodifferential equation for the current in this circuit is obtained from the Kirchhoff voltage law, which gives

$$x = Ri + L\frac{di}{dt} + \frac{1}{C}\int i\,dt. \tag{3.3-1}$$

Also the desired output voltage is

$$y = \frac{1}{C}\int i\,dt. \tag{3.3-2}$$

(Note that x, y, and i are all functions of time; the time dependence is suppressed to simplify the appearance of the equations.) We can eliminate the variable i from the two equations by differentiating Eq. (3.3-2). This gives

$$i = C\frac{dy}{dt}, \tag{3.3-3}$$

$$\frac{di}{dt} = C\frac{d^2y}{dt^2}. \tag{3.3-4}$$

Substituting these back into Eq. (3.3-1) results in the second-order differential equation

$$LC\frac{d^2y}{dt^2} + RC\frac{dy}{dt} + y = x. \tag{3.3-5}$$

The differential equations for more complicated circuits are obtained in a similar way, and all such circuits can be represented by a differential equation of the general form

$$\sum_{i=0}^{n} a_i \frac{d^i y}{dt^i} = \sum_{i=0}^{k} b_i \frac{d^i x}{dt^i}, \tag{3.3-6}$$

where y is the output and x the input. The *degree* of the equation is n, and for most practical systems $k \leq n$. (A reason for this will appear later.)

Solution in Terms of Exponential Functions

The general solution of differential equations of the form of Eq. (3.3-6) consists of two parts: the complementary or transient solution and the particular or driven solution. The transient solution is obtained by setting the right-hand side of Eq. (3.3-6) equal to zero and assuming a solution in the form

$$y(t) = Ce^{st}, \tag{3.3-7}$$

where C and s are arbitrary complex numbers. Substitution of Eq. (3.3-7) into the equation results in

$$C\sum_{i=0}^{n} a_i s^i e^{st} = 0. \tag{3.3-8}$$

If this is to be true for all t, then $e^{st} \neq 0$. Also, $C = 0$ yields $y(t) = 0$, which, although perfectly valid, is usually not of interest. Hence, for nontrivial solutions

$$\sum_{i=0}^{n} a_i s^i = a_n s^n + \cdots + a_0 = 0. \tag{3.3-9}$$

This algebraic equation is called the characteristic equation. It has n roots $s = \lambda_1, s = \lambda_2, \ldots, s = \lambda_n$, called characteristic roots. The function $Ce^{\lambda_i t}$ is a solution of the differential equation if λ_i is one of the characteristic roots. Since the differential equation is linear, any linear combination of functions $C_i e^{\lambda_i t}$, where the λ_i, $i = 1, \ldots, n$, are different characteristic roots, will also be a solution. Hence, the complete complementary solution is

$$y = C_1 e^{\lambda_1 t} + C_2 e^{\lambda_2 t} + \cdots + C_n e^{\lambda_n t}. \tag{3.3-10}$$

The coefficients C_i, $i = 1, \ldots, n$, are arbitrary and are not directly determined by the differential equation. They are instead determined by the initial or "boundary" conditions. If there is to be a unique solution for all the n coefficients $\{C_i\}$, there must be n independent boundary conditions.

The nature of the complementary solution of the differential equation is determined by the characteristic roots. If any one root λ_i has a positive real part, the corresponding $e^{\lambda_i t}$ will grow exponentially with time. Although practical systems cannot permit such growth indefinitely and eventually saturate, any system in which a small initial disturbance grows with time is termed *unstable*. Thus in a *stable* system none of the characteristic roots can have positive real values.

The coefficients a_i and b_i of real systems are real, e.g., combinations of R, L, or C as in Eq. (3.3-5). Hence any complex characteristic roots must appear in complex conjugate pairs: If λ_i is a complex root, λ_i^* is also a root. The corresponding coefficients $\{C_i\}$ must also occur as complex conjugate pairs if the boundary conditions and the subsequent system response are to be real functions of time. Suppose that a pair of complex conjugate characteristic roots is $\alpha \pm j\beta$ and that the respective coefficients are $A \pm jB$. The corresponding component of the response will be

$$(A + jB)e^{(\alpha + j\beta)t} + (A - jB)e^{(\alpha - j\beta)t} = 2e^{\alpha t}(A \cos \beta t - B \sin \beta t)$$

$$= 2\sqrt{A^2 + B^2}\, e^{\alpha t} \cos \left(\beta t + \tan^{-1} \frac{B}{A} \right).$$

$$(3.3\text{-}11)$$

If α is positive, this kind of response represents an oscillation whose amplitude grows exponentially with time, and it constitutes a form of instability, as already mentioned. If $\alpha = 0$, the output will contain an oscillating component with constant amplitude, which is generally also regarded as a form of instability. Hence one requires all the roots to have negative real parts for stability, or equivalently the roots should lie in the *left half* of the complex plane.

3.4   The Transfer Function

The response of the system to the input is given by the *particular solution* of the differential equation. For stable systems the transient response given by the complementary solution eventually dies out, and therefore the particular solution is also the *steady-state* solution. In most applications to communications, one is interested mainly in the steady-state solution.

If the Fourier transforms of the functions $x(t)$ and $y(t)$ exist, the particular solution can be obtained by Fourier-transforming both sides of Eq. (3.3-6). This results in†

$$Y(j\omega) \sum_{i=0}^{n} a_i(j\omega)^i = X(j\omega) \sum_{i=0}^{k} b_i(j\omega)^i$$

or

$$\frac{Y(j\omega)}{X(j\omega)} \equiv H(j\omega) = \frac{\sum_{i=0}^{k} b_i(j\omega)^i}{\sum_{i=0}^{n} a_i(j\omega)^i}. \qquad (3.4\text{-}1)$$

†In this chapter we frequently write Fourier transforms with arguments of $j\omega$ rather than ω, i.e., $H(j\omega)$ instead of $H(\omega)$, because we often switch between variables s and $j\omega$. Use of ω would necessitate switching between ω and $-js$. Also, instead of $H(j2\pi f)$ we frequently simply use $H(f)$. The meaning of the notation can generally be inferred from the context, and there should be no confusion.

In communication theory $H(j\omega)$ is frequently taken as the transfer function of the system. More generally, however, the transfer function is defined in terms of the complex variable $s = \sigma + j\omega$; i.e.,

$$\frac{Y(s)}{X(s)} \equiv H(s) = \frac{\sum\limits_{i=0}^{k} b_i s^i}{\sum\limits_{i=0}^{n} a_i s^i}. \tag{3.4-2}$$

Note that this result can be obtained directly by taking Laplace [Eq. (3.2-1)] rather than Fourier transforms of $x(t)$ and $y(t)$ in Eq. (3.3-6). For linear, time-invariant, lumped-parameter systems the transfer function is always in the form of a ratio of two polynomials in the variable $s = j\omega$. The denominator polynomial can be represented in the form

$$\sum\limits_{i=0}^{n} a_i s^i = a_n(s - \lambda_1)(s - \lambda_2), \ldots, (s - \lambda_n), \tag{3.4-3a}$$

where the numbers $\lambda_1, \lambda_2, \ldots, \lambda_n$ are the n characteristic roots already referred to. Similarly, the numerator polynomial can be put into the form

$$\sum\limits_{i=0}^{k} b_i s^i = b_k(s - \mu_1)(s - \mu_2)\cdots(s - \mu_k), \tag{3.4-3b}$$

where the numbers $\mu_1, \ldots, \mu_k$ are the roots of the equation $\sum_{i=0}^{k} b_i s^i = 0$. Since $H(s) = 0$ for $s = \mu_i$, the μ's are referred to as the *zeros* of the transfer function. Similarly, $H(s)$ will be infinite if s takes on one of the values $\lambda_1, \ldots, \lambda_n$; these are referred to as the *poles* of the transfer function. The poles are identical to the characteristic roots.

If we substitute Eqs. (3.4-3a) and (3.4-3b) with $s = j\omega$ into Eq. (3.4-1), we find that

$$H(j\omega) = \frac{b_k}{a_n} \frac{(j\omega - \mu_1) \cdots (j\omega - \mu_k)}{(j\omega - \lambda_1) \cdots (j\omega - \lambda_n)}.$$

It follows that except for the scale factor b_k/a_n the transfer function, and therefore the steady-state behavior of the system, is completely determined by the zeros and poles.

Qualitatively we see that if a pole λ_i is complex, i.e., $\lambda_i = \sigma_i + j\omega_i$, and if $\sigma_i \ll \omega_i$, then $|H(j\omega)|$ will tend to peak for $\omega = \omega_i$. Similarly, a complex zero in which the imaginary part is much larger than the real part tends to cause a notch in the frequency response. The effect of pole and zero location on frequency response is discussed in more detail in the next section.

We have already seen that if the poles of a transfer function of a real system are complex they must occur as complex conjugate pairs. This is true also of the zeros, and for the same reason. A stable system must have all of its poles in the left half-plane. There is no such restriction on the location of the zeros, but if none of the zeros is in the right half-plane, the transfer function is of the *minimum phase* type. The meaning of this notion will be explained shortly.

Properties of the Transfer Function

The transfer function of a real system (i.e., one where the coefficients of the differential equation are real) evaluated for real frequencies ($s = j\omega$) must have Hermitian symmetry; i.e.,

$$|H(j\omega)| = |H(-j\omega)| \qquad (3.4\text{-}4)$$

and

$$\angle H(j\omega) = -\angle H(-j\omega). \qquad (3.4\text{-}5)$$

This follows directly from the fundamental definition, Eq. (3.4-1). Hence, if we set

$$H(j\omega) = A(j\omega) + jB(j\omega), \qquad A, B \text{ real}, \qquad (3.4\text{-}6)$$

then

$$A(j\omega) = A(-j\omega), \qquad B(j\omega) = -B(-j\omega), \qquad (3.4\text{-}7)$$

and

$$|H(j\omega)| = [A^2(j\omega) + B^2(j\omega)]^{1/2}. \qquad (3.4\text{-}8)$$

The transfer function $H(j\omega)$ is identical with the frequency response of the system. We show this formally by considering an input $x(t) = \cos 2\pi f_0 t$. For simplicity we write $H(f)$ instead of $H(j2\pi f)$. Then, as shown in Chapter 2,

$$X(f) = \tfrac{1}{2}[\delta(f - f_0) + \delta(f + f_0)], \qquad (3.4\text{-}9)$$

and therefore

$$Y(f) = \frac{H(f)}{2}[\delta(f - f_0) + \delta(f + f_0)]$$

$$= \frac{H(f_0)}{2}\delta(f - f_0) + \frac{H(-f_0)}{2}\delta(f + f_0). \qquad (3.4\text{-}10)$$

Because of the Hermitian symmetry of the transfer function, this can be put into the form

$$Y(f) = \frac{A(f_0)}{2}[\delta(f - f_0) + \delta(f + f_0)] - \frac{B(f_0)}{2j}[\delta(f - f_0) - \delta(f + f_0)].$$

$$(3.4\text{-}11)$$

Finally, to get the steady-state time response we inverse-transform to get

$$y(t) = \mathcal{F}^{-1}[Y(f)] = A(f_0) \cos 2\pi f_0 t - B(f_0) \sin 2\pi f_0 t$$

$$= |H(f_0)| \cos [2\pi f_0 t + \angle H(f_0)]. \qquad (3.4\text{-}12)$$

Thus the steady-state output is a sinusoid of the same frequency as the input but with amplitude and phase determined by the magnitude and phase, respectively, of the transfer function evaluated at the frequency of the applied signal.

The fact that the transfer function and the impulse response of a linear time-invariant system are Fourier transforms of each other, i.e., that

$$H(j\omega) = \mathcal{F}[h(t)] \qquad (3.4\text{-}13a)$$

and

$$h(t) = \mathcal{F}^{-1}[H(j\omega)], \qquad (3.4\text{-}13b)$$

follows directly from Eq. (3.4-1) by assuming $X(j\omega)$ to be the Fourier transform of the impulse function. This gives $X(j\omega) = 1$, $Y(j\omega) = H(j\omega)$, and therefore

$$y(t) = h(t) = \mathcal{F}^{-1}[H(j\omega)].$$

Equation (3.4-13a) follows by the uniqueness of the Fourier transform.

Bode Plots

There are a number of simple methods for obtaining the frequency response from the poles and zeros of the transfer function. Perhaps the most convenient is the asymptotic log-amplitude or Bode plot.[†] To develop this we write the transfer function in the form

$$
\begin{aligned}
H(j\omega) &= \frac{b_k(j\omega - \mu_1)(j\omega - \mu_2)\cdots(j\omega - \mu_k)}{a_n(j\omega - \lambda_1)(j\omega - \lambda_2)\cdots(j\omega - \lambda_n)} \\
&= K \frac{\left(\dfrac{j\omega}{-\mu_1} + 1\right)\left(\dfrac{j\omega}{-\mu_2} + 1\right)\cdots\left(\dfrac{j\omega}{-\mu_k} + 1\right)}{\left(\dfrac{j\omega}{-\lambda_1} + 1\right)\left(\dfrac{j\omega}{-\lambda_2} + 1\right)\cdots\left(\dfrac{j\omega}{-\lambda_n} + 1\right)},
\end{aligned} \qquad (3.4\text{-}14)
$$

where

$$K = \frac{b_k(-\mu_1)(-\mu_2)\cdots(-\mu_k)}{a_n(-\lambda_1)(-\lambda_2)\cdots(-\lambda_n)}$$

is a constant.

Because Eq. (3.4-14) involves a ratio of products of simple factors, it is convenient to use a logarithmic representation of the form

$$H(j\omega) = e^{\alpha(\omega) + j\beta(\omega)}$$

so that

$$\alpha(\omega) = \ln|H(j\omega)| \qquad (3.4\text{-}15)$$

and

$$\beta(\omega) = \angle H(j\omega).[‡] \qquad (3.4\text{-}16)$$

---

†After H. W. Bode [3-2].

‡In transmission-line theory the function $\gamma(\omega) = \alpha(\omega) + j\beta(\omega)$ is called the *propagation function*; $\alpha(\omega)$ is the *attenuation function* and $\beta(\omega)$ the *phase function*.

If we take the logarithm of both sides of Eq. (3.4-14), we find that

$$\alpha(\omega) = \ln |H(j\omega)| = \ln |K| + \sum_{i=1}^{k} \ln \left| \frac{j\omega}{-\mu_i} + 1 \right| - \sum_{i=1}^{n} \ln \left| \frac{j\omega}{-\lambda_i} + 1 \right| \quad (3.4\text{-}17)$$

and

$$\beta(\omega) = \angle H(j\omega) = \sum_{i=1}^{k} \angle \left(\frac{j\omega}{-\mu_i} + 1 \right) - \sum_{i=1}^{n} \angle \left(\frac{j\omega}{-\lambda_i} + 1 \right). \quad (3.4\text{-}18)$$

Suppose initially that the μ_i and λ_i are negative real numbers. A typical term of the form $\ln |(j\omega/-\mu_i) + 1|$ approaches zero as $|\omega/\mu_i| \rightarrow 0$, and it approaches $\ln |\omega/\mu_i| = \ln |\omega| - \ln |\mu_i|$ asymptotically for very large $|\omega/\mu_i|$. The two asymptotes and the actual magnitude are shown in Fig. 3.4-1. If

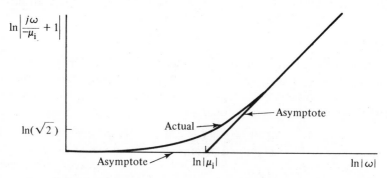

Figure 3.4-1. Asymptotic and actual behavior of the term $\ln |(j\omega/|\mu_i|) + 1|$.

logarithmic scales are used for both the magnitude and the frequency, the asymptotic plot consists of two straight lines that meet at the point $\omega = |\mu_i|$. For frequencies greater than $|\mu_i|$ the line has a slope of $+1$. For frequencies below $|\mu_i|$ the slope is zero. The frequency $\omega = |\mu_i|$ is called the *break frequency*. If $-\mu_i$ is a positive real number, the actual magnitude of the term $[(j\omega/-\mu_i) + 1]$ never departs very far from the asymptotes, and at the break point it is equal to $\sqrt{2}$. Hence for many purposes a plot of only the asymptotes gives an adequate picture of the magnitude of the frequency response. Such a plot is easily sketched on log-log paper by adding lines of the type shown in Fig. 3.4-1, paying due attention to sign. As an example we show a Bode diagram for the function

$$H(j\omega) = 10 \frac{(j\omega/1) + 1}{[(j\omega/2) + 1][(j\omega/10) + 1]} \quad (3.4\text{-}19)$$

in Fig. 3.4-2.

For negative real μ_i the phase shift contributed by a typical term $[(j\omega/-\mu_i) + 1]$ is $\tan^{-1}(\omega/|\mu_i|)$. Thus the phase-shift curve can easily be sketched on semilog paper (logarithmic abscissa, linear ordinate scale) by adding typical terms. The phase curve for the function $H(j\omega)$ in Eq. (3.4-19)

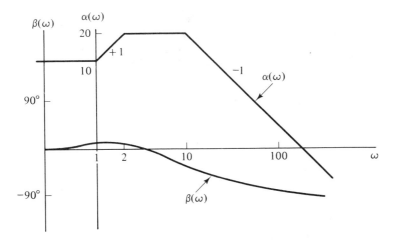

Figure 3.4-2. Asymptotic log-magnitude plot (Bode diagram) and the corresponding minimum phase lag curve for the transfer function $H(j\omega) = 10[j\omega/1) + 1]/\{[(j\omega/2) + 1][(j\omega/10) + 1]\}$.

is also shown in Fig. 3.4-2. We see that real zeros in the transfer function cause the magnitude of the frequency response to go up, while real poles cause it to go down. The frequencies where the increases or decreases take place are substantially equal to the magnitude of the zeros or poles, respectively.

Complex Poles and Zeros

The basic argument leading to the Bode diagram is unaffected if the zeros and poles are complex. As noted earlier, complex poles or zeros must occur in complex conjugate pairs if the impulse response is real. A typical pair of factors with complex conjugate poles is

$$\left(\frac{j\omega}{-\lambda} + 1\right)\left(\frac{j\omega}{-\lambda^*} + 1\right) = -\left|\frac{\omega}{\lambda}\right|^2 + 2j\zeta\left|\frac{\omega}{\lambda}\right| + 1, \qquad (3.4\text{-}20)$$

where

$$\zeta = \frac{Re\,(-\lambda)}{|\lambda|}. \qquad (3.4\text{-}21)$$

The parameter ζ is called the damping constant, and for complex roots, ζ is restricted to $0 < \zeta < 1$. The log-magnitude and phase diagram for a transfer function having only a single pair of such complex poles is shown in Fig. 3.4-3 for various values of ζ. The asymptotic log-magnitude diagram has a second-order break point at $\omega = |\lambda|$; i.e., the slope changes from 0 to -2. For small ζ the actual magnitude curve departs considerably from the asymptotic curve. This is due to the resonance phenomenon. Observe that complex

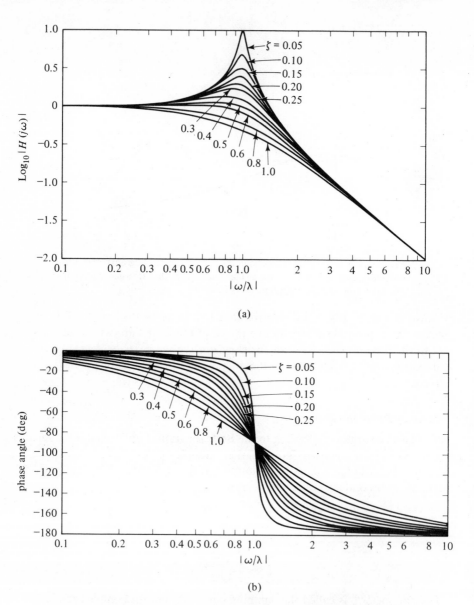

Figure 3.4-3. Frequency-response diagram for the transfer function $H(j\omega) = (-|\omega/\lambda|^2 + 2j\zeta|\omega/\lambda| + 1)^{-1}$: (a) actual log-magnitude curves for various values of ζ; (b) corresponding phase characteristic.

poles in the transfer function generally result in resonant peaks in the frequency response for frequencies close to the pole magnitude. If the real part of the pole is small, the peak is large.

By way of example, the transfer function of the series RLC circuit considered in the last section (Fig. 3.3-1) has two poles

$$\lambda = -\frac{R}{2L} \pm \sqrt{\left(\frac{R}{2L}\right)^2 - \frac{1}{LC}}. \qquad (3.4\text{-}22)$$

These poles are complex if $(R/2L)^2 < 1/LC$, and in this case $|\lambda| = \omega_0 = 1/\sqrt{LC}$, the resonant frequency of the circuit. The parameter ζ is given by $\zeta = R/(2L\omega_0)$.

The Quality Factor Q

For values of $\zeta \ll 1$, the peak of the magnitude of

$$H(j\omega) = \frac{1}{1 + 2j\zeta(\omega/\omega_0) - (\omega/\omega_0)^2} \qquad (3.4\text{-}23)$$

occurs near $\omega = \omega_0$ and has a value very nearly equal to $1/(2\zeta)$. The sharpness of the peak is often a measure of the selectivity of filters, tuned circuits, and oscillators. A measure of the sharpness of the peak is the quality factor Q, defined by [3-3]

$$Q = 2\pi \frac{\text{energy stored in the circuit}}{\text{energy dissipated per cycle}}. \qquad (3.4\text{-}24)$$

This is also proportional to the peak of the resonance curve divided by the half-power bandwidth (which is determined by the frequency at which the peak has decayed to $1/\sqrt{2}$ times its maximum). Clearly, the smaller the energy loss, the sharper the resonance and the larger is Q. It can be shown (Prob. 3-8) that the series RLC circuit has a Q given by $Q = \omega_0 L/R = (2\zeta)^{-1}$. Hence Eq. (3.4-23) can also be written as

$$H(j\omega) = \frac{1}{1 + (j/Q)(\omega/\omega_0) - (\omega/\omega_0)^2}. \qquad (3.4\text{-}25)$$

The peak of the resonance curve is then seen to be approximately equal to Q (for $\zeta \ll 1$).

Decibels

As we have already seen, logarithmic scales are often used in plots of $|H(j\omega)|$. The commonly used unit for measuring power ratios is the *decibel*. The gain or loss of power expressed in decibels is 10 times the common logarithm of the power ratio. If P is the power level to be compared to the reference level P_0, the number of decibels is given by $10 \log_{10} (P/P_0)$. Thus, a power ratio of 10 is 10 dB, a power ratio of 2 is 3 dB (since $\log_{10} 2 = 0.30103$). When the decibel measure is applied to transfer-function magnitude levels one identifies $|H(j\omega)|^2$ with power, and therefore the number of decibels corresponding to a magnitude ratio $|H(j\omega)/H(j\omega_0)|$ is $20 \log_{10} |H(j\omega)/H(j\omega_0)|$.

For instance, the magnitude of the transfer function $H(j\omega) = \alpha/(j\omega + \alpha)$ is $1/\sqrt{2}$ at the break frequency α; this is 3 dB down from the value at zero frequency. For this reason one sometimes refers to the break frequency as the 3-dB point. A slope of ± 1 in the Bode diagram means that the magnitude doubles (or halves) for each doubling of the frequency; therefore this is sometimes referred to as 6 dB per octave. Also a 10% change in transfer-function magnitude is commonly referred to as a 1-dB change; this is an approximation since 1 dB is about 12.2%.

The decibel is one tenth of a larger unit, the *bel*,† named after Alexander Graham Bell, who in addition to inventing the telephone also was an early pioneer in sound, speech, and hearing research. The unit is appropriate to measuring sound, because the ear responds approximately logarithmically, and equal decibel increments are perceived as equal increments in sound. One decibel is the smallest change in sound intensity that the ear can normally detect. In applications to sound measurement, 0 dB is a reference power of 10^{-16} watts, supposedly the least detectable sound for a normal ear, but other reference levels are also used.

Minimum Phase Property

The magnitude vs. frequency characteristic of the transfer function $H(j\omega)$ is unaffected if a zero is moved from the left half-plane to its mirror location in the right half-plane, i.e., if only the sign of the real part is changed. This is made clear by Fig. 3.4-4, where μ_1 is a zero in the left half-plane and μ_2 the corresponding zero in the right half-plane. The magnitude of the "vector" $j\omega - \mu_1$ is simply its length, and this is seen to be the same as the length of $j\omega - \mu_2$. However, the phase change $d\theta_1/d\omega$, contributed by μ_1, is positive, while that contributed by μ_2, i.e., $d\theta_2/d\omega$, is negative. Thus the zero in the right half-plane contributes more negative phase (i.e., phase lag) to the transfer function than the corresponding left half-plane zero. The minimum amount of phase lag for a given magnitude characteristic is obtained if all the zeros lie in the left half-plane. For this reason a transfer function having no zeros in the right half-plane is called a minimum phase-lag (minimum phase for short) function.

For a minimum phase transfer function the phase function is completely determined by the magnitude function. To see why this is so, let $H(s) = e^{\alpha(s) + j\beta(s)}$. Then $\gamma(s) \equiv \alpha(s) + j\beta(s) = \ln H(s)$ is a complex function of the complex variable s, and it will be analytic in the right half-plane if $H(s)$ is stable and minimum phase, i.e., if all the zeros and poles of $H(s)$ are in the left half-plane. We saw in Sec. 2.13 that if a complex function is analytic in

---

†The decibel, rather than the bel, is widely used because for most applications the bel is too large a unit.

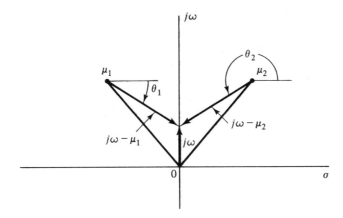

Figure 3.4-4. Effect of moving a zero from the left half-plane to its mirror location in the right half-plane. Only one set of the complex conjugate pair is shown.

one of the half-planes formed by the axis of reals then its real and imaginary parts for real argument are related by a Hilbert transform. A very similar argument holds if the complex function is analytic in one of the half-planes formed by the imaginary axis, as is the case here. Then the real and imaginary parts evaluated for $s = j\omega$ are related by a Hilbert transform; in fact we have

$$\beta(\omega) = \frac{1}{\pi} \int_{-\infty}^{\infty} \frac{\alpha(\lambda)}{\omega - \lambda} \, d\lambda, \tag{3.4-26}$$

$$\alpha(\omega) = -\frac{1}{\pi} \int_{-\infty}^{\infty} \frac{\beta(\lambda)}{\omega - \lambda} \, d\lambda. \tag{3.4-27}$$

Equations (3.4-26) and (3.4-27) can be used to establish further transformations and lead to the famous Bode phase-integral theorem ([3-2] and [3-4], Chap. 6):

$$\beta(\omega_0) = \frac{1}{\pi} \int_{-\infty}^{\infty} \left(\frac{d\alpha}{du}\right) \ln \coth \frac{|u|}{2} \, du, \tag{3.4-28}$$

where $u = \ln(\omega/\omega_0)$ is the frequency on a logarithmic scale so that $d\alpha/du$ is the slope of the α curve on a log-log scale. Equation (3.4-28) therefore is a statement of the fact that for a minimum phase transfer function the phase depends on the slope of the α curve. The factor $\ln \coth |u/2|$ is sharply peaked at $u = 0$ (or $\omega = \omega_0$) and shows that the phase at a given frequency ω_0 is most heavily influenced by the slope of the α curve in the immediate vicinity of ω_0.

Bode's phase integral is one of the few mathematical formulas known to the authors that is covered by a patent; it is U.S. Patent No. 2,123,178.

Behavior of $|H(\omega)|$ for Infinite ω

If a transfer function has more zeros than poles, the magnitude of the transfer function goes to infinity for infinite frequency. Practical circuits do not behave this way; in fact $|H(\omega)|$ generally goes to zero for infinite ω. At most one encounters systems having finite response at very high frequency. Thus the transfer function of a simple resistive voltage divider circuit is $R_2/(R_1 + R_2)$, where R_2 is the shunt resistor and R_1 the series resistor; this remains constant with increasing frequency as long as shunt capacitances generally present in such circuits have negligible effect. For this reason one generally assumes that the number of zeros of the transfer function is no larger than the number of poles.†

3.5   Input-Output

From the definition of the transfer function, Eq. (3.4-1), we obtain immediately the important formula

$$Y(j\omega) = H(j\omega)X(j\omega), \tag{3.5-1}$$

where

$$X(j\omega) = \int_{-\infty}^{\infty} x(t)e^{-j\omega t}\, dt. \tag{3.5-2}$$

Thus we can compute the output $y(t)$ of a linear time-invariant system resulting from any arbitrary input $x(t)$ by first Fourier-transforming the input, multiplying the transform by the transfer function, and then inversely transforming to obtain $y(t)$.

In Sec. 3.2 we found that the output can also be obtained by the superposition integral:

$$y(t) = \int_{-\infty}^{\infty} h(t, \tau)x(\tau)\, d\tau, \tag{3.5-3}$$

where $h(t, \tau)$ is the response of the system at time t to an impulse applied at time τ. For time-invariant systems this becomes the convolution

$$y(t) = \int_{-\infty}^{\infty} h(t - \tau)x(\tau)\, d\tau. \tag{3.5-4}$$

From the convolution-multiplication property of the Fourier transform (property 12 in Table 2.9-1), we see that Eq. (3.5-4) can be transformed to

$$Y(j\omega) = \mathcal{F}[h(t)]X(j\omega), \tag{3.5-5}$$

which is, of course, identical with Eq. (3.5-1).

†Since a zero or a pole at infinity has no effect on the transfer function at finite frequencies, one sometimes regards transfer functions with fewer zeros than poles as having enough infinite zeros to make the total number of zeros equal to the number of poles. With this convention all transfer functions have the same number of zeros as poles.

We see that we have two different methods for obtaining the output of a linear, time-invariant system to an arbitrary input. One is the direct method, using the superposition integral; the other is an indirect method where we first Fourier-transform the input, multiply by the transfer function, and then inverse-transform the product. It would seem that the direct method should be preferable, but this is not necessarily so. For many standard signals (e.g., sine waves) the Fourier transform of the input is known, and therefore the first step in the procedure is not necessary—it has, in effect, been done once and for all. Also it frequently turns out that the output in the form $Y(j\omega)$ is as useful and informative as $y(t)$; thus the inverse transformation is also not needed. Then the advantage of the transfer-function method is the substitution of a simple multiplication for an integration. Another point in favor of the transfer-function method is that it is usually much easier to derive the transfer function from a circuit diagram, whereas derivation of the impulse response for other than the simplest circuits is not straightforward. The impulse response is in fact best obtained as the inverse Fourier transform of the transfer function.

The superposition integral is easier to use when the input signal and impulse response have simple forms for which the Fourier transforms may be complicated. Also, the superposition integral can accommodate time-varying systems, at least in principle. As should be clear from our derivation here, the transfer function is, strictly speaking, defined only for time-invariant systems. (Time-varying transfer functions $H(\omega, t)$ can be defined, but they give only approximate results, and generally work only if the time variation is slow compared to the frequencies ω of interest [3-5].)

EXAMPLE

To illustrate the two methods for calculating the output we consider the simple *RC* circuit shown in Fig. 3.5-1, driven by a square pulse.

The transfer function of this circuit can be obtained by the application of Kirchhoff's law as explained in Sec. 3.3. However, with simple circuits such as the

Figure 3.5-1. *RC* circuit with square pulse input signal.

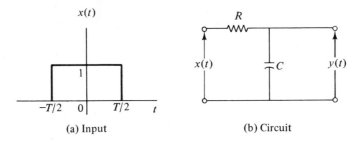

(a) Input   (b) Circuit

one in this example one can obtain the transfer function by inspection, using the principle of the voltage divider:

$$H(j\omega) = \frac{Z_2(j\omega)}{Z_1(j\omega) + Z_2(j\omega)}, \tag{3.5-6}$$

where $Z_2(j\omega)$ is the impedance of the shunt branch and $Z_1(j\omega)$ the impedance of the series branch. In our example $Z_2(j\omega) = 1/j\omega C$ and $Z_1(j\omega) = R$; hence

$$H(j\omega) = \frac{1/j\omega C}{1/j\omega C + R}, \tag{3.5-7}$$

or, if we use the frequency variable f and, for simplicity, write $H(f)$ instead of $H(j2\pi f)$,

$$H(f) = \frac{a}{a + j2\pi f}, \tag{3.5-8}$$

where

$$a \equiv \frac{1}{RC}. \tag{3.5-9}$$

The input signal is

$$x(t) = \text{rect}\left(\frac{t}{T}\right), \tag{3.5-10}$$

and its Fourier transform is

$$X(f) = T \text{ sinc } (Tf). \tag{3.5-11}$$

To get the output by means of the transfer-function method, we multiply $X(f)$ by $H(f)$:

$$Y(f) = H(f)X(f) = \frac{aT \text{ sinc } (Tf)}{a + j2\pi f}. \tag{3.5-12}$$

The input spectrum, transfer function, and output spectrum are shown in Fig. 3.5-2 for $aT/2\pi = 0.5$.

To obtain the output as a time function, we perform the inverse transformation of $Y(f)$:

$$\begin{aligned}
y(t) &= \int_{-\infty}^{\infty} \frac{aT \text{ sinc } (Tf)e^{j2\pi ft}}{a + j2\pi f} \, df \\
&= a\int_{-\infty}^{\infty} \frac{(e^{j\pi fT} - e^{-j\pi fT})e^{j2\pi ft}}{(a + j2\pi f)j2\pi f} \, df \\
&= a\int_{-\infty}^{\infty} \frac{e^{j2\pi(t+T/2)f} - e^{j2\pi(t-T/2)f}}{j2\pi f(a + j2\pi f)} \, df \\
&= \int_{-\infty}^{\infty} \left(\frac{1}{j2\pi f} - \frac{1}{a + j2\pi f}\right) e^{j2\pi(t+T/2)f} \, df \\
&\quad - \int_{-\infty}^{\infty} \left(\frac{1}{j2\pi f} - \frac{1}{a + j2\pi f}\right) e^{j2\pi(t-T/2)f} \, df. \tag{3.5-13}
\end{aligned}$$

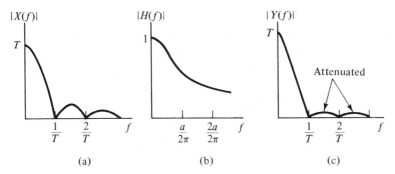

Figure 3.5-2. (a) Fourier spectrum of the input signal; (b) magnitude of the transfer function; (c) output spectrum.

In the last step the partial fraction expansion

$$\frac{a}{j2\pi f(a + j2\pi f)} = \frac{1}{j2\pi f} - \frac{1}{a + j2\pi f} \qquad (3.5\text{-}14)$$

is used.

The inverse transformation is now easily completed by using some of the Fourier pairs in Table 2.12-1; thus from pairs 6 and 4 we have

$$\frac{1}{j2\pi f} \longleftrightarrow \frac{1}{2} \operatorname{sgn}(t)$$

$$\frac{1}{a + j2\pi f} \longleftrightarrow e^{-at} u(t). \qquad (3.5\text{-}15)$$

Therefore

$$y(t) = \frac{1}{2} \operatorname{sgn}\left(t + \frac{T}{2}\right) - e^{-a(t+T/2)} u\left(t + \frac{T}{2}\right)$$

$$- \frac{1}{2} \operatorname{sgn}\left(t - \frac{T}{2}\right) + e^{-a(t-T/2)} u\left(t - \frac{T}{2}\right)$$

$$= \operatorname{rect}\left(\frac{t}{T}\right) - e^{-a(t+T/2)} u\left(t + \frac{T}{2}\right) + e^{-a(t-T/2)} u\left(t - \frac{T}{2}\right), \qquad (3.5\text{-}16)$$

where the second step follows by combining the two sgn functions. The three components and the result are shown in Fig. 3.5-3.

An example of the use of the superposition integral has already been presented in Sec. 3.2. In the present example the impulse response is given by

$$h(t) = ae^{-at}u(t),$$

and therefore

$$y(t) = \int_{-\infty}^{\infty} x(\tau)h(t - \tau)\, d\tau$$

$$= \int_{-\infty}^{\infty} \operatorname{rect}\left(\frac{\tau}{T}\right) ae^{-a(t-\tau)} u(t - \tau)\, d\tau. \qquad (3.5\text{-}17)$$

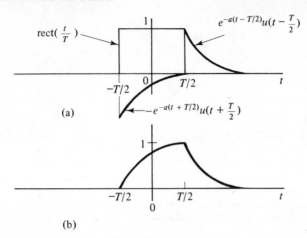

Figure 3.5-3. Output of the *RC* circuit by the transfer-function method: (a) the three components of Eq. (3.5-16); (b) the complete response.

The factors making up the integrand are shown in Fig. 3.5-4 for three different values of *t*. The output for the three values is as follows:

$$\text{for } t < -\frac{T}{2}: \qquad y(t) = 0,$$

$$\text{for } -\frac{T}{2} \le t \le \frac{T}{2}; \qquad y(t) = ae^{-at} \int_{-T/2}^{t} e^{a\tau}\, d\tau = 1 - e^{-a(t+T/2)},$$

$$\text{for } \frac{T}{2} \le t: \qquad y(t) = ae^{-at} \int_{-T/2}^{T/2} e^{a\tau}\, d\tau = e^{-a(t-T/2)}(1 - e^{-aT}). \qquad (3.5\text{-}18)$$

The three solutions are put together in Fig. 3.5-5. The result is identical with the one shown in Fig. 3.5-3. The same result would be obtained by using the alternative form of the superposition integral:

$$y(t) = \int_{-\infty}^{\infty} x(t - \tau)\, h(\tau)\, d\tau. \qquad (3.5\text{-}19)$$

The reader is invited to try this for himself.

While this example has been presented mainly to illustrate two methods of input-output calculations, we point out that it illustrates several other things as well. The *RC* circuit in the example is an example of a low-pass filter since it selectively attenuates the higher frequencies. This is shown in Fig. 3.5-2. The effect is to convert the discontinuous steps of the input into a more gradually rising and falling function. Note that for $aT \ll 1$ this effect is very pronounced, and the square pulse is converted approximately into a triangular pulse. On the other hand, if $aT \gg 1$, the filter decay time is much shorter than the duration of the input, and therefore it only causes a small amount of rounding of the corners of the input function.

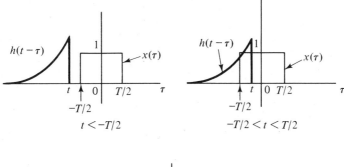

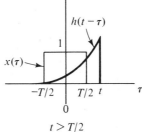

Figure 3.5-4. Input signal $x(\tau) = \text{rect}\,(\tau/T)$ and reversed and shifted impulse response $h(t - \tau) = ae^{-a(t-\tau)}u(t - \tau)$ for three different values of t.

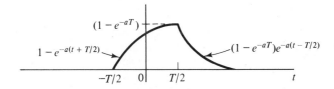

Figure 3.5-5. Output via superposition integral.

3.6 Ideal Linear Filters

We have defined a filter as a circuit that passes certain desired frequencies of the input and rejects or attenuates unwanted frequencies. An ideal filter performs this function perfectly; i.e., it passes the desired parts of the input spectrum with no change, and it rejects the unwanted parts completely. We shall see that ideal filters can only be approximated by practical circuits.

In communication systems small time delays between input and output are generally of no consequence;† therefore if $x(t)$ is the input, the output is ideally

$$y(t) = Kx(t - t_0), \qquad (3.6\text{-}1)$$

where t_0 is the time delay and K is a gain constant. Fourier-transforming this expression results in

$$Y(f) = KX(f)e^{-j2\pi f t_0}. \qquad (3.6\text{-}2)$$

Thus, if $x(t)$ has frequency components only in a certain band, then the ideal filter should have a constant amplitude and linear phase shift in that band. Any filter that does not satisfy these conditions introduces distortion into the signal. Nonuniform magnitude of the filter frequency response gives rise to *amplitude distortion*; departure from linearity of the phase causes *phase distortion*. The effects of these two kinds of distortion on the output signal are similar; however, since the human ear is relatively insensitive to phase, phase distortion is less serious in audio systems than amplitude distortion.

A *low-pass* filter is a filter that passes all frequencies whose magnitudes lie between zero and a certain cutoff frequency B. The frequency range $-B \leq f \leq B$ is called the *passband* of the filter. The transfer function of the *ideal* low-pass filter is given by

$$H(f) = K \operatorname{rect}\left(\frac{f}{2B}\right)e^{-j2\pi f t_0}. \qquad (3.6\text{-}3)$$

This is illustrated in Fig. 3.6-1(a). This filter characteristic is clearly noncausal, as can be seen from the impulse response:

$$h(t) = \mathcal{F}^{-1}[H(f)] = 2KB \operatorname{sinc}\left[2B(t - t_0)\right], \qquad (3.6\text{-}4)$$

†Large delays are frequently of no consequence either. For instance, the unavoidable delays in the signals transmitted from deep-space probes are generally of little consequence. On the other hand, in telephony large delays are intolerable.

Figure 3.6-1. (a) Frequency response and (b) impulse response of an ideal low-pass filter.

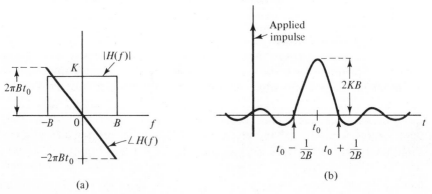

(a)

(b)

shown in Fig. 3.6-1(b). The tails of the sinc function go to infinity in both directions and therefore extend beyond $t < 0$. However, if $Bt_0 \gg 1$, the tail extending to negative t is very small and can be neglected. Thus, although the ideal low-pass characteristic can never be exactly causal, it can be fairly well approximated by a causal function if one makes the delay t_0 large enough.

The ideal *band-pass filter* is given by

$$H(f) = K\left[\text{rect}\left(\frac{f-f_0}{B}\right) + \text{rect}\left(\frac{f+f_0}{B}\right)\right]e^{-j2\pi f t_0}. \qquad (3.6\text{-}5)$$

This characteristic is shown in Fig. 3.6-2. The bandwidth of the filter is B Hz. If this is to be a band-pass transfer function, it is necessary that $f_0 > B/2$; generally one assumes that $f_0 \gg B/2$.

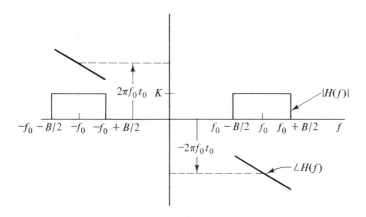

Figure 3.6-2. Ideal band-pass characteristic.

The impulse response of the ideal band-pass filter is the inverse Fourier transform of Eq. (3.6-5):

$$h(t) = 2KB \text{ sinc } [B(t - t_0)] \cos 2\pi f_0 (t - t_0). \qquad (3.6\text{-}6)$$

This is shown in Fig. 3.6-3 for $f_0 \gg B/2$. The 3-dB width ($\simeq B^{-1}$) of the main lobe may be regarded as the *response time* τ_c of the filter. The envelope of the response, i.e., the locus of the peaks, is similar to the response of the low-pass filter, except that the envelope is superimposed on the high-frequency signal $\cos 2\pi f_0(t - t_0)$. As will be seen in Chapter 6, this is a form of amplitude modulation. The response is noncausal, just like the low-pass response; in fact, except for the high-frequency carrier it has the same characteristics as the low-pass response. In particular, for very large Bt_0 the response can be approximately realized with a causal system.

Two other generic filter types are the *high-pass* filter and the *band-stop* filter. These are complementary forms of the low-pass and band-pass filters,

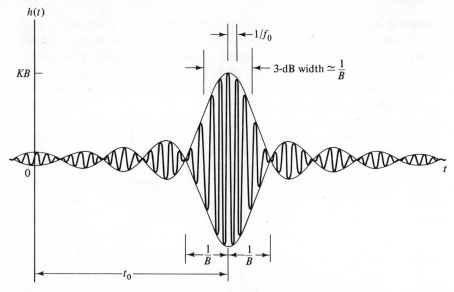

Figure 3.6-3. Impulse response of a band-pass filter with $f_0 \gg B/2$.

respectively. Thus the ideal high-pass transfer function is given by

$$H(f) = K\left[1 - \text{rect}\left(\frac{f}{2B}\right)\right]e^{-j2\pi f t_0}, \tag{3.6-7}$$

and the ideal band-stop transfer function is

$$H(f) = K\left[1 - \text{rect}\left(\frac{f - f_0}{B}\right) - \text{rect}\left(\frac{f + f_0}{B}\right)\right]e^{-j2\pi f t_0}. \tag{3.6-8}$$

These are illustrated in Fig. 3.6-4. Both of these filters are modeled as having

Figure 3.6-4. (a) Ideal high-pass characteristic and (b) ideal band-stop characteristic.

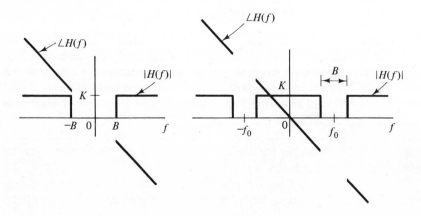

constant gain, K, even as f goes to infinity. Therefore, besides being noncausal they are also not realizable in practice because small capacitances and inductances that may be negligible at normal operating frequencies will generally reduce the gain at sufficiently high frequencies. Nevertheless, in this as well as the other cases, approximations to ideal behavior can be realized that are satisfactory for most applications.

One other filter type that should be mentioned is the *all-pass* filter. This has constant gain at all frequencies (and hence is also unrealizable) but an arbitrary phase shift. All-pass filters are used as *phase equalizers*, i.e., to correct the phase characteristic of some other filter to give it a desired form. Since the phase shift associated with a minimum phase transfer function having constant gain is zero, all-pass networks are necessarily nonminimum phase, and they therefore can only produce phase lag.

3.7   Causal Filters

None of the ideal filters considered in the previous section are causal because the sharp cutoff of the frequency response results in an impulse response containing sinc functions. The general question of whether a particular form of $|H(\omega)|$ can be realized by a causal filter is answered by the Paley-Wiener criterion [3-6]. This says that an even-magnitude function $|H(\omega)|$ can be realized by a causal filter if and only if it satisfies

$$\int_{-\infty}^{\infty} \frac{|\ln|H(\omega)||}{1 + \omega^2} \, d\omega < \infty. \tag{3.7-1}$$

If the integral is bounded, a phase function $\theta(\omega)$ exists such that the impulse response associated with $|H(\omega)| e^{j\theta(\omega)}$ has zero response for $t < 0$. A proof of this relation is given in [3-7]; it is not trivial.

If one attempts to generate a causal response from noncausal responses such as those given in Eqs. (3.6-4) or (3.6-6) by simply setting $h(t) = 0$ for $t < 0$, then the resulting frequency response will extend beyond the passband and will generally have ripples in the passband. This is illustrated in Fig. 3.7-1; for computational details, see Prob. 3-27.

The ripples in the transfer function result in amplitude distortion. We can study this distortion by considering a transfer function of the form

$$H(f) = K(1 + \gamma \cos 2\pi\alpha f)e^{-j2\pi f t_0}. \tag{3.7-2}$$

This is a constant with a superimposed cosinusoidal ripple of frequency α and relative amplitude γ, together with a linear phase-lag term. Equation (3.7-2) can be written in the expanded form

$$H(f) = K\left(e^{-j2\pi f t_0} + \frac{\gamma}{2} e^{j2\pi f(\alpha - t_0)} + \frac{\gamma}{2} e^{-j2\pi f(\alpha + t_0)}\right). \tag{3.7-3}$$

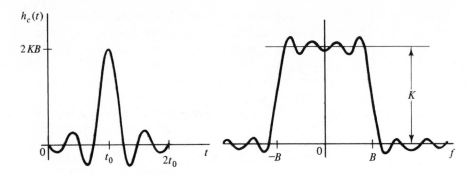

Figure 3.7-1. (a) Causal impulse response obtained by symmetric truncation of the response of Figure 3.6-1; (b) corresponding frequency response ($Bt_0 = 2$).

Then if the input spectrum is $X(f)$, the output spectrum is

$$Y(f) = X(f)H(f)$$

$$= KX(f)\left(e^{-j2\pi f t_0} + \frac{\gamma}{2}e^{-j2\pi f(t_0-\alpha)} + \frac{\gamma}{2}e^{-j2\pi f(t_0+\alpha)}\right). \qquad (3.7\text{-}4)$$

The resulting time function is

$$y(t) = Kx(t - t_0) + \frac{K\gamma}{2}x(t - t_0 + \alpha) + \frac{K\gamma}{2}x(t - t_0 - \alpha). \qquad (3.7\text{-}5)$$

Note that the amplitude ripple results in the generation of echoes. If $x(t)$ is a short pulse, these echoes may appear as two additional distinct pulses, but for an input of duration long compared to the time α, the effect of the echoes is simply to distort the output. This is illustrated in Fig. 3.7-2.

This method for analyzing amplitude distortion is referred to as the *paired-echo* method. It was invented by H. A. Wheeler in 1939 [3-8] and can be extended to more complicated transfer functions. It can be shown that

Figure 3.7-2. Distortion resulting from ripple in the magnitude of the frequency response.

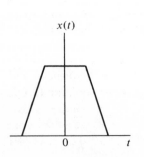

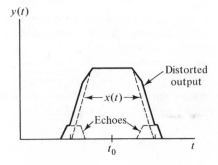

phase distortion can be treated in a similar fashion and gives rise to similar kinds of distortion (Prob. 3-20).

3.8 Realizable Low-Pass Filters

The transfer function of a filter constructed of linear lumped-parameter circuit elements has the form

$$H(j\omega) = \frac{N(j\omega)}{D(j\omega)} = \frac{b_0 + b_1(j\omega) + \cdots + b_k(j\omega)^k}{a_0 + a_1(j\omega) + \cdots + a_n(j\omega)^n}. \tag{3.8-1}$$

However, it can be shown from complex-variable theory that the ideal transfer characteristic given in Eq. (3.6-3) *cannot* be represented by a ratio of polynomials. The main problem in practical filter design is therefore to determine, for all i, the coefficients a_i and b_i in Eq. (3.8-1) so that the transfer function best approximates the ideal characteristic in some sense. One usually imposes the constraint that the *order n* of the resulting filter, that is, the number of poles, should be fixed. The order of a filter is directly related to its complexity, to the number of components needed in its construction, and to the difficulty of making it work as designed. Recall that in general there are at least as many poles as zeros; hence the zeros do not affect the order of the filter.

Network synthesis was at one time one of the most challenging and interesting problems facing communications engineers, and many thick books have been written on the subject. A list of some of the foremost names associated with this effort, with photographs, is given in Van Valkenburg [3-9]. It is clear that in our very brief treatment we can present only a small selection of some of the many results that have been obtained in all of this work. We hope to have chosen the most important, but our judgment is necessarily subjective.

Butterworth Filters

One of the first questions that arises in the design of practical filters is in what sense the approximation to the ideal is to be "best." One common criterion is that the transfer function is to be *maximally flat*. This means that as many derivatives of $|H(j\omega)|$ as possible should go to zero as ω goes to zero. This criterion leads to the *Butterworth* characteristic (Prob. 3-21):

$$|H(j\omega)| = \frac{1}{\sqrt{1 + (\omega/\omega_0)^{2n}}}. \tag{3.8-2}$$

This is shown in Fig. 3.8-1 for several values of n. The Butterworth magnitude characteristic is monotonically decreasing both inside the passband and outside the passband (i.e., the stopband). It has relatively small attenuation in the passband, particularly for $\omega \ll \omega_0$, but the transition between passband

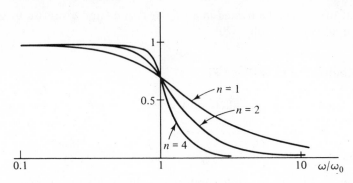

Figure 3.8-1. Butterworth amplitude characteristic for several values of n.

and stopband is relatively slow, and the attenuation in the stopband is not so complete as in some other filter types to be described later.

The transfer function $H(j\omega)$ of Eq. (3.8-2) has n poles. To find them, consider the function $|H(j\omega)|^2$, and replace $j\omega$ by the complex frequency s. Since $|H(j\omega)|^2 = H(j\omega)H(-j\omega)$, replacing $j\omega$ by s results in $H(s)H(-s)$. If s_i is a pole of $H(s)$ lying in the left half-plane, then $-s_i$, a pole of $H(-s)$, is its mirror image in the right half-plane. The function $H(s)H(-s)$ has $2n$ poles of which n lie in the left half-plane and give rise to a stable filter. Hence the desired poles are the values of s lying in the left half-plane for which $H(s)H(-s)$ is infinite. The function

$$H(s)H(-s) = \frac{1}{1 + (-1)^n (s/\omega_0)^{2n}}$$

goes to infinity if s is one of the $2n$ roots of the equation

$$\left(\frac{s}{\omega_0}\right)^{2n} = \begin{cases} +1, & n \text{ odd,} \\ -1, & n \text{ even.} \end{cases}$$

The roots are

$$s_i = \omega_0 e^{j\pi(2i/n)/2} \qquad \text{for odd } n$$

$$s_i = \omega_0 e^{j\pi[(2i-1)/n]/2} \qquad \text{for even } n, \quad i = 1, \ldots, 2n.$$

These roots are equally spaced points on a circle of radius ω_0, as shown in Fig. 3.8-2. The poles of interest are the roots lying in the left half-plane. The first few Butterworth polynominals [i.e., denominators of $H(s)$] are given in Table 3.8-1.

Tchebycheff Filters

By permitting the magnitude of the transfer function to have small ripples in the passband one can obtain a narrower transition between the pass- and

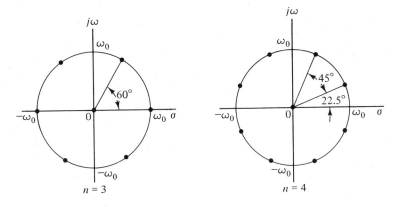

Figure 3.8-2. Roots of the equation $(s/\omega_0)^{2n} + (-1)^n = 0$ for $n = 3$ and $n = 4$. The poles of the Butterworth transfer function are the roots lying in the left half-plane.

TABLE 3.8-1

n	Butterworth polynominal $(x = s/\omega_0)$
1	$1 + x$
2	$\left(x + \dfrac{\sqrt{2} + j\sqrt{2}}{2}\right)\left(x + \dfrac{\sqrt{2} - j\sqrt{2}}{2}\right) = x^2 + x\sqrt{2} + 1$
3	$\left(x + \dfrac{1 + j\sqrt{3}}{2}\right)\left(x + \dfrac{1 - j\sqrt{3}}{2}\right)(x + 1) = (x + 1)(x^2 + x + 1)$
4	$(x^2 + 0.765x + 1)(x^2 + 1.848x + 1)$
5	$(x + 1)(x^2 + 0.618x + 1)(x^2 + 1.618x + 1)$

stopbands than that achievable with a Butterworth characteristic of the same order. One can also increase the attenuation in the stopband. The usual constraint is that the maximum extent of the ripples should be less than some predetermined amount, say 10%. This leads to the *Tchebycheff* filter characteristic illustrated in Fig. 3.8-3.

The magnitude of the transfer function oscillates in the passband between two limits shown as $|H_{max}|$ and $|H_{min}|$ in Fig. 3.8-3 and then drops off monotonically outside the passband. The Tchebycheff characteristic has the property that for a *given maximum peak-to-peak ripple inside the passband and monotonic attenuation outside the passband*, the transition from passband to stopband is *fastest for any nth order filter*. The magnitude characteristic is given by

$$|H(j\omega)|^2 = \frac{1}{1 + \epsilon^2 T_n^2(\omega/\omega_0)},\qquad(3.8\text{-}3)$$

where ϵ is a constant and $T_n(\cdot)$ is the Tchebycheff polynominal of the first kind of degree n, defined for $|x| < 1$ by

$$T_n(x) = \cos{(n\cos^{-1}x)}.$$

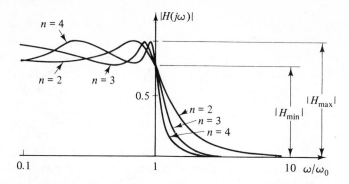

Figure 3.8-3. Tchebycheff magnitude responses for $n = 2, 3, 4$.

Although it may not be obvious at first glance, this is indeed a polynomial because one can express $\cos n\theta$ as a polynomial in $\cos \theta$ [3-10]. For example, $\cos 2\theta = 2\cos^2 \theta - 1$, $\cos 3\theta = 4\cos^3 \theta - 3\cos\theta$, etc. Thus by identifying θ with $\cos^{-1} x$ one can readily find the entries for the Tchebycheff polynomials in Table 3.8-2. The polynomials are defined for all x even though $\cos^{-1}x$

TABLE 3.8-2

n	Tchebycheff polynomials
1	x
2	$2x^2 - 1$
3	$4x^3 - 3x$
4	$8x^4 - 8x^2 + 1$
5	$16x^5 - 20x^3 + 5x$
6	$32x^6 - 48x^4 + 18x^2 - 1$

is defined only for $|x| \leq 1$. Observe that $-1 \leq T_n(x) \leq 1$ for $|x| \leq 1$ and that therefore $|H(j\omega)|^2$ oscillates between $(1 + \epsilon^2)^{-1}$ and 1 for $|\omega/\omega_0| \leq 1$. Hence ϵ is determined by the specified value of $|H_{max}| - |H_{min}|$. At the cutoff frequency $x = \omega/\omega_0 = 1$ and $T_n(x) = 1$; hence $|H(j\omega_0)|$ is exactly equal to $|H_{min}|$ for all orders, i.e., the characteristics for all n go through the same point. It can be shown [3-11] that the poles of the Tchebycheff transfer function lie on an ellipse whose major and minor diameters depend on n and ϵ.

Another standard class of filters is the class of *elliptic filters*, so called because elliptic functions are used in their design. The transfer function of elliptic filters can be put into the form of Eq. (3.8-3) except that the polynomial $T_n^2(\omega/\omega_0)$ is replaced by a ratio of polynomials. The transfer function of the elliptic filter therefore has zeros as well as poles. The zeros are located in the stopband and result in nulls in the frequency response. The transfer function therefore has ripples in the stopband as well as in the passband. The advantage is that this permits a still more rapid transition between the pass- and stopbands than is achieved by a Tchebycheff filter. In fact elliptic filters have the most rapid transition between pass- and stopbands of any filter of order n. Elliptic filters and other equiripple filters are discussed in some detail in [3-11] and [3-12].

There are a number of other filter types that are used for somewhat more special applications. For instance, one can specify a filter to have a maximally linear phase characteristic. This leads to a class of filters called Bessel filters since their design is based on Bessel polynomials. These filters have excellent transient response, but their amplitude response is generally poorer than any of the other filter types mentioned above. Additional filter types are given in Lubkin [3-13].

3.9 Active Filter Circuits

Practical analog filters are constructed from standard circuit elements: resistors, capacitors, inductors, and operational amplifiers (op-amps). Thus there still remains the task of showing how to translate a given realizable arrangement of poles and zeros into a hardware configuration. A convenient procedure for doing this is to establish a small number of canonical circuits that provide certain standard combinations of poles and zeros. Then the desired arrangement of the poles and zeros in the filter is obtained by cascading or otherwise combining a number of these canonical circuits. The proper location of the poles and zeros is achieved by adjusting the parameters in the canonical circuits.

As a simple example we can regard the RC and RLC filters shown in Fig. 3.9-1 as canonical circuits. There are numerous other simple circuits that could be added to this list. The transfer functions of the circuits given in (a) and (b) have been obtained earlier, and those in (c) and (d) are easily derived. In principle any transfer function having real and complex poles and real or complex zeros can be synthesized by cascading such circuits. In making up such a cascade one has to make sure that the assumption of zero source impedance and infinite load impedance is justified for each element of the cascade. This can be accomplished by raising the impedance level of each successive element of the cascade by about an order of magnitude over that of its predecessor, but this procedure can generally be used only in cascades of two or three elements. Alternatively the elements of the cascade can be separated by isolating amplifiers.

Simple passive low-pass and band-pass filters employing mainly LC resonant circuits are commonly used in radio and television sets. However, for more demanding applications passive circuits have a number of serious drawbacks. For one thing the range of permissible pole and zero locations achievable with practical passive circuits is fairly limited. Also the use of

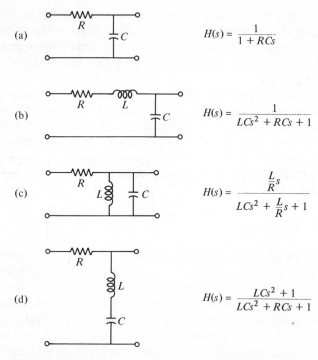

$$H(s) = \frac{1}{1 + RCs}$$

$$H(s) = \frac{1}{LCs^2 + RCs + 1}$$

$$H(s) = \frac{\frac{L}{R}s}{LCs^2 + \frac{L}{R}s + 1}$$

$$H(s) = \frac{LCs^2 + 1}{LCs^2 + RCs + 1}$$

Figure 3.9-1. Canonical circuits: (a) *RC* low-pass filter; (b) *RLC* low-pass filter; (c) *RLC* band-pass filter; (d) *RLC* notch filter.

inductors is generally undesirable because they are bulky, they cannot easily be implemented in integrated circuits, and their characteristics are often far from ideal.

For this reason modern filter designs frequently use active circuits. In an active circuit a high-gain op-amp is used in a feedback configuration together with resistors and capacitors. Active circuits are easily designed with transfer functions having poles and zeros located anywhere in the complex plane. A large variety of op-amps is available as inexpensive integrated circuits, and therefore active circuits are actually simpler and cheaper to construct than passive circuits.

Operational Amplifiers

A symbol for an op-amp is shown in Fig. 3.9-2(a). It is basically a very high-gain linear amplifier, normally used with output feedback, having two input terminals and one output terminal. Practical op-amps have additional terminals to provide power supply, frequency compensation, and dc compensation, but we need not be concerned with these. The terminal marked "−" is the inverting input, and that marked "+" the noninverting input terminal.

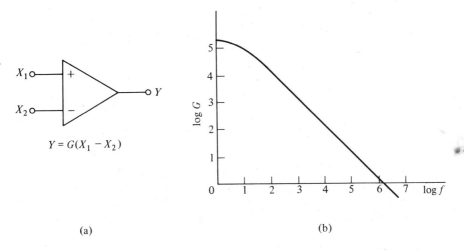

(a)   (b)

Figure 3.9-2. Operational amplifier: (a) elementary circuit operation; (b) open-loop frequency response.

Hence for low-frequency input signals† X_1 and X_2 connected as shown in Fig. 3.9-2, the output is ideally $Y = G(X_1 - X_2)$. In practice, zero output voltage may require a small nonzero value of $X_1 - X_2$, called the offset voltage. In many IC op-amps the offset is negligible, while in others it can be compensated. We shall therefore neglect it in our discussion below.

A typical gain vs. frequency curve is shown in Fig. 3.9-2(b). For stable operation in a feedback loop, op-amps require frequency compensation to control the high-frequency roll-off characteristic. Some IC op-amps such as the 741 are internally compensated.‡ Others, such as the 101, require external compensation but can be used to higher frequencies. Typically the low-frequency open-loop gain for these units is 2×10^5. The crossover frequency (i.e., the frequency where the gain is down to unity) for the 741 is about 1 MHz; for the 101 it may go as high as 3 MHz. There are special wideband units that generally have lower dc gain (about 3000) but crossover frequencies as high as 30 MHz.

A simple op-amp circuit with negative feedback is shown in Fig. 3.9-3. Assume that the input frequency is low enough so that the open-loop gain of the amplifier is $G \gg 1$. Also we assume that the amplifier has infinite input impedance and zero output impedance.

The input-output relation for the circuit is most easily obtained by using the fact that, except for a constant offset (which we ignore), the input voltage

---

†Capital letters indicate the frequency-domain representations of signals, i.e., $X(j\omega)$ or $X(s)$.

‡The complete number may be something like SN72 741 or MCCI 741; the prefix identifies the manufacturer. Sometimes a suffix is added for further description.

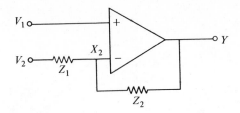

Figure 3.9-3. Operational amplifier with negative feedback.

$V_1 - X_2 \approx 0$. This follows since, for linear operation of the circuit, the absolute value of the output $|Y|$ cannot exceed some limit $|Y_{max}|$, which is fixed by supply voltages but is typically around 5 V. This means that $|V_1 - X_2| \leq |Y_{max}|/G$; hence for typical values of G, $|V_1 - X_2|$ is in the microvolt range.

We assume, therefore, that

$$V_1 = X_2. \qquad (3.9\text{-}1)$$

Also, if V_2 is assumed to be a zero-impedance source,

$$X_2 = \frac{Z_1 Y + Z_2 V_2}{Z_1 + Z_2}. \qquad (3.9\text{-}2)$$

Solving these equations simultaneously results in

$$Y = \frac{Z_1 + Z_2}{Z_1} V_1 - \frac{Z_2}{Z_1} V_2. \qquad (3.9\text{-}3)$$

Observe particularly that the gain depends only on the ratios of passive impedances Z_1 and Z_2 and is essentially independent of the open-loop gain G. Therefore large variations in open-loop gain have almost no effect on the circuit.

If the input signal V_1 is grounded, the signal X_2 must also be essentially at ground potential. For this reason the X_2 terminal is sometimes referred to as a *virtual ground*. The principle of the virtual ground can be used to show that the circuit shown in Fig. 3.9-4 acts as an adding circuit for the input signals

Figure 3.9-4. Op-amp used as an adding circuit.

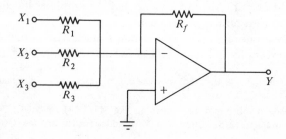

X_1, X_2, X_3. Since the inverting terminal is a virtual ground and the sum of the currents flowing into the inverting terminal must be zero, we have

$$\frac{X_1}{R_1} + \frac{X_2}{R_2} + \frac{X_3}{R_3} = -\frac{Y}{R_f}$$

or

$$Y = -\frac{R_f}{R_1} X_1 - \frac{R_f}{R_2} X_2 - \frac{R_f}{R_3} X_3.$$

If all the R's are equal, Y is the sum of the X's; if the R's are not equal, Y is a weighted sum of the X's.

A general active filter prototype is shown in Fig. 3.9-5. We see from Eq. (3.9-3) that the gain from the noninverting input X_1 to the output Y is given by

$$\frac{Y}{X_1} = K = \frac{R_a + R_b}{R_b}. \tag{3.9-4}$$

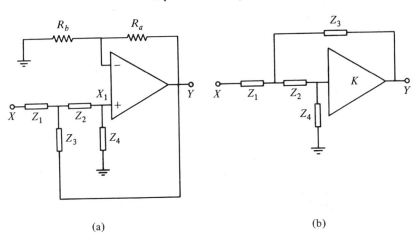

(a)   (b)

Figure 3.9-5. (a) Active filter prototype; (b) equivalent circuit where $K = (R_a + R_b)/R_b$.

To obtain this result, set $V_2 = 0$, $Z_1 = R_b$, and $Z_2 = R_a$ in Eq. (3.9-3). Hence an equivalent circuit can be drawn in the form shown in Fig. 3.9-5(b). The transfer function of this circuit is easily obtained by standard methods and can be put into the form

$$H(s) = \frac{Y}{X} = \frac{KZ_3Z_4}{Z_1Z_2 + Z_3(Z_1 + Z_2 + Z_4) + Z_1Z_4(1 - K)}. \tag{3.9-5}$$

The general prototype is converted into a low-pass filter by making Z_1 and Z_2 resistances and Z_3 and Z_4 capacitances. A simple choice of values is $Z_1 = Z_2 = R$ and $Z_3 = Z_4 = 1/Cs$, as in Fig. 3.9-6. If we substitute these values into

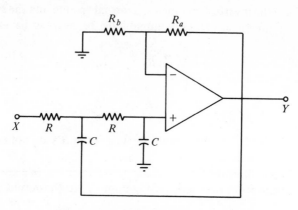

Figure 3.9-6. Second-order low-pass filter.

Eq. (3.9-5) and let $\omega_0 = 1/RC$, we obtain

$$H(s) = \frac{K\omega_0^2}{s^2 + s\omega_0(3 - K) + \omega_0^2}. \tag{3.9-6}$$

We see that the transfer function has two poles that will be complex if $1 < K < 5$ and that can be made purely imaginary if $K = 3$ (for $K = 3$, the circuit will oscillate at $\omega = \omega_0$). For a second-order (i.e., $n = 2$) Butterworth response we see from Table 3.8-1 that $3 - K = \sqrt{2}$ or $K = 1.586$. Any even-order Butterworth filter can be constructed by cascading several sections having the same RC product. Thus, for a fourth-order filter we see from Table 3.8-1 that we cascade a section in which $K = 2.235$ with another one in which $K = 1.152$. Odd-order filters require a first-order prototype of the type shown in Fig. 3.9-7. A fifth-order Butterworth filter is shown in Fig. 3.9-8. The cutoff frequency $\omega_0 = 1/RC$. If R is set to 1000 ohms, then $C = 1000/\omega_0 \ \mu\text{F}$.

Figure 3.9-7. (a) First-order low-pass filter; (b) equivalent circuit where $K = (R_b + R_a)/R_b$.

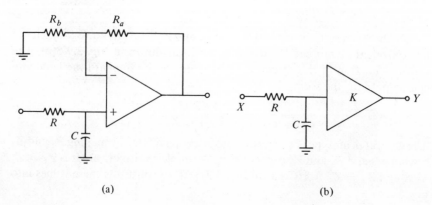

(a)   (b)

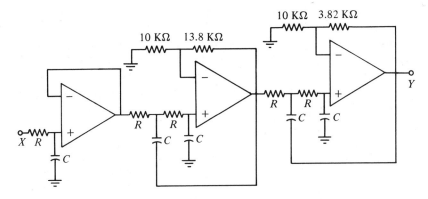

Figure 3.9-8. Fifth-order Butterworth low-pass filter.

A feature of this design is that the filter does not have unity gain. If unity gain is desired, the gains of all the op-amps can be made equal to unity by connecting the output terminal directly to the inverting input terminal and omitting R_b. The filter parameters can then be adjusted by using either two different capacitances or two different resistors. We leave it to the reader to show that if the resistors are both equal and if the capacitors are C_1 and C_2 (from left to right in Fig. 3.9-6), then $\omega_0^2 = 1/R^2 C_1 C_2$, and the denominator polynomial of Eq. (3.9-6) is changed to $s^2 + 2\sqrt{C_2/C_1}\,\omega_0 s + \omega_0^2$. Thus any positive value of the coefficient of the middle term can be obtained by properly adjusting the ratio C_2/C_1.

Although active filter circuits will work well with a wide range of parameter values, it is important that the impedance levels of the R's are such that the input impedance of the op-amp is much larger and the output impedance much smaller than R. A value for R of about 1000 ohms is usually satisfactory. Also, the filter will not perform in accordance with the expressions derived here if the open-loop gain of the amplifier is not very large. Since this gain decreases with increasing frequency, there is always a maximum frequency for which these circuits should be used. A typical upper cutoff frequency is 20 kHz, but higher frequencies can be used.

Since the transfer functions of Tchebycheff and Butterworth filters are very similar, the active prototypes presented in Figs. 3.9-6 and 3.9-7 can also be used to construct Tchebycheff filters by appropriate choice of the feedback resistors R_a and R_b. The values of these resistors depend on the choice of the ripple-amplitude parameter ϵ, and although the calculations needed for their determination are basically straightforward, they are also quite tedious and are therefore omitted. In fact the easiest way to construct any of these filters is by the use of standardized charts or nomographs such as are given in several publications ([3-14] and [3-15]). These references also give design charts for some of the other filter types mentioned in Sec. 3.8, and they contain ranges of parameter values and frequencies for which the circuits perform properly.

3.10 Frequency Transformations

The discussion thus far has dealt exclusively with low-pass filters, but the results can be generalized to some of the other filter types discussed in Sec. 3.6 by simple transformations. For example, the transfer function of a high-pass filter can be obtained from the low-pass transfer function by using the transformation ([3-2], p. 209)

$$\frac{j\omega}{\omega_0} \longrightarrow \frac{\omega_0}{j\omega}. \tag{3.10-1}$$

Thus the second-order low-pass Butterworth transfer function

$$H(j\omega) = \frac{1}{1 + j\sqrt{2}(\omega/\omega_0) - (\omega/\omega_0)^2} \tag{3.10-2}$$

is transformed into the high-pass form:

$$H(j\omega) = \frac{-(\omega/\omega_0)^2}{1 + j\sqrt{2}(\omega/\omega_0) - (\omega/\omega_0)^2}. \tag{3.10-3}$$

Also, the first- and second-order low-pass prototype given in Figs. 3.9-7 and 3.9-6 can be converted to high-pass prototypes by simply exchanging R's and C's in the noninverting branch. This is shown in Fig. 3.10-1.

A transformation that converts a low-pass transfer function to band-pass is ([3-2], p. 209)

$$\omega \longrightarrow \frac{\omega^2 - \omega_c^2}{\omega} \tag{3.10-4a}$$

or, when complex frequencies are used,

$$s \longrightarrow \frac{s^2 + \omega_c^2}{s}. \tag{3.10-4b}$$

Figure 3.10-1. Active high-pass filters: (a) first order; (b) second order.

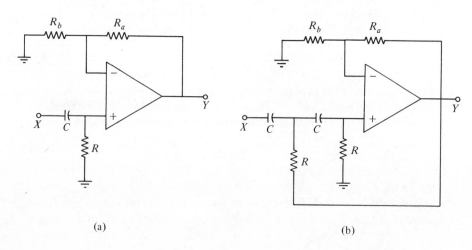

(a)   (b)

If the original transfer function had a pole on the real axis at, say, $s = -\omega_0$, and $\omega_c > \omega_0/2$, then the effect of the transformation is to replace the pole at $s = -\omega_0$ with a set of poles at $s = -(\omega_0/2) \pm j\sqrt{\omega_c^2 - (\omega_0/2)^2}$. If $\omega_c \gg \omega_0$, then $s \simeq -(\omega_0/2) \pm j\omega_c$. The configuration is shown in Fig. 3.10-2.

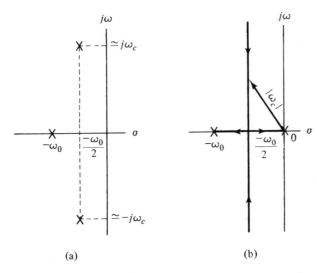

(a)   (b)

Figure 3.10-2. (a) Example of a low-pass to band-pass transformation: The original poles have been transformed to a set of poles at $s \simeq -(\omega_0/2) \pm j\omega_c$ if $\omega_0 \ll \omega_c$. (b) As ω_c decreases, the locus of the poles is a line intersecting the negative real axis at $\sigma = -(\omega_0/2)$. When $\omega_c \leq \omega_0/2$, the poles are purely real and move toward the points $\sigma = 0$ and $\sigma = -\omega_0$ as $\omega_c \to 0$. In practice, the band-pass property refers to the situation in (a) when $\omega_c \ll \omega_0$.

The simplest example illustrating this transformation is the first-order low-pass transfer function

$$H(j\omega) = \frac{1}{j(\omega/\omega_0) + 1}. \tag{3.10-5}$$

After we apply the transformation of Eq. (3.10-4) this becomes

$$H(j\omega) = \frac{j\omega\omega_0}{\omega_c^2 + j\omega\omega_0 - \omega^2} = \frac{j(\omega/\omega_c)(\omega_0/\omega_c)}{1 + j(\omega/\omega_c)\cdot(\omega_0/\omega_c) - [\omega/\omega_c]^2}. \tag{3.10-6}$$

This is a typical second-order resonant filter response with $Q = \omega_c/\omega_0$ [cf. Eq. (3.4-25)].

For large Q the resonant peak is very close to $\omega = \omega_c$ and the 3-dB bandwidth is the frequency interval for which $|\omega - \omega_c| \leq \omega_0/2$. The width of this interval along the positive-frequency axis is ω_0, and there is another interval of the same length along the negative-frequency axis. Thus the total interval on the whole frequency axes for which $|H(j\omega)|$ exceeds $1/\sqrt{2}$ is $2\omega_0$, just as for the low-pass case.

The low-pass transfer function can be converted to the *band-stop* form by applying the high-pass and band-pass transformation in sequence ([3-2], p. 209). The transformation is therefore

$$\frac{j\omega}{\omega_0} \longrightarrow \frac{\omega\omega_0}{j(\omega^2 - \omega_c^2)}. \tag{3.10-7}$$

The effect of this transformation on the first-order low-pass transfer function of Eq. (3.10-5) is to produce

$$H(j\omega) = \frac{\omega_c^2 - \omega^2}{\omega_c^2 + j\omega_0\omega - \omega^2} = \frac{1 - (\omega^2/\omega_c^2)}{1 - (\omega^2/\omega_c^2) + j(\omega\omega_0/\omega_c^2)}. \tag{3.10-8}$$

This is seen to have zeros on the imaginary axis and therefore a null at $\omega = \pm\omega_c$. It has not so much a band-stop as a single-frequency-stop characteristic, and filters with this kind of transfer function are sometimes called *notch filters*. They are very commonly used as *frequency traps* in radio or television in order to eliminate interference from carrier-frequency components near the channel of interest.† The magnitude $|H(j\omega)|$ of a second-order notch filter is plotted in Fig. 3.10-3 for various values of ω_0/ω_c.

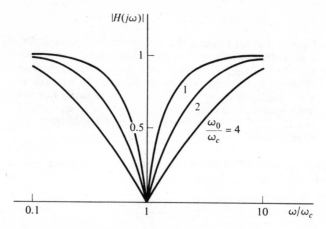

Figure 3.10-3. First-order band-stop characteristic.

There are a number of active circuits having band-pass or band-stop characteristics. Perhaps the simplest arrangement is to place a high-pass filter in series with a low-pass filter to produce a band-pass filter and to use a parallel combination of high-pass and low-pass filters to obtain a band-stop characteristic. These two methods are illustrated in Fig. 3.10-4. An advantage

†In commercial TV sets simple *LC* traps rather than circuits involving op-amps are ordinarily used.

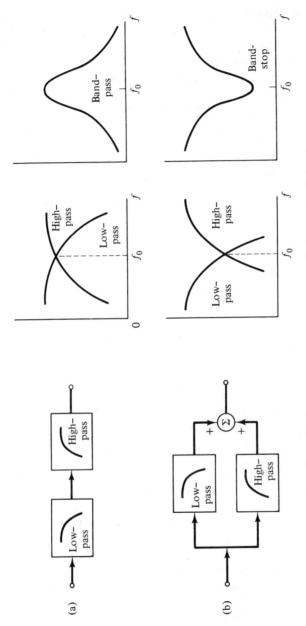

Figure 3.10-4. Combining high-pass and low-pass filters to achieve (a) band-pass and (b) band-stop filters.

of this method is its inherent flexibility. The high-pass and low-pass sections can be designed separately and need not be "matched" in any sense. By separating the cutoff frequencies of the two sections one can also easily obtain filters that pass or reject relatively wide frequency bands. On the other hand, the complexity, i.e., number of op-amps, resistors, and capacitors, of these combination filters is generally greater than for filters specifically designed for band-pass or band-stop.

An example of a cascadable second-order band-pass filter is shown in Fig. 3.10-5(a). This is one version of a standard class of related circuits called single amplifier biquad (SAB) circuits. A more general circuit is shown in Fig. 3.10-5(b). We leave it to the reader to show that the transfer function for this circuit is in the form of a ratio of two quadratic polynomials in s:

$$H(s) = K\frac{b_0 + b_1 s + b_2 s^2}{a_0 + a_1 s + a_2 s^2};$$ (3.10-9)

hence the designation biquad (for biquadratic). In the band-pass version of this circuit [Fig. 3.10-5(a)] resistors R_3 and R_4 are eliminated, and this causes b_0 and b_2 in Eq. (3.10-9) to vanish. If resistance R_3 is present but R_4 is left disconnected, if can be shown (Prob. 3-31) that $a_0 = b_0$, $a_2 = b_2$. By proper choice of component values one can then obtain zero b_1 and finite positive a_1 with the result that the transfer function is that of the notch filter, Eq. (3.10-8). With R_4 connected one can obtain a combination of notch and low-pass characteristics. SAB circuits are discussed in detail in [3-16] and [3-17]. The latter reference describes a general biquad circuit contained entirely on a hybrid integrated circuit (HIC) chip. Changes in the circuit topography are made by changes in the external pin connections, and resistance values are adjusted by burning away certain sections with a laser beam.

In this necessarily brief treatment of the subject of active filters we have been able to do little more than scratch the surface. We hope, however, that the examples presented will have given the reader a general overview of the subject. We have already given several references to more extensive treatments; additional ones are given at the end of the chapter ([3-18]–[3-20]).

3.11 Summary

In this chapter we have examined some of the properties of linear lumped-parameter electric circuits and filters. We have shown that the response of such circuits can be obtained both in the time domain, i.e., via the superposition integral, and in the frequency domain using the transfer function. We have investigated some of the properties of transfer functions, such as the poles and zeros, Bode plots, phase-shift curves, etc. We then defined a number of ideal filter characteristics: low-pass, high-pass, band-pass, and band-stop.

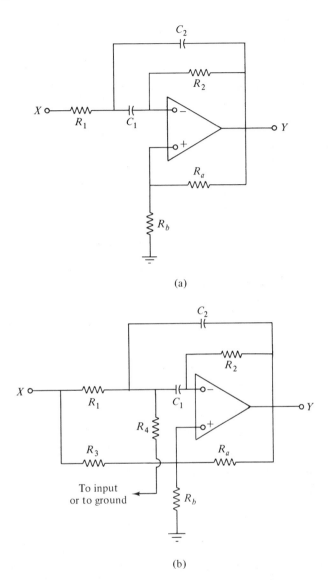

(a)

(b)

Figure 3.10-5. Single amplifier biquad (SAB) circuits: (a) band-pass filter; (b) general form.

We showed that the ideal forms are noncausal, and we considered several practical transfer functions such as the Butterworth, Tchebycheff, and elliptic forms that are commonly used to approximate the ideal filter response functions. We showed finally how to design hardware realizations in the form of active filters.

PROBLEMS

3-1. ([3-21], p. 4) Consider a linear, time-invariant system $\mathcal{H}$ that furnishes an output $y(t)$ when the input is $x(t)$.

a. Which properties of the system are involved to argue that the response to $(1/\Delta t)[x(t + \Delta t) - x(t)]$ is $(1/\Delta t)[y(t + \Delta t) - y(t)]$?

b. Extend the result in part a to argue that the response to dx/dt is dy/dt.

c. Take $x(t) = Ae^{j\omega t}$. Use the result in part b and the fact that the system is linear to show that $y(t)$ satisfies the differential equation

$$\frac{dy(t)}{dt} = j\omega y(t).$$

Show that the solution is $y(t) = Be^{j\omega t}$, where B is an arbitrary constant.

d. Compute the transfer function $H(\omega)$ defined by

$$H(\omega) = \frac{\mathcal{H}(Ae^{j\omega t})}{Ae^{j\omega t}};$$

i.e., $H(\omega)$ is the response of $\mathcal{H}$ to $Ae^{j\omega t}$ divided by $Ae^{j\omega t}$.

3-2. Assume that $x(t)$ is a continuous function over $[t_0, t_0 + T]$ and continuous at $t = t_0$. If $\Delta\xi$ is a small increment of time, then to a good approximation, $x(t)$, for $t_0 < t < T + t_0$, can be approximated by a finite number of unit steps according to

$$x(t) \simeq x(t_0)u(t - t_0) + \sum_{i=1}^{N} \{x(t_0 + i\Delta\xi)$$

$$- x[t_0 + (i - 1)\Delta\xi]\}u(t - t_0 - i\Delta\xi), \qquad t_0 \leq t \leq t_0 + T,$$

where $\Delta\xi = T/N$. Derive this result.

3-3. Extend the result in Prob. 3-2, by letting $\Delta\xi \rightarrow 0$ and $N \rightarrow \infty$ with $N\Delta\xi = T$, to

$$x(t) = x(t_0)u(t - t_0) + \int_0^T \dot{x}(t_0 + \xi)u(t - t_0 - \xi)\,d\xi.$$

3-4. In Prob. 3-3, assume that $x(t_0) = 0$, and let $\lambda = t_0 + \xi$. Show that $x(t)$ can be written, with $T \rightarrow \infty$, as

$$x(t) = \int_{t_0}^{\infty} \dot{x}(\lambda)u(t - \lambda)\,d\lambda, \qquad t_0 > 0.$$

3-5. Show that if $\mathcal{H}$ is a causal system and $a(t)$ is defined to be the response of $\mathcal{H}$ to a unit step applied at $t = 0$, then the output $y(t)$ can be written as

$$y(t) = \int_0^{\infty} \dot{x}(t - \lambda)a(\lambda)\,d\lambda.$$

3-6. Use the result in Prob. 3-5 to compute the response $y(t)$ to an input $x(t) = \text{rect}\,[(t - T/2)/T]$ when $a(t) = (1 - e^{-t/T})u(t)$.

3-7. Use Eq. (3.2-9) to compute the response to the input $x(t) = \text{rect}\,[(t - T/2)/T]$ when $h(t)$ is $e^{-t}u(t)$.

3-8. Show that the parameter Q, as defined in Eq. (3.4-24), is given by $Q = \omega_0 L/R$ for the series RLC circuit in Fig. P3-8. *Hint:* Take $v(t) = V\cos\omega_0 t$,

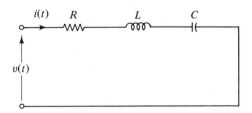

Figure P3-8

and use the definition

$$Q = \frac{2\pi(\text{energy stored in the circuit})}{\text{energy dissipated per cycle}}.$$

3-9. Consider the parallel RLC circuit shown in Fig. P3-9.

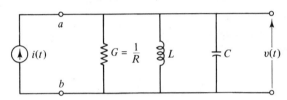

Figure P3-9

a. Show that the differential equation relating the voltage $v(t)$ to the input current $i(t)$ is

$$C\frac{d^2v}{dt^2} + G\frac{dv}{dt} + \frac{1}{L}v(t) = \frac{di}{dt}.$$

b. Show that when $v(t)$ and $i(t)$ are replaced by their Fourier transforms, the impedance between a-b is

$$Z(j\omega) = \frac{V(j\omega)}{I(j\omega)} = \frac{j(\omega/\omega_0)}{QG[1 - (\omega/\omega_0)^2 + j(\omega/\omega_0)(1/Q)]},$$

where $Q = [\omega_0 LG]^{-1}$ and $\omega_0 = (LC)^{-1/2}$.

c. Let $s = \sigma + j\omega$, and consider $Z(s)$. Discuss the locations of the poles in the complex s plane when $R \leq \omega_0 L/2$ and when $R > \omega_0 L/2$.

3-10. In this problem we consider some of the implications of Q. Assume where needed that $R > \omega_0 L/2$.

a. Show that $Z(j\omega)$ in Prob. 3-9 can be written as

$$Z(j\omega) = \frac{R}{1 + jQ[(\omega^2 - \omega_0^2)/\omega\omega_0]}.$$

The half-power frequencies ω_1, ω_2 are determined from the points where $|Z(j\omega)|^2 = \frac{1}{2}|Z(j\omega)|_{\max}^2$. Show that ω_1, ω_2 can be found from the equation

$$Q\frac{\omega^2 - \omega_0^2}{\omega\omega_0} = \begin{cases} 1, & \text{for } \omega = \omega_1, \\ -1, & \text{for } \omega = \omega_2, \end{cases}$$

and that $\omega_1 - \omega_2 = [RC]^{-1}$ and $\omega_1\omega_2 = \omega_0^2$.

b. Let $\alpha \equiv [2RC]^{-1}$. Show that the 3-dB bandwidth B satisfies

$$\frac{B}{\omega_0} = \frac{2\alpha}{\omega_0} = \frac{1}{Q} \qquad (B \text{ in radians per second})$$

or, equivalently,

$$Q = \frac{f_0}{B} \qquad (B \text{ in hertz}).$$

For extracting the fundamental component of a square wave, would you choose a small or a large Q circuit?

c. As Q increases, what happens to the poles of $Z(s)$? How would you expect the circuit to deal with transients?

d. Take $i(t)$ in Prob. 3-9 as $i(t) = I\cos\omega_0 t$. Show that

$$\frac{2\pi \text{ (peak energy stored)}}{\text{energy dissipated per cycle}} = \frac{2\pi(\frac{1}{2}CV_1^2)}{(2\pi/\omega_0)(V_1^2/2R)} = \omega_0 CR = Q,$$

where $V_1 \equiv IR$.

3-11. Compute the Bode diagram (amplitude and phase curve) for the transfer function given by

$$H(j\omega) = \frac{j\omega + 100}{(j\omega + 10)(j\omega + 20)}.$$

3-12. Show that for a causal system whose transfer function is $H(j\omega) = A(\omega) + jB(\omega)$

$$h(t) = \frac{2}{\pi}\int_0^\infty A(\omega)\cos\omega t\,d\omega$$

$$= -\frac{2}{\pi}\int_0^\infty B(\omega)\sin\omega t\,d\omega.$$

Hint: For a causal system $h(t) = 0$ for $t < 0$. Therefore, by using the dual of the problem considered in Sec. 2.13 it can be shown that $A(\omega)$ and $B(\omega)$ are Hilbert transforms of each other.

3-13. Find the transfer function of the system shown in Fig. P3-13. Sketch the output for a rectangular pulse input of duration $\tau = T$, $\tau \ll T$, and $\tau \gg T$.

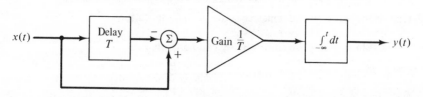

Figure P3-13

3-14. Compute the response of the circuit shown in Fig. P3-14 to an input $x(t) = \text{rect}(t/T)$. Obtain the solution with the help of the transfer function and the Fourier inversion theorem.

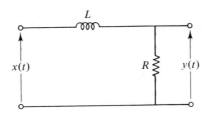

Figure P3-14

3-15. A series LC network with losses is sometimes modeled as shown in Fig. P3-15. Compute $Y(j\omega)/X(j\omega)$. Take $L/R_1 = R_2 C$; show that the transfer function is then given by

$$H(s) = \frac{\alpha/(1 + \alpha)}{(s^2/\omega_0^2) + (2/\sqrt{1 + \alpha})(s/\omega_0) + 1},$$

where $\alpha = R_2/R_1$ and $\omega_0^{-2} = R_2 LC/(R_1 + R_2)$. Plot the loci of the poles of $H(s)$ as α varies from ∞ to 0.

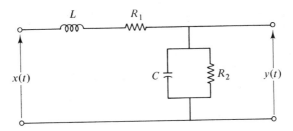

Figure P3-15

3-16. The input $x(t)$ (Fig. P3-16) is applied in turn to two systems with impulse responses as shown in Figs. P3-16(a) and (b), respectively. Sketch the output

Figure P3-16

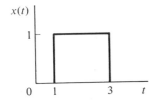

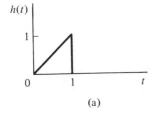

(a)

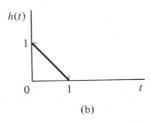

(b)

time function in each case, indicating the amplitude of the output pulse and the quantitative shapes of leading and trailing edges.

3-17. A linear system has the impulse response shown in Fig. P3-17(a) and is subjected to the indicated input. Find the output time function
a. By formal evaluation of the convolution integral.
b. By use of Fourier transforms.

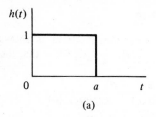
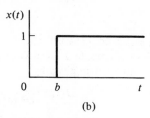

(a)   (b)

Figure P3-17

3-18. Use the alternative form of the convolution theorem given in Eq. (3.5-19) to compute the waveform shown in Fig. 3.5-5 in the text.

3-19. Assume that a time-invariant system is stable if it satisfies

$$\int_{-\infty}^{\infty} |h(t - \lambda)| \, d\lambda < \infty.$$

a. Is the ideal integrator, i.e., a system that furnishes an output $y(t)$ when the input is $x(t)$ according to

$$y(t) = \int_{-\infty}^{t} x(\lambda) \, d\lambda,$$

stable?

b. Is the system in Fig. P3-19 stable? If the integration time is T, what inequality must the product RC satisfy in order for the circuit to behave as an integrator?

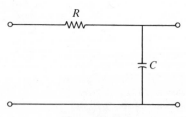

Figure P3-19

3-20. A network whose phase shift has ripples can produce echoes. Show this by considering

$$|H(f)| = 1, \qquad \theta(f) = -\omega t_0 + b \sin \omega t_d, \qquad \text{where } b \ll 1.$$

Hint: Use a series expansion for $\exp(jb \sin \omega t_d)$.

3-21. A Butterworth low-pass filter has

$$|H(f)| = \left[1 + \left(\frac{f}{f_0}\right)^{2n}\right]^{-1/2},$$

where n is the number of reactive components (i.e., inductors or capacitors).
   a. Show that as $n \to \infty$ the amplitude response of the Butterworth filter approaches that of the ideal low-pass filter.
   b. Find n so that $|H(f)|^2$ is constant to within 1 dB over the range $|f| \le 0.8f_0$. Repeat for $|f| \le 0.9f_0$.

3-22. A Gaussian filter has

$$H(f) = \exp(-af^2).$$

Calculate the 3-dB bandwidth and

$$B_{eq} = \frac{1}{2H(0)} \int_{-\infty}^{\infty} |H(f)| \, df.$$

3-23. Compute and sketch the poles of a Butterworth filter of order 5. Repeat for a Butterworth filter of order 6.

3-24. Derive the transfer functions given in the text for the circuits shown in Fig. 3.9-1.

3-25. Derive Eq. (3.9-5), which gives the transfer function of the general prototype in Fig. 3.9-5(b).

3-26. Prove that for the second-order low-pass filter whose transfer function is given in Eq. (3.9-6) the roots will be complex when $1 < K < 5$.

3-27. It is possible to convert the noncausal low-pass impulse response given in Eq. (3.6-4) to a causal response by the simple expedient of multiplying $h(t)$ by the step function $u(t)$ and thereby removing the response for negative t. An almost equivalent procedure leading to a much simpler transfer function is to use a square pulse instead of a step; i.e., consider the causal impulse response

$$h_c(t) = 2KB \text{ sinc } [2B(t - t_0)] \text{ rect} \left(\frac{t - t_0}{2t_0}\right).$$

   a. Sketch this function for $Bt_0 = 2$.
   b. Show that the corresponding transfer function is

$$H_c(f) = \frac{K}{\pi} e^{-j2\pi f t_0}\{\text{Si } [2\pi t_0(f + B)] - \text{Si } [2\pi t_0(f - B)]\},$$

   where Si $(x) = \int_0^x [(\sin t)/t] \, dt$ is the sine-integral function.
   c. Sketch $|H_c(f)|$ for $Bt_0 = 2$; observe that instead of the square pulse corresponding to the noncausal impulse response we now have an oscillatory approximation to a square pulse.
   d. Show that for large Bt_0, $H_c(t)$ approximates the noncausal frequency response except at the transition frequencies.

3-28. Consider the transformation in Eq. (3.10-4b), i.e.,

$$s \longrightarrow \frac{s^2 + \omega_c^2}{s},$$

in the transfer function of the low-pass filter shown in Fig. P3-28. Compare the position of the original poles with the position of the new poles.

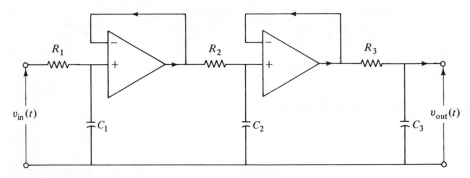

Figure P3-28

3-29. Derive Eq. (3.10-6) and show that the 3-dB frequencies are indeed $\omega_c + (\omega_0/2)$ and $\omega_c - (\omega_0/2)$ when Q is much larger than unity.

3-30. Show that the transfer function for the SAB band-pass circuit of Fig. 3.10-5(a) is given (for infinite op-amp gain) by

$$H(s) = \frac{-(1 + b)R_2 C_1 s}{R_1 R_2 C_1 C_2 s^2 + [R_1(C_1 + C_2) - bR_2 C_1]s + 1},$$

where $b = R_b/R_a$.

3-31. Show that the transfer function for the SAB circuit of Fig. 3.10-5(b), with R_4 disconnected, has the general form of Eq. (3.10-9). (Assume that the op-amp has infinite input impedance, zero output impedance, and infinite gain.)

REFERENCES

[3-1] RALPH J. SCHWARZ and BERNARD FRIEDLAND, *Linear Systems*, McGraw-Hill, New York, 1965, p. 12.

[3-2] H. W. BODE, *Network Analysis and Feedback Amplifier Design*, Van Nostrand Reinhold, New York, 1945.

[3-3] S. RAMO, J. R. WHINNERY, and T. VAN DUZER, *Fields and Waves in Communication Electronics*, Wiley, New York, 1967, p. 8.

[3-4] E. A. GUILLEMIN, *The Mathematics of Circuit Analysis*, Wiley, New York, 1949.

[3-5] L. A. ZADEH, "Frequency Analysis of Variable Networks," *Proc. IRE*, **38**, pp. 291–299, March 1950.

[3-6]   RAYMOND E. A. C. PALEY and NORBERT WIENER, "Fourier Transforms in The Complex Domain," *American Mathematical Society Colloquium Publication 19*, New York, 1934.

[3-7]   A. N. PAPOULIS, *The Fourier Integral and Its Applications*, McGraw-Hill, New York, 1962, pp. 215–217.

[3-8]   H. A. WHEELER, "The Interpretation of Amplitude and Phase Distortion in Terms of Paired Echos," *Proc. IRE*, **27**, pp. 359–385, June 1939.

[3-9]   M. E. VAN VALKENBURG, *Introduction to Modern Network Synthesis*, Wiley, New York, 1960.

[3-10]   B. O. PEIRCE, *A Short Table of Integrals*, Ginn, Boston, 1929, p. 75.

[3-11]   L. WEINBERG, *Network Analysis and Synthesis*, McGraw-Hill, New York, 1962, Chap. 11.

[3-12]   D. A. CALAHAN, *Modern Network Synthesis*, Vol. 1: Approximation, Hayden, New York, 1964.

[3-13]   Y. J. LUBKIN, *Filter Systems and Design*, Addison-Wesley, Reading, Mass., 1970.

[3-14]   JOHN L. HILBURN and DAVID E. JOHNSON, *Manual of Active Filter Design*, McGraw-Hill, New York, 1973.

[3-15]   ROBERT R. SHEPARD, "Active Filters: Part 12, Shortcuts to Network Design," *Electronics*, **42**, No. 17, pp. 82–91, Aug. 18, 1969.

[3-16]   P. E. FLEISCHER, "Sensitivity Minimization in a Single Amplifier Biquad Circuit," *IEEE Trans. Circuits Sys.*, **CAS23**, No. 1, pp. 46–55, Jan. 1976.

[3-17]   J. J. FRIEND, C. A. HARRIS, and D. HILBERMAN, "STAR: An Active Biquadratic Filter Section," *IEEE Trans. Circuits Sys.*, **CAS22**, No. 2, pp. 115–121, Feb. 1975.

[3-18]   L. P. HUELSMAN, *Active Filters: Lumped, Distributed, Integrated, Digital and Parametric*, McGraw-Hill, New York, 1970.

[3-19]   S. K. MITRA, ed., *Active Inductorless Filters*, IEEE Press, New York, 1971.

[3-20]   R. W. NEWCOMB, *Active Integrated Circuit Synthesis*, Prentice-Hall, Englewood Cliffs, N.J., 1968.

[3-21]   W. R. BENNETT, *Introduction to Signal Transmission*, McGraw-Hill, New York, 1970.

Sampling and Pulse Modulation

4.1 Introduction

When one observes a typical signal, such as the output of a microphone picking up someone's speech, on an oscilloscope, one sees a very complicated looking random wave apparently able to take on an infinity of different values and shapes. However, closer examination of the waveform reveals that there are limits to its variability. One finds, for instance, that the wave does not have any abrupt breaks; i.e., it is continuous. Also, on still closer examination (which may involve expansion of the time scale) one can see that all slopes and curvatures are finite. Thus, although quite random and undisciplined in first appearance, the wave is seen to be subject to some important constraints.

These constraints are imposed in part by the limitations of the physical apparatus, e.g., the human vocal tract, that generates the signal. These constraints translate into a limit on the lower and upper frequencies in the signal. For speech this ranges from about 100 to about 4000 Hz; for television signals it ranges from 0 to about 4 MHz, but it is always finite.

One of the important consequences of finite signal bandwidth is that signals can be represented by samples taken at discrete instants called sampling instants. As we shall show in the following sections, such sampling introduces essentially no loss, and it is possible to find interpolation functions which accurately reconstruct the original signal from the samples. The fact that band-limited signals can be accurately represented by discrete samples is of major importance in signal theory. It means that the undenumerable infinity of a continuous signal representation can be reduced to a countable infinity. Countability implies that the signal can be represented by a vector, and therefore the tools of vector-space mathematics can be utilized in the analysis of signals and of signal processing.

From the practical point of view, sampling makes it possible to deal with a signal only at isolated instants of time. This permits, for instance, time sharing of transmission facilities by a procedure referred to as time-division multiplexing. Perhaps even more importantly the waveform samples can be digitized, i.e., converted to numbers having only a finite number of digits. Although such a conversion is generally approximate, it can be made as exact as desired by using a large enough number of digits. By sampling and digitizing, signals can be converted into a series of numbers that can be handled by a digital computer or any other digital circuit. In fact this is the only way in which signals can be processed by digital computers. Hence sampling of signals is an essential preprocessing step in all applications of digital systems to signal processing.

In the following sections we shall consider some of the consequences of the bandwidth constraint on signals. We derive several versions of the sampling theorem and show how the signal can be reconstructed from the samples. We study time-division multiplexing (TDM) of several signals and some of the problems, such as intersymbol interference, that have to be solved if TDM is to be practical. We then consider a number of pulse-modulation schemes that are used to transmit the sampled data.

Before getting into a detailed discussion of sampling we must briefly consider a general property of the frequency limitation. We saw in Chapter 2 (Sec. 2.9) that the Fourier spectrum of a time-limited signal such as the square pulse extends to infinite frequencies. By the duality principle we conclude that a signal having a finite bandwidth must extend to infinity in time. Apparently a signal cannot be finite in both time and frequency. This observation is more precisely enunciated in the following rule ([4-1], p. 70):

A strictly band-limited signal cannot be simultaneously time limited, and vice versa.

By strictly band limited we mean

$$X(f) = 0 \qquad \text{for } |f| > W, \tag{4.1-1}$$

and, similarly, strictly time limited means

$$x(t) = 0 \qquad \text{for } |t| > T. \tag{4.1-2}$$

The proof of this statement rests on the Paley-Wiener condition [Eq. (3.7-1)]. According to this criterion, if Eq. (4.1-2) holds, $x(t - T)$ can be regarded as the output of a causal system, and therefore $e^{j2\pi fT}X(f)$ cannot vanish for any finite (or infinite) frequency interval. The converse follows by duality.

Although the strict time and frequency limitations given in Eqs. (4.1-1) and (4.1-2) cannot be imposed simultaneously, we can deal with signals that are *essentially* time and frequency limited. This means, for instance, that the spectrum of a time-limited signal has only very small magnitude outside some frequency band. (For a good discussion of this subject, see [4-2].)

4.2 Implications of the Frequency Limitation

We now consider a strictly band-limited signal, i.e., one that satisfies Eq. (4.1-1). We assume also that $|X(f)|$ remains finite throughout the interval $-W < f < W$. It then follows that $x(t)$ and all of its time derivatives exist, since the nth derivative of $x(t)$ is

$$\frac{d^n x(t)}{dt^n} = (2\pi j)^n \int_{-W}^{W} f^n X(f) e^{j2\pi ft} \, df, \qquad (4.2\text{-}1)$$

and by hypothesis the integral on the right exists for any finite n. Formally, $x(t)$ can therefore be expanded in a Taylor series about any (finite) particular value of $t = t_0$:

$$x(t) = x(t_0) + \frac{dx(t_0)}{dt}(t - t_0) + \frac{1}{2!}\frac{d^2 x(t_0)}{dt^2}(t - t_0)^2 + \cdots$$

$$= \sum_{n=0}^{\infty} \frac{x^{(n)}(t_0)}{n!}(t - t_0)^n. \qquad (4.2\text{-}2)$$

This series converges for all t. To show this, we calculate the remainder in Taylor's formula

$$x(t) = \sum_{n=0}^{N} \frac{x^{(n)}(t_0)}{n!}(t - t_0)^n + R_N(t).$$

As is shown in any standard calculus text,

$$R_N(t) = \frac{x^{(N+1)}(\xi)(t - t_0)^{N+1}}{(N+1)!},$$

where ξ lies between t and t_0. Assume that the maximum absolute value of $X(f)$ in the interval $-W \le f \le W$ is X_m; such a maximum exists because we required $X(f)$ to be finite. Then $|x^{(N+1)}(\xi)|$ is bounded from above according to

$$|x^{(N+1)}(\xi)| \le 2WX_m(2\pi W)^{N+1},$$

and, therefore,

$$|R_N(t)| \le \frac{2WX_m(2\pi W|t - t_0|)^{N+1}}{(N+1)!}. \qquad (4.2\text{-}3)$$

It is easily shown (for instance by examining the ratio R_{N+1}/R_N) that $\lim_{N\to\infty} R_N(t) = 0$. Therefore by the definition of analyticity ([4-3], p. 128) $x(t)$ is analytic at t. This means that if $x(t)$ and all derivatives are known at a particular value of time t_0, $x(t)$ can be completely predicted for all future times. However, such predictability is inconsistent with the requirement that an information-bearing signal must be unpredictable to the receiver. Thus, in a sense, we find that a strictly band-limited signal cannot be used to convey information.

The paradox here again lies in the requirement that $x(t)$ be strictly band limited. This requirement is violated if the signal contains only a tiny amount of random noise. As shown in Chapter 10, the spectrum of *white* noise extends over a very broad band—theoretically infinite—and therefore the combination of signal and noise is not strictly band limited in the sense used here. The practical effect of noise is to make the measurement of $x(t_0)$ and the derivatives of $x(t)$ uncertain, and therefore the distance over which $x(t)$ can be extrapolated becomes, in fact, quite small.

Sampling Theorem

The Taylor series expansion can be thought of as a means of expressing a continuous waveform in terms of a countable set of measurements. Although the Taylor series is not a particularly useful way to do this, the fact that it converges for band-limited signals suggests that there may be other expansions that would be more useful.

Probably the most useful of these is the sampling theorem. In a somewhat general form this theorem states that a signal whose Fourier transform vanishes for all frequencies above W Hz is completely specified by samples taken at the rate of at least $2W$ samples per second.

The general principle was demonstrated by Nyquist [4-4] by considering a section of length T of a signal $x(t)$ having a bandwidth of W and regarding it as one period of a periodic signal of period T. The T-second section can be expanded in a Fourier series. The fundamental frequency in this expansion is $2\pi/T$ radians per second. Because $x(t)$ has a bandwidth of W, it is argued, albeit not quite correctly, that the Fourier expansion has a maximum frequency of $2\pi W = 2\pi n_{max}/T$ radians per second. Then if WT is assumed to be an integer, we find that $n_{max} = WT$.

This argument is not completely correct, because as we have seen earlier a time function of finite duration has an infinite bandwidth. It therefore cannot be exactly represented by a finite Fourier series. Hence the Fourier series

$$x_s(t) = c_0 + \sum_{n=1}^{WT} \left[a_n \cos \frac{2\pi nt}{T} + b_n \sin \frac{2\pi nt}{T} \right] \qquad (4.2\text{-}4)$$

can be only approximately equal to the T-second section of $x(t)$. The approximation can, however, be very good, especially in regions away from the end points of the T-second interval and if WT is very large.

Suppose we know the values of $x_s(t) \simeq x(t)$ at the points $0 \leq t_0 < t_1 <$

$t_2 < \cdots < t_{2TW} \leq T$. Then Eq. (4.2-4) can be regarded as a set of $2WT + 1$ simultaneous equations in the unknowns c_0, a_n, and b_n. These equations can be solved if they are all independent, i.e., if the coefficient matrix is non-singular. [The elements of this matrix are in the form $\cos(2\pi n t_i/T)$, $\sin(2\pi n t_i/T)$, or 1, depending on whether they multiply a_n's, b_n's, or c_0.] There are many ways of choosing the sample times t_i, $i = 0, 1, 2, \ldots, 2WT$, to make the coefficient matrix nonsingular; in fact almost any random choice will do. Once c_0, a_n, and b_n are known, $x_s(t)$ is, in principle, specified.

We see that $2WT + 1$ sample values are required, or if $WT \gg 1$, approximately $2WT$ values. We say that the signal has $2WT$ *degrees of freedom*. Clearly, if the sample values are uniformly spaced, the sampling rate is $2W$ samples per second. There is, however, no need for the samples to be uniformly spaced. Also note that we can differentiate Eq. (4.2-4) once or several times and generate different equations in which some derivative of $x(t)$ appears on the left. Thus some of the $2WT$ samples can be of $x(t)$, others of $\dot{x}(t)$, etc.

Beyond having a theorem which states that $x(t)$ is specified by a countable number of samples, one would like some simple formula for interpolating $x(t)$ for values of t other than the sampling times. Although Eq. (4.2-4) is potentially such a formula, at least when $2WT \gg 1$, it is not particularly simple. For one thing, it does not provide an explicit connection between the sample values and the function. Also, it provides no information about how well a small number of terms of the summation approximates the function. Finally, it is not exact.

The commonly used version of the sampling theorem is the uniform sampling theorem, also referred to as the Whittaker-Shannon sampling theorem.

Uniform Sampling Theorem

If $X(f)$, the Fourier transform of $x(t)$, is band limited such that $X(f) = 0$ for $|f| \geq W$, then $x(t)$ is completely specified by instantaneous samples uniformly spaced in time, with a uniform intersample period T_s such that $T_s \leq 1/(2W)$. At times other than the sampling instants, $x(t)$ is given by the interpolation formula

$$x(t) = \sum_{n=-\infty}^{\infty} x(nT_s) \operatorname{sinc}\left(\frac{t - nT_s}{T_s}\right). \tag{4.2-5}$$

As we shall show, the summation on the right is an exact representation of $x(t)$. Note that the sample values $x(nT_s)$ appear explicitly in this formula. Because of the orthogonality of the sinc functions only one term of the summation is nonzero at the sampling instants. Although interpolation to other instants requires theoretically an infinite number of terms, the fact that the sinc function goes to zero for large arguments makes it possible to approximate $x(t)$ quite well by a relatively small number of samples in the neighborhood of t. This is shown in Fig. 4.2-1. Note that the contributions from samples other than the ones in the immediate neighborhood of a particular point t tend to cancel.

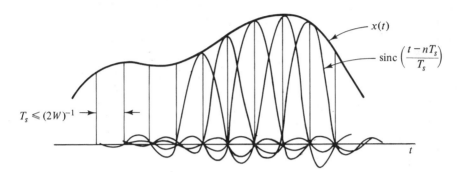

Figure 4.2-1. Between-sample interpolating property of the sinc functions.

The maximum sampling interval is $T_s = 1/(2W)$. This interval is known as the Nyquist interval, and the corresponding sampling rate as the Nyquist rate. In practice a shorter interval is often used to account for the fact that signals are never exactly band limited. A commonly used value is $T_s = 0.7/(2W)$.

To prove the sampling theorem, suppose that $X(f) = 0$ for $|f| \geq W$. It is then possible to regard $X(f)$ as a single cycle of a spectrum that repeats periodically with a period (in frequency) of $2W$. (See Fig. 4.2-2.) This can be compactly expressed by the identity

$$X(f) = \text{rect}\left(\frac{f}{2W}\right) \sum_{n=-\infty}^{\infty} X(f - 2nW). \tag{4.2-6}$$

It is shown in Chapter 2 [Eq. (2.12-20)] that the inverse Fourier transform of the summation in Eq. (4.2-6) is the "comb"

$$\frac{1}{2W} \sum_{n=-\infty}^{\infty} x\left(\frac{n}{2W}\right) \delta\left(t - \frac{n}{2W}\right).$$

Figure 4.2-2. Periodic spectrum.

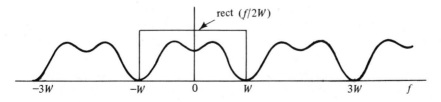

Therefore, if we inversely Fourier-transform Eq. (4.2-6), we get

$$x(t) = 2W \operatorname{sinc} 2Wt * \left[\frac{1}{2W} \sum_{n=-\infty}^{\infty} x\left(\frac{n}{2W}\right) \delta\left(t - \frac{n}{2W}\right) \right]$$

$$= \sum_{n=-\infty}^{\infty} x\left(\frac{n}{2W}\right) \left[\operatorname{sinc} 2Wt * \delta\left(t - \frac{n}{2W}\right) \right], \qquad (4.2\text{-}7)$$

and this is identical with Eq. (4.2-5) if T_s is set equal to $1/(2W)$.

It is worth examining Eq. (4.2-7) more carefully, since it actually describes the ideal sampling and recovery process in some detail. Thus, consider the comb expression

$$\sum_{n=-\infty}^{\infty} x\left(\frac{n}{2W}\right) \delta\left(t - \frac{n}{2W}\right).$$

This can also be written in the form of the product $x(t) \sum_{n=-\infty}^{\infty} \delta[t - (n/2W)]$ and can therefore be thought of as the operation of a delta-function sampler operating on the original signal $x(t)$. (See Fig. 4.2-3.) Although delta-function

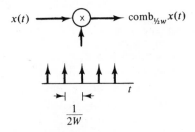

Figure 4.2-3. Ideal sampler.

samplers are not practical, a real sampler consisting of a switch (electronic or otherwise) that periodically connects the signal to the output can approximate such samplers. The approximation can be made very good by making the "on" time of the switch very short. This, in effect, replaces the delta-function sequence by a sequence of narrow pulses with a very short duty cycle. Since $T_s x(t) \sum_{n=-\infty}^{\infty} \delta(t - nT_s)$ and $\sum_{n=-\infty}^{\infty} X[f - (n/T_s)]$ are Fourier pairs, we see that the sampling operation results in a periodic spectrum with period $1/T_s$. Thus the periodic spectrum which we introduced somewhat artificially in the proof of the sampling theorem actually exists in the system— it is the spectrum at the output of the sampler. It then follows that multiplication of the spectrum by rect $(f/2W)$ or the equivalent—convolution of the sampled signal by $2W \operatorname{sinc} 2Wt$—can be regarded as the operation of a reconstruction filter which converts the sampler output back to the original signal. By Eqs. (4.2-6) and (4.2-7) this filter is an ideal low-pass filter (in the sense discussed in Chapter 3) with passband W. (See Fig. 4.2-4.)

We now have another reason the sampling interval must be less than $1/(2W)$. The generation of the periodic spectrum will take place no matter

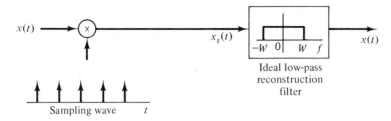

Figure 4.2-4. Complete sampling and reconstruction system.

what the sampling interval is, but if the sampling interval exceeds $1/(2W)$, the period of the spectrum is less than $2W$, and therefore there will be spectral overlap. This is shown in Fig. 4.2-5. Note that this overlap—also called *spectral folding* or *aliasing*—causes distortions in the original spectrum.

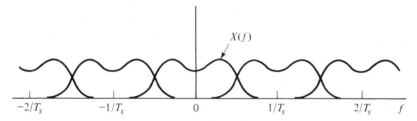

Figure 4.2-5. Spectral overlap.

Frequency components at the high-frequency edge of the spectrum are reflected down to lower frequencies. For example, if a signal containing the typical 60-Hz line-frequency noise is sampled at 100 samples/sec, corresponding to a folding frequency of 50 Hz, there will appear a nonzero spectral component at 40 Hz even though the original signal had no energy at 40 Hz.

The effect is illustrated in another way in Fig. 4.2-6, which shows a sine wave sampled at a rate slightly larger than half of its frequency (actually considerably less than the Nyquist rate). The signal reconstruction from the sample values looks like a low-frequency sine wave; it has taken an "alias."

Since aliasing distortion causes high-frequency spectral components to appear at lower frequencies, it seriously affects intelligibility of speech signals. The kind of spectral reversal produced by aliasing is in fact the basis for a form of speech scrambler used to preserve confidentiality in telephone conversations. For this reason a low-pass filter is frequently used *ahead* of the sampler to prevent aliasing. The cutoff frequency of this low-pass filter is set at something like 35 or 40 % of the sampling frequency to make certain that there are no significant spectral components at half the sampling frequency or above. The elimination of the high-frequency components in the signal

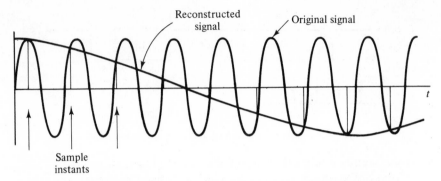

Figure 4.2-6. Showing the effect of sampling below the Nyquist rate.

obviously degrades the fidelity of transmission to some extent, but the loss of intelligibility is much less than it would be if aliasing were permitted. As an example, in time-division multiplexing of telephone conversations (discussed in Sec. 4.4) the telephone signals are passed through a low-pass filter with cutoff frequency of 3.2 kHz before being sampled at an 8-kHz sampling frequency.

It should be noted that although the mapping from $x(t)$ to the sample values $x(nT_s)$ is unique, the inverse mapping from the sequence of samples to the continuous time function is not. In fact the sequence of samples can yield any function $y(t) = x(t) + z(t)$, where $x(t)$ is given by Eq. (4.2-5) and $z(t)$ is an arbitrary function that is zero at all the sampling instants (see Fig. 4.2-7). Note that $z(t)$ has a minimum frequency of $W = 1/2T_s$ Hz. This

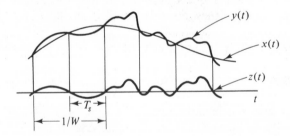

Figure 4.2-7. Illustration of nonuniqueness of inverse mapping of sample values.

phenomenon was first noted by E. T. Whittaker in 1915 [4-5], and it is clear that the function $x(t)$ given by Eq. (4.2-5) is only the *lowest frequency* function that can be passed through the set of sample points. The statement of the theorem guarantees that this function has a bandwidth of *less than W*, whereas the function $y(t)$ formed by adding a nonzero $z(t)$ to $x(t)$ must have a bandwidth of *at least W*. Thus there is no inconsistency.

Sampling of Band-pass Signals

For band-pass signals we assume that $X(f)$ is nonzero only in the bands $|f - f_0| < W/2, |f + f_0| < W/2$, as shown in Fig. 4.2-8. Although such a signal can be regarded as having a spectrum extending from $-f_0 - W/2$ to $f_0 + W/2$, one can reduce the sampling rate greatly by taking advantage of the fact that the effective bandwidth is much smaller.

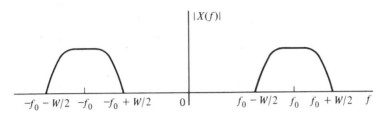

Figure 4.2-8. Band-pass spectrum.

As in previous discussions of band-pass signals we find it convenient to deal with the *complex analytic form* $\xi(t)$ of the signal $x(t)$ (see Sec. 2.13) for which Re $\xi(t) = x(t)$, Im $\xi(t) = \hat{x}(t)$. The spectrum of $\xi(t)$ exists only for positive frequencies and is given there by $\Xi(f) = 2X(f)$.

We can then use the same procedure as in the baseband case. The band-limited spectrum is regarded as one cycle of a periodic spectrum and is written in the form

$$\Xi(f) = \text{rect}\left(\frac{f - f_0}{W}\right) \sum_{n=-\infty}^{\infty} \Xi(f - nW). \tag{4.2-8}$$

Fourier-transforming and using Eq. (2.12-20), we obtain

$$\xi(t) = \sum_{n=-\infty}^{\infty} \xi\left(\frac{n}{W}\right)\left[\text{sinc}(Wt)e^{j2\pi f_0 t} * \delta\left(t - \frac{n}{W}\right)\right]$$

$$= \sum_{n=-\infty}^{\infty} \xi\left(\frac{n}{W}\right) \text{sinc}(Wt - n)e^{j2\pi f_0[t-(n/W)]}. \tag{4.2-9}$$

To get back the original signal, take the real part of Eq. (4.2-9):

$$x(t) = \text{Re} \sum_{n=-\infty}^{\infty} \left[x\left(\frac{n}{W}\right) + j\hat{x}\left(\frac{n}{W}\right)\right] \text{sinc}(Wt - n)e^{j2\pi f_0[t-(n/W)]}$$

$$= \sum_{n=-\infty}^{\infty} \text{sinc}(Wt - n)\left[x\left(\frac{n}{W}\right)\cos\omega_0\left(t - \frac{n}{W}\right)\right.$$

$$\left. - \hat{x}\left(\frac{n}{W}\right)\sin\omega_0\left(t - \frac{n}{W}\right)\right] \tag{4.2-10}$$

$$= \sum_{n=-\infty}^{\infty} \text{sinc}(Wt - n)\left|\xi\left(\frac{n}{W}\right)\right|\cos\left\{\omega_0\left(t - \frac{n}{W}\right)\right.$$

$$\left. + \arg\left[\xi\left(\frac{n}{W}\right)\right]\right\}. \tag{4.2-11}$$

Note that the Nyquist sampling interval is now $1/W$; i.e., it is twice as long as for baseband sampling. However, at each point it is necessary to make a measurement of a complex quantity, which means either the real and imaginary part, or the amplitude and phase. Thus at each sampling point there are two measurements, and therefore the minimum number of uniform samples in time T is $2WT$, which is the same as for the baseband signal of bandwidth W.

Here again it is possible, and usually desirable, to sample the signal at a higher rate to account for the fact that the spectrum is not precisely band limited.

4.3 Practical Aspects of Sampling

In practice, sampling is performed by high-speed switching circuits. Typical circuits utilize field-effect transistors (FET's) as the switching element, together with suitable drivers to apply the sampling pulses. The effect is to connect the input to the output during the time that the sampling signal is "high" and to disconnect the output when it is "low." Since the switching pulses have finite width, this circuit approximates the delta-function sampling pulses by short rectangular pulses. An equivalent circuit employing a mechanical switch and the resulting output signal are shown in Fig. 4.3-1.

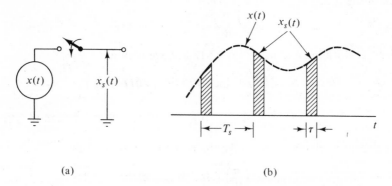

(a)   (b)

Figure 4.3-1. (a) Elementary sampling circuit and (b) sampled wave.

The effect of the finite width of the sampling pulses is easily investigated. The sampled wave $x_s(t)$ can be written as

$$x_s(t) = x(t) \cdot g(t), \qquad (4.3\text{-}1)$$

where $g(t)$ is the sampling function given by

$$g(t) = \sum_{n=-\infty}^{\infty} \text{rect}\left(\frac{t - nT_s}{\tau}\right). \qquad (4.3\text{-}2)$$

The Fourier series expansion of the sampling function was derived in Chapter 2 [Eq. (2.4-20)] and is given by

$$g(t) = \sum_{n=-\infty}^{\infty} d \text{ sinc } (nd)e^{j2\pi nt/T_s}$$

$$= d[1 + 2 \sum_{n=1}^{\infty} \text{sinc } (nd) \cos n\omega_s t], \qquad (4.3\text{-}3)$$

where $d = \tau/T_s$ is the duty cycle and $\omega_s = 2\pi/T_s$ is the sampling frequency. Substituting Eq. (4.3-3) into Eq. (4.3-1) gives

$$x_s(t) = d[x(t) + 2 \sum_{n=1}^{\infty} \text{sinc } (nd)x(t) \cos n\omega_s t], \qquad (4.3\text{-}4)$$

and therefore the Fourier spectrum of the sampled signal is

$$X_s(\omega) = dX(\omega) + d \sum_{n=1}^{\infty} \text{sinc } (nd)[X(\omega - n\omega_s) + X(\omega + n\omega_s)]. \qquad (4.3\text{-}5)$$

This spectrum is illustrated in Fig. 4.3-2 under the assumption that $x(t)$ is sampled at a sufficiently high frequency so that there is no aliasing, i.e., such that $X(\omega)$ does not extend beyond $\pm\omega_s/2$.

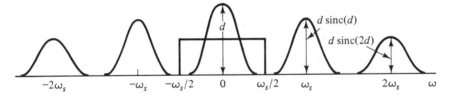

Figure 4.3-2. Effect of finite-width sampling.

Note that the finite width of the sampling pulses causes each lobe of the spectrum to be multiplied by d sinc (nd). It does not alter the shape of the central lobe of the spectrum. An ideal low-pass filter with a passband of $\omega_s/2$ can therefore be used to recover the spectrum of the original signal. Thus the finite width of the sampling pulses is seen to have no important effects on the sampling process.

Shape of the Sampling Pulses

Electronic switching circuits typically will not produce perfect rectangular sampling pulses; thus we consider arbitrary pulse shapes. Again, let

$$x_s(t) = x(t) \cdot g(t) \qquad (4.3\text{-}6)$$

but with

$$g(t) = \sum_{n=-\infty}^{\infty} p(t - nT_s), \qquad (4.3\text{-}7)$$

where $p(t)$ is some arbitrary pulse shape. Fourier transforming gives

$$X_s(f) = X(f) * \frac{1}{T_s} \sum_{n=-\infty}^{\infty} P\left(\frac{n}{T_s}\right)\delta\left(f - \frac{n}{T_s}\right)$$

$$= \frac{1}{T_s} \sum_{n=-\infty}^{\infty} X\left(f - \frac{n}{T_s}\right)P\left(\frac{n}{T_s}\right)$$

$$= \left(\frac{1}{T_s}\right)X(f)\cdot P(0) + \frac{1}{T_s} \sum_{\substack{n=-\infty \\ n\neq 0}}^{\infty} X\left(f - \frac{n}{T_s}\right)P\left(\frac{n}{T_s}\right). \qquad (4.3\text{-}8)$$

The Fourier transform of $g(t)$ was obtained by using the Poisson sum formula [Eq. (2.12-20)]. We see that the lobes of the sampled spectrum are now multiplied by $P(n/T_s)$; however, since the central lobe is multiplied by the constant $P(0)$, sampling with arbitrary pulses that multiply the original signal is essentially the same as ideal sampling.

Instantaneous Sampling

Certain practical sampling circuits such as the sample-and-hold (S/H) circuit, to be discussed shortly, sample the signal over a very short time but deliver pulses that are stretched out in time. One reason for doing this is that the transmission of very narrow pulses requires an excessive bandwidth in all the circuits making up the transmission channel—for delta-function pulses this bandwidth would be infinite. The stretching out of the pulses may therefore be intentional, right in the sampler, or it may simply be a consequence of the filtering action of the circuits that handle the pulses after they are generated. If we suppose that the impulse response of the stretching circuit is $h(t)$, then the sampled output is

$$x_s(t) = [x(t)\cdot \sum_{n=-\infty}^{\infty} \delta(t - nT_s)] * h(t) = \sum_{n=-\infty}^{\infty} x(nT_s)h(t - nT_s). \qquad (4.3\text{-}9)$$

Such a sampled wave for $h(t)$ having the form of a square pulse is shown in Fig. 4.3-3.† Note the difference between this figure and Fig. 4.3-1, which represents a type of sampling sometimes referred to as "natural sampling" [4-6].

†As shown in Fig. 4.3-3, this is not causal. Why?

Figure 4.3-3. Instantaneous sampling.

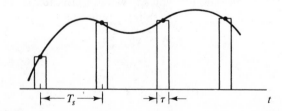

The Fourier spectrum of the sampled signal of Eq. (4.3-9) is given by

$$X_s(f) = \left[X(f) * \frac{1}{T_s} \sum_{n=-\infty}^{\infty} \delta\left(t - \frac{n}{T_s}\right) \right] H(f)$$

$$= \frac{1}{T_s} \sum_{n=-\infty}^{\infty} X\left(f - \frac{n}{T_s}\right) H(f), \tag{4.3-10}$$

where $H(f)$ is the Fourier transform of $h(t)$. As before, only the central lobe of this spectrum is passed by the reconstruction filter, but this is now given by $X(f)H(f)$. Thus instantaneous sampling alters the spectrum of the signal, and in order to recover the original spectrum it would be necessary to use a compensating or equalization filter having the transfer function $1/H(f)$.

In practical circuits, sampling cannot be truly instantaneous; in fact the output of the sampler is generally a weighted average of the input during a short but finite sampling period. If $h(t)$ is the impulse response of such a weighting circuit, the sampled output and its spectrum are also given by Eqs. (4.3-9) and (4.3-10). Both effects, i.e., the weighting over a finite time and the subsequent holding or stretching, may occur together, and the resulting output will again be given by Eqs. (4.3-9) and (4.3-10), but with a more complicated $h(t)$. Thus Eqs. (4.3-9) and (4.3-10) represent a general form of *filtered sampling* characterized by the fact that the signal is convolved with the sampling function rather than being multiplied as in natural sampling.

Sample-and-Hold Circuit

A special case of filtered sampling occurs in the very frequently used sample-and-hold (S/H) circuit, shown in ideal form in Fig. 4.3-4. During the time that the sampling pulse is "high" the switch is on, and the capacitance C charges up to the value of the input signal. During the remainder of the sampling period the switch is off, and the signal is held on the capacitor. Ideally the switch is "on" for only an instant during which the capacitor

Figure 4.3-4. (a) Sample-and-hold (S/H) circuit; (b) S/H output.

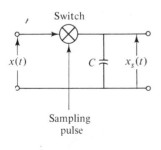
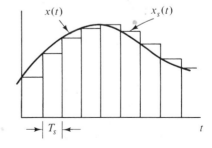

(a)   (b)

charges up to the exact value of the input. Then the S/H output is the stair-case signal shown in Fig. 4.3-4(b). This is described by

$$x_s(t) = \sum_{n=-\infty}^{\infty} x(nT_s) \, \text{rect} \left[\frac{t - (n + 1/2)T_s}{T_s} \right], \qquad (4.3\text{-}11)$$

which is in the form of Eq. (4.3-9) with

$$h(t) = \text{rect} \left[\frac{t - (T_s/2)}{T_s} \right]. \qquad (4.3\text{-}12)$$

The "hold" time of the S/H circuit need not extend over the entire sampling period T_s; in fact it is frequently much shorter to permit sampling pulses from several signals to be interleaved. This is discussed more fully below. Clearly, if the hold time is $T \leq T_s$, the transfer function of the S/H circuit is

$$H(f) = T \, \text{sinc} \, (Tf) e^{-j\pi Tf}. \qquad (4.3\text{-}13)$$

Practical S/H circuits differ from the ideal in that the switch has to be "on" for a finite time. Since the charging current is limited by the combined switch resistance and source output impedance, the capacitor voltage can reach only some fraction of the input during the "sample" part of the cycle. There is also a certain amount of leakage from the capacitor during the "hold" period because of finite load resistance. The size of the capacitor used in any given application is, in fact, generally chosen as a best compromise between minimizing the charging time constant and maximizing the leakage time constant.

Nonideal Reconstruction Filters

Practical filters don't have an infinitely sharp frequency cutoff characteristic, and therefore a certain amount of signal energy from the adjacent lobes of the sampled spectrum is always passed. This is shown in Fig. 4.3-5; the

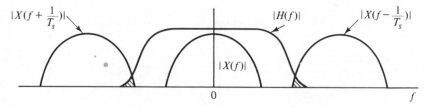

Figure 4.3-5. Reconstruction filter with a passband that is too large.

part shown shaded represents undesirable leakage of signal from adjacent spectral lobes. The effect is a high-frequency hiss or whistle in the reconstruction. It can be eliminated by narrowing the bandwidth of the filter or, more effectively, by using a higher sampling frequency to provide more separation between the spectral lobes. Note that the effect discussed here is different from aliasing, where significant portions of the spectral lobes overlap. The distortion caused by aliasing can only be eliminated by using a higher sampling frequency or by reducing the signal bandwidth before sampling, while the high-frequency noise considered here can at least in theory also be eliminated by using a sharper cutoff reconstruction filter.

4.4 Time-Division Multiplexing

There are many applications of the principle of sampling in communications systems. One of the most important is time-division multiplexing (TDM), which is commonly used to simultaneously transmit several different signals over a single channel. Each of the signals is sampled at a rate in excess of the Nyquist rate. The samples are interleaved, and a single composite signal consisting of all the interleaved pulses is transmitted over the channel. At the receiving end the interleaved samples are separated by a synchronous switch or *demultiplexer*, and then each signal is reconstructed from the appropriate set of samples.

The samples can be coded into pulses suitable for transmission in a number of different ways. The most obvious is probably pulse amplitude modulation (PAM) where the signal is converted into a train of pulses whose amplitude is proportional to the amplitude of the signal at the sample points. However, it is also possible to keep the pulse amplitudes constant and to vary their width or their position relative to some reference point in proportion to the values of the signal sample. This results in pulse duration modulation (PDM) and pulse position modulation (PPM), respectively. Other schemes include pulse code modulation (PCM) where the pulses represent the signal in some form of digital code. For the purpose of our discussion of TDM we assume that PAM is used. We consider the various other types of pulse modulation in the next two sections.

Time-division multiplexing is widely used in telephony, telemetry, radio, and data processing. Data about temperature, magnetic intensity, vehicle attitude, etc., measured on various instruments aboard a spacecraft are almost always transmitted this way. All of the instrument outputs are sampled at regular intervals, and the samples are impressed on a single radio-frequency carrier which is beamed to earth. In telephony both frequency-division multiplexing (FDM) and TDM are jointly used to permit many hundreds of different conversations to utilize the same microwave link. When a computer is used to monitor several different experiments the outputs of various instruments are multiplexed so that the computer can deal with them sequentially, one at a time.

The PAM-TDM system is illustrated in Fig. 4.4-1, where we show an elementary switching circuit that acts to multiplex the N signals $s_1(t)$, $s_2(t)$, $\ldots, s_N(t)$. Instead of the mechanical commutator shown in that figure,

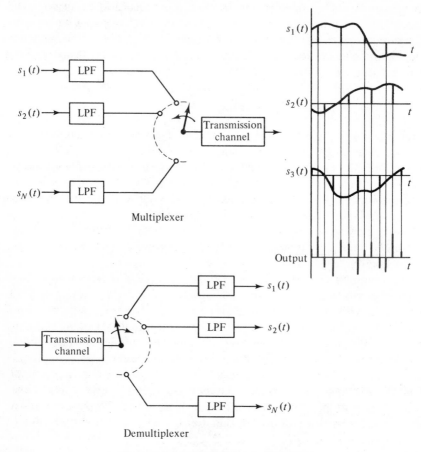

Figure 4.4-1. Time-division multiplexing.

practical multiplexers are generally all electronic and are available in the form of integrated-circuits (IC's).

Figure 4.4-1 shows how the signals are sampled and how the samples are interleaved before being transmitted. At the receiver a demultiplexer separates the interleaved samples and sends them on to appropriate reconstruction filters. It is obvious that the multiplexer and demultiplexer must operate in precise synchronism. Synchronization is, in fact, one of the major problems, especially with high-speed data systems and where the transmitter and receiver are physically far apart. In such systems synchronization is generally achieved by utilizing very stable local *clock* oscillators. These, in turn, are kept in synchronism by transmitting an occasional special synchronizing signal. It is also possible to derive timing information from the signal pulses themselves by averaging over long periods of time. The synchronization problem and the various methods used in practice to solve it are discussed in more detail in the literature ([4-7] and [4-8]).

For channels involving a radio link the pulses are themselves modulated onto a carrier using AM or FM modulation, discussed in Chapters 6 and 7. For this purpose it is often undesirable to have both positive and negative pulses. Also synchronization may be simplified if the pulses never go to zero, i.e., so that a positive pulse is always present at each pulse time. Both of these objectives can be easily achieved by adding a dc level larger than the largest expected signal amplitude to all signals before sampling. The resulting pulse train is referred to as unipolar PAM, while the kind of system shown in Fig. 4.4-1 is bipolar. For very long-distance communication, where transmitter power may be at a premium, bipolar PAM has the advantage over unipolar PAM of requiring less signal power for the same signal-to-noise ratio at the receiver.†

An important consideration in TDM is the bandwidth required for the channel. An estimate of this is easily obtained if it is assumed that the individual signals $s_1(t), \ldots, s_N(t)$ all have the same bandwidth W. (See Prob. 4-12 for an example of unequal bandwidths.) Then the sampling rate for each signal must be at least $2W$ samples/sec, but in order to provide for the less-than-perfect cutoff characteristics of practical low-pass filters the sampling rate should be $2W_1$, where W_1 exceeds W by a small amount called a *guard band*. For instance, if the $\{s_i(t)\}$ are telephone conversations, $W = 3.2$ kHz and the sampling rate might be 8 kHz, providing a 1.6-kHz guard band. The number of revolutions per second at which the rotary switch in Fig. 4.4-1 rotates is equal to the sampling rate. If there are N signals, the length of each sampling pulse cannot exceed $1/N$ of a revolution or $1/(2W_1N)$ sec.

Suppose for the moment that the pulses are square (as in Fig. 4.3-3). The Fourier spectrum for a square pulse of length τ is τ sinc (τf). The bandwidth may be defined in terms of the first zero crossing of the sinc function and is therefore given by

$$B = \frac{1}{\tau}, \qquad (4.4\text{-}1)$$

and for $\tau = 1/(2W_1N)$ the channel bandwidth is

$$B = 2W_1N \text{ Hz}. \qquad (4.4\text{-}2)$$

This is the bandwidth for pulses with maximum duty cycle, shorter pulses requiring a larger bandwidth.

At the expense of somewhat poorer separation between adjacent pulses one can define B as the 3-dB bandwidth which for square pulses is approxi-

†This is discussed in more detail in Chapter 11.

mately $1/(2\tau)$. This gives

$$B_{3\,dB} = \frac{1}{2\tau} = W_1 N \text{ Hz.} \qquad (4.4\text{-}3)$$

The bandwidth requirements appear to be more or less what one would expect: proportional to the signal bandwidth and to the number of signals to be multiplexed. This result is very similar to that obtained for frequency-division multiplexing (FDM) where different signals are sent on different frequency carriers (see Sec. 6-16). Thus insofar as required channel bandwidth is concerned these two techniques are approximately equivalent.

Intersymbol Interference

The question of whether the bandwidth should be calculated according to Eq. (4.4-2) or (4.4-3), or some other way, can be answered much more precisely by considering the spillover from one pulse into the adjacent time slot and setting some upper limit on the permissible amount. Such spillover is called intersymbol interference, and in a TDM system it results in cross talk; i.e., a signal in one channel can be heard in an adjacent channel. Cross talk is generally highly objectionable. The amount of interference clearly depends on the channel bandwidth because if the bandwidth is too small, pulses that are well separated at the point of origin will be spread out by the time they reach the receiver. However, additional factors are the pulse shape and the method used by the receiver to detect the pulses.

Thus, suppose that the receiver samples the signal using instantaneous sampling. Also, suppose that the pulse shape at the receiver has its maximum at the sampling point and goes through zero at all adjacent sampling points (see Fig. 4.4-2); then there will be no intersymbol interference, even though the received pulses persist over several time slots. In fact this result is obtained if the received pulses have the form $\text{sinc}\,[(t - nT)/T]$, where $n = 0, \pm 1, \pm 2, \ldots$ and where T is the time between samples. This shape is obtained by using delta-function pulses at the transmitter and giving the remainder of the channel the characteristic of an ideal low-pass filter with a passband $W = 1/(2T)$. (See Fig. 4.4-3.) Since by our previous discussion $T = 1/(2W_1 N)$, this argument gives a channel bandwidth of $W_1 N$, i.e., as in Eq. (4.4-3).

In practice a somewhat larger bandwidth has to be used because the ideal low-pass filter characteristic is difficult to realize and also because any inexactness in the sampling instants would again result in intersymbol interference. In fact since the sinc function decays very slowly, it is even possible (although not likely) that the sum of pulse tails at times other than integral multiples of T could diverge. The object is to retain the property that the pulse goes through zero at adjacent sample points, but in addition the tails should also be small so that small jitter in sampling time at the receiver does not cause large intersymbol interference. It was shown by Nyquist [4-4] that

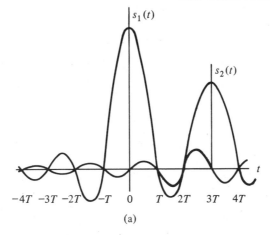

(a)

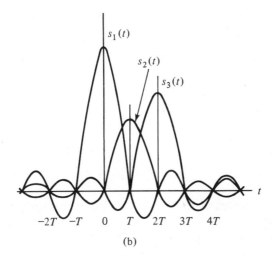

(b)

Figure 4.4-2. Zero intersymbol interference with sinc pulses: (a) two sinc pulses whose maxima are separated by an integral number of sample points; (b) three sinc pulses as close together as possible without interfering with each other.

this objective can be obtained quite easily (albeit at the expense of increasing the channel bandwidth) by using a channel filter characteristic having a more gradual falloff characteristic than the rect (·) shown in Fig. 4.4-3. Nyquist's result has been extended by Gibby and Smith [4-9], whose conclusions we briefly outline here.

Suppose that the channel output signal in Fig. 4.4-3 is $h(t)$. For zero

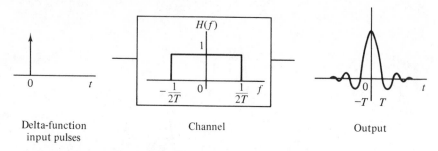

Delta-function input pulses   Channel   Output

Figure 4.4-3. Output signal of a channel having the ideal low-pass filter characteristic when the input is an impulse.

intersymbol interference from transmitted pulses spaced T seconds apart we must have

$$h(mT) \equiv h_m = \begin{cases} h_0, & \text{for } m = 0, \\ 0, & 0 \quad m \neq 0. \end{cases} \tag{4.4-4}$$

To translate this constraint into a requirement on the channel transfer function $H(f)$ we consider the periodic (in frequency) function $\sum_{n=-\infty}^{\infty} H[f - (n/T)]$. By use of the Poisson sum formula [actually the dual of Eq. (2.12-18)], we can write

$$\sum_{n=-\infty}^{\infty} H\left(f - \frac{n}{T}\right) = T \sum_{m=-\infty}^{\infty} h_m e^{-j2\pi mTf}. \tag{4.4-5}$$

Hence, substituting the constraint equation (4.4-4) into Eq. (4.4-5), we find that $H(f)$ must satisfy

$$\sum_{n=-\infty}^{\infty} H\left(f - \frac{n}{T}\right) = Th_0. \tag{4.4-6}$$

In other words, the sum of $H(f)$ and all of its replicas shifted by intervals of n/T in frequency must be a constant. In general $H(f)$ is complex, and therefore Eq. (4.4-6) should be written in the form

$$\sum_{n=-\infty}^{\infty} \operatorname{Re} H\left(f - \frac{n}{T}\right) = Th_0$$

$$\sum_{n=-\infty}^{\infty} \operatorname{Im} H\left(f - \frac{n}{T}\right) = 0. \tag{4.4-7}$$

We may now choose any function $H(f)$ having a gradual roll-off and satisfying Eq. (4.4-6) or (4.4-7). The example considered by Nyquist was the *raised cosine*

$$H(f) = \begin{cases} \frac{1}{2}(1 + \cos \pi fT), & |f| < \frac{1}{T} \\ 0, & \text{otherwise.} \end{cases} \tag{4.4-8}$$

This satisfies Eq. (4.4-5), as can be seen in Fig. 4.4-4. The corresponding impulse response is

$$h(t) = \frac{1}{T}\left[\operatorname{sinc}\frac{2t}{T} + \frac{1}{2}\operatorname{sinc}\left(\frac{2t}{T} + 1\right) + \frac{1}{2}\operatorname{sinc}\left(\frac{2t}{T} - 1\right)\right]. \qquad (4.4\text{-}9)$$

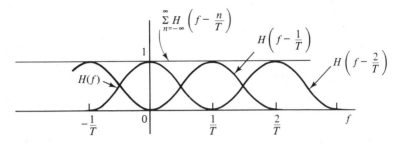

Figure 4.4-4. The sum of raised cosine functions shifted by intervals of n/T is constant.

This function is shown in Fig. 4.4-5(b). Note that $h(t)$ is exactly zero at all adjacent sample points, but it also is quite small at all other $|t| > T$. Practical filters whose transfer functions satisfy Eq. (4.4-7) are relatively easy to construct.

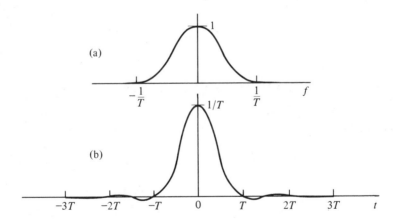

Figure 4.4-5. (a) Raised cosine transfer function and (b) its inverse Fourier transform.

Observe that the bandwidth for the raised cosine spectrum is $1/T = 2W_1N$, as given in Eq. (4.4-2).

In practice the received pulse shape is determined not only by shaping filters at the transmitter but also by characteristics of the transmission medium. The latter may not always be the same; for instance, in telephony different trunk lines generally have different transmission loss functions. To prevent intersymbol interference from this source adaptive equalizing filters are inserted at the transmission line terminals ([4-8] and [4-10]).

4.5 Pulse Duration and Pulse Position Modulation

Besides PAM two additional analog methods for impressing the sample value on a pulse-train carrier are pulse duration modulation† (PDM) and pulse position modulation (PPM). Typical pulse shapes for the three techniques are shown in Fig. 4.5-1. If $s(t)$ is the signal which modulates square

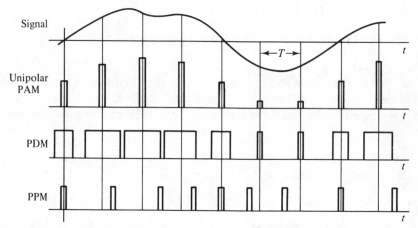

Figure 4.5-1. Analog pulse modulation methods.

pulses, mathematical expressions for the three forms of pulse modulation are as follows:

$$\text{PAM:} \qquad x(t) = \sum_{n=-\infty}^{\infty} [s(nT) + K]\,\text{rect}\left(\frac{t - nT}{\tau}\right), \qquad (4.5\text{-}1)$$

where $K \geq |s(t)|_{\max}$ is a constant added to the signal to prevent the pulses from becoming negative, τ is the pulse width, and T is the pulse period.

$$\text{PDM:} \qquad x(t) = \sum_{n=-\infty}^{\infty} \text{rect}\left(\frac{t - nT}{\tau}\right)$$
$$\tau = \tau(s) = as(nT) + K. \qquad (4.5\text{-}2)$$

†PDM is similar to the halftone process used for printing photographs in books and newspapers.

Here the pulse width τ is made proportional to $s(nT)$,‡ and it is necessary to choose a and K such that $0 < \tau < T$; i.e.,

$$K > a\,|\,s(t)|_{\max}$$

$$K + a\,|\,s(t)|_{\max} < T.$$

PPM: $$x(t) = \sum_{n=-\infty}^{\infty} \text{rect}\left(\frac{t - nT - \alpha}{\tau}\right) \qquad (4.5\text{-}3)$$

$$\alpha = as(nT).$$

The pulse position is proportional to $s(nT)$,‡ and the maximum shift must be less than $\frac{1}{2}(T - \tau)$, i.e.,

$$a\,|s(t)|_{\max} < \tfrac{1}{2}(T - \tau).$$

There are a number of variations of these pulse trains that are frequently encountered. Bipolar PAM has already been mentioned. The PDM train shown is symmetric around the sampling point, but frequently the leading edge is fixed and only the trailing edge is modulated. The PPM pulse train shown is bipolar in the sense that negative signal levels result in pulses that precede the sample instant. In practice, unipolar PPM, where the pulse always trails the sampling instant, may be easier to generate.

The main advantage of PDM and PPM over PAM is the fixed amplitude of the pulses, which makes them relatively immune from additive noise. Also, amplifiers and repeaters need not be linear; in fact limiting or clipping amplifiers which eliminate all amplitude variation and transmit only the rise and fall times of the pulses are commonly used.

Greater noise immunity could be achieved if the pulses were square, since additive disturbances would leave the rise and fall times unchanged. In practice the pulses cannot be square because of finite channel bandwidths, and therefore they are affected by noise to some extent. This is shown in Fig. 4.5-2. We assume that the receiver determines the pulse arrival time by observing the instant at which the trailing edge of the pulse passes a certain threshold (usually set at half of the pulse amplitude). Such threshold circuits are referred to as *slicing circuits*, and the threshold is called the *slicing level*. Note that for a square pulse the arrival time is unaffected by the noise as long as the noise amplitude is less than the slicing level. However, for a pulse with a finite slope there is an error which is proportional to the reciprocal of the slope of the trailing edge. It is easily demonstrated that the slope is proportional to the bandwidth. For example, consider a unit amplitude Gaussian pulse,

$$s(t) = e^{-\pi B^2 t^2}, \qquad (4.5\text{-}4)$$

‡Although proportional modulation is simplest, other monotonic functions $f[s(nT)]$ could be used as well.

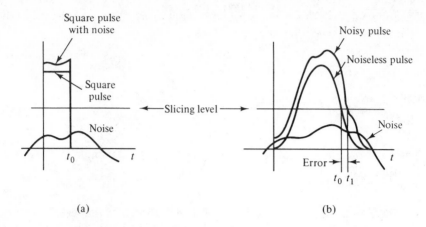

Figure 4.5-2. Noise effect in pulse demodulation: (a) square pulse; (b) pulse with finite bandwidth.

with Fourier transform

$$S(f) = \frac{1}{B} e^{-\pi f^2/B^2}. \tag{4.5-5}$$

We can loosely define the bandwidth by B. The slope is given by

$$s'(t) = \frac{ds(t)}{dt} = -2\pi B^2 t e^{-\pi B^2 t^2}, \tag{4.5-6}$$

and if t is evaluated, say, for a slicing level of $e^{-1/2}$, we get

$$t_0 = \pm \frac{1}{\sqrt{2\pi} B} \tag{4.5-7}$$

and

$$|s'(t_0)| = \sqrt{2\pi} B e^{-1/2}. \tag{4.5-8}$$

Thus for a particular noise amplitude the error will be inversely proportional to B, or the mean-square error is inversely proportional to B^2.

There is an inconsistency in this argument because we have assumed tacitly that the noise will remain unchanged as the bandwidth of the channel is increased. Actually if we increase the bandwidth in order to make the pulses more nearly square, we generally also increase the noise power. The analysis of this problem is therefore more complicated. The general problem of the relation between bandwidth and signal-to-noise ratio appears again in our discussion of FM (Chapter 7), and is discussed in more detail in Chapter 10.

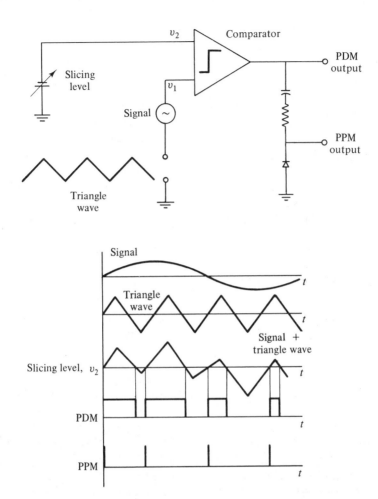

Figure 4.5-3. Generation of PDM and PPM.

Generation of PDM and PPM

A simple circuit that will generate both PDM and PPM is shown in Fig. 4.5-3. A sawtooth or triangular wave whose fundamental frequency is the desired sampling rate is added to the signal, and the sum is passed into a slicing circuit. The slicing circuit is simply a comparator which puts out a "high" signal when $v_1 > v_2$ and a "low" signal otherwise. The slicing level is the value of the voltage v_2. The use of a triangular wave results in symmetrical PDM, whereas a sawtooth wave generates either leading-edge or trailing-edge modulation. The capacitor in the output differentiates the comparator output voltage, producing a positive spike when this voltage goes positive and a negative spike when it goes negative. The diode eliminates the negative spikes.

As shown in Fig. 4.5-3, the sampling produced by this circuit is nonuniform, because in effect sampling occurs whenever the combined wave crosses the slicing threshold in the positive direction and not when the signal level matches the level of the biasing waves. This kind of sampling is referred to as *natural* sampling and gives rise to some distortion [4-11], [4-12]. Observe that the term *natural sampling* has a different meaning for PDM and PPM than for PAM (see Fig. 4.3-1).

Uniform sampling can be achieved with this circuit if the input is first passed through a sample-and-hold circuit. The S/H circuit performs the actual sampling, while the combination of triangle wave and slicing circuit determines the length or location of the respective PDM or PPM pulse.

Because of the complicated way in which signal information is transferred to pulse length or pulse position, the computation of the Fourier spectrum of PDM and PPM is not straightforward. Much of the early work on this problem is due to W. R. Bennett, and a fairly complete treatment can be found in books by Schwarz et al. [4-12] and Rowe [4-6]. Simplified analyses leading to approximate forms of the spectrum can be found in Carlson ([4-13] and [4-14]). One of the most important properties of the spectrum of PDM is that if the pulse-repetition frequency is much higher than the highest signal frequency and if the modulation index a in Eq. (4.5-2) is small, then the lowest harmonic contains the signal. This can be understood by observing that under these conditions the width of, and therefore the energy in, each pulse is proportional to the signal. As a consequence the signal can be recovered by simply passing the PDM wave through a low-pass filter which responds to the slowly changing average value of the pulses but rejects frequency components at the pulse-repetition frequency.

The Fourier spectrum of PPM can be shown to contain a term proportional to the derivative of the baseband signal. Hence PPM can be demodulated by passing the pulses through a low-pass filter which passes the signal and rejects harmonic components at the pulse-repetition frequency, followed by an integrator. In practice PPM signals are often demodulated by first converting them to PDM. This can be done by using each pulse of the PPM signal to trigger a bistable flip-flop and resetting the flip-flop with pulses from a clock oscillator running at a constant pulse-repetition rate equal to the average rate of the PPM pulse stream. The resulting PDM is then demodulated by use of a low-pass filter.

An important feature of the Fourier spectrum of PDM and PPM is that the harmonics generally spread into the baseband even if the baseband signal bandwidth is considerably less than one half of the average pulse-repetition frequency. This gives rise to a form of distortion not present in

PAM; however, it can be minimized by using a low value of modulation index.

4.6 Pulse Code Modulation

Pulse code modulation (PCM) is a method of transmitting a signal in terms of a series of numbers. The signal is sampled as in PAM, but then each sample is quantized, i.e., rounded to the nearest one of a set of levels referred to as quantum levels. If there are 50 such levels, quantizing will produce a set of integers between 0 and 49. The numbers are usually coded into binary pulse strings. One way of doing this is to represent the number in its binary form. Thus if a particular message sample has the quantized level 22, the string 010110 would be transmitted. Other binary coding schemes can also be used, particularly the so-called *Gray* codes [4-15] in which adjacent integers differ by only one binary digit. The zeros and ones can be represented by pulses of opposite polarity, or by pulses of different amplitudes, phases, or frequencies. The output need not be binary but may have more than two levels. For instance, if there are four levels, the code transmitted might be the base 4 representation of the quantum level. The number 22 would then be transmitted as 112.

The fact that a continuous quantity, such as the instantaneous value of a speech waveform, can be adequately represented by a number having only a few digits is, of course, well known. The advantage of digitizing is that it is much easier to handle discrete numbers than continuous variables. Requirements on linearity, signal-to-noise ratio, and fidelity can be relaxed greatly, and therefore lower-cost channels can be used with no real sacrifice in quality. Although analog-to-digital and digital-to-analog converters are needed at the terminals, the advent of large-scale integrated circuitry has made these relatively simple and inexpensive.

Pulse code modulation appears to have been first invented in 1926 by Paul M. Rainey (U.S. Patent No. 1,608,527, Nov. 30, 1926). Rainey's patent dealt with the transmission of a facsimile signal over telegraph channels by a process of sampling, quantizing, and coding which is essentially PCM. PCM was reinvented by A. H. Reeves (French Patent No. 853183, Oct. 23, 1939). Reeves' patent contains specific electronic circuits and proposed 32 quantizing levels as suitable for constant-volume speech. During World War II PCM was invented once more at Bell Laboratories, this time for the purpose of providing secret telephoning by means of an enciphering method developed by Vernam in World War I [4-16]. Vernam's method was developed for use with binary telegraphy and consists of adding a random key k to the message M, modulo 2; i.e., $0 + 0 = 0$, $0 + 1 = 1$, $1 + 0 = 1$, $1 + 1 = 0$. The resulting telegraphic message was completely scrambled and unintelligible but could be easily unscrambled at the receiving end by adding the key again, since $(M + k + k) \bmod 2 = M$. The application of this method to speech converted into a binary code via PCM is in principle straightforward. It was only the first of many other investigations into the use of PCM to permit various discrete transmission and detection systems to be used for analog signals.

In long-distance telephony it is necessary to use repeaters at regular intervals to amplify the signal which would otherwise disappear because of transmission losses. Analog signals are inevitably degraded in this process because of the ever-present noise, which is, of course, also amplified to some extent. Furthermore, it is impossible to make a perfect repeater, and by the time the signal has passed through a number of them the small distortions inserted by each repeater may build up to serious proportions.

The situation is quite different if binary signals are transmitted. In binary communication the repeaters can act as regenerators. During each pulse interval the repeater decides whether a pulse is present or not. Depending on the outcome of this decision, the repeater sends on a clean new pulse or zero. Noise affects such a system only if it is large enough to cause the repeaters to make erroneous decisions. With signal-to-noise ratios achievable in existing installations and reasonable repeater spacings (2 km) the probability of errors can easily be made vanishingly small. This important advantage makes it worthwhile to transmit analog signals in digital form.

Another important application is in time-division multiplexing. The problem of intersymbol interference in TDM was discussed in Sec. 4.4. In theory intersymbol interference can be completely eliminated by using specially shaped pulses and very precise synchronization of transmitter and receiver. In practice some cross talk is inevitable with PAM or any other analog system. However, in a discrete system, finite amounts of intersymbol interference that are not large enough to cause erroneous level decisions are perfectly acceptable and cause no cross talk at all. Thus cross-talk-free TDM almost requires some form of digital transmission.

The operation of PCM is shown schematically in Fig. 4.6-1, where we have assumed eight equally spaced quantum levels coded into their binary number form using pulses of opposite polarity. The quantum level assigned to each sample is the nearest one *above* the continuous sample value and is shown by the heavy dot at each sample point. The corresponding pulse signal is shown at the bottom.

It should be clear that for the 2-level, i.e., binary, output code shown in Fig. 4.6-1 the number of pulses per sample is given by

$$N = \lceil \log_2 q \rceil, \tag{4.6-1}$$

where q is the number of quantum levels and where the symbol $\lceil \ \rceil$ is defined by

$$\lceil A \rceil = A \qquad\qquad \text{if } A \text{ is an integer}$$

$$\lceil A \rceil = \text{the next integer greater than } A \qquad \text{if } A \text{ is not an integer.}$$

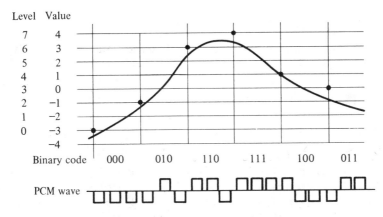

Figure 4.6-1. Pulse code modulation.

Thus 8 levels require 3 pulses, and any number of levels between 32 and 63 requires 6 pulses.

For a multilevel output code having m levels the number of pulses per sample is

$$N = \lceil \log_m q \rceil. \qquad (4.6\text{-}2)$$

The fact that several pulses are generally needed to code each sample increases the bandwidth, as has already been shown in the discussion of TDM. Specifically, suppose that the effective signal bandwidth (i.e., inclusive of guard band, etc.) is W; then a minimum bandwidth for pulses short enough so that N of them will fit into a sampling period is in accordance with Eq. (4.4-3):

$$B = WN$$
$$= W\lceil \log_m q \rceil. \qquad (4.6\text{-}3)$$

Since for fixed q, $\log_m q$ decreases as m increases (e.g., $\log_2 64 = 6$, $\log_4 64 = 3$), we see that binary PCM requires the largest bandwidth. The advantage of binary PCM is that a decision between two states is easier and therefore less prone to error than one between several levels. For this reason binary PCM is commonly used. However, the signal-to-noise ratios available on commercial channels are often so high that the error rate of a code having as many as 8 or 16 levels is still extremely low. In such cases a binary code is inefficient, and a higher-level code can be used. In this way the requirement on bandwidth is reduced and the channel capacity is better utilized.

It is in principle possible to use as many levels in the code as there are quantizing levels. Such a system is essentially quantized PAM, and the bandwidth is reduced to the baseband width.† In fact it is even possible to

---

†Actually a wave consisting of many small steps will occupy a larger bandwidth, but since one is not ordinarily interested in preserving the steps, it is not necessary to provide extra bandwidth in the channel.

go beyond this point and to code k samples of the original signal into a code having q^k levels. This is referred to as *hyperquantization*. It theoretically permits the bandwidth to be made less than baseband, but the extremely high signal-to-noise ratio needed makes such systems impractical. Note that the required signal-to-noise ratio increases exponentially with reduction in bandwidth.

Encoding and Decoding

Conversion of the continuous signal into a digital code is usually done by an analog-to-digital converter (ADC). At the receiver a digital-to-analog converter (DAC) reverses the process. There are a large number of different types of converters, and we shall describe here only a few representative ones. The reader is referred to [4-17] for more details.

Since several commonly used ADC circuits employ a DAC, we consider the latter first. One of the most popular techniques for D/A conversion is the R-$2R$ ladder method. This consists of a resistive ladder network having series values of R and shunt values of $2R$ as shown in Fig. 4.6-2. The digital signal to be converted is stored in an auxiliary register not shown in the figure so that all of its bits are simultaneously and continuously available. The switches shown at the bottom of the ladder are electronic gating devices such as transistors or FET's that are activated by the digital input. If a

Figure 4.6-2. D/A converter using R-$2R$ ladder.

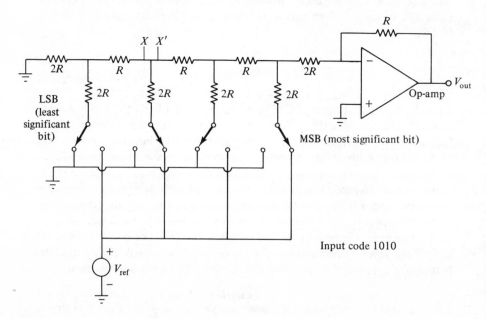

particular bit of the input is a "one," the corresponding switch connects its shunt resistor to the positive reference voltage; if a bit is a "zero," the corresponding shunt resistor is grounded. In Fig. 4.6-2 the binary number 1010 is applied to the converter input. Observe that the most significant bit (MSB) is applied to the switch nearest to the operational amplifier.

The operation of the network is based on the binary division of current as it flows down the ladder. This can be seen by examining the points X and X' in the ladder. As explained in Chapter 3, the operational amplifier input is a *virtual ground*; i.e., the effect of the large gain and negative feedback is to permit essentially no voltage to exist at this point. Also, we can assume that the internal impedance of the reference voltage is zero. For these reasons we find that the resistance looking to the left from point X is $R + (2R/2) = 2R$. Similarly, the resistance looking to the right from X is R. At point X' the resistance looking to the right is $2R$, and looking to the left it is R. This is true at any point on the ladder. If a shunt resistor is switched to the reference voltage source, the source sees a resistance of $2R$ in series with two $2R$ resistances in parallel, i.e., $3R$, and therefore the current in the shunt branches is either $V_{ref}/3R$ or zero. At the junction the current divides equally, with half flowing to the left and half flowing to the right. The right-hand current goes to the next junction where it again divides in half, and so on to the right end of the ladder. Thus the current from the source feeding into the junction marked with X and X' will be halved three times before reaching the operational amplifier, and the signal produced by it will be proportional to $\frac{1}{8}$. Similarly, the current flowing upward in the last branch to the right is halved once, and therefore its contribution to the total is $\frac{1}{2}$. The currents from all the "on" switches add in the output because of superposition. Hence with the switch setting shown, the analog output is proportional to $\frac{1}{2} + \frac{1}{8}$, which is exactly the value of the binary number 1010 if the binary point is assumed to be to the left of the number (i.e., 0.1010). The ladder can be extended at will to accommodate as many bits as desired. An important advantage of the circuit is that it uses only two resistance values so that matching and temperature tracking is very simple. The accuracy of the circuit can therefore be made very high.

One of the most widely used A/D converters is the successive approximation type, because it combines high resolution and high speed. This converter operates with a fixed conversion time per bit, independent of the value of the analog input. The system is illustrated in Fig. 4.6-3 and operates by comparing the input voltage with the DAC output. For the sake of concreteness we consider an 8-bit converter.

The input signal is assumed to be held in a *hold* circuit and is therefore constant during the entire conversion process.

At the start of the conversion cycle the logic circuit generates the digital number 10000000 (i.e., the most significant bit is 1, and all other bits are

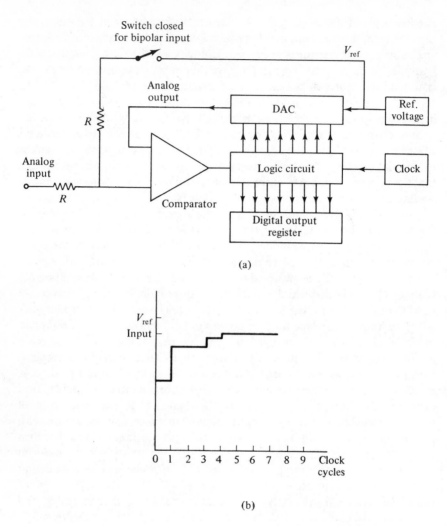

(a)

(b)

Figure 4.6-3. Analog-to-digital converter using the successive approximation method: (a) block diagram; (b) output for a typical input signal.

0) and applies it to the D/A converter. The resulting analog output will be
one half of the reference voltage, V_{ref}, and this is compared to the input
signal. If the input signal is larger, the logic circuit next applies 11000000 to
the D/A converter, resulting in an output of $\frac{3}{4}V_{ref}$. Otherwise the logic
circuit turns off the MSB and applies 01000000, giving a D/A output of
$\frac{1}{4}V_{ref}$. This process repeats down to the least significant bit (LSB) after which
the output register contains the digital output. The way in which the input
signal is approximated and the resulting digital output are shown in Fig.
4.6-3(b).

The operation described thus far is for unipolar signals; that is, the range of input signals is between zero and V_{ref}. The circuit is, however, easily converted to bipolar operation by connecting the two resistors R shown in the figure. If one assumes that both the signal and the reference are zero-impedance sources, then the input to the comparator will be $\frac{1}{2}(V_{in} + V_{ref})$. This varies between 0 for $V_{in} = -V_{ref}$ to V_{ref} for $V_{in} = V_{ref}$.

Analog-to-digital conversion can also be accomplished by applying the analog signal to a precision integrator and turning on a clock until the integrator output reaches a certain fixed level. The clock pulses can be used to step up a counter, so that the digital output is the number appearing in the counter when the clock is stopped. Still another, somewhat similar approach is to have the analog signal control the frequency of a voltage-controlled oscillator, whose output is applied to a counter for a certain fixed time. Both of these methods have the disadvantage that the precision of conversion depends on the accuracy of analog devices such as an integrator or a voltage-controlled oscillator.

With all A/D conversion methods it is best to use a sample-and-hold circuit before the ADC to keep the analog signal fixed during the conversion time. Otherwise the output will be the digital equivalent of a not-too-clearly defined weighted average of the analog signal existing during the conversion period.

In PCM, after each input sample is digitized it must be converted into a sequential string of pulses for transmission. This can be done by use of a simple binary multiplexer connected to the A/D converter output register. At the receiver a demultiplexer reverses the process and loads the group of pulses corresponding to each input sample into the D/A input register.

All the A/D and D/A systems described here are for binary codes, but the principle is easily adapted to higher-level codes. In the D/A converter the switches at the bottom of the ladder could be replaced by a precision operational amplifier which would apply one of several levels to the shunt resistances. The binary output of the A/D converter can always be converted to a multilevel output by feeding each group of m bits to a small DAC.

Both D/A and A/D converters are available today in large-scale-integrated-circuit (LSI) packages and at relatively moderate cost. Thus the complexity, lack of reliability, and expense of early coding systems are no longer factors against the use of PCM.

Delta Modulation

A method for converting analog signals to a string of binary digits that requires much simpler circuitry than PCM is delta modulation (DM) [4-18]. A simple DM system is shown in Fig. 4.6-4. When the output $v(t)$ of the integrator is less than $s(t)$ the comparator output is positive, and therefore the modulator output consists of positive pulses. Otherwise it consists of negative

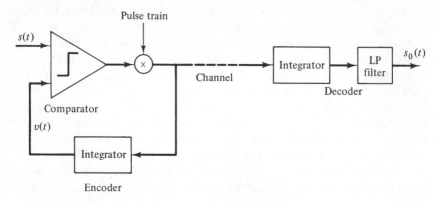

Figure 4.6-4. Delta-modulation system.

pulses. If $v(t) < s(t)$, the positive pulses entering the integrating network cause $v(t)$ to increase. The integrator output therefore follows $s(t)$. The sequence of positive and negative pulses constitutes the digital output code. At the receiver, an integrating circuit similar to the one used in the transmitter reconstitutes the analog signal from the pulses and the low-pass filter removes some of the quantizing noise.

The integrating circuit may be simply an *RC* low-pass filter with a time constant much larger than the period of the lowest signal frequency. Thus the DM encoder consists of only a few very simple components. However, because of the simple coding scheme used, a relatively large pulse repetition rate and correspondingly large bandwidth is needed for acceptable fidelity. There are a number of variations of the simple scheme described here that have somewhat better performance characteristics at the expense of a somewhat more complicated implementation. For details see, for instance, Panter [4-19].

Noise Effects in PCM

There are two different sources of error in PCM: those due to random, additive noise and those due to the quantization process. The random noise will cause an occasional error in the receiver so that in the case of binary PCM a high pulse will be decoded into a low pulse or vice versa. Similarly, in a multilevel PCM system, random noise may cause the receiver to put out the wrong level.

The error due to random noise can be made extremely small in a properly operating PCM system; in fact its elimination is one of the main reasons for using PCM. We analyze this error here for the case of bipolar binary PCM and under the assumption that the receiver works in the following way (see Fig. 4.6-5):

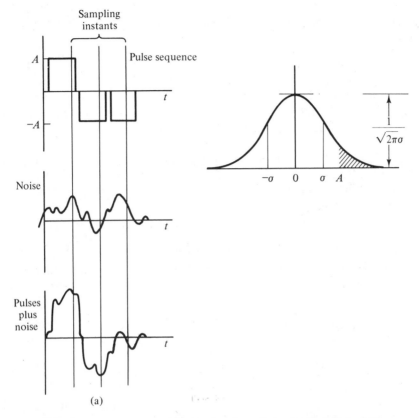

Figure 4.6-5. Noise in PCM: (a) showing effect of added noise; (b) probability of error is proportional to the shaded area.

1. The received signal is sampled instantaneously at a fixed point in each pulse period (usually close to the end of the period).
2. With a bipolar signal the pulse level in the absence of noise is either A or $-A$. With additive noise the height fluctuates, but if the signal is positive at the sample instant, it is called high, and otherwise low.†

We see that there will be a decoding error if at the sampling instant the noise is of the opposite sign from the signal pulse and its amplitude exceeds that of the signal. Note that we are interested in the noise only at the sampling instant; its value at other times is immaterial.

---

†A different kind of bipolar signal is used in telephony. A 0 bit is transmitted as a zero voltage. A 1 bit is transmitted as a nonzero voltage of constant magnitude whose sign alternates to keep the average of the transmitted signal zero.

Suppose that the probability† of the noise amplitude at the sampling point falling between the values n and $n + dn$ is given by the Gaussian function

$$\frac{dn}{\sqrt{2\pi\sigma^2}} \exp\left(-\frac{n^2}{2\sigma^2}\right). \tag{4.6-4}$$

This is shown in Fig. 4.6-5(b). The *standard deviation*, σ, is the rms noise amplitude; for the most important type of electrical noise, σ^2 can be calculated from the time average of the square of $n(t)$; i.e.,

$$\sigma^2 = \lim_{T\to\infty} \frac{1}{2T} \int_{-T}^{T} n^2(t)\, dt. \tag{4.6-5}$$

The probability of an error, $P(\epsilon)$, in any one digit can then be expressed in the following way:

$$P(\epsilon) = P(\text{high sent and } n < -A)$$
$$+ P(\text{low sent and } n > A). \tag{4.6-6}$$

It is reasonable to assume that the noise is independent from the signal, so the probability of the joint events involving signal and noise is the product of the probabilities of the individual events. Hence

$$P(\text{high sent and } n < -A) = P(\text{high sent})\, P(n < -A). \tag{4.6-7a}$$

Similarly,

$$P(\text{low sent and } n > A) = P(\text{low sent})\, P(n > A). \tag{4.6-7b}$$

Also it is most reasonable to assume that high and low pulses are equiprobable. In fact, it is shown in classical information theory that coding of the signal to achieve equiprobability maximizes the channel capacity [4-20]. Hence $P(\text{high sent}) = P(\text{low sent}) = \frac{1}{2}$. The probability of the noise amplitude being greater than A is the area under the probability density curve shown shaded in Fig. 4.6-5. Because of the symmetry of the curve, this is also the probability of the noise being less than $-A$. Thus Eq. (4.6-6) becomes

$$P(\epsilon) = \frac{1}{2} P(n < -A) + \frac{1}{2} P(n > A)$$
$$= P(n > A)$$
$$= \frac{1}{\sqrt{2\pi}\sigma} \int_{A}^{\infty} e^{-n^2/2\sigma^2}\, dn$$
$$= \frac{1}{\sqrt{\pi}} \int_{A/\sqrt{2}\sigma}^{\infty} e^{-z^2}\, dz = \frac{1}{2}\,\text{erfc}\left(\frac{A}{\sqrt{2}\,\sigma}\right). \tag{4.6-8}$$

---

†We apologize to the reader for jumping the gun and using probabilistic concepts before a proper introduction (Chapters 8 and 9). However, the notions used here are fairly simple, and we hope that they will be understood.

In the last step we have made the change of variable $z = n/\sqrt{2}\sigma$ in order to put the integral into the form tabulated as the complementary error function. Values of $P(\epsilon)$ are given in the following table:

A/σ	$P(\epsilon)$
0	0.5
1	0.16
2	0.023
3	0.00135
4	0.000032

Observe that the probability of error can be made negligibly small by using A/σ on the order of 4 or higher. Hence for practical purposes the effects of random noise can be neglected.

Quantizing Noise

The quantization error is the roundoff error resulting from the fact that a continuous signal is converted into a finite set of discrete levels. This means that the output differs from the input, and the difference can be regarded as noise. This noise will typically appear as shown in Fig. 4.6-6 and results in a somewhat gritty sound in sound systems. The amplitude of this noise is directly related to the size of the steps between the quantizing levels. If the differences between the levels are all equal, and if the maximum excursion of the analog input is fixed, step size and therefore noise amplitude can be decreased by increasing the number of quantizing levels. However, more

Figure 4.6-6. Quantizing noise.

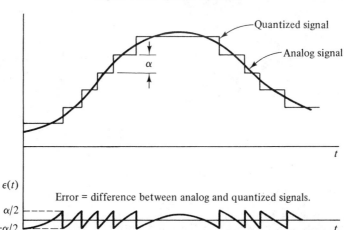

Error = difference between analog and quantized signals.

levels require a larger bandwidth for transmission, and therefore the amount of noise reduction is limited mainly by bandwidth considerations.

Computations of quantizing noise resulting from a sinusoidal input signal and giving the amplitudes of various harmonics have been performed by Bennett and are described in [4-12]. These show a very complicated behavior which is very sensitive to small changes in any input parameter and therefore not too useful. A simpler approach is to assume that the quantizing error is random and uniformly distributed within each quantization level.

Suppose that the width of all the quantization steps is the same (uniform quantization) and equal to α. Then the randomness assumption is equivalent to saying that the quantization error ϵ at any instant t is a random variable with

$$p(\epsilon) = \begin{cases} \dfrac{1}{\alpha} & \text{for } |\epsilon| < \dfrac{\alpha}{2} \\ 0 & \text{for } |\epsilon| > \dfrac{\alpha}{2}. \end{cases} \tag{4.6-9}$$

With this assumption the mean-square error is easily calculated; it is given by

$$\overline{\epsilon^2} = \int_{-\alpha/2}^{\alpha/2} \epsilon^2 p(\epsilon)\, d\epsilon = \frac{1}{\alpha} \int_{-\alpha/2}^{\alpha/2} \epsilon^2\, d\epsilon = \frac{\alpha^2}{12}. \tag{4.6-10}$$

The signal-to-noise ratio can be computed by assuming a sinusoidal input of the form

$$s(t) = A \cos 2\pi f_m t, \qquad f_m = \frac{1}{T}, \quad T \text{ is the period.} \tag{4.6-11}$$

If the peak-to-peak signal amplitude encompasses q quantization levels, then

$$A = \frac{q\alpha}{2}. \tag{4.6-12}$$

The averaged-over-time signal power is

$$P_s = \frac{1}{T} \int_{-T/2}^{T/2} s^2(t)\, dt = \frac{A^2}{2} = \frac{q^2\alpha^2}{8}. \tag{4.6-13}$$

The signal-to-noise ratio, in this case, is given by

$$\text{SNR} = \frac{\text{ave. signal power per cycle}}{\text{mean-square error due to noise}}$$

$$= \frac{q^2\alpha^2/8}{\alpha^2/12} = \frac{3}{2} q^2. \tag{4.6-14}$$

This shows quantitatively that the signal-to-noise ratio can be increased at will by increasing q.

The relation between q and the bandwidth is given in Eq. (4.6-3). For simplicity, assume that $\log_m q$ is an integer†; then

$$q = m^{B/W}, \tag{4.6-15}$$

where m is the size of the pulse alphabet (2 for binary PCM), W the signal bandwidth, and B the required channel bandwidth. Substituting into Eq. (4.6-14) results in

$$\text{SNR} = \tfrac{3}{2}m^{2(B/W)}. \tag{4.6-16}$$

For $m = 2$, this specializes to

$$\text{SNR} = 3 \cdot 2^{(2B/W)-1}.$$

The SNR is seen to increase exponentially with the quantity $V \equiv B/W$, the *bandwidth expansion ratio*. A similar trade-off of bandwidth for signal-to-noise ratio is observed for PDM, and it is also an important feature of FM communication. Because of the exponential relation between bandwidth expansion ratio and SNR, PCM is superior in this respect to these two methods. However, in practice, large values of V are rarely used because bandwidth conservation is usually the highest-priority requirement. Therefore this advantage of PCM can usually not be realized.

It is not necessary that all the quantization steps be equal. In fact it is frequently desirable to use small steps for low signal levels and larger steps for larger levels to compensate for the reduction in quantizing SNR that always occurs for small-signal amplitudes. In practice unequal quantization is conveniently accomplished by *companding*, i.e., compressing the signal at the transmitter input and expanding it again at the receiver output. This is done by passing the signal through a nonlinear device having the general characteristic of the form shown in Fig. 4.6-7(a) and then using the reciprocal nonlinear device shown in Fig. 4.6-7(b) at the receiver. Compression can be done just prior to the A/D converter that generates the digital signal and expansion just after the D/A converter. If the compression and expansion

†Otherwise use $\lceil \cdot \rceil$.

Figure 4.6-7. Companding.

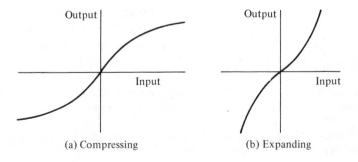

(a) Compressing   (b) Expanding

functions are properly matched (i.e., do inverse operations), the process introduces no distortion since its only effect is to change the A/D and D/A coding process. Companding is used also in analog transmission channels, but there serious distortion can occur if the channel has any amplitude or phase errors. In this respect PCM is therefore also much superior to other forms of transmission.

With all of the many apparent advantages of PCM one may ask why it has thus far found only relatively limited commercial application.† One of the reasons for not using PCM in its early days was the relative complexity of its physical implementation compared to more conventional systems. This is no longer a serious disadvantage since extremely complicated systems that operate very reliably are now available as a result of large-scale integration (LSI). Another disadvantage of PCM is its relatively large bandwidth requirement which translates directly into more coaxial cables or microwave links to handle a given number of simultaneous transmissions. This disadvantage is probably the major one that prevents more widespread use of PCM at the present time. A very good discussion of the trade-offs involved and of the reasons the Bell system utilized one system in preference to another is given by Bennett [4-21].

Pulse Transmission

Pulses generated by the various pulse-modulation schemes discussed in this chapter, as well as data pulses produced by computer terminals and related devices, are usually not transmitted over any great distance in the simple form shown, say, in Fig. 4.6-1. Instead these pulses are used to *modulate* a carrier whose frequency matches the medium being used. For instance, for a radio link the carrier frequency might be a few megahertz; in a microwave link the frequency will be in the gigahertz range. For the transmission of data pulses over telephone circuits, voice frequencies (1–2 kHz) are used. The device used for this purpose is called a *modem* (for *mo*dulation and *dem*odulation).

There are three basic modulation types: (1) amplitude-shift keying (ASK), (2) phase-shift keying (PSK), and (3) frequency-shift keying (FSK). In ASK modulation a fixed-frequency carrier wave is switched between several levels. For binary pulses the 0 bit is generally transmitted as a zero voltage (*off* state). The 1 bit is transmitted as a nonzero voltage of constant amplitude (*on* state). This kind of ASK is therefore also called off-on keying (OOK). The phase of the on-state voltage may vary from pulse to pulse; it carries no information. In PSK the phase of a fixed-amplitude carrier is switched between several values. If binary pulses are used, the *on* state might

---

†PCM *is* used in space communications and in computer communication nets. It is now also used in optical communication systems.

correspond to the carrier being in phase with some reference, and the *off* state to the carrier being 180° out of phase. In four-level PSK the carrier might be switched to 0°, 90°, 180°, and 270° relative to the reference. Finally, in FSK the frequency of a constant-amplitude carrier is switched between several frequencies or tones. The three systems are illustrated in Fig. 4.6-8.

In addition to these basic modulation types there are numerous variations and combinations. One of these is differential phase-shift keying (DPSK) in which the pulse information is contained in the phase change between adjacent pulse intervals. For instance, in the commonly used ±45°, ±135° system a phase change from one interval to the next of +45° encodes the binary code 11, +135° encodes 01, −135° encodes 00, and −45° encodes 10. The advantage of DPSK over PSK is that there is no need to transmit a phase reference; all that is necessary at the receiver is a short-term memory device to store the phase for one pulse interval.

A modification of the PAM technique is quadrature amplitude modulation (QAM) in which the pulses are amplitude modulated as in PAM but

Figure 4.6-8. Three methods of modulating a high-frequency carrier with pulses.

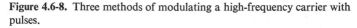
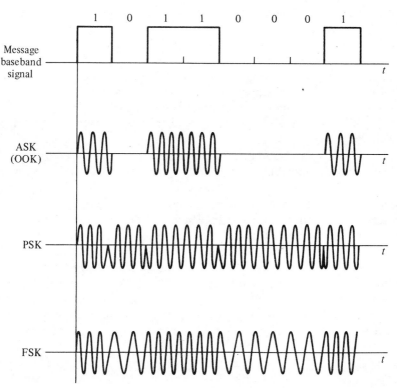

using two carriers at the same frequency that are 90° out of phase. Thus, if $s(t)$ and $\hat{s}(t)$ are two pulse streams, the QAM signal has the form $s(t) \cos \omega_0 t + \hat{s}(t) \sin \omega_0 t$. Quadrature signal modulation is also used in color television and for some forms of stereo and quadraphonic sound transmission and is explained in more detail in Chapter 6. Its main advantage is the saving in channel bandwidth achievable by transmitting two signal streams at the same carrier frequency. A similar bandwidth reduction can be achieved by use of vestigial sideband modulation (VSB), which is also explained in Chapter 6.

Because of its simplicity and economy FSK is most commonly used in low-speed (up to about 1800 bits/sec) data transmission where there is ample signal-to-noise ratio and where bandwidth minimization is not a critical requirement. The frequency shift in hertz is typically between one half and three quarters of the maximum pulse rate, and the channel bandwidth is almost double this value. FSK signals are easily demodulated by passing the signal through narrowband filters tuned to the expected frequencies and observing the magnitude of their output. Phase-locked loops (discussed in Sec. 6.10) are sometimes used for this purpose.

PSK requires a coherent demodulator; i.e., a carrier signal with the correct reference phase must be available at the receiver so that the instantaneous phase of the PSK signal can be determined. This is both an advantage and a disadvantage. As will be seen later (Chapter 11), the advantage of coherent demodulation is that it results in the lowest error probability for a given signal-to-noise ratio. Equivalently, a given error probability can be achieved with somewhat less signal power in a coherent system than in an incoherent system.† It can be shown [4-22] that given a peak power constraint the required signal-to-noise ratio in PSK is about 5 dB less than that needed for (incoherent) FSK.

The obvious disadvantage of a coherent system is the need to have the reference phase available. Although there are methods for handling this problem, they tend to be somewhat complicated. (We shall meet this problem again in the detection of double-sideband-modulated signals in Sec. 6.4.) PSK is therefore not ordinarily used unless its lower signal-to-noise ratio requirement is essential. This is true, for instance, in space communications, and there PSK as well as other coherent signaling methods are widely used. A combination of PSK and FSK called minimum shift keying (MSK) and having many of the advantages of PSK and FSK is described in [4-23].

The three basic modulation schemes briefly mentioned here, as well as some of their variants, are discussed in considerably more detail in some of the references. Particularly complete treatments can be found in Bennett and

---

†An example of the use of PSK modulation in deep-space communication is given in the Appendix.

Davey [4-7] and Davey [4-24]. For an analysis of QAM, see [4-25]. Several other systems are described in [4-23] and [4-26].

4.7 Summary

In this chapter we have dealt with some of the signal properties arising from the fact that signals are generally band limited. Although signals cannot be strictly confined to a finite time and frequency interval, in practice one can define reasonable confinement intervals. In particular, frequency confinement was shown to permit representation of a continuous signal in terms of a sequence of discrete samples. We introduced the sampling theorem, which says that the minimum required number of samples for signal reconstruction is $2W$ per second, where W is the signal bandwidth. From these samples the signal can, in theory, be exactly reproduced for all values of time, and we discussed some of the reconstruction methods used in practice. Sampling underlies a variety of pulse-transmission systems used in practice; we discussed PAM, PDM, PPM, and PCM systems. PCM employs an additional quantization of signal amplitudes at each sampling instant, which permits analog signals to be transmitted in digital form. A number of ways of modulating digital pulses was then briefly considered.

PROBLEMS

4-1. Use the identity

$$\sum_{n=-\infty}^{\infty} \delta(t + nT) = \frac{1}{T} \sum_{n=-\infty}^{\infty} e^{jn\omega_0 t}, \qquad \omega_0 \equiv \frac{2\pi}{T},$$

and the fact that for a linear system, the responses are given by

Input	Response
$\delta(t + nT)$	$h(t + nT)$
$e^{jn\omega_0 t}$	$H(n\omega_0)e^{jn\omega_0 t}$

to prove that

$$\sum_{n=-\infty}^{\infty} h(nT) = \frac{1}{T} \sum_{n=-\infty}^{\infty} H(n\omega_0).$$

This result is one version of *Poisson's summation formula*.

4-2. a. Show that Eq. (4.3-5) can be written as

$$X_s(\omega) = d \sum_{n=-\infty}^{\infty} \text{sinc}\,(nd)X(\omega - n\omega_s),$$

where $X(\omega)$ is the spectrum of $x(t)$ and $X_s(\omega)$ is the spectrum of the pulse-sampled signal.

b. Compute $X_s(\omega)$ when $x(t) = 1 + m \cos \omega_m t$, and identify the term that represents the spectrum of the sampling pulses.

4-3. Consider the sampling theorem with $T_s = 1/(2W)$. Then Eq. (4.2-5) takes the form

$$x(t) = \sum_{n=-\infty}^{\infty} x\left(\frac{n}{2W}\right) \phi_n(t),$$

where $\phi_n(t) \equiv \operatorname{sinc} 2W[t - n/(2W)]$. Show that the set of functions $\phi_n(t)$, $n = \pm 1, \pm 2, \ldots$, are orthogonal over the interval $-\infty < t < \infty$. *Hint:* Parseval's theorem

$$\int_{-\infty}^{\infty} v(t)u^*(t)\, dt = \int_{-\infty}^{\infty} V(f)U^*(f)\, df$$

shows that if two functions are orthogonal in one domain, their transforms will be orthogonal in the other domain.

4-4. Let the energy, E, in $x(t)$ be given by

$$E = \int_{-\infty}^{\infty} |x(t)|^2\, dt.$$

Show that if $x(t)$ is strictly band-limited to B Hz, then

$$E = \frac{1}{2B} \sum_{n=-\infty}^{\infty} \left|x\left(\frac{n}{2B}\right)\right|^2.$$

4-5. Assume a more general form of the "sampling" theorem in which $x(t)$ has the representation

$$x(t) = \sum_n x_n \phi_n(t),$$

where x_n are the "samples" and the set $\{\phi_n(t)\}$ is orthonormal over $a < t < b$. Show that

$$E \equiv \int_a^b |x(t)|^2\, dt = \sum_n |x_n|^2.$$

4-6. The output, y, of a square-law device is $y = x^2$ if x is the input. How can $y(t)$ be written in terms of the sampled values of $x(t)$? At what rate must $x(t)$ be sampled to enable the reconstruction of $y(t)$? Assume that $x(t)$ is band-limited to B Hz.

4-7. (Frequency sampling theorem) Show that if $x(t) = 0$ for $|t| > T$, then

$$X(f) = \sum_{n=-\infty}^{\infty} X\left(\frac{n}{2T}\right) \operatorname{sinc} 2T\left(f - \frac{n}{2T}\right).$$

4-8. In the sample-and-hold circuit of Fig. 4.3-4(a), show that the sampler/ capacitor system is represented by a transfer function

$$H(\omega) = \frac{1 - e^{-j\omega T_s}}{j\omega}$$

if the hold time is equal to the sampling interval. Derive Eq. (4.3-13) from this result.

4-9. N baseband signals, each limited to W Hz, are time-multiplexed. Assume the minimum adequate sampling rate. What is the minimum required bandwidth required to transmit the multiplexed signal on a PAM system?

4-10. Thirty baseband channels, each band-limited to 3.2 kHz, are sampled and multiplexed at an 8-kHz rate.
    a. What is the required bandwidth for transmission of the multiplexed samples on a PAM system?
    b. Explain the function of the "excess" sampling rate. What problems are encountered when the theoretical minimum rate is used?

4-11. It is suggested that the S/H circuit described by Eq. (4.3-13) be used as a low-pass filter to reconstruct the signal from its samples. Assume that the hold time is equal to the sampling period. Let the normalized response of the S/H circuit be

$$H(f) = \text{sinc}\left(\frac{f}{f_s}\right)e^{-j2\pi(f/f_s)}, \qquad f_s \equiv \frac{1}{T_s}.$$

Consider a baseband signal with highest frequency B Hz.
    a. Let $f_s = 2B$. How effective is the filter in keeping out the higher lobes in the periodic spectrum?
    b. Let $f_s = 6B$. Indicate how the performance of the filter has improved by computing the response for the wanted (baseband) spectrum and the unwanted (first-order) spectrum.

4-12. A PAM telemetry system multiplexes four signals: $s_i(t)$, $i = 1, \ldots, 4$. Two of the signals $s_1(t)$, $s_2(t)$ have $B = 80$ Hz, while $s_3(t)$, $s_4(t)$ have $B = 1000$ Hz. The sampling rate for $s_3(t)$ and $s_4(t)$ is $f_{sm} = 2400$ samples/sec. Assume that other sampling rates can be derived from f_{sm} by dividing by powers of 2. Design a multiplexing system that does a preliminary multiplexing of $s_1(t)$ and $s_2(t)$ into a single sequence, $s_5(t)$, and a final multiplexing of $s_5(t)$, $s_3(t)$, and $s_4(t)$. What is the required minimum sampling rate for the two-step multiplexing? How does it compare with the required sampling rate if all signals were simultaneously multiplexed? Sketch a block diagram for a two-step demultiplexer. How many more signals of 80 Hz bandwidth could this system accommodate without an increase in sampling rate?

4-13. Show that the impulse response of the raised cosine filter [Eq. (4.4-9)] can be written in a somewhat more revealing form as

$$h(t) = \frac{\sin(2\pi t/T)}{2\pi t(1 - 4t^2/T^2)}.$$

This function has its first nulls at $t = \pm T$ and is very small *outside* the range $|t| < T$. Hence intersymbol interference is reduced.

4-14. The circuit shown in Fig. P4-14 (adapted from Reference [4-12], p. 249) is a simplified PPM generator. The signal $s(t)$ is applied to the base of the transistor, in series with a periodic sweep signal described by

$$V(t) = \frac{2V_0}{T}(t - nT), \qquad nT - \frac{T}{2} < t < nT + \frac{T}{2}, \qquad n = 0, \pm 1, \ldots.$$

Assume that collector current flows when the base-emitter voltage exceeds zero.
    a. Explain how this circuit works.

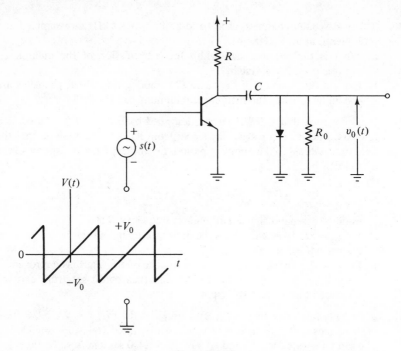

Figure P4-14

b. Let $g(t)$ describe the output pulse when the latter is produced at $t = 0$. Let $s(t)$ be described by

$$s(t) = A \cos (\omega_c t + \theta),$$

where A, ω_c, and θ are constant. Show that the output signal $v_0(t)$ is given by

$$v_0(t) = \sum_{n=-\infty}^{\infty} g(t - t_n),$$

where t_n represents the time at which the nth pulse occurs and satisfies

$$A \cos (\omega_c t_n + \theta) + \frac{2V_0}{T}(t_n - nT) = 0.$$

(See Reference [4-12] for a discussion of this problem.)

4-15. It is desired to design an FSK system using two separate and phase-incoherent crystal oscillators. Let the oscillator outputs be given by

$$x_1(t) = \cos (\omega_1 t + \theta_1)$$

$$x_2(t) = \cos (\omega_2 t + \theta_2), \qquad \omega_1, \omega_2 \gg \frac{2\pi}{T},$$

where T is the *mark* or *space* interval. To ensure that switching between *marks* and *spaces* is done at instants of phase coherence, the system shown in Fig. P4-15 is used.

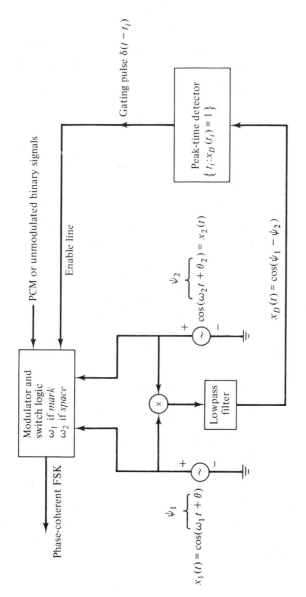

Figure P4-15

a. Explain how the system works.

b. The effect of phase continuity between *mark* and *space* signals can be investigated by considering the signal

$$x(t) = [1 - u(t - t_1)] \cos(\omega_1 t + \theta_1) + u(t - t_1) \cos(\omega_2 t + \theta_2),$$

where t_1 is the time at which the output switches from x_1 to x_2. Calculate the frequency spectrum of $x(t)$. What condition on t_1 assures phase continuity?

c. Show that the spectrum decreases as $1/\omega^2$ if the phase is continuous, compared to a decrease of $1/\omega$ if the phase is discontinuous.

4-16. a. Show that 0°–180° PSK and ASK waves can be written as

$$x(t) = f(t) \cos \omega_c t,$$

where ω_c is the carrier frequency. What is $f(t)$ in the two cases?

b. What happens to this kind of PSK signal in a synchronous detection system if the local carrier has the right frequency but the wrong phase?

4-17. To reduce intersymbol interference due to pulse distortion by the channel, *equalizing* filters in the form of tapped delay lines are inserted between the A/D converter at the sender and the receiver. In Fig. P4-17 is shown a

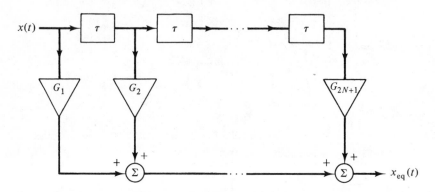

Figure P4-17

transversal equalizer consisting of $2N + 1$ taps on a delay line, $2N + 1$ adjustable gain elements, and $2N$ delay elements.

a. Show that the output of such a filter can be written as

$$x_{eq}(t) = \sum_{i=1}^{2N+1} G_i x[t - (i - 1)\tau]$$

$$= \sum_{i=-N}^{N} F_i x[t - (i + N)\tau],$$

where $F_i \equiv G_{i+N+1}$.

b. Suppose that $x(t)$ has a peak of 1 at $t = 0$ and intersymbol interference on either side. Let $x_{eq}(t)$ be sampled at $t_k = (k + N)\tau, k = \ldots, -2, -1, 0,$

1, ... How should the gains $\{F_i\}_{i=-N}^{N}$ be adjusted so that

$$x_{eq}(t_k) = \begin{cases} 1, & k = 0, \\ 0, & k \neq 0. \end{cases}$$

For a discussion of this problem see [4-10].

4-18. Use the method of Sec. 4.6 to compute the probability of error per digit, $P(\epsilon)$, if unipolar PCM is used; i.e., the pulse heights switch between zero and A. Assume that the noise obeys the Gaussian function of Eq. (4.6-4).

4-19. Show that in symmetrical bipolar PCM, $P(\epsilon)$ is independent of the *a priori* probabilities that a high or low was sent.

4-20. Consider a three-alphabet PCM system. A 10-volt peak-to-peak signal is quantized into 20 uniform levels. Assume that the signal bandwidth is 3.2 kHz.
a. Compute the bandwidth expansion ratio.
b. Compute the SNR. What assumption about the signal is being made in this calculation?
c. Why is the expression SNR $= \frac{3}{2}q^2$ somewhat misleading in the sense that unbounded SNR is not assured by unbounded quantizing?

REFERENCES

[4-1] A. PAPOULIS, *The Fourier Integral and Its Applications*, McGraw-Hill, New York, 1962.

[4-2] D. SLEPIAN, "On Bandwidth," *Proc. IEEE*, **64**, No. 3, pp. 292–300, March 1976.

[4-3] R. C. BUCK, *Advanced Calculus*, McGraw-Hill, New York, 1965.

[4-4] H. NYQUIST, "Certain Topics in Telegraph Transmission Theory," *Trans. AIEE*, **47**, pp. 617–644, April 1928.

[4-5] E. T. WHITTAKER, "On the Functions Which Are Represented by the Expansions of the Interpolation Theory," *Proc. Roy. Soc. Edinburgh*, **35**, pp. 181–194, 1915.

[4-6] H. E. ROWE, *Signals and Noise in Communication Systems*, Van Nostrand Reinhold, New York, 1965, p. 225.

[4-7] W. R. BENNETT and J. R. DAVEY, *Data Transmission*, McGraw-Hill, New York, 1965, Chap. 14.

[4-8] R. W. LUCKY, J. SALZ, and E. J. WELDON, *Principles of Data Communications*, McGraw-Hill, New York, 1968.

[4-9] R. A. GIBBY and J. W. SMITH, "Some Extensions of Nyquist's Telegraph Transmission Theory," *Bell System Tech. J.*, **44**, pp. 1487–1510, Sept. 1965.

[4-10] R. W. LUCKY, "Techniques for Adaptive Equalization of Digital Communications Systems," *Bell System Tech. J.*, **45**, pp. 255–286, 1966.

[4-11]   W. R. BENNETT, "Statistics of Regenerative Digital Transmission," *Bell System Tech. J.*, **37**, pp. 1501–1542, Nov. 1958.

[4-12]   M. SCHWARZ, W. R. BENNETT, and S. STEIN, *Communication Systems and Techniques*, McGraw-Hill, New York, 1966, Chap. 6.

[4-13]   A. B. CARLSON, *Communication Systems*, McGraw-Hill, New York, 1968, pp. 293–294.

[4-14]   A. B. CARLSON, *Communication Systems*, 2nd ed., McGraw-Hill, New York, 1975, pp. 312–314.

[4-15]   M. P. MARCUS, *Switching Circuits for Engineers*, Prentice-Hall, Englewood Cliffs, N.J., 1962, p. 155.

[4-16]   G. S. VERNAN, "Cipher Printing Telegraph Systems for Secret Wire and Radio Telegraphic Communications," *AIEE Trans.*, **65**, pp. 295–301, Feb. 1926.

[4-17]   H. SCHMIDT, *Analog/Digital Conversion*, Van Nostrand Reinhold, New York, 1970.

[4-18]   E. M. DELORAINE, S. VAN MIERLO, and B. DERJAVICH, French Patent No. 932,140, Aug. 1946, U.S. Patent No. 2,629,857, Feb. 24, 1953.

[4-19]   P. F. PANTER, *Modulation, Noise and Spectral Analysis*, McGraw-Hill, New York, 1965, Chap. 22.

[4-20]   C. E. SHANNON and W. WEAVER, *The Mathematical Theory of Communication*, University of Illinois Press, Urbana, Ill., 1949.

[4-21]   W. R. BENNETT, *Introduction to Signal Transmission*, McGraw-Hill, New York, 1970, Chap. 10.

[4-22]   J. M. WOZENCRAFT and I. M. JACOBS, *Principles of Communication Engineering*, Wiley, New York, 1965, pp. 248–252.

[4-23]   S. A. GRONEMEYER and A. L. MCBRIDE, "MSK and Offset QPSK Modulation," *IEEE Trans. Commun.*, **COM 24**, No. 8, pp. 809–820, Aug. 1976.

[4-24]   J. R. DAVEY, "Modems," *Proc. IEEE*, **60**, pp. 1284–1292, Nov. 1972.

[4-25]   D. D. FALCONER and G. J. FOSCHINI, "Theory of Minimum Mean-Square-Error. QAM Systems Employing Decision Feedback Equalization," *Bell System Tech. J.*, **52**, No. 10, pp. 1821–1849, Dec. 1973.

[4-26]   M. K. SIMON, "An MSK Approach to Offset QASK," *IEEE Trans. Commun.*, **COM 24**, No. 8, pp. 921–923, Aug. 1976.

5

Discrete System Theory

5.1 Introduction

The linear system theory presented in Chapter 3 dealt essentially with continous-time signals such as the waveforms generated in speech or in television. These signals are defined for every instant in a given interval and can, in principle, take on any value. In Chapter 4 we showed that band-limited signals can be specified by instantaneous samples taken at discrete times. Sampling converts continuous signals into a sequence of numbers. These numbers can be further processed and transmitted, as is done, for example, in pulse code modulation. Such a sequence of numbers is called a discrete-time signal, and the systems that deal with them are discrete-time systems, or more briefly, discrete systems. If, in addition, the signal levels are discrete (i.e., quantized), the discrete signal is said to be digital.

The sampling of continuous waveforms is only one means of generating discrete signals. Another obvious source of discrete signals is the digital computer, or, in fact, any of the vast variety of digital circuits that are in use today. These devices generate streams of numbers that may not have any very clear relation to analog signals. A good deal of our present-day communication equipment and circuitry must be designed to handle such number streams.

5.2 Discrete Linear Systems

The definition of linearity given in Chapter 2 and its relation to superposition carries over directly to the discrete case. Generally, the input to a discrete linear system is a *sequence* of numbers. Then if y_n is the response to an input sequence u_n and z_n is the response to an input sequence v_n, the response to $a_1 u_n + a_2 v_n$ is $a_1 y_n + a_2 z_n$.

If the input to a discrete linear system consists of the sequence x_1, $x_2, \ldots, x_n$, the output consists of a weighted sum of the inputs:

$$
\begin{aligned}
y_1 &= h_{11} x_1 + h_{12} x_2 + \cdots + h_{1k} x_k + \cdots \\
y_2 &= h_{21} x_1 + h_{22} x_2 + \cdots + h_{2k} x_k + \cdots
\end{aligned}
$$

$$
\cdot \tag{5.2-1}
$$

$$
y_n = h_{n1} x_1 + h_{n2} x_2 + \cdots + h_{nk} x_k + \cdots .
$$

The set of coefficients $\{h_{nk}\}$ characterizes the system.

Equation (5.2-1) can be written in several equivalent and more convenient ways. Allowing for input strings that may be, at least in theory, infinitely long, we can write

$$
y_n = \sum_{k=-\infty}^{\infty} h_{nk} x_k, \qquad n = \ldots, -1, 0, +1, \ldots . \tag{5.2-2}
$$

This form is analogous to the superposition integral

$$
y(t) = \int_{-\infty}^{\infty} h(t, \tau) x(\tau) \, d\tau \tag{5.2-3}
$$

if we identify the indices n and k with the time variables t and τ, respectively, and the constants h_{nk} with the impulse response $h(t, \tau)$. In fact, consider the special input consisting of the discrete impulse function:

$$
x_k = \begin{cases} 1, & k = i \\ 0, & \text{otherwise}; \end{cases}
$$

then the response is just h_{ni}. Hence for any k, h_{nk} lends itself to the interpretation: h_{nk} is the response at "time" n to a unit function applied at "time" k. We put quotes around the word *time* because the indices do not necessarily have to refer to real time, although in communication systems they often do.

Because of the close analogy between Eqs. (5.2-2) and (5.2-3), the former is referred to as a *discrete superposition summation* or just *(discrete) superposition*. For a time-discrete system where the subscripts n, k do refer to time, *causality* requires that y_n cannot depend on future values of x_k, i.e., values of x_k such that $k > n$. Also, the input sequence is typically of finite duration and can be assumed to begin at $k = 0$ or $k = 1$. The choice of time origin is arbitrary. Hence Eq. (5.2-2) for *causal* systems with inputs that begin at $k = 1$

takes the form

$$y_n = \sum_{k=1}^{n} h_{nk} x_k, \tag{5.2-4}$$

which is in direct analogy to the continuous case where

$$y(t) = \int_0^t h(t, \tau) x(\tau) \, d\tau. \tag{5.2-5}$$

Another convenient description of the system of equations in Eq. (5.2-1) results from using vector notation. In this representation the input and output sequences are represented by column vectors $\mathbf{x}$ and $\mathbf{y}$, where $\mathbf{x}$ has the K elements $x_1, \ldots, x_K$ and $\mathbf{y}$ has, say, the N elements $y_1, \ldots, y_N$. Then, through the use of basic matrix multiplication, Eq. (5.2-1) is converted to

$$\mathbf{y} = \mathbf{Hx}, \tag{5.2-6}$$

where $\mathbf{y} = (y_1, \ldots, y_N)^T$, $\mathbf{x} = (x_1, \ldots, x_K)^T$, the T denotes transpose, and $\mathbf{H}$ is given by

$$\mathbf{H} = \begin{bmatrix} h_{11} & h_{12} & h_{13} & \cdots & h_{1K} \\ h_{21} & h_{22} & h_{23} & \cdots & h_{2K} \\ \cdot & & & & \cdot \\ \cdot & & & & \cdot \\ \cdot & & & & \cdot \\ h_{N1} & & \cdots & & h_{NK} \end{bmatrix}. \tag{5.2-7}$$

We see that a discrete linear system can be characterized by a weighting matrix $\mathbf{H}$ instead of the impulse or step-function response. The matrix $\mathbf{H}$ is referred to as the *transmission matrix* of a system. It can contain a finite or infinite number of elements. The matrix for a causal system is triangular (Prob. 5-1).

We said earlier that discrete systems need not necessarily be indexed by real time. Often the discrete input data are already collected on magnetic tape or disk, and outputs can be computed using stored "future" values of the input. For this reason concepts such as causality or even time invariance play a less important role in discrete system theory than in continuous time systems. On the other hand when $\mathbf{y} = \mathbf{Hx}$ is used to model a real-time sampled-data system as shown in Fig. 5.2-1, then causality and time invariance are notions of principal importance.

We have already mentioned causality. The concept of time invariance in discrete systems is a direct carry-over from the continuous case. For the latter a system is said to be time-invariant if $h(t, \tau) = h(t - \tau)$. In a discrete, linear, time-invariant (DLTI) system the impulse response satisfies

$$h_{nk} = h_{n-k} \tag{5.2-8}$$

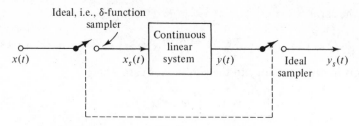

Figure 5.2-1. Real-time ideal sampled-data system.

and Eq. (5.2-2) reduces to

$$y_n = \sum_{k=-\infty}^{\infty} h_{n-k} x_k. \tag{5.2-9}$$

Equation (5.2-9) is referred to as a *discrete convolution*. Although it may not be immediately apparent, the transmission matrix for a DLTI system has far fewer *distinct* entries than is the case for a noninvariant system (Prob. 5-2). The transmission matrix for a causal DLTI system has the form

$$\mathbf{H} = \begin{bmatrix} h_0 & & & & \\ & h_0 & & & \\ h_1 & & h_0 & & \textbf{0} \\ & h_1 & & h_0 & \\ h_2 & & h_1 & & h_0 \\ & h_2 & & h_1 & & h_0 \\ & & h_2 & & h_1 & & h_0 \end{bmatrix}. \tag{5.2-10}$$

Note that h_{nk} is a function of only one subscript, i.e., $n - k \equiv m$, and hence the elements along diagonals are all equal.

For DLTI systems that are also causal, with inputs that begin at $k = 1$, the expression equivalent to Eq. (5.2-4) is

$$y_n = \sum_{k=1}^{n} h_{n-k} x_k. \tag{5.2-11}$$

If in Eq. (5.2-11) we let $m \equiv n - k$, we obtain

$$y_n = \sum_{m=0}^{n-1} h_m x_{n-m}. \tag{5.2-12}$$

It is convenient to denote a discrete convolution as $y_n = h_n * x_n$; the subscript n on both h and x is standard usage.

Difference Equations

The description of DLTI systems by the discrete impulse response h_m is not the only one in use. A powerful approach is a representation by linear, constant-coefficient difference equations. This is analogous to the representation of continuous systems by differential equations. The coefficients of the difference equation can be related to the discrete impulse response, and this also is analogous to the relation that exists between the impulse response and the coefficients of the differential equation describing a continuous system. We confine ourselves to a brief discussion of the difference-equation method.

A Kth order difference equation can be written as

$$\sum_{k=0}^{K} a_k y_{n-k} = \sum_{k=0}^{M} b_k x_{n-k}, \qquad a_0 = 1, \qquad (5.2\text{-}13)$$

where the $\{a_i\}$ and $\{b_i\}$ are coefficients that characterize the system and $a_K \neq 0$ if the system is Kth order. The order of the equation is K because the present value of y, i.e., y_n, depends on the previous K values $y_{n-1}, \ldots, y_{n-K}$. For most systems of engineering interest, $M < K$.

There is a somewhat subtle notational paradox in Eq. (5.2-13). The subscripts on the a_i and b_i identify the coefficients and imply their position in the equation but *do not* refer to time. On the other hand, the subscripts on the input and output variables x_i and y_i, respectively, often *do* refer to time, and therefore it has become conventional to write Eq. (5.2-13) as

$$\sum_{k=0}^{K} a_k y(n-k) = \sum_{k=0}^{M} b_k x(n-k), \qquad a_0 = 1, \qquad (5.2\text{-}14)$$

which clearly differentiates between the two meanings. When $y(\cdot)$ and $x(\cdot)$ are written as functions of integers, the implication usually is that the sampling is uniform in time. It is convenient to write Eq. (5.2-14) as

$$y(n) = -\sum_{k=1}^{K} a_k y(n-k) + \sum_{k=0}^{M} b_k x(n-k), \qquad (5.2\text{-}15)$$

which directly shows the dependence of $y(n)$ on its past values and on the input and its past values.†

As an example, consider the first-order difference equation

$$y(n) = -5y(n-1) + x(n) \qquad (5.2\text{-}16)$$

with initial condition $y(-1) = 0$. Let $x(n) = n$. The solution can be obtained recursively as follows:

$$y(0) = -5y(-1) + x(0) = 0$$
$$y(1) = -5y(0) + x(1) = 1$$
$$y(2) = -5y(1) + x(2) = -3$$
$$y(3) = -5y(2) + x(3) = 18 \qquad (5.2\text{-}17)$$

$$\cdot$$
$$\cdot$$
$$\cdot$$

etc.

---

†The notation $y(n-k)$, $x(k)$, etc., rather than y_{n-k}, x_k, etc., will be adopted for the remainder of this chapter.

Another approach is to compute the solution directly as the sum of the driven solution $y_d(n)$ and the transient solution $y_t(n)$. For $y_t(n)$ we write

$$y_t(n) + 5y_t(n-1) = 0 \qquad (5.2\text{-}18)$$

and assume a solution of the form $A\gamma^n$, where A and γ are unknown. Then, substituting $A\gamma^n$ directly into Eq. (5.2-18) enables us to write

$$A\gamma^{n-1}(\gamma + 5) = 0,$$

for which the only nontrivial solution is

$$y_t(n) = A(-5)^n. \qquad (5.2\text{-}19)$$

To compute $y_d(n)$ we *assume* a solution $y_d(n) = Bn + C$. When this result is substituted into Eq. (5.2-16) we obtain

$$Bn + C = -5[B(n-1) + C] + n. \qquad (5.2\text{-}20)$$

By matching coefficients of like powers of n we get

$$B = \tfrac{1}{6}, \qquad C = \tfrac{5}{36}.$$

The complete solution, with A still unknown, is therefore

$$y(n) = \tfrac{1}{6}n + \tfrac{5}{36} + A(-5)^n. \qquad (5.2\text{-}21)$$

When the initial condition $y(-1) = 0$ is inserted in Eq. (5.2-21) we get

$$A = -\tfrac{5}{36}.$$

The completely determined solution is finally

$$y(n) = \frac{6n + 5}{36} - \frac{5}{36}(-5)^n, \qquad (5.2\text{-}22)$$

which has an advantage over Eq. (5.2-17) in that $y(n)$ can be computed for any n without first recursively computing $y(n-1)$, $y(n-2)$, ..., etc. A very simple filter that is represented by Eq. (5.2-16) is shown in Fig. 5.2-2.

We have already said that invariant systems require far fewer descriptors

Figure 5.2-2. First-order system described by the difference equation $y(n) = -5y(n-1) + x(n)$.

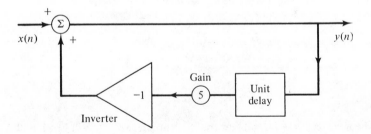

than other systems. But that isn't the only advantage offered by the property of invariance; by direct analogy to the continuous case, input-output calculations for DLTI systems can be conveniently done with the discrete Fourier transform. With the aid of the latter, convolution is replaced by multiplication, and the transform technique offers a real savings in the amount of labor involved provided that a convenient way of computing the discrete Fourier transform exists. Fortunately such an algorithm does exist and is known as the fast Fourier transform. It is discussed in Sec. 5.6. First, however, we shall discuss the theory behind the discrete Fourier transform.

5.3 The Discrete Fourier Transform

In the following discussion we consider discrete signals resulting from sampling of a continuous time function $x(t)$ at a uniform sampling rate $1/\Delta t$. This time function will in general have a Fourier transform, and we can define a discrete Fourier transform (DFT) for the sampled time function that approximates the continuous Fourier transform (CFT) in the same way as the sequence $x(1) \cdots x(N)$ approximates the function $x(t)$. Since digital computers can only work with discrete signals, Fourier transforms obtained with a digital computer are always discrete Fourier transforms.

To develop the discrete Fourier transform, consider the expression for the CFT [Eq. (2.8-7)]:

$$X(f) = \int_{-\infty}^{\infty} x(t)e^{-j2\pi ft} \, dt. \tag{5.3-1}$$

For discrete signals, the time variable t becomes $i \, \Delta t$, and the function $x(t)$ is replaced by $x(i)$. Also, because we are interested only in discrete frequencies, the frequency variable f is replaced by $k \, \Delta f$, and $X(f)$ becomes $X(k)$. In practice the number of time or frequency samples that can be handled is a finite number N, determined by the size of the computer. If the signal is sampled at the Nyquist rate, then Δt has the value

$$\Delta t = \frac{1}{2W}, \tag{5.3-2}$$

where W is the bandwidth. However, with N frequency components separated by a frequency increment Δf, the total frequency range that can be considered is $N \, \Delta f$, so that

$$2W = N \, \Delta f. \tag{5.3-3}$$

Combining Eqs. (5.3-2) and (5.3-3), we see that

$$\Delta t \, \Delta f = \frac{1}{N}. \tag{5.3-4}$$

The infinite limits of integration must be replaced by finite limits, and the integral itself is replaced by a sum. The infinitesimal dt becomes $\Delta t = 1/(2W)$ $= T/N$, where T is the length of time over which the N samples of $x(t)$ have been obtained. Then finally Eq. (5.3-1) is transformed to

$$X(k) = \frac{T}{N} \sum_{i=1}^{N} x(i)e^{-j2\pi ki/N}, \qquad j = \sqrt{-1}, \tag{5.3-5}$$

and by direct analogy the inverse transform is

$$x(i) = \frac{2W}{N} \sum_{k=1}^{N} X(k)e^{j2\pi ki/N}. \tag{5.3-6}$$

The validity of these formulas is easily demonstrated by inserting the definition for $X(k)$ from Eq. (5.3-5) into Eq. (5.3-6). This gives

$$x(m) = \frac{2W}{N} \sum_{k=1}^{N} \frac{T}{N} \sum_{i=1}^{N} x(i)e^{-j2\pi ki/N}e^{j2\pi km/N}$$

$$= \frac{2WT}{N^2} \sum_{k=1}^{N} \sum_{i=1}^{N} x(i)e^{j2\pi k(m-i)/N}. \tag{5.3-7}$$

Note the use of the index m to prevent confusion with the index i used in the second summation; this is equivalent to the use of a dummy variable τ in the demonstration of the Fourier identity in Sec. 2.8. Since both sums in Eq. (5.3-7) are finite, their order can be interchanged, giving

$$x(m) = \frac{2WT}{N^2} \sum_{i=1}^{N} x(i) \sum_{k=1}^{N} e^{j2\pi k(m-i)/N}. \tag{5.3-8}$$

The second summation has the form of a geometric series:

$$\sum_{k=1}^{N} G^k = \frac{G(1 - G^N)}{1 - G}, \tag{5.3-9}$$

where $G \equiv e^{j2\pi(m-i)/N}$. Since $m - i$ is an integer, $G^N = 1$, and therefore the series sums to zero unless $G = 1$, i.e., when $m = i \bmod (N)$. In this case it has the form $\sum_{i=1}^{N} (1)$ and sums to N. Thus Eq. (5.3-8) becomes

$$x(m) = \frac{2WT}{N} x(m). \tag{5.3-10}$$

However, it is easily shown that $2WT = N$. This follows from the fact that $\Delta t = 1/(2W)$ and $T = N \Delta t$. Thus Eq. (5.3-10) is an identity.

The discrete Fourier pair is often given in the form

$$X(k) = \sum_{i=1}^{N} x(i)U_N^{-ik}, \tag{5.3-11}$$

$$x(i) = \frac{1}{N} \sum_{k=1}^{N} X(k)U_N^{ik}, \tag{5.3-12}$$

where $U_N \equiv e^{j2\pi/N}$. This form eliminates all reference to T and W since these constants are only scale factors and are essentially irrelevant to the basic operation. The validity of this pair is as easily demonstrated as that of the previous one; in fact the discussion subsequent to Eq. (5.3-10) can be omitted.

Because the series of Eq. (5.3-9) sums to N for all $m = i \bmod (N)$, we see that the inverse transform of the sequence $\{X(k)\}$ gives not only $x(i)$ but also all $x(i \pm cN)$, where c is any integer. Similarly, the discrete Fourier transform of $\{x(i)\}$ results in $X(k \pm cN)$. Hence we have

$$x(i) = x(i + N) = x(i + 2N) \cdots, \qquad i = 1, \ldots, N, \quad (5.3\text{-}13)$$

$$X(k) = X(k + N) = X(k + 2N) \cdots, \qquad k = 1, \ldots, N. \quad (5.3\text{-}14)$$

Thus both the transform and its inverse yield periodic sequences. In most applications one is interested in only a single period. For the time series, the period is $T = N \Delta t$, and in frequency, the period is $f_s = N \Delta f = 2W$.

When the $\{x(i)\}$ sequence is real, the real part of $X(k)$ is symmetric about the folding frequency $f_s/2$. This is easily demonstrated by computing, say, $[X(N/2) + 1]$ and $X[(N/2) - 1]$ using Eq. (5.3-11) (assume N is even). We have

$$X\left(\frac{N}{2} + 1\right) = \sum_{i=1}^{N} x(i) e^{-j\pi i - j2\pi i/N} = \sum_{i=1}^{N} x(i)(-1)^i e^{-j2\pi i/N}$$

$$X\left(\frac{N}{2} - 1\right) = \sum_{i=1}^{N} x(i) e^{-j\pi i + j2\pi i/N} = \sum_{i=1}^{N} x(i)(-1)^i e^{j2\pi i/N}.$$

Hence $X[(N/2) + 1] = X^*[(N/2) - 1]$, where the upper asterisk denotes complex conjugate and

$$\text{Re}\left[X\left(\frac{N}{2} + 1\right)\right] = \sum_{i=1}^{N} x(i)(-1)^i \cos \frac{2\pi i}{N} = \text{Re}\left[X\left(\frac{N}{2} - 1\right)\right]. \quad (5.3\text{-}15)$$

Thus the real part is symmetric, and the imaginary part of $X(k)$ is antisymmetric about $N/2$. But since $X(k)$ is a periodic sequence with period N, we also have that $X[(N/2) + k] = X[(-N/2) + k] = X^*[(N/2) - k]$ so that $X(k)$ has Hermitian symmetry about $k = 0$. Furthermore, we see from Eq. (5.3-11) that when $k = N$, $X(N)$ is proportional to the dc value of the $\{x(i)\}$. Therefore we associate $k = N$ with dc. The lowest positive frequency above dc is $k = 1$; the nearest negative frequency to dc is $N - 1$. Hence we identify the Fourier coefficients $X(k)$ for $1 \leq k \leq N/2$ as positive-frequency components while those for $N/2 < k < \text{N}$ are negative-frequency components. The maximum frequency corresponds to $k = +N/2$. For the time series $x(i)$ we similarly identify $N/2 < i < N$ as negative time and $1 \leq i \leq N/2$ as positive time. See Fig. 5.3-1.

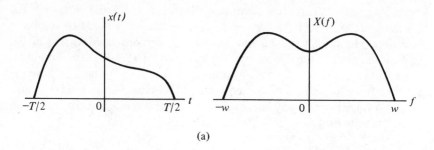

(a)

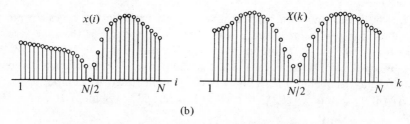

(b)

Figure 5.3-1. (a) Continuous signal and its Fourier transform; (b) discrete signal and its Fourier transform.

5.4 Properties of the Discrete Fourier Transform

Most of the elementary properties of the continuous Fourier transform presented in Sec. 2.9 have their counterpart in the discrete Fourier transform. We consider some of them here.

Parseval's Theorem

This takes the form

$$\sum_{i=1}^{N} |x(i)|^2 = \frac{1}{N} \sum_{k=1}^{N} |X(k)|^2 \tag{5.4-1}$$

and is proved by writing

$$\sum_{i=1}^{N} |x(i)|^2 = \sum_{i=1}^{N} x^*(i) \cdot \frac{1}{N} \sum_{k=1}^{N} X(k) e^{j2\pi i k/N}$$

$$= \frac{1}{N} \sum_{k=1}^{N} X(k) \sum_{i=1}^{N} x^*(i) e^{j2\pi i k/N} = \frac{1}{N} \sum_{k=1}^{N} |X(k)|^2. \tag{5.4-2}$$

Linearity

If $X(k)$ and $Y(k)$ are the DFT of $x(i)$ and $y(i)$, respectively, the DFT of $ax(i) + by(i)$ is $aX(k) + bY(k)$.

Shifting Property

Let $X(k)$ be the DFT of $x(i)$, and consider the DFT of $x(i + m)$, where m is an integer. This is given by

$$\sum_{i=1}^{N} x(i + m)e^{-j2\pi ki/N} = \sum_{l=1+m}^{N+m} x(l)e^{-j2\pi(l-m)k/N} \tag{5.4-3}$$

$$= e^{j2\pi mk/N} \sum_{l=1+m}^{N+m} x(l)e^{-j2\pi kl/N}$$

$$= e^{j2\pi mk/N}X(k). \tag{5.4-4}$$

In the first step we have substituted $l \equiv i + m$ so that instead of counting from 1 we start counting from $m + 1$. In the last step we make use of the fact that the $x(l)$ are periodic, i.e., $x(l) = x(l + N)$, etc. Hence, as long as we sum N successive terms of the form $x(l)e^{-j2\pi kl/N}$ we get the same $X(k)$ no matter where we start the summing process.

Convolution-Multiplication

Discrete convolution was discussed in Sec. 5.2. When the sequences being convolved are obtained by sampling continuous time functions, the operation takes the form

$$y(m) = \sum_{i=1}^{N} x(i)h(m - i). \tag{5.4-5}$$

Taking the discrete Fourier transform of both sides yields

$$Y(k) = \sum_{m=1}^{N} y(m)e^{-j2\pi mk/N} = \sum_{m=1}^{N}\sum_{i=1}^{N} x(i)h(m - i)e^{-j2\pi mk/N}$$

$$= \sum_{i=1}^{N} x(i)e^{-j2\pi ik/N} \sum_{m=1}^{N} h(m - i)e^{-j2\pi(m-i)k/N} \tag{5.4-6}$$

or

$$Y(k) = X(k)H(k). \tag{5.4-7}$$

In the second step we exchange orders of summation and multiply by $e^{-j2\pi(ik-ik)/N} = 1$. The last step uses the periodicity of the $h(m)$, which implies that summing from $1 - i$ to $N - i$ is the same as summing from 1 to N. Just as the CFT enjoys the convolution-multiplication property for continuous waveforms, so does the DFT for discrete signals; this property is, in fact, one of the most important reasons for using the DFT.

Some of the other Fourier properties discussed in Chapter 2 also have their discrete counterparts. Duality is easily proved and yields an additional set of simple relations. The scaling property of the continuous Fourier transform is not used very much in the discrete case since scale information generally does not appear in the discrete formulation.

5.5 Pitfalls in the Use of the DFT

The discrete Fourier transform is in some ways conceptually simpler than the continuous transform because it deals with finite sets of numbers. Questions of convergence or continuity never arise. Also, the various Fourier relations are exact, and there is no need to consider concepts such as the delta function that are sometimes regarded as being of doubtful validity.

Problems arise when the discrete Fourier transform is used to approximate the continuous Fourier transform. The DFT is a mapping of one periodic number sequence into another, while the CFT usually maps a finite interval in time (or frequency) into an infinite interval in frequency (or time). The problems are then mainly due to truncation effects and sampling approximations, and some of them have already been discussed in Chapter 4.

Consider the Fourier transformation of a cosine wave of frequency f_0 Hz, as shown in Fig. 5.5-1. Line (a) shows that the cosine, which extends over the infinite interval $-\infty < t < \infty$, transforms into two delta functions at $f = \pm f_0$. If the cosine is observed over only a finite time period, the frequency function is the convolution of the delta functions with a sinc function as shown in line (b). Sampling of the time-limited cosine is equivalent to multiplication in the time domain by a comb of delta functions, or a convolution in the frequency domain by a related comb of delta functions [see Eq. (2.12-24)]. The sampling comb and its transform are shown in line (c) and the sampled time function and its continuous Fourier transform in line (d). As discussed in Chapter 4, there will be aliasing or overlap in the spectra unless the sampling rate is large enough. Since the spectrum of the truncated cosine, or of any time function extending over a finite time interval, extends to infinity on the frequency axis, a certain amount of aliasing will always take place, even though it can be minimized by using a sufficiently large sampling rate.

In any case, it is clear that the spectrum is no longer exactly that of the truncated cosine wave, and therefore retransforming will result in a somewhat altered wave.

Sampling of the spectrum results in convolution of the time function with a delta-function comb so that the single cosine-pulse sequence is transformed into a periodic sequence. This is shown in line (e). Here again there will generally be some aliasing if the rate at which the spectrum is sampled is not high enough. Finally, if both the time and frequency function are reconstructed from N samples, we see that neither one will in general be exact.

In addition to the aliasing errors, truncation causes an effect referred to as *leakage*. This is illustrated in Fig. 5.5-1, line (b), where we show that the delta function spectrum of the infinite cosine is converted into a sinc spectrum because of truncation. The word leakage refers to the fact that energy in the original spectral components at $f = \pm f_0$ leaks to other frequencies after truncation in time. The truncation shown in Fig. 5.5-1 is rather severe in order

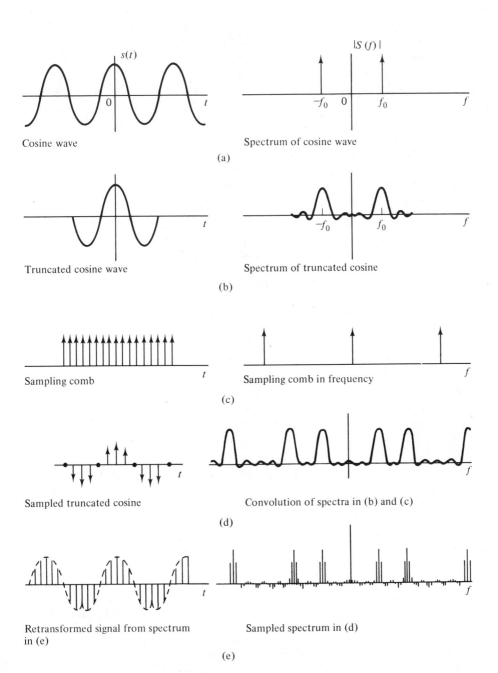

Cosine wave

Spectrum of cosine wave

(a)

Truncated cosine wave

Spectrum of truncated cosine

(b)

Sampling comb

Sampling comb in frequency

(c)

Sampled truncated cosine

Convolution of spectra in (b) and (c)

(d)

Retransformed signal from spectrum in (e)

Sampled spectrum in (d)

(e)

Figure 5.5-1. Effects of truncation and sampling. *After Bergland* [5-1].

197

to clearly show the effect. However, if we wanted to compute the Fourier spectrum of a cosine, or any other function, we would always have to perform the transformation on a finite piece of the function because of the limitations in computer memory. Thus truncation is usually necessary (unless the function is naturally time limited), and it will result in the smearing effect shown.† Leakage can be reduced by using truncation functions that reduce the time function more gradually near the ends of the interval than the rect function illustrated in Fig. 5.5-1. Such functions are referred to as *data windows*, and much work has gone into the design of windows that reduce leakage without distorting the spectrum too much. A commonly used window is the cosine bell window, shown in Fig. 5.5-2. If the original time function is $x(t)$, then the truncated function is $x_T(t) = w(t)x(t)$, where

$$w(t) = \begin{cases} \dfrac{1}{2} - \dfrac{1}{2}\cos 2\pi \dfrac{t}{\alpha T}, & 0 < t \le \dfrac{\alpha T}{2} \\[2mm] 1, & \dfrac{\alpha T}{2} < t < T - \dfrac{\alpha T}{2} \\[2mm] \dfrac{1}{2} - \dfrac{1}{2}\cos 2\pi \dfrac{T-t}{\alpha T}, & T - \dfrac{\alpha T}{2} < t < T. \end{cases} \quad (5.5\text{-}1)$$

Typical values of α are 0.1–0.2.

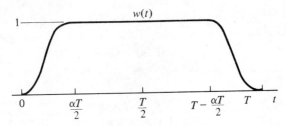

Figure 5.5-2. Data window with cosine roll-off.

Another frequently used window function is the generalized Hamming‡ window, defined by

$$w(t) = \begin{cases} \alpha - (1 - \alpha)\cos \dfrac{2\pi t}{T}, & 0 \le t \le T \\[2mm] 0, & \text{otherwise.} \end{cases} \quad (5.5\text{-}2)$$

This is the raised cosine pulse already encountered several times in this book. Note that for $\alpha \ne \frac{1}{2}$ there is a jump in $w(t)$ at $t = 0$ and $t = T$. When $\alpha = 0.54$, the window is called the (no-longer-generalized) Hamming window;

---

†There is no leakage if the input spans exactly one fundamental period ([5-2], p. 92). However, in most applications the truncation of the input cannot be arranged to guarantee this.

‡After Richard W. Hamming.

when $\alpha = 0.50$, it is called the Hanning† window [5-3]. An advantage of the Hanning window is that multiplication by $\frac{1}{2}$ is a particularly simple operation on the computer since it involves merely a right shift of the binary representation of the multiplicand. The Hamming window is optimum in the sense that of all raised cosine windows it has the smallest maximum side lobe level ([5-2], p. 99).

Instead of multiplying the input signal by $w(t)$ one can, equivalently, convolve the spectrum of the input with the Fourier transform of the data window. The latter is called a *spectral window*. For the generalized Hamming window, the spectral window has the form (delay terms omitted)

$$W(f) = \alpha T \text{ sinc } fT + \frac{1-\alpha}{2} T \text{ sinc } (Tf - 1) + \frac{1-\alpha}{2} T \text{ sinc } (Tf + 1).$$

$$(5.5\text{-}3)$$

The DFT equivalent of Eq. (5.5-3) is

$$W(k) = \alpha \text{ sinc } k + \frac{1-\alpha}{2} \text{ sinc } (k-1) + \frac{1-\alpha}{2} \text{ sinc } (k+1), \qquad (5.5\text{-}4)$$

and the discrete convolution of $W(k)$ with $X(k)$ results in replacing each point $X(k)$ in the original DFT by $\alpha X(k) + [(1-\alpha)/2][X(k+1) + X(k-1)]$ (see Prob. 5-13). Thus the operation of the window function is nothing more than a weighted moving average of three adjacent spectral lines.

Data and spectral windows are discussed further in Chapter 10 where they are considered in relation to the problem of spectral estimation.

A third problem in the use of the DFT arises from the improper use of Eq. (5.4-7) to obtain the output of a linear system. The impulse response of such a system is generally defined only over a finite time interval T and is assumed to be zero outside of this interval. Similarly, the input signal may exist over some time T_1. A typical input signal and impulse response are shown in Fig. 5.5-3. Convolution of these two signals results in a signal such as shown in Fig. 5.5-3(c), which typically extends over a time interval $T + T_1$.

†After the Austrian mathematician Julius von Hann.

Figure 5.5-3. (a) Finite-duration impulse response; (b) finite-duration signal; (c) output.

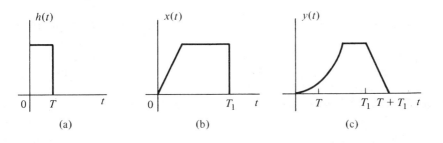

Suppose that instead of actually performing the convolution, the output is obtained by multiplying the discrete Fourier transforms of $x(t)$ and $h(t)$, using Eq. (5.4-7). To get as short a sampling interval as possible, one might naively decide to obtain the N uniformly spaced samples from the longer of the two intervals T or T_1. However, since the DFT treats all functions as being periodic, this results in fact in the convolution of two periodic signals such as the ones shown in Figs. 5.5-4(a) and (b). The output would be as shown in Fig. 5.5-4(c). Note that the result, even over one period, is quite different from that shown in Fig. 5.5-3.

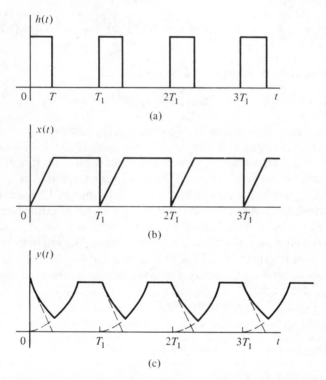

Figure 5.5-4. Periodicities induced by the DFT: (a) impulse response; (b) input; (c) distorted output.

A simple cure for this problem is to lengthen the sampling interval to be at least equal to $T + T_1$. Then even though the signals are still periodic, the nonzero parts will be separated by intervals of zero signals such that the correct output is obtained in any one of the extended periods. See Fig. 5.5-5.

In practice one frequently wants to convolve signals having widely differing periods. This is true, for instance, when the input is a random signal

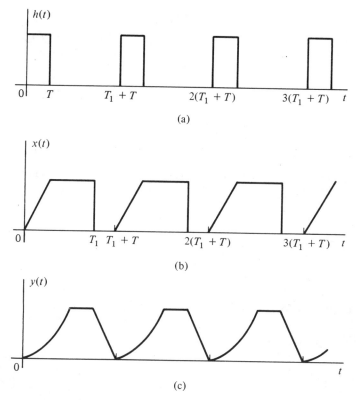

Figure 5.5-5. Periodicities induced by the DFT when a sufficiently large interval is used to produce the correct output: (a) impulse response; (b) input; (c) undistorted output appears as a periodic signal.

extending over indefinite time, and the impulse response can be assumed to have only a finite length. Such convolutions are most rapidly performed by using the fast Fourier transform (FFT), described below, to transform the impulse response and adjacent intervals of the signal, using Eq. (5.4-7), and then retransforming. For this operation to yield valid results one must use overlapping intervals as indicated in Fig. 5.5-6. The length of each interval, containing N samples, is $2T$, where T is the time over which $h(t)$ is nonzero. Convolution of this $h(t)$ with an equally long interval of signal results in incorrect results for the first $N/2$ spectral components but in valid results for the last $N/2$ components. We therefore retain only the last $N/2$ components from each set of N that is generated, and after each complete computation we advance the signal T seconds, i.e., $N/2$ points, and repeat the process. For a complete discussion of this procedure, see [5-1].

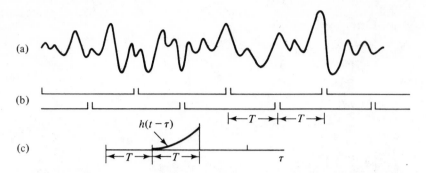

Figure 5.5-6. Use of overlapping intervals for processing signals of indefinite time duration with the DFT of a time-limited impulse response: (a) input; (b) overlapping interval of length $2T$; (c) impulse response.

5.6 The Fast Fourier Transform

The discrete Fourier transform has in recent years become one of the most widely used computational tools because of the development in 1965 of the fast Fourier transform (FFT) algorithm by J. W. Cooley and J. W. Tukey [5-4]. This algorithm reduces the number of computations required to perform the DFT from something on the order of N^2 to $N/2 \log_2 N$. For $N = 1024$ the reduction factor is about $10^6/5000$ or 200. Thus if a typical operation takes 10 microseconds, the time required for the transformation is reduced from 10 seconds to 50 milliseconds. Machines are now available in which the FFT is hard-wired so that a 1024-point FFT may be obtained in less than 10 milliseconds. This is fast enough so that fairly extensive signal processing can be done on line in many cases.

To explain the algorithm, consider the expression

$$X(k) = \sum_{i=0}^{N-1} x(i)U^{ik}. \tag{5.6-1}$$

This is essentially Eq. (5.3-11), with the indexing changed to run from 0 to $N - 1$ and with $U \equiv U_N^{-1}$. The algorithm is particularly simple if N is a power of 2, although this is not essential [5-5]. For the sake of our discussion, let $N = 8$. The indices i and k can be expressed as binary numbers in the form

$$i = (i_2, i_1, i_0)$$
$$k = (k_2, k_1, k_0), \tag{5.6-2}$$

where i_2, i_1, etc., take on only the values 0 or 1. Only three binary digits are needed since $N = 8$; for example, if $i = 5$, it becomes 101 in binary form. The decimal equivalent of the number (i_2, i_1, i_0) is $4i_2 + 2i_1 + i_0$ and similarly for the k's.

With the binary number representation Eq. (5.6-1) can be rewritten in the form

$$X(k_2, k_1, k_0) = \sum_{i_0=0}^{1} \sum_{i_1=0}^{1} \sum_{i_2=0}^{1} x(i_2, i_1, i_0) U^{(4i_2+2i_1+i_0)(4k_2+2k_1+k_0)}$$

$$= \sum_{i_0=0}^{1} \sum_{i_1=0}^{1} \sum_{i_2=0}^{1} x(i_2, i_1, i_0) U^{k_0(4i_2+2i_1+i_0)} U^{2k_1(2i_1+i_0)} U^{4k_2 i_0}, \quad (5.6\text{-}3)$$

where the second line is obtained from the first by multiplying out the exponent and observing that terms such as $U^{16k_2i_2}$ or $U^{8k_1i_2}$, etc., can be dropped since they are equal to 1.

The equation can now be solved recursively as follows. We let

$$A_1(k_0, i_1, i_0) \equiv \sum_{i_2=0}^{1} x(i_2, i_1, i_0) U^{k_0(4i_2+2i_1+i_0)}$$

$$A_2(k_0, k_1, i_0) \equiv \sum_{i_1=0}^{1} A_1(k_0, i_1, i_0) U^{2k_1(2i_1+i_0)}$$

$$A_3(k_0, k_1, k_2) \equiv \sum_{i_0=0}^{1} A_2(k_0, k_1, i_0) U^{4k_2 i_0}$$

$$X(k) = X(k_2, k_1, k_0) = A_3(k_0, k_1, k_2). \quad (5.6\text{-}4)$$

Note that each sum contains only two terms, each of which involves multiplication by some power of U. For each k there are therefore 6 complex multiplications and 3 complex additions, or since there are 8 k's, a total of 48 multiplications and 24 additions. Half of the multiplications can be eliminated immediately by noting that the exponent of U is zero for the 4 A_1's for which k_0 is zero in the first line,† and similarly for the 4 A_2's in the second line for which $k_1 = 0$ and for the 4 A_3's in the third line for which $k_2 = 0$. This follows since $U^0 = 1$. A further halving of the number of multiplications is possible if half of the additions are replaced by subtractions. This follows because $U^4 = -U^0$, $U^5 = -U^1$, etc. This has the effect in the first line, for instance, of making $A_1(1, i_1, i_0) = [x(0, i_1, i_0) - x(1, i_1, i_0)] U^{(2i_1+i_0)}$, etc.

A signal flow diagram illustrating these factors is shown in Fig. 5.6-1 for the 8-point algorithm. In this diagram the junction of two arrows represents an addition, and a minus sign near one line at a junction means that this line should have its sign reversed before addition; i.e., it should be subtracted. Numbers such as U^0, U^1, etc., placed next to lines mean multiplication by that amount. Note that there are 12 additions, 12 subtractions, and 12 multiplications. The numbers in this example indicate exactly each of $N/2 \log_2 N$ additions, subtractions, and multiplications. More generally this can be shown to be true whenever the number of points is a power of 2 ([5-6], Chap. 6). Actually 7 of the multiplications are by U^0 and could be omitted, leaving

---

†That is, $A_1(0, 0, 0)$, $A_1(0, 0, 1)$, $A_1(0, 1, 0)$, $A_1(0, 1, 1)$.

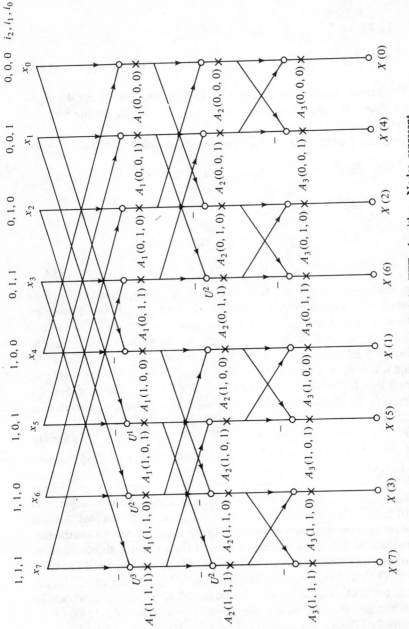

Figure 5.6-1. Signal flow diagram for the FFT algorithm. Nodes represent algebraic addition, and the numbers U^k next to some links represent multiplication by that amount. The values at the crosses are the A's computed at that level.

only 5 multiplications.† It is easily shown that for an N-point algorithm this remaining reduction in multiplications results in $(N/2) \log_2 N - (N - 1)$ multiplications.

Multiplications are in general an order of magnitude more time-consuming than additions or subtractions. The amount of time required to perform an N-point FFT is therefore on the order of $(N/2) \log_2 N - (N - 1)$ times the time required to perform a single complex multiplication, or since a complex multiplication is equivalent to four real multiplications, the time is $2N \log_2 N - 4(N - 1)$ real multiplication times. This contrasts with the roughly N^2 complex multiplications that would be needed if the DFT were directly evaluated using Eq. (5.3-11). Figure 5.6-2 shows the substantial time saving afforded by the FFT for large N [5-1].

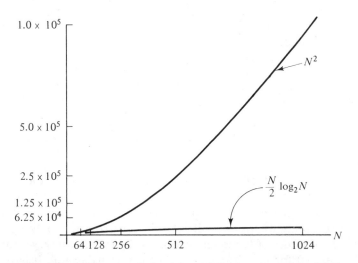

Figure 5.6-2. Required number of multiplications in a direct DFT calculation vs. the number required in the FFT algorithm. *After Bergland [5-1].*

Examination of the flow chart of Fig. 5.6-1 shows that the transform can be done "in place," that is, by writing all intermediate results over the original data sequence and writing the final answer over the intermediate results. Thus no storage is needed beyond that required for the original N complex numbers. This is an additional important advantage of the FFT. Also, since the FFT is essentially a series of nested 2-point DFT's, it is in principle possible to do a transform for N larger than the available storage space by doing several smaller FFT's and then combining the results. The algorithm is discussed in much more detail in some of the references ([5-4]–[5-7]). References

---

†The multiplications by U^0 are omitted in Fig. 5.6-1.

[5-2] and [5-6] contain short FORTRAN programs that implement the algorithm.

5.7 The Z Transform

While the DFT and FFT have their principal applications in the numerical *computation* of Fourier transforms by computer, the Z transform has its principal application as an *analytical tool* in the *analysis* of discrete-time systems. The Z transform and the DFT are closely related and share many properties. However, unlike the DFT, the frequency variable in the Z transform is a continuous complex variable, z, instead of a discrete real variable k. In a sense, one may regard the DFT as a weighted average of the input data, while the Z transform may be regarded as a *resolution* of the input data into signals of the form z^{-n}, where z is a complex variable. The one-sided Z transform of a discrete signal $x(n)$ is defined by

$$Z_1[x(n)] = X_1(z) = \sum_{n=0}^{\infty} x(n)z^{-n}, \qquad (5.7\text{-}1)$$

while the two-sided Z transform is defined by

$$Z_2[x(n)] = X_2(z) = \sum_{n=-\infty}^{\infty} x(n)z^{-n}. \qquad (5.7\text{-}2)$$

We shall consider only the one-sided Z transform here; therefore expressions such as $X(z)$ or $Z[x(u)]$ all refer to one-sided Z transforms in the sequel.

The variable z is an arbitrary complex variable, but for the Z transform to exist, z must lie inside a region of convergence that depends on the particular sequence $x(n)$ being transformed. If $x(n)$ has nonzero values only over some finite range of values of n, the summation in Eq. (5.7-1) converges for all z. Also if $x(n) = 0$ for $n < 0$ and is otherwise reasonably well behaved, the series $X_1(z)$ converges if $|z|$ is larger than some radius of convergence r.†

One of the main reasons for using the Z transform is that it possesses the convolution-multiplication property of all Fourier-type transformations. This permits the replacement of the relatively complex procedure of convolution by the much simpler one of multiplication. Convolution of two discrete-time signals was discussed in Sec. 5.2. However, instead of assuming that $x(m)$ begins at $m = 1$, as we did in Eq. (5.2-11), we shall for convenience assume that $x(m) = 0$ for $m < 0$. In that case we write

$$y(n) = \sum_{m=0}^{n} x(m)h(n - m), \qquad (5.7\text{-}3)$$

---

†For the one-sided Z transform the conditions for convergence are (1) $|x(n)| < \infty$ for all finite n, and (2) $|x(n)| \leq kr^n$ if $n \geq N$, for some finite k, r, and N. The smallest value of r for which this inequality is satisfied is the radius of convergence.

where $h(\cdot)$ is a causal response. Taking the Z transform of both sides of Eq. (5.7-3) gives

$$
\begin{aligned}
Y(z) &= \sum_{n=0}^{\infty} y(n)z^{-n} = \sum_{n=0}^{\infty} \sum_{m=0}^{\infty} x(m)h(n-m)z^{-n} \\
&= \sum_{m=0}^{\infty} x(m)z^{-m} \sum_{k=-m}^{\infty} h(k)z^{-k} \\
&= \sum_{m=0}^{\infty} x(m)z^{-m} \sum_{k=0}^{\infty} h(k)z^{-k} \\
&= X(z)H(z),
\end{aligned}
\tag{5.7-4}
$$

where

$$
H(z) = \sum_{k=0}^{\infty} h(k)z^{-k}.
\tag{5.7-5}
$$

In the second equality of the first line of this development we have changed the upper limit on the summation from n to infinity; this doesn't affect anything since $h(\cdot)$ is zero for negative arguments. In the second line we interchanged orders of summation and substituted the index $k \equiv n - m$. In the third line we again used the fact that $h(\cdot)$ is causal and is therefore zero for negative arguments.

The usefulness of the Z transform depends in part on the possibility of finding simple *closed-form expressions* for commonly used sequences $x(n)$ and $h(n)$. Then the resulting $Y(z)$ can be analyzed or inverted to study the properties of the response. In this respect the Z transform differs from the DFT, where closed-form expressions are mostly irrelevant. We shall consider the Z transforms of some common functions.

The Unit Step Function

Assuming that the step takes place at $n = 0$, we have that

$$
x(n) = \begin{cases} 1 & \text{for } n = 0, 1, 2, \ldots \\ 0 & \text{for } n < 0. \end{cases}
$$

Then

$$
X(z) = \sum_{n=0}^{\infty} z^{-n} = \frac{1}{1 - z^{-1}}.
\tag{5.7-6}
$$

Note that the summation is just a geometric series in z^{-1}. It converges if $|z| > 1$; therefore the radius of convergence is 1.

The Delayed Unit Step

Suppose the step takes place at $n = k$; i.e.,

$$
x(n) = \begin{cases} 0, & n < k \\ 1, & n \geq k. \end{cases}
$$

Then

$$X(z) = \sum_{n=k}^{\infty} z^{-n} = \frac{z^{-k}}{1 - z^{-1}}. \qquad (5.7\text{-}7)$$

The radius of convergence is again 1. This example is a particular application of the *shifting property* of the Z transform:

If $\qquad\qquad X(z) = Z[x(n)],$

then $\qquad Z[x(n - k)] = z^{-k}X(z) + z^{-k}\sum_{m=-k}^{m=-1} x(m)z^{-m}. \qquad (5.7\text{-}8)$

The proof is left as an exercise (Prob. 5-14). The second term vanishes if $x(m) = 0$ for negative m.

The Linearly Increasing Sequence

Here

$$x(n) = \begin{cases} n, & n \geq 0 \\ 0, & n < 0. \end{cases} \qquad (5.7\text{-}9)$$

Then

$$X(z) = \sum_{n=0}^{\infty} nz^{-n}. \qquad (5.7\text{-}10)$$

This summation, as well as others in which n appears to higher powers, is easily evaluated by differentiating Eq. (5.7-6) with respect to z, which results in $-\sum_{n=0}^{\infty} nz^{-n-1}$. Thus the Z transform desired here is

$$Z[n] = -z\frac{d}{dz}\left(\frac{1}{1 - z^{-1}}\right) = \frac{z^{-1}}{(1 - z^{-1})^2}. \qquad (5.7\text{-}11)$$

The Exponential Sequence

If

$$x(n) = \begin{cases} a^n, & n \geq 0 \\ 0, & n < 0, \end{cases}$$

then

$$Z[a^n] = \sum_{n=0}^{\infty} a^n z^{-n} = \frac{1}{1 - az^{-1}}. \qquad (5.7\text{-}12)$$

The radius of convergence is now seen to be $|a|$. If $|a| < 1$, the sequence $x(n)$ converges to zero, and this permits a smaller radius of convergence for z. On the other hand, if $|a| > 1$, a^n diverges, and then a larger z magnitude is needed for convergence. Note that by making the identification $a = e^{-c}$ this result is seen to be analogous to the Fourier transform of the exponential function.

The Discrete Impulse

This is the discrete equivalent of the Dirac delta function discussed in Sec. 2.11. It is defined by

$$\delta_{mn} = \begin{cases} 1, & n = m \\ 0, & n \neq m. \end{cases} \tag{5.7-13}$$

This function is also called the unit function, or the Kronecker delta. It is sometimes written $\delta(n - m)$. If $m = 0$, $\delta(n)$ is 1 for $n = 0$, and it is zero otherwise. The corresponding Z transform is

$$\mathcal{Z}[\delta_{mn}] = z^{-m}, \tag{5.7-14}$$

and for the particular value $m = 0$

$$\mathcal{Z}[\delta_{0n}] = 1. \tag{5.7-15}$$

Again note the analogy to the continuous Fourier transform, especially Eq. (5.7-15). The discrete impulse has the advantage over the Dirac delta function of being finite and requiring no limiting arguments. This is another illustration of the relative analytic simplicity of discrete systems compared to continuous ones.

The Finite Square Pulse

As a final example we consider the sequence

$$x(n) = \begin{cases} 1, & n = 0, 1 \ldots, m - 1 \\ 0, & \text{otherwise.} \end{cases} \tag{5.7-16}$$

Then

$$X(z) = \sum_{n=0}^{m-1} z^{-n} = \frac{1 - z^{-m}}{1 - z^{-1}}.$$

This can be put into the more symmetric form

$$X(z) = z^{-(m-1)/2} \left[\frac{z^{m/2} - z^{-m/2}}{z^{1/2} - z^{-1/2}} \right], \tag{5.7-17}$$

where the factor $z^{-(m-1)/2}$ can be regarded as resulting from a shift of $(m - 1)/2$ sample points, so that the result in brackets is the Z transform of a symmetrical pulse. By using $z = e^{j\omega}$, Eq. (5.7-17) can, for small ω, be identified as an approximation to the sinc function; this demonstration is left as an exercise. Because of the finite number of terms, $X(z)$ converges everywhere.

5.8   The Inverse Z Transform

The one-sided Z transform can be computed by several methods, but the most general is the method of contour integration using the calculus of residues. Since

$$X(z) = \sum_{n=0}^{\infty} x(n)z^{-n},$$

the Z transform is recognized *to be a Laurent* series [5-8] about the point $z = 0$. Therefore the coefficients are given by

$$x(n) = \frac{1}{2\pi j} \int_C X(z)z^{n-1}\, dz, \tag{5.8-1}$$

where C is any contour that encloses all the singularities of $X(z)$. The function $X(z)$ must be analytic on C. The residue theorem states that if C is a closed contour within which an arbitrary function $F(z)$ is analytic except for a finite number of singular points $z_1, z_2, \ldots, z_N$ interior to C, then

$$\frac{1}{2\pi j} \int_C F(z)\, dz = \text{sum of residues of } F \text{ at } z_1, z_2, \ldots, z_N. \tag{5.8-2}$$

To use the residue theorem to compute $x(n)$ we let $F(z) \equiv X(z)z^{n-1}$.

Since the Laurent expansion is unique, any other way of getting $X(z)$ into a form in which the coefficients can be recognized is acceptable as an inversion method. A simple method applicable to Z transforms that are ratios of polynomials in z is just to divide the denominator into the numerator. Still another method of performing the inversion is to use partial fractions to put the transform into a form where the inverses can be recognized. Some examples are given below.

EXAMPLES

Consider

$$X(z) = \frac{z^2 - 9z}{z^2 - 6z + 5}.$$

1. *Partial fractions.*

$$X(z) = \frac{z^2 - 9z}{z^2 - 6z + 5} = \frac{z^2 - 9z}{(z - 1)(z - 5)} = \frac{2z}{z - 1} - \frac{z}{z - 5}$$

$$= \frac{2}{1 - z^{-1}} - \frac{1}{1 - 5z^{-1}}.$$

By Eq. (5.7-6) the first term corresponds to the step function sequence with a coefficient of 2. The second term is in the form of Eq. (5.7-12) with $a = 5$. Thus

$$x(n) = 2 - 5^n, \qquad n = 0, 1, 2, \ldots. \tag{5.8-3}$$

2. *Residues.* The function $X(z)z^{n-1}$ has poles at $z = 1$ and $z = 5$. The residue at $z = 1$ is

$$\lim_{z \to 1} \frac{(z - 1)(z^2 - 9z)z^{n-1}}{(z - 1)(z - 5)} = 2,$$

and the residue at 5 is

$$\lim_{z \to 5} \frac{(z - 5)(z^2 - 9z)z^{n-1}}{(z - 1)(z - 5)} = -5^n.$$

Therefore $x(n) = 2 - 5^n$ as before. Note that the function $X(z)z^{n-1}$ may have a pole at the origin for $n = 0$; in fact this will generally happen if the numerator of $X(z)$ does not contain z as a factor. Then the residue for $z = 0$ must also be computed. As an example, if $X(z) = 10/(z - 1)(z - 5)$, the residues of $X(z)z^{n-1}$ for $n > 0$ are -2.5 and $5^n/2$, but for $n = 0$ there are three poles with residues 2, -2.5, and 0.5. Therefore for this function $x(0) = 0$, $x(n) = 2.5(5^{n-1} - 1)$ for $n > 0$.

3. *Long division.* The function $X(z)$ is first converted into a function of z^{-1} by dividing the numerator and denominator by the highest power of z. In our example we divide by z^2. This results in

$$X(z) = \frac{1 - 9z^{-1}}{1 - 6z^{-1} + 5z^{-2}}.$$

The denominator is then divided into the numerator as follows:

$$\begin{array}{r}
1 - 3z^{-1} - 23z^{-2} - 123z^{-3} \cdots . \\
\hline
1 - 6z^{-1} + 5z^{-2}\overline{)1 - 9z^{-1}} \\
\underline{1 - 6z^{-1} + 5z^{-2}} \\
- 3z^{-1} - 5z^{-2} \\
\underline{- 3z^{-1} + 18z^{-2} - 15z^{-3}} \\
- 23z^{-2} + 15z^{-3} \\
\underline{- 23z^{-2} + 138z^{-3} - 115z^{-4}} \\
- 123z^{-3} + 115z^{-4}
\end{array}$$

Thus

$$x(0) = 1$$
$$x(1) = -3$$
$$x(2) = -23$$
$$x(3) = -123$$
$$\vdots$$
$$x(n) = 2 - 5^n,$$

as before. Note that the method of long division yields numerical values for the $x(n)$ from which a functional form has to be inferred.

Properties of the Z Transform

The properties of the Z transform are very similar to those of the DFT, already discussed in some detail. We shall therefore confine ourselves here to a listing of the most important properties without proof (Table 5.8-1). By way of example, consider the computation of the Z transform of na^n. Use property 5 with $k = 1$. The transform for a^n is $1/(1 - az^{-1})$, and therefore the desired

TABLE 5.8-1   PROPERTIES OF THE ONE-SIDED Z TRANSFORM

Property	Time sequence	Z Transform
1. Linearity	$ax(n) + by(n)$	$aX(z) + bY(z)$
2. Shifting	$x(n - k)$	$z^{-k}X(z) + z^{-k}\sum\limits_{m=-k}^{m=-1} x(m)z^{-m}$
3. Scale change	$x(an)$ (a is an integer)	$X(z^{a^{-1}})$
4. Multiplication by a^n	$a^n x(n)$	$X(a^{-1}z)$
5. Multiplication by n^k	$n^k x(n)$	$\left(z^{-1}\dfrac{d}{dz^{-1}}\right)^k X(z)$
6. Forward difference	$\Delta x(n) = x(n+1) - x(n)$	$(z-1)X(z) - zx(0)$
7. Backward difference	$\nabla x(n) = x(n) - x(n-1)$	$(1 - z^{-1})X(z) - z^{-1}x(-1)$
8. Sums	$\sum\limits_{k=-\infty}^{n} x(k)$	$\dfrac{X(z) + \sum\limits_{k=-\infty}^{-1} x(k)}{1 - z^{-1}}$
9. Convolution	$y(n) = \sum\limits_{m=0}^{\infty} h(n-m)x(m)$	$Y(z) = H(z)X(z)$
10. Product of two functions	$y(n) = x(n)\cdot g(n)$	$Y(z) = \dfrac{1}{2\pi j}\displaystyle\int_C X(\lambda)G(z\lambda^{-1})\dfrac{d\lambda}{\lambda}$ All poles of $X(\lambda)$ inside C, all poles of G outside C

transform is

$$z^{-1}\frac{d}{dz^{-1}}\left(\frac{1}{1 - az^{-1}}\right) = \frac{az^{-1}}{(1 - az^{-1})^2}.$$

The Z transform is discussed in considerable detail in books dealing with discrete signal processing. See particularly References [5-9]–[5-12].

5.9   Relation Between the Z Transform and the DFT

In Eq. (5.7-1), consider a sequence $x(n)$ having nonzero values for only a finite number of n's; i.e.,

$$x_N(n) = \begin{cases} x(n), & 0 \leq n < N \\ 0, & \text{otherwise.} \end{cases} \qquad (5.9\text{-}1)$$

By Eq. (5.7-1) the Z transform of this truncated sequence is

$$X(z) = \sum_{n=0}^{N-1} x_N(n)z^{-n}. \qquad (5.9\text{-}2)$$

The DFT of the sequence $\tilde{x}(n)$ is given in Eq. (5.3-11). With a minor change of notation this can be written in the form†

$$\tilde{X}(k) = \sum_{n=0}^{N-1} x_N(n)U_N^{-kn}, \qquad (5.9\text{-}3)$$

†The $\tilde{\ }$ over $\tilde{X}(k)$ is used to designate the DFT.

where $U_N = e^{j2\pi/N}$. Thus

$$\tilde{X}(k) = X(e^{j2\pi k/N}); \qquad (5.9\text{-}4)$$

that is, the DFT is the Z transform evaluated for equally spaced values of z on the unit circle.

Since the sequence $x_N(n)$ and its DFT $\tilde{X}(k)$ uniquely define each other, it should be possible to obtain $X(z)$ for all values of z from the DFT. To obtain this relation we start with Eq. (5.9-2), but for $x_N(n)$ we use the inverse DFT relation:

$$x_N(n) = \frac{1}{N} \sum_{k=0}^{N-1} \tilde{X}(k) U_N^{kn}. \qquad (5.9\text{-}5)$$

Then Eq. (5.9-2) becomes

$$X(z) = \frac{1}{N} \sum_{n=0}^{N-1} \sum_{k=0}^{N-1} \tilde{X}(k) (U_N^{k} z^{-1})^n,$$

which, by a change in the order of summation, becomes

$$X(z) = \sum_{k=0}^{N-1} \tilde{X}(k) \frac{1}{N} \sum_{n=0}^{N-1} (U_N^{k} z^{-1})^n$$

$$= \sum_{k=0}^{N-1} \tilde{X}(k) \frac{1 - U_N^{kN} z^{-N}}{N(1 - U_N^{k} z^{-1})}$$

$$= \sum_{k=0}^{N-1} \frac{\tilde{X}(k)(1 - z^{-N})}{(1 - U_N^{k} z^{-1})N}. \qquad (5.9\text{-}6)$$

In going from the first line to the second we have summed the sum over n using the formula for a finite geometric series, and in the second step we used the fact that $U_N^{kN} = 1$.

Equation (5.9-6) is the desired relation and can be regarded as an interpolation formula by which $X(z)$ is expressed in terms of its values at N equally spaced points on the unit circle. In this respect it is somewhat similar to Eq. (4.2-5), which relates a function of time to its values at discrete-time points. The similarity can be made somewhat more evident by considering $X(e^{j\omega})$, i.e., $X(z)$ evaluated on the unit circle. By Eq. (5.9-6) this is given by

$$X(e^{j\omega}) = \sum_{k=0}^{N-1} \frac{\tilde{X}(k)(1 - e^{-j\omega N})}{(1 - e^{-j[\omega - (2\pi k/N)]})N}. \qquad (5.9\text{-}7)$$

By factoring a factor $e^{-j\omega N/2}$ out of the numerator and $e^{-j[\omega - (2\pi k/N)]/2}$ from the denominator this can be put into the form

$$X(e^{j\omega}) = \sum_{k=0}^{N-1} \tilde{X}(k) \Phi\left(\omega - \frac{2\pi k}{N}\right), \qquad (5.9\text{-}8)$$

where

$$\Phi(\omega) = \frac{\sin \omega N/2}{N \sin \omega/2} e^{-j\omega(N-1)/2} \qquad (5.9\text{-}9)$$

is the interpolation function. Note that it is somewhat similar to the function sinc $[(t - n\tau_s)/\tau_s]$ appearing in Eq. (4.2-5). In particular, for $\omega = 2\pi l/N$, i.e., on the DFT sample points, $\Phi[\omega - (2\pi k)/N]$ is zero for all $l \neq k$ and it is equal to 1 for $l = k$. Thus Eq. (5.9-7) is seen to give the exact value of $X(z)$ on the sample points. Equation (5.9-6) is the basis for one form of digital filter, as will be seen in Sec. 5.11.

5.10   Digital Filters

Digital filters or digital signal processors are small special-purpose digital computers designed to implement an algorithm that converts an input sequence $x(n)$ into a desired output sequence $y(n)$. Such filters employ devices such as adders, multipliers, shifters, and delay elements rather than resistors, capacitors, or operational amplifiers. As a result they are generally unaffected by factors such as component accuracy, temperature stability, long-term drift, etc., that afflict analog filter circuits. Also many of the circuit restrictions imposed by physical limitations of analog devices can be removed or at least circumvented in a digital processor. On the other hand, digital filter designs have to take into account such things as finite word size, roundoff errors, aliasing, and other factors.

Of the large variety of possible digital processors the only ones considered in any detail here are the linear time-invariant (LTI) [also called linear shift invariant (LSI)] systems. Linearity was discussed at the beginning of this chapter; time (or shift) invariance means that if the input sequence $x(n)$ produces the output sequence $y(n)$, then the input $x(n - n_0)$ produces the output $y(n - n_0)$ for all n_0. The input-output relation for LTI systems is the discrete convolution

$$y(n) = \sum_{m=-\infty}^{\infty} x(m)h(n - m), \tag{5.10-1}$$

where $h(n)$ is the discrete impulse response, i.e., the sequence generated by the filter when the input is the discrete impulse $\delta(n)$, given by†

$$\delta(n) = \begin{cases} 1, & n = 0 \\ 0, & n \neq 0. \end{cases} \tag{5.10-2}$$

Recursive Structures

LTI systems are further subdivided into recursive and nonrecursive structures. A recursive system is one in which the current output value depends on preceding values of both the output and the input; i.e.,

$$y(n) = \mathcal{H}[y(n - 1), y(n - 2), \ldots, x(n), x(n - 1), \ldots], \tag{5.10-3}$$

†See Eq. (5.7-13).

where $\mathcal{H}$ is some linear operator. Recursive LTI systems are describable by linear constant-coefficient difference equations:

$$y(n) = \sum_{m=0}^{k} a_m x(n - m) - \sum_{m=1}^{k} b_m y(n - m). \tag{5.10-4}$$

They are therefore fairly direct analogs of ordinary continuous filters that are described by linear constant-coefficient differential equations, and their properties are similar. Their frequency response (actually the Z transform) has poles and zeros, and the response will be unstable if the poles lie outside the unit circle.

We can show this in a somewhat heuristic manner by writing the Z transform relationship between input and output:

$$Y(z) = H(z)X(z), \tag{5.10-5}$$

where $H(z)$ and $X(z)$ are transforms of the impulse response and input, respectively, and $Y(z)$ is the transform of the output. If the output sequence is computed from an inversion of a partial fraction expansion of $Y(z)$, such an expansion will contain terms generated by the poles of $H(z)$. These terms represent the transient response and will be of the form

$$a_1 z_1^k + a_2 z_2^k + \cdots + a_n z_n^k, \tag{5.10-6}$$

where the a_i are coefficients, the z_i are the poles of $H(z)$, and k is the discrete-time index. Clearly, if the output is to remain bounded for a bounded input, $|z_i|$ for $i = 1, \ldots, n$ should be less than unity. This in turn requires that *the poles be inside the unit circle*. Therefore the poles of a stable system must satisfy

$$|z_i| < 1 \qquad \text{for } i = 1, \ldots, n. \tag{5.10-7}$$

Also the amplitude and phase characteristics of $H(z)$ must satisfy some of the constraints in Chapter 3.

The location of the poles determines not only the stability but also the inherent character of the transient response of the system. For example, suppose that a typical term in the partial fraction expansion of $Y(z)$ is

$$\frac{a_i z}{z - z_i} = \frac{a_i}{1 - z_i z^{-1}}.$$

From Eq. (5.7-12) we know that this term gives rise to a sequence $x(n) = z_i^n$, $n = 0, 1, \ldots$. We already know that if $|z_i| < 1$, the transient response decays with increasing k and the system is stable. If z_i is real, the algebraic sign of z_i determines whether the sequence alternates or behaves monotonically. If the poles appear as complex conjugate pairs, i.e., $z^{(1)} = \alpha + j\beta$ and $z^{(2)} = \alpha - j\beta$, stability requires that $(\alpha^2 + \beta^2)^{1/2} < 1$. Figure 5.10-1 shows some typical sequences and their corresponding pole locations.

Recursive filters are frequently referred to as IIR (infinite impulse response) systems to distinguish them from nonrecursive FIR (finite impulse

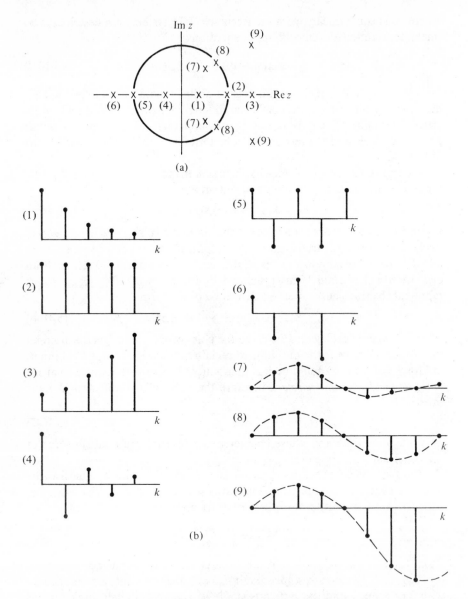

Figure 5.10-1. (a) Typical pole locations of $H(z)$; (b) the corresponding discrete-time sequences generated by such poles.

response) systems whose impulse response is finite by design and equal to the number of delay elements employed in their construction. We discuss FIR filters below. Note that in practice the impulse response time of stable IIR filters is actually also finite because of finite-word-size effects. That is, when the response has decayed to a value smaller than the smallest number that can be represented in the special-purpose digital computer that implements the filter the response is effectively zero.

There are a number of standard realizations for IIR systems. A realization which implements Eq. (5.10-4) directly is the so-called direct form 1 realization, in which separate delays are used to generate both the lagged inputs and outputs. An example of a third-order, direct form 1 realization is shown in Fig. 5.10-2. It is called a third-order system because it realizes a third-order difference equation in $y(n)$. In other words, the present value of $y(n)$ depends on $y(n-1)$, $y(n-2)$, and $y(n-3)$.

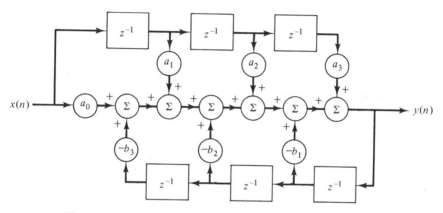

Figure 5.10-2. Example of a direct form 1 digital filter realization.

From the direct form 1 realization it is easy to derive $H(z)$. First, Eq. (5.10-4) is rewritten as

$$\sum_{m=0}^{k} a_m x(n-m) = \sum_{m=0}^{k} b_m y(n-m), \qquad (5.10\text{-}8)$$

where $b_0 = 1$. The Z transform then yields

$$X(z) \sum_{m=0}^{k} a_m z^{-m} = Y(z) \sum_{m=0}^{k} b_m z^{-m},$$

or, equivalently,

$$H(z) = \frac{Y(z)}{X(z)} = \frac{\displaystyle\sum_{m=0}^{k} a_m z^{-m}}{\displaystyle\sum_{m=0}^{k} b_m z^{-m}}, \qquad (5.10\text{-}9)$$

where the a_m and b_m are the gain parameters in the direct form 1 realization.

In Fig. 5.10-2, the blocks labeled z^{-1} are unit delays; in practice these would be implemented by *shift registers*. The "gains" $a_1, a_2, \ldots, b_1, b_2, \ldots$ represent multiplication by constant numbers, and the blocks with the

$\sum$ label are adders. Thus the circuit employs shift-register delays, multipliers, and adders.

A somewhat different form requiring fewer delays can be developed from Eq. (5.10-9) by writing

$$H(z) = \frac{\displaystyle\sum_{m=0}^{k} a_m z^{-m}}{\displaystyle\sum_{m=0}^{k} b_m z^{-m}}$$

$$= \left(\frac{1}{\displaystyle\sum_{m=0}^{k} b_m z^{-m}}\right) \cdot \sum_{m=0}^{k} a_m z^{-m}$$

$$= H_1(z) \cdot H_2(z). \tag{5.10-10}$$

$H_1(z)$ can be represented by a block diagram similar to the lower part of Fig. 5.10-2. $H_2(z)$ is simply a weighted sum of delayed versions of the input. Hence, for a third-order system as in Fig. 5.10-2, we obtain the block diagram of Fig. 5.10-3. Inspection of the circuit shows that the two sets of delay elements do the same thing and can therefore be replaced by a single set. The resulting canonical form referred to as direct form 2 is shown in Fig. 5.10-4. Note that half as many delay elements are needed as in the filter shown in Fig. 5.10-2. In fact, since the filter shown is third order and contains just three delay elements, it has a minimum number of delay elements. However, this is not the only realization having this feature. For instance, it is always possible to write $H(z)$ as a product of factors:

$$H(z) = H_1(z)H_2(z) \cdots H_k(z), \tag{5.10-11}$$

Figure 5.10-3.   Realization of Eq. (5.10-10) for a third-order system.

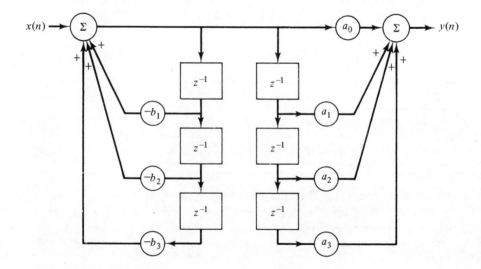

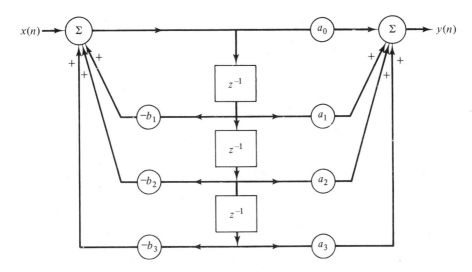

Figure 5.10-4. Direct form 2 realization.

where the $H_i(z)$ are first- or second-order filters. The factored form suggests a cascade connection as shown in Fig. 5.10-5. Each of the elements of this cascade may evidently be realized as a first- or second-order filter having the direct form 2 shown in Fig. 5.10-4, and then the cascade will have a minimum number of delay elements as well. Other possibilities are parallel combinations [obtained by expressing $H(z)$ as a sum of terms] or any combination of series and parallel connections.

Figure 5.10-5. Cascade connection.

Although all these different forms are described by the same equations, they do not perform the same way in practice because of the effects of finite word size.† For instance, there is a severe coefficient sensitivity problem with the two direct forms shown in Figs. 5.10-2 and 5.10-4 when the poles of $H(z)$ are close together or near the unit circle. Conversely, it turns out that cascade connections in which poles and zeros are distributed among the various sections so that poles and zeros that are close to each other in the z plane are placed in the same section of the cascade tend to be less sensitive to word-size noise effects [5-13]. The general problem of how best to arrange a digital

---

†This is a much less severe problem with analog filters where different configurations of the block diagram generally result in essentially the same output.

filter so as to minimize finite-word-size effects is clearly very important, but it is beyond the scope of this necessarily very brief treatment. Readers interested in more details are referred to one of a number of excellent references ([5-2], [5-6], [5-7]).

5.11 Nonrecursive (FIR) Filters

A widely used class of filters does not use output feedback. These filters are therefore nonrecursive and have a finite impulse response (FIR). One simple form of FIR structure is obtained by direct implementation of the discrete convolution:

$$y(n) = \sum_{m=-\infty}^{\infty} h(m)x(n-m). \tag{5.11-1}$$

Suppose that $h(m) = 0$ for $m < 0$, and also that for $m \geq N$, $h(m)$ has decayed to a value small enough to be neglected. Because of finite word size there is always such an N if $h(m)$ is a stable impulse response, but in practice a smaller number may be used to reduce the length of the filter. Then Eq. (5.11-1) becomes

$$y(n) = \sum_{m=0}^{N-1} h(m)x(n-m) = \sum_{m=0}^{N-1} a_m x(n-m). \tag{5.11-2}$$

This is a difference equation just like Eq. (5.10-4), but it is nonrecursive since $y(n-m)$ does not appear on the right. The Z transform of Eq. (5.11-2) has all its poles at the origin; hence it cannot represent an unstable system. The length of the impulse response is exactly N samples.

The structure suggested by Eq. (5.11-2) is shown in Fig. 5.11-1 and is often referred to as a *transversal filter*. Another designation is *tapped delay-line filter*; this name comes from the fact that the string of delays shown in Fig. 5.11-1 acts like a discrete tapped delay line. In fact analog versions of this filter that use a tapped delay line and continuously adjustable gains have been used [5-14]. In principle the transversal filter can realize any arbitrary causal impulse response. Questions of stability or realizability do not arise.

Figure 5.11-1. Transversal filter.

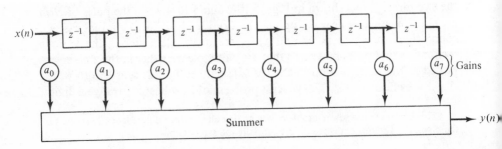

There are no difficulties in realizing impulse responses that have arbitrary jumps or that have radically different forms for different values of time n. Also by defining the input sequence as the signal somewhere near the middle of the delay line (i.e., by considering the input only after a fixed delay) it is possible to get an impulse response with finite values for negative argument. The resulting "noncausality" is essentially mathematical; the filter is still causal in real-time operation.

Transversal filters are easily made time-variable by varying the a's. There are simple algorithms (see Chapter 11) for automatically adjusting the a's so that some overall performance criterion is minimized. Such systems find considerable use in adaptive equalizers† ([5-14], [5-15]).

The major disadvantage of the transversal filter is the large number of delay elements and multipliers that are required. For instance, an IIR filter can be designed to have a fourth-order Butterworth response with no more than eight delay elements and about the same number of multipliers and adders. The number of delay elements and multipliers needed for an FIR design tends to be very much greater. This is particularly true for filters having a sharp cutoff characteristic, since they have a long impulse response. The advent of large-scale integration has, however, greatly lessened the disadvantages of complexity, and it is possible to construct special-purpose digital circuits for FIR filters that occupy only a single integrated circuit chip and are therefore physically no larger than a corresponding IIR filter. A simple digital circuit involving only two shift-register memories and a very minimum of additional circuitry has, in fact, been designed to implement the FIR algorithm ([5-6], p. 543). The arrangement for $N = 8$ is shown in Fig. 5.11-2. At the moment shown, the accumulator has been cleared and $a_7 x(n - 7)$ is entered. During the first shift cycle a new value of $x(n)$ enters at the left, and $x(n - 7)$ is shifted off at the right. During the remaining seven shift cycles both shift registers circulate, and it is left as an exercise to show that after a total of eight shifts $y(n) = \sum_{m=0}^{7} a_m x(n - m)$ is in the accumulator register and can be shifted out. After $y(n)$ is shifted out a new sample of $x(n)$ enters, and a new sample of $y(n)$ leaves the system only once per complete shift cycle. Thus for an N-stage filter the shift-register speed must be N times the sampling frequency of the signal. With current technology it is possible to obtain shift-register memories with more than 1000 stages, switchable at rates in excess of 20 MHz in a single IC chip. Thus a 1000-point FIR filter capable of sampling rates of 20 kHz is quite feasible. Evidently higher speeds can be attained at the expense of greater complexity by paralleling additional units.

The circuit shown in Fig. 5.11-2 using shift-register memories can, of course, also be implemented with random-access memories, i.e., with a small general-purpose computer, possibly involving a microprocessor. The advan-

---

†Adaptive filters are discussed in Chapter 11.

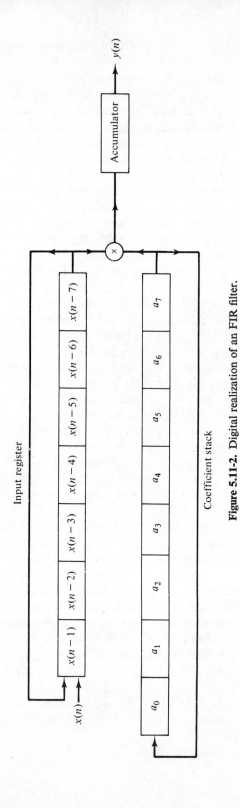

Figure 5.11-2. Digital realization of an FIR filter.

tage of the shift-register memory is, at the moment, its considerably lower cost. Similar simple circuits can be designed for IIR filters. (See [5-6], Chap. 11.)

FIR filters can be designed to have precise linear phase characteristics. As pointed out in Chapter 3, this means that except for a delay corresponding to the slope of the phase vs. frequency curve of the transfer function the input is reproduced exactly at the output. All that is needed for a filter to have a linear phase response is that for $n = 0, 1, \ldots, (N/2 - 1)$

$$h(n) = h(N - 1 - n), \tag{5.11-3}$$

where, for the sake of simplicity, we assume N to be even. (A slightly different result holds for odd N.) To show that Eq. (5.11-3) implies linear phase, write the Z transform:

$$
\begin{aligned}
H(z) &= \sum_{n=0}^{(N/2)-1} h(n)z^{-n} + \sum_{N/2}^{N-1} h(n)z^{-n} \\
&= \sum_{n=0}^{(N/2)-1} [h(n)z^{-n} + h(n)z^{-(N-1-n)}],
\end{aligned}
\tag{5.11-4}
$$

where the second step follows if one makes the change of variable $n' = N - 1 - n$ in the second summation and then uses Eq. (5.11-3). Evaluating $H(z)$ for $z = e^{j\omega}$ gives

$$
\begin{aligned}
H(e^{j\omega}) &= e^{-j\omega(N-1)/2} \sum_{n=0}^{(N/2)-1} h(n)\{e^{j\omega[n-(N-1)/2]} + e^{-j\omega[n-(N-1)/2]}\} \\
&= 2e^{-j\omega(N-1)/2} \sum_{n=0}^{(N/2)-1} h(n) \cos \omega \left(n - \frac{N-1}{2}\right).
\end{aligned}
\tag{5.11-5}
$$

Since the summation is real, phase shift is contributed only by the factor $\exp\left[-j\omega(N-1)/2\right]$, and it is seen to be linear with slope $-(N-1)/2$. An FIR filter structure suggested by Eq. (5.11-4) is given in Fig. 5.11-3 (for even N).

The single delay line filter shown in Fig. 5.11-1 is the FIR counterpart of one of the two direct forms for IIR filters shown in Fig. 5.10-1, and when the Z transform of the filter response is written as a product or sum, the resultant expressions suggest filter realizations by cascade or parallel structures here as well. It has been shown [5-16] that, as with IIR filters, the cascaded form of the FIR filter tends to be less sensitive to coefficient errors and finite-word-size effects than the direct form.

Frequency Sampling Structure

This is another form of nonrecursive filter structure. It is suggested by the relation between the Z transform and the discrete Fourier transform, developed in Sec. 5.9 [Eq. (5.9-6)]. If instead of $X(z)$ we write $H(z)$ in this

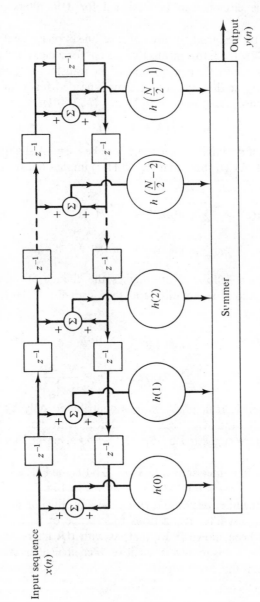

Figure 5.11-3. FIR filter having a linear phase characteristic.

equation we obtain

$$H(z) = \frac{1 - z^{-N}}{N} \sum_{k=0}^{N-1} \frac{\tilde{H}(k)}{1 - z^{-1}e^{j2\pi k/N}}, \tag{5.11-6}$$

where the $\tilde{H}(k)$ are the DFT coefficients corresponding to the impulse response sequence $h(n)$ of the desired filter. This expression suggests the filter structure shown in Fig. 5.11-4.

Since the zeros of the function $1 - z^{-N}$ are the N roots of 1, i.e., $e^{j2\pi n/N}$ for $n = 0, 1, \ldots, N - 1$, each of the poles in the parallel branches is in theory exactly cancelled by a zero contributed by the series part of the circuit, and therefore the transfer function has no nonzero poles. (It does have N poles at the origin as the transfer functions of all FIR filters do, however.) In practice, because of finite-word-length effects, this cancellation is not perfect. Also it is undesirable to have poles right on the unit circle since this results in stability problems. Therefore the multipliers $e^{j2\pi k/N}$ shown in Fig. 5.11-4 are in prac-

Figure 5.11-4. Frequency sampling filter.

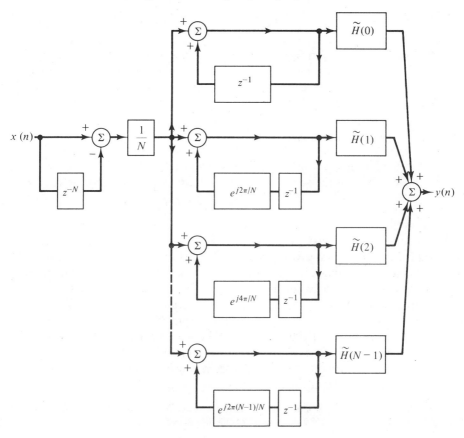

tice replaced by $re^{j2\pi k/N}$, where r is a real number slightly smaller than unity. For both of these reasons the filter is not exactly an FIR filter. It is referred to as a frequency sampling filter because the basic coefficients $\tilde{H}(k)$ are the values of the filter frequency response $H(e^{j\omega})$ sampled at N equally spaced points around the unit circle.

The chief advantage of this filter is that it can result in a much simpler structure than the direct, or transversal, form, especially when the desired filter is a low-pass, high-pass, or band-pass filter with one or more stopbands. Since the frequency samples in the various stopbands are zero, the corresponding parallel branches can be omitted. Thus a narrowband filter may require only a small number of branches—in contrast to the transversal form where a narrowband filter with a sharp cutoff characteristic tends to be especially long. Another possible advantage is that if several filters with the same number N of frequency samples are used in a filter bank, the $-z^{-N}$ feedforward and the feedbacks in the individual branches need to be implemented only once, with different sections of the bank differing only in the multipliers $\tilde{H}(k)$.

Since the $\tilde{H}(k)$ and also the factors $re^{j2\pi k/N}$ are complex, the hardware implementation of the circuit shown in Fig. 5.11-4 requires complex multipliers and adders. By making use of some of the symmetry properties of the Z transforms of real impulse responses, the structure can be modified so that only real operations are needed. For details of this procedure, see [5-7], Sec. 4.5.

5.12   Fast Convolution

The convolution equation (5.10-1) is referred to as "slow" convolution to distinguish it from the so-called fast convolution symbolized by the relation

$$y(n) = \mathcal{F}^{-1}\{\mathcal{F}[h(n)] \cdot \mathcal{F}[x(n)]\}. \tag{5.12-1}$$

The symbols $\mathcal{F}$ and $\mathcal{F}^{-1}$ refer here† to discrete Fourier transformation (DFT) and inverse DFT, respectively, and the reason Eq. (5.12-1) is called "fast" is that the FFT is used to perform the actual transformation. The block diagram of a filter employing fast convolution is shown in Fig. 5.12-1. The input sequence goes into a buffer memory holding N words. When this is full it dumps the N values of $x(n)$ into the FFT processor and then commences to store the next N input samples. The FFT output can be manipulated in any arbitrary way by the digital processor. There are no realizability, phase, or stability constraints. The result is then inverse-transformed in the second FFT box and loaded into the output buffer from which the output samples can be taken at some constant rate. Note that even if the two FFT processors and

---

†In other places in this book $\mathcal{F}$ refers to the *continuous* Fourier transform operator. The usage is made self-evident by the context of the discussion.

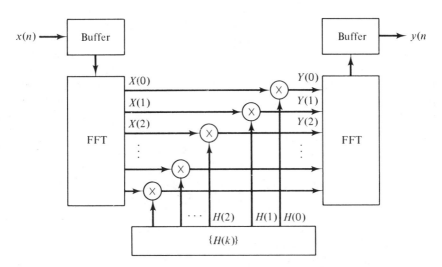

Figure 5.12-1. Digital filter employing fast convolution.

the multiplication by $H(k)$ were instantaneous, there would be a minimum delay between input and output of N samples because of the requirement of filling the input buffer before applying the FFT. Thus the elimination of the realizability constraints is bought at the price of a delay, just as with the FIR filter. In other words, "there is no free lunch." However, because of the general availability of FFT processors either in hardware or in software, the fast convolution approach is widely used in practice.

Certain precautions must be observed in the application of the fast convolution technique. These have to do with sampling errors, aliasing errors, the fact that the DFT implicitly deals with harmonic sequences, etc. All of these considerations have already been discussed in Sec. 5.5, where the DFT was discussed in detail.

5.13 Summary

In this chapter we have studied the properties of discrete linear systems (DLS), and we have shown that a DLS can be characterized by a transmission matrix **H** instead of the impulse or step-function responses which are commonly used for continuous systems. For systems that are time-invariant, we saw that a constant-coefficient *difference* equation can be used to relate output to input.

We investigated the properties of the DFT and found that the convolution-multiplication property, so useful in the continuous case, carried over to the discrete case as well. We saw that the DFT can be computed rapidly and efficiently with the FFT algorithm.

We found that the Z transform, which is essentially analogous to the Laplace transform in the continuous case, was a useful tool for the analysis of discrete filters. With respect to the latter, we studied both recursive and nonrecursive structures, and we showed that the former will be unstable if its Z-transform poles lie outside the unit circle. Nonrecursive, finite impulse response filters are fundamentally stable since the impulse response is N unit delays long.

We ended the chapter by considering the technique known as fast convolution, which uses the FFT algorithm for rapid signal processing.

PROBLEMS

5-1. Write, or at least indicate, the form of the most succinct transmission matrix **H** for the following cases:
a. **H** noncausal, **x** extends from $k = -\infty$ to $k = +\infty$.
b. **H** noncausal, **x** is nonzero for $1 \le k \le K$.
c. **H** causal, **x** extends from $k = -\infty$ to $k = +\infty$.
d. **H** causal, **x** nonzero for $1 \le k \le K$.

5-2. Consider a discrete system with a square transmission matrix **H**. How many independent elements are there in **H** when the system is
a. Noncausal, non-time-invariant.
b. Causal, non-time-invariant.
c. Causal, time-invariant.

5-3. Use discrete convolution to compute the sequence $\{y_n\}$ when x_k and h_k are as given below:

k	0	1	2	3	4	5	6	7	8→
x_k	0	1	1	1	1	1	0	all zeros	
h_k	1	1	1	1	0	0	0	all zeros	

5-4. a. Write a difference equation that describes the system in Fig. P5-4.

Figure P5-4

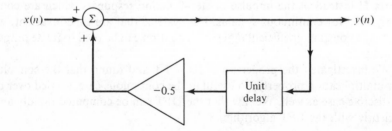

    b. Compute $y(n)$, by recursion, when $x(n) = n^2$.

    c. Compute $y(n)$ by solving the difference equation in part a.

5-5. a. Write a difference equation that describes the second-order system shown in Fig. P5-5.

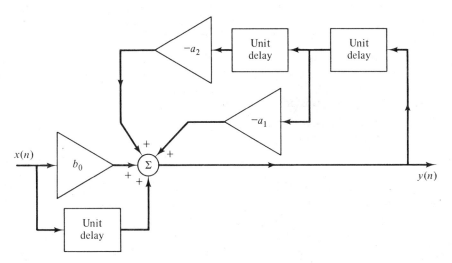

Figure P5-5

    b. Let $y(-2) = y(-1) = 0$. Solve for $y(n)$ by recursion when

$$x(n) = \begin{cases} 1, & 0 \le n \le 3 \\ 0, & \text{otherwise,} \end{cases}$$

    and $b_0 = 1$, $a_1 = \frac{1}{2}$, and $a_2 = 2$. Plot the output.

5-6. Demonstrate that Eqs. (5.3-11) and (5.3-12) form a valid discrete Fourier transform pair.

5-7. Demonstrate for arbitrary k that

    a. $X(-k) = X^*(k)$

    b. $X\left(\dfrac{N}{2} - k\right) = X^*\left(\dfrac{N}{2} + k\right),$

    where $X(\cdot)$ is the DFT and the sequence $\{x(i)\}$ is real. Assume N is even.

5-8. The discrete system counterpart of the derivative is the *difference*. For the sequence $\{x(i)\}$ the first forward difference [written $\Delta x(i)$] is the sequence $\{x(i + 1) - x(i)\}$. The backward difference [written $\nabla x(i)$] is $\{x(i) - x(i - 1)\}$.

    a. Find the DFT for the forward difference $\Delta x(i)$.

    b. Find the DFT for the backward difference $\nabla x(i)$.

5-9. Find the DFT for the product $x(i)y(i)$.

5-10. A waveform $x(t) = \cos{(2\pi/T_0)}t$ is sampled by the comb function

$$g(t) = \sum_{n=-\infty}^{\infty} \delta(t - nT).$$

a. Sketch $x(t)g(t)$, assuming that $T \simeq T_0/10$.
b. Assume a rectangular truncation function

$$p(t) = \text{rect}\left[\frac{t - (T_0 - T)/2}{T_0}\right],$$

and sketch $|P(f)|$.
c. Sketch $h(t) \equiv x(t)g(t)p(t)$ and $H(f) = \mathfrak{F}[h(t)]$.
d. Sketch $H(f) \sum_{n=-\infty}^{\infty} \delta[f - (n/T_0)]$, and compare with $X(f)$. What does the inverse transform of the sampled spectrum look like?

5-11. Repeat Prob. 5-10, except choose a truncation interval that covers slightly more than one period. For example, choose

$$p(t) = \text{rect}\left[\frac{t - (T_0 - T/2)/2}{T_0 + T/2}\right].$$

5-12. Write expressions for each of the waveforms shown in Fig. 5.5-1, and carefully label all significant parameters such as zero crossing, sampling widths, etc. Justify that neither the DFT nor the retransformed sampled wave need resemble the original spectrum and signal.

5-13. Show that convolving a DFT signal spectrum with the discrete generalized Hamming window is equivalent to replacing every $X(k)$ in the original sequence with the weighted average

$$\alpha X(k) + \frac{1 - \alpha}{2}[X(k + 1) + X(k - 1)].$$

5-14. Establish the shifting property of the Z transform given in Eq. (5.7-8).

5-15. The one-sided Z transform for a function $x(n)$ is given by

$$F(z) = \frac{1 + 2z}{z^2 + z}.$$

Determine the region of convergence, and compute $x(n)$ for all n.

5-16. Compute the discrete-time signal associated with

$$X(z) = \frac{1}{(z - 1)(z - 2)}$$

if $X(z)$ represents the one-sided Z transform. Describe the regions of convergence.

5-17. Starting with Eq. (5.9-7), prove that $X(e^{j\omega})$ can be written as in Eqs. (5.9-8) and (5.9-9). *Hint*: Use the fact that $(-1)^{2k} = 1$ for $k = 0, 1, 2, \ldots$.

5-18. Prove that

$$H(z) = \frac{\text{response to } z^n}{z^n},$$

where $H(z) \equiv Z[h_k]$. State your assumptions.

5-19. Prove that the realizations in Fig. 5.10-3 or 5.10-4 are equivalent to the one shown in Fig. 5.10-2. *Hint*: In Fig. 5.10-3, define an auxiliary variable $w(n)$ as the signal in the direct (undelayed) link.

5-20. Prove that the Z transform of an FIR filter with impulse response $h(m)$ that satisfies

$$h(m) = \begin{cases} =0, & m < 0, m \geq N \\ \neq 0, & \text{otherwise,} \end{cases}$$

has its poles at the origin. Why is stability therefore not a problem?

REFERENCES

[5-1] G. D. BERGLAND, "A Guided Tour of the Fast Fourier Transform," *IEEE Spectrum*, pp. 41–52, July 1969.

[5-2] SAMUEL D. STEARNS, *Digital Signal Analysis*, Hayden, New York, 1975.

[5-3] R. B. BLACKMAN and J. W. TUKEY, *The Measurement of Power Spectra*, Dover, New York, 1958.

[5-4] J. W. COOLEY and J. W. TUKEY, "An Algorithm for the Machine Calculation of Complex Fourier Series," *Math. Comput.*, **19**, p. 297, April 1965.

[5-5] G.A.E. SUBCOMMITTEE ON MEASUREMENT CONCEPTS, "What is the Fast Fourier Transform?," *IEEE Trans. Audio Electroacoustics*, **AU-15**, pp. 45–55, June 1967.

[5-6] LAWRENCE R. RABINER and BERNARD GOLD, *Theory and Application of Digital Signal Processing*, Prentice-Hall, Englewood Cliffs, N.J., 1975.

[5-7] ALAN V. OPPENHEIM and RONALD W. SCHAFER, *Digital Signal Processing*, Prentice-Hall, Englewood Cliffs, N.J., 1975.

[5-8] R. V. CHURCHILL, *Introduction to Complex Variables and Applications*, McGraw-Hill, New York, 1948, p. 102.

[5-9] RALPH J. SCHWARZ and BERNARD FRIEDLAND, *Linear Systems*, McGraw-Hill, New York, 1965, Chap. 8.

[5-10] J. R. RAGAZZINI and G. F. FRANKLIN, *Sampled Data Control Systems*, McGraw-Hill, New York, 1958, Chap. 4.

[5-11] H. FREEMAN, *Discrete Time Systems*, Wiley, New York, 1965.

[5-12] E. I. JURY, *Theory and Application of the Z-Transform Method*, Wiley, New York, 1964.

[5-13] L. B. JACKSON, "Roundoff Noise Analysis for Fixed-Point Digital Filters Realized in Cascade or Parallel Form," *IEEE Trans. Audio Electroacoustics*, **AU-18**, pp. 107–122, June 1970.

[5-14] R. W. LUCKY, "Automatic Equalization for Digital Communication," *Bell System Tech. J.*, **44**, pp. 547–588, April 1965.

[5-15]   J. H. CHANG and F. B. TUTEUR, "A New Class of Adaptive Array Processors," *J. Acoust. Soc. Am.*, **49**, No. 3 (Part 1), pp. 639–649, March 1971.

[5-16]   O. HERMANN and H. W. SCHUESSLER, "On the Accuracy Problem in the Design of Nonrecursive Digital Filters," *Arch. Elek. Ubertragung*, **24**, pp. 525–526, 1970.

6

Amplitude Modulation Systems and Television

6.1 Introduction

It is no exaggeration to state that in the communication sciences, modulation holds a central place. The terms amplitude modulation and frequency modulation are used by every layperson at one time or another, even if their only associations with these words are popular music for the former and classical music for the latter. Nevertheless, the word modulation has a distinct technical meaning and was defined by an appropriate committee of the IRE [6-1] as "the process . . . whereby some characteristic of a wave is varied in accordance with another wave." We shall be concerned with *controlled* modulation, i.e., the *desired* and controlled shifting of the spectrum of a message wave which is usually baseband and contains information of interest to humans. The term baseband refers generally to a low-pass wave such as simple speech or, as in more complicated systems, a multiplex wave consisting of many low-pass waves.

There are two fundamental reasons we modulate. One has to do with the laws of electromagnetic propagation which require that the size of the radiating element be a significant fraction of the wavelength of the signal to be transmitted. Thus the transmission of a 1000-Hz signal by a quarter-wave antenna would require a radiating element 75 kilometers long if we didn't modulate the 1000-Hz signal onto a high-frequency carrier. There exists a mismatch or gap between the frequencies that the human ear can detect and the frequencies at which electromagnetic energy can be efficiently radiated. Modulation bridges this gap.

The other main reason for modulating is the need for simultaneous transmission of different signals. The signals of interest to a human are primarily in a frequency band that spans from tens of hertz to several thousand hertz. If we didn't modulate, we could only broadcast one baseband signal in any locality at a given time; simultaneous transmission of more than one signal would cause the signals to overlap without hope of separation. Through use of modulation, however, we can transmit many messages over the same medium and still ensure their separability at the receiver. It is the multichannel capability furnished by modulation which enables us to have many radio and television channels in the same locality at the same time. The same principles are in use when transmitting many telephone messages over the same cable. The separation of the messages by different carrier assignment is sometimes called frequency-division multiplexing. In the case of telephoning, signals are also separated in time by a process called time-division multiplexing (TDM), which is not a carrier-modulation technique. TDM was discussed in Sec. 4.4.

There are other, secondary, reasons we modulate that have to do with obtaining noise reduction, limiting the size and weight of circuits and components, and other factors. Spurious modulation such as occurs to transmitted signals during electrical storms and modulation due to faulty practices and components are of course also commonplace, but we shall not discuss them further.

Since modulation involves the generation of new frequency components (i.e., shifting of spectral bands), we cannot modulate by using linear, time-invariant systems. Modulation is generally done by using time-varying linear systems or systems using one or more nonlinear elements.

In this chapter we shall discuss amplitude modulation (AM) and related modulation methods. AM is very easy to understand and is essentially a direct translation of the message spectrum to the carrier frequency. Closely related schemes are double-sideband (DSB), single-sideband (SSB), and vestigial sideband (VSB). All of these modulation methods are closely related; either both sidebands, or one sideband, or fractions of both, with or without a carrier signal, are transmitted between source and receiver. At the receiver, we must demodulate the signal and recover the baseband message $s(t)$, also sometimes called the information.† A block diagram of an ideal communication system is shown in Fig. 6.1-1.

Perhaps the most widely used communication system of all and one that uses all of these modulation methods is color television. Black/white and color TV are therefore discussed at length at the end of this chapter.

---

†Also sometimes loosely called *intelligence*.

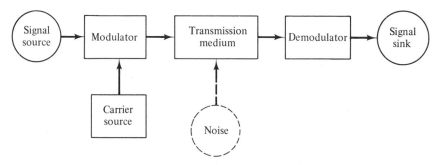

Figure 6.1-1. Ideal system that ignores all undesirable effects except channel noise.

6.2 Amplitude Modulation

An ordinary AM wave can be constructed by adding to the baseband signal $s(t)$ a constant and multiplying the sum by a sine wave. Our set of messages $\{s(t)\}$ will be assumed normalized and band-limited so that the set is described by $\{s(t): |s(t)|_{\max} \leq 1; S(f) = 0, |f| > W\}$. W is the bandwidth of the signal. The normalization $|s(t)|_{\max} \leq 1$ enables us to write for the AM signal

$$x(t) = A_c[1 + ms(t)]\cos(2\pi f_c t + \phi), \tag{6.2-1}$$

where $x(t)$ is the AM signal, m is the modulation index, f_c is the carrier frequency, A_c is the carrier amplitude, and ϕ is a constant-phase term that can be set to zero by appropriate choice of the time origin. The number m should not exceed the value of unity; for $m \geq 1$, $ms(t) + 1$ can go negative if $s(t)$ takes its most negative value. This means that $x(t)$ has undergone a 180° phase reversal and will lead to distortion in the recovery of $s(t)$ *if the simplest AM detection, i.e., envelope detection, is used.* When $m = 1$, we have 100% modulation; for $m > 1$, we have *overmodulation,* which implies distortion. Figure 6.2-1 shows AM waveforms for various values of m and a sinusoidal modulating signal.

The Fourier transform of Eq. (6.2-1) gives the spectrum of $x(t)$. It is (with ϕ set to zero for convenience)

$$X(f) = \frac{A_c}{2}\delta(f - f_c) + \frac{A_c}{2}\delta(f + f_c) + \frac{mA_c}{2}S(f - f_c) + \frac{mA_c}{2}S(f + f_c). \tag{6.2-2}$$

Figure 6.2-2(a) shows a possible spectrum for $s(t)$; Fig. 6.2-2(b) shows the spectrum of $x(t)$. Several interesting results are now apparent. The bandwidth of the AM wave is

$$B = 2W, \tag{6.2-3}$$

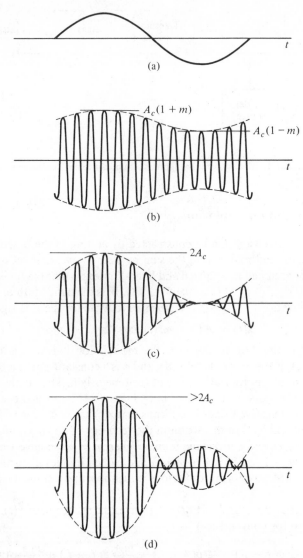

Figure 6.2-1. AM waveforms: (a) $s(t)$; (b) $x(t)$ for $m < 1$; (c) $x(t)$ for $m = 1$; (d) $x(t)$ for $m > 1$. Note 180° phase reversal in carrier.

i.e., twice as great as the bandwidth of the original signal. Also the AM wave consists of a *pair* of sidebands each of width W. Foldover distortion occurs when the carrier frequency is not high enough (Fig. 6.2-3). The foldover phenomenon also occurs in sampling theory, i.e., when the sampling rate is too low (Sec. 4.2). For example, in Fig. 4.2-5 it can be seen that the high-

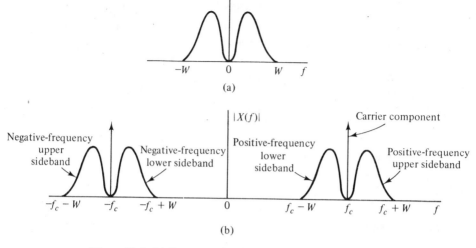

Figure 6.2-2. (a) Spectrum of $s(t)$; (b) spectrum of the AM wave.

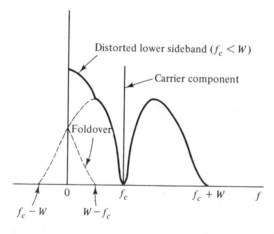

Figure 6.2-3. Sideband foldover if $f_c < W$. The low frequencies of the lower sideband (which contain the high-frequency information in the baseband) add to the lower frequencies and produce distortion.

frequency components of a displaced spectrum fold into the lower frequencies of its neighbor. For no foldover distortion in AM we must have

$$f_c > W.$$

We can now describe some of the other modulation schemes that are closely related to AM. If the carrier component in Eq. (6.2-2) is absent, we obtain a wave of the form

$$x(t) = Bs(t) \cos 2\pi f_c t, \qquad (6.2\text{-}4)$$

where m has been dropped. The wave in Eq. (6.2-4) is known as double-sideband suppressed carrier (DSBSC) or just double-sideband (DSB) for short. Its spectrum is identical with the spectrum in Fig. 6.2-2 except for the missing carrier. If, in addition, we remove either of the sidebands, we obtain single-sideband modulation (SSB). In either case, a small carrier component (sometimes called a vestigial carrier) may be transmitted to synchronize a *local* oscillator to the carrier frequency to simplify demodulation. If a large portion of one sideband is sent with a small portion of the other sideband, we eliminate the abrupt frequency cutoff required in SSB and still obtain many of its advantages. This scheme is called vestigial sideband (VSB) and is in use in TV broadcasting.

6.3 Modulators

The generation of an AM wave can be conceptually described as in Fig. 6.3-1. The key operation is the ideal product between the carrier signal and the biased signal $1 + ms(t)$. The product of two signals can be obtained in a number of ways. One way is to use one signal to control the gain of an active device and use the other signal as input. Another way is to use one or more nonlinear elements that furnish the desired product but also, frequently, generate additional terms that need to be removed by filtering.

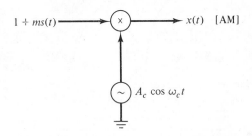

Figure 6.3-1. Ideal generation of an AM wave.

Multipliers

An analog multiplier is a three-port device that produces at the output port a signal that is proportional to the product of the two input signals present at its input ports. The variable transconductance multiplier (VTM) uses one signal to control the gain (transconductance) of an active device which then amplifies the instantaneous value of the other signal in proportion to the instantaneous value of the control signal. The active device has the configuration of an emitter-coupled differential amplifier except that one input is used to control the current in the common emitter. A practical realization of a VTM is the monolithic unit shown in Fig. 6.3-2 (No. AD 530, Analog Devices, Inc.). A discussion of this circuit is given in [6-2] and [6-3]. It is not too difficult to show that the output signal E_0 is given by

$$E_0 = KV_XV_Y, \qquad (6.3\text{-}1)$$

where K is a constant and V_X and V_Y are the two input signals to be multiplied.

Another type of multiplier in common use is the logarithmic multiplier ([6-3], p. 233) which uses logarithmic amplifiers to multiply together two positive signals. This operation is called one-quadrant operation. A bipolar signal can be multiplied by another bipolar signal if these signals are first

Figure 6.3-2. Practical two-quadrant variable transconductance multiplier. (*Adapted from [6-3], by permission.*)

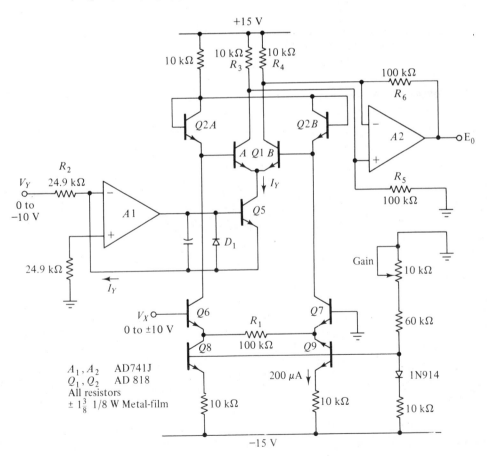

added to positive dc levels that exceed the magnitude of the largest negative excursion of the signals. Figure 6.3-3 shows a logarithmic amplifier. The purpose of the op-amp is to facilitate the addition of the signals generated by the logarithmic amplifiers and to provide isolation from the $\log^{-1}$ circuit.

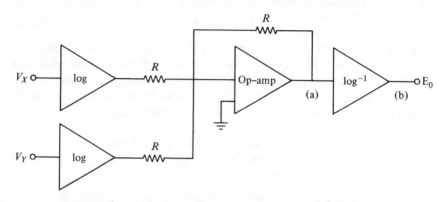

Figure 6.3-3. Logarithmic multiplier. At (a) we have $\log V_X + \log V_Y$. At (b) we have $E_0 = K V_X V_Y$, where K is a scale factor.

Square-Law Modulators

Semiconductor diodes and certain other types of nonlinear devices have a volt-ampere characteristic not unlike that shown in Fig. 6.3-4. When such a device is connected to a resistor, the voltage developed across the resistor can be described as a power series of the form

$$e_0 = \sum_{i=1}^{\infty} a_i e_i^i, \qquad (6.3\text{-}2)$$

Figure 6.3-4. Volt-ampere characteristic of *P-N* diodes and certain other nonlinear devices.

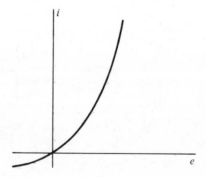

where e_0 is taken as the output voltage and e_i is the voltage impressed across the device. If $|e_i|$ is small enough we can generally approximate the transfer characteristic by retaining only the first two terms. Then we say that we have a square-law device. For such a device,

$$e_0 = a_1 e_i + a_2 e_i^2. \qquad (6.3\text{-}3)$$

Let the input consist of the sum of modulating signal plus carrier; i.e.,

$$e_i(t) = s(t) + \cos 2\pi f_c t.$$

The output is then

$$e_0(t) = a_1 s(t) + a_2 s^2(t) + a_2 \cos^2 2\pi f_c t$$

$$+ a_1 \left[1 + \frac{2a_2}{a_1} s(t) \right] \cos 2\pi f_c t, \qquad (6.3\text{-}4)$$

which can be written as

$$e_0(t) = A_c s(t) + a_2 s^2(t) + \frac{a_2}{2}(1 + \cos 4\pi f_c t)$$

$$+ \underbrace{A_c[1 + ms(t)] \cos 2\pi f_c t}_{\text{AM wave}}, \qquad (6.3\text{-}5)$$

where $A_c \equiv a_1$ and $m \equiv 2a_2/a_1$. The unscaled spectrum of $e_0(t)$ is shown in Fig. 6.3-5. A band-pass filter of bandwidth $2W$ centered at f_c will isolate the

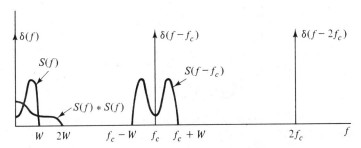

Figure 6.3-5. Spectrum of $e_0(t)$.

AM wave. The filter bandwidth can, of course, be larger provided it doesn't admit the other spectra. Note that for modulation with a square-law device we must have

$$f_c > 3W \qquad \text{(square-law modulation)}. \qquad (6.3\text{-}6)$$

A square-law diode modulator is shown in Fig. 6.3-6. The voltage e_i is assumed to be very small in order for the square-law model to be valid.

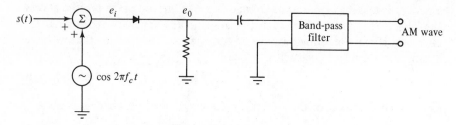

Figure 6.3-6. Square-law diode modulator.

Switch-Type Modulators

The square-law transfer characteristic assumed for diodes and other nonlinear devices such as nonlinear resistors, reactors, etc., is only an approximation and, in practice, it is possible to get distortion terms that fall within the passband and that cannot be removed by filtering. This problem can, to some extent, be overcome by switch-type modulators.

In a switch-type modulator, the modulating signal is switched either in polarity or on-off at the carrier rate. A polarity switch is shown in Fig. 6.3-7.

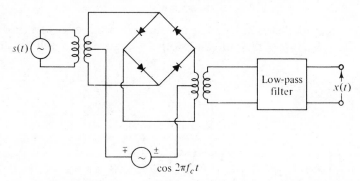

Figure 6.3-7. Switch-type modulator consisting of a ring diode arrangement.

The signal flow paths for the two carrier polarities are shown in Fig. 6.3-8. We assume that the amplitude of the switching signal is much greater than $|s(t)|_{\max}$.

The output signal $e(t)$ can be written as

$$e(t) = s(t)g(t), \qquad (6.3-7)$$

where

$$g(t) = \begin{cases} 1 & \text{for } \cos 2\pi f_c t > 0 \\ -1 & \text{for } \cos 2\pi f_c t < 0; \end{cases} \qquad (6.3-8)$$

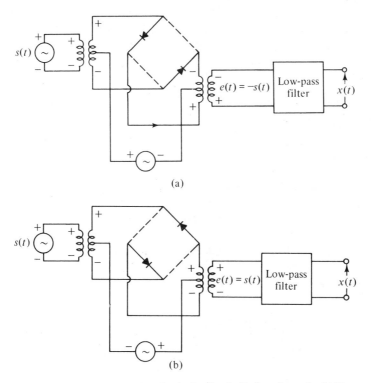

Figure 6.3-8. (a) Odd signal parity during first half of carrier cycle. (b) Even signal parity during second half of carrier cycle.

i.e., $g(t)$ is a square wave of unit amplitude and fundamental frequency f_c. We note that this (idealized) system is linear since

$$g(t)[s_1(t) + s_2(t)] = g(t)s_1(t) + g(t)s_2(t)$$
$$= e_1(t) + e_2(t). \qquad (6.3\text{-}9)$$

However, the system is *time-varying* since the output depends on the dynamical state of the system at time t. By using the Fourier series expansion of the square wave, the output signal $e(t)$ can be written as

$$e(t) = s(t)\left[\frac{4}{\pi} \cdot \sum_{n=0}^{\infty} (-1)^n(2n + 1)^{-1} \cos 2\pi(2n + 1)f_c t\right], \qquad (6.3\text{-}10)$$

and the positive frequency spectrum is as shown in Fig. 6.3-9. To separate the unwanted sidebands from the signal of interest we require that $3f_c - W > f_c + W$ or $f_c > W$. A filter centered at f_c with bandwidth $2W$ can be used to separate the modulated signal from the other terms. The output $x(t)$ has the form

$$x(t) = Bs(t) \cos 2\pi f_c t, \qquad (6.3\text{-}11)$$

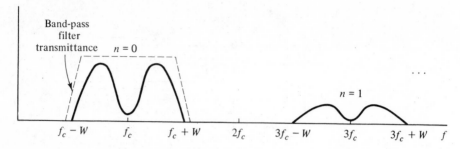

Figure 6.3-9. Spectrum of the output of a ring modulator.

which is identical with Eq. (6.2-4). Thus we see that balanced switching produces a DSB wave which can be converted to an AM wave by the addition of a signal $A_c \cos 2\pi f_c t$ with A_c sufficiently large (Fig. 6.3-10).

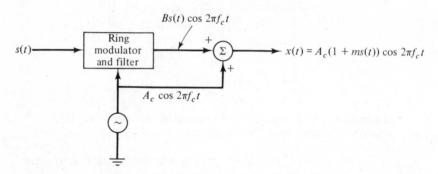

Figure 6.3-10. Generation of AM wave from DSB wave.

Balanced Modulators

The ring-diode modulator is but one example of a type of modulator known as a *balanced modulator*. The output of such a modulator is typically DSB. When the modulating element is a nonlinear device with transfer characteristic given by Eq. (6.3-3), with $a_1 \neq 0$, it is possible to produce a DSB wave and simulate an ideal multiplier through the use of the balanced configuration shown in Fig. 6.3-11.

The signal e_1 is proportional to $a_1(s + \xi) + a_2(s^2 + 2s\xi + \xi^2)$, while the signal e_2 is proportional to $a_1(-s + \xi) + a_2(s^2 - 2s\xi + \xi^2)$. The difference is proportional to $2a_1 s + 4a_2 s\xi$, and the $2a_1 s$ term is removed by the filter. The output is a pure DSB wave and simulates the product of $s(t)$ and the carrier $\xi(t) = \cos 2\pi f_c t$. If the constant $a_1 = 0$, there would be no need for band-pass filtering. However, very few nonlinear devices satisfy the equation

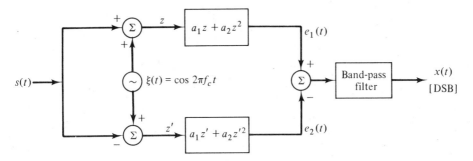

Figure 6.3-11. Use of a balanced modulator to generate DSB.

$y = x^2$. Note that if the two nonlinear devices in Fig. 6.3-11 are not matched, potentially troublesome terms such as $s^2(t)$ remain after summation.

A balanced modulator circuit available as an integrated-circuit package is shown in Fig. 6.3-12(a). In this circuit the carrier signal is applied to the bases of the four transistor switches Q_1, Q_2, Q_3, and Q_4, and it acts to switch on either pair Q_1, Q_4 or pair Q_2, Q_3. The modulating signal is applied to the emitters of the same four transistors via the two transistors Q_5 and Q_6. If transistors Q_5, Q_6 are operating in the linear region, then their collector voltages are proportional to their base voltages. When line 8 is positive, the Q_3 and Q_2 switches are closed, and the positive input appears at $V_0(+)$ and the negative input at $V_0(-)$. During the next half-cycle of the carrier the positive signal input goes to $V_0(-)$ and the negative signal input to $V_0(+)$. Thus the signal is in effect multiplied by a balanced square wave. Typical input and output wave shapes for this circuit are shown in Fig. 6.3-12(b).

Class C Amplifier Modulation

The modulators discussed so far are low-level (i.e., low-power modulation). For AM, high-level modulation (i.e., high-power modulation) is achieved with the aid of tuned class C amplifiers.

A class C amplifier is an amplifier for which the operating point is chosen so that the output current or voltage is zero for more than one half of an input sinusoidal signal cycle. A simplified version of a class C amplifier is shown in Fig. 6.3-13.

With the carrier signal off, the value of V_{BB} and R_B are so chosen that the transistor is well beyond cutoff (for an *n-p-n* transistor this requires that $V_{BE} < 0.1$ V for germanium and < 0.0 V for silicon). The transistor goes into conduction only near the positive peak portion of the carrier cycle. The transistor is thus switched on and off by the carrier at the carrier frequency, f_c. In the absence of $s(t)$, the output is a series of pulses which are filtered by the tuned *RLC* circuit (sometimes called a tank circuit) to produce a constant

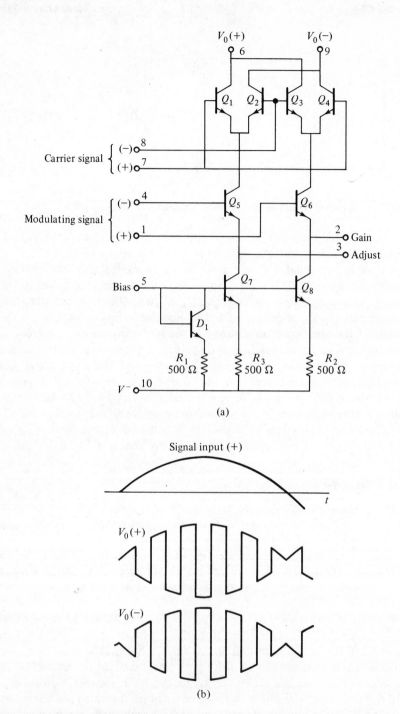

Figure 6.3-12. (a) Integrated circuit for balanced modulator-demodulator (Signetics 5596); (b) typical input and output wave shapes.

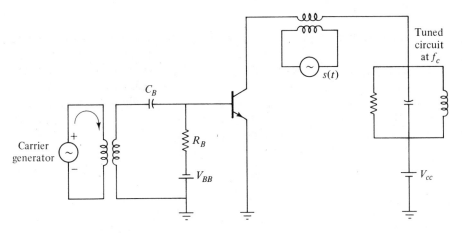

Figure 6.3-13. Simplified circuit of a class C amplifier.

amplitude collector current at the frequency f_c. When the collector supply voltage is varied by $s(t)$, the envelope of the collector current follows the variations in $s(t)$ and an AM wave is produced. Class C amplifiers are widely used for the generation of high-level AM signals. A simplified block diagram for an AM broadcast system is shown in Fig. 6.3-14. The level of $s(t)$ in class C modulation is generally much higher than in the circuits described earlier.

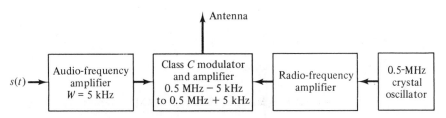

Figure 6.3-14. AM high-level broadcast system tuned to 500 kHz.

6.4 Detection of AM Waves

Synchronous Detection

From a conceptual point of view, synchronous detection is perhaps the simplest, although it is almost never used to detect AM waves. Nevertheless we shall illustrate how synchronous detection might be used to detect AM waves. With $x(t) = A_c[1 + ms(t)] \cos 2\pi f_c t$ representing the incoming AM wave, the product of $x(t)$ with a carrier gives

$$x(t) \cos 2\pi f_c t = \frac{1}{2} A_c[1 + ms(t)] \cos (4\pi f_c t) + \frac{A_c}{2}$$

$$+ \frac{1}{2} A_c ms(t). \tag{6.4-1}$$

The last term is the desired signal. A block diagram for the synchronous detector is shown in Fig. 6.4-1. In principle the generation of the carrier signal at the receiver can be done with a crystal oscillator† tuned to precisely the same frequency as the incoming carrier, by separation of the incoming carrier by narrowband filtering, or by tuning a local oscillator. Practical methods for obtaining a carrier wave at the receiver will be discussed later. With few exceptions, some form of synchronous detection is generally required to demodulate DSB, SSB, VSB, or some other derived form of AM. However, AM itself is detected with an *envelope detector*.

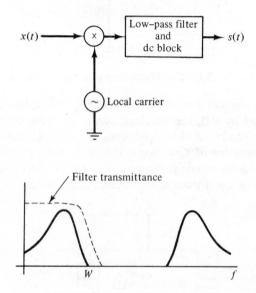

Figure 6.4-1. Synchronous detection of AM waves.

Homodyne techniques (i.e., where signals of the same frequencies are multiplied) are a special case of the more general *heterodyne* principle, which is widely used in AM receivers. Heterodyning is the process of generating new frequencies by multiplying carrier signals of different frequencies. Heterodyning leads to a frequency shift of the carrier signals; i.e., $\cos 2\pi f_1 t \cdot \cos 2\pi f_2 t = \frac{1}{2}[\cos 2\pi (f_1 + f_2) + \cos 2\pi (f_2 - f_1)t]$. One of the two signals thus produced is then filtered out. If the sum frequency is rejected, we say that the signal has been down-converted. If the difference frequency is rejected, the signal is said to be up-converted. Thus homodyning is down conversion to baseband (or up conversion to double the frequency). A block diagram of a

†However, there will be, in general, a phase difference between the local demodulating carrier and the incoming carrier signal. This could result in significant loss of detected power.

heterodyne system is shown in Fig. 6.4-2. To enable separation of the up-converted signal from the down-converted signal without sideband foldover we require that (assume $f_2 > f_1$)

$$\text{(i)} \quad f_2 - f_1 > W$$
$$\text{(ii)} \quad f_1 > W \qquad\qquad (6.4\text{-}2)$$
$$\text{(iii)} \quad f_2 > 2W.$$

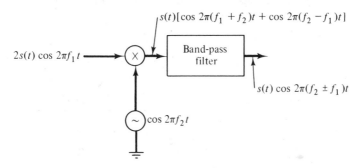

Figure 6.4-2. Heterodyning or frequency conversion.

The last condition follows from the first two. The product of the two signals can be achieved by the techniques discussed in Sec. 6.3.

An interesting modification of a heterodyne system, discussed by Carlson ([6-4], p. 191), is the so-called regenerative frequency divider, which takes an input frequency f_0 and produces an output frequency f_0/n. The system is shown in Fig. 6.4-3. We leave it to the reader (Prob. 6-12) to show that if the band-pass filter rejects the upper frequency $(2n - 1)f_0/n$, then the output is indeed f_0/n.

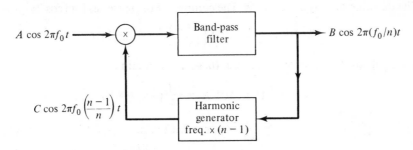

Figure 6.4-3. Regenerative frequency divider.

Frequency conversion of the form f/N where N is a multiple of 2 can be done with digital circuitry. For example, a four-stage ripple counter ([6-5], p. 301) can be used to produce the frequencies $f_0/2, f_0/4, f_0/8, f_0/16$ from an input pulse sequence of frequency f_0. Although the intermediate signals are square waves, appropriate band-pass filtering will generate sinusoids of the same frequency.

Detection with Rectifiers

A nonlinear operation on the AM signal can produce a signal which, after filtering, furnishes the message $s(t)$. The linear half-wave and linear full-wave rectifiers shown in Fig. 6.4-4 are devices useful for such operations. The

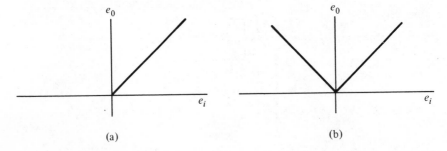

Figure 6.4-4. (a) Half-wave and (b) full-wave rectifier transfer characteristics.

linear full-wave rectifier has transfer characteristic $e_0 = |e_i|$. Hence denoting an AM wave by

$$x(t) = r(t) \cos 2\pi f_c t,$$

where $r(t) \equiv A_c[1 + ms(t)] \geq 0$, we can write for the output of the full-wave rectifier

$$z(t) = |x(t)| = r(t) |\cos 2\pi f_c t|. \qquad (6.4\text{-}3)$$

The dc value of $|\cos 2\pi f_c t|$ is $2/\pi$. The complete Fourier series for $|\cos 2\pi f_c t|$ is

$$|\cos 2\pi f_c t| = \frac{2}{\pi} - \frac{4}{\pi} \sum_{n=1}^{\infty} \frac{(-)^n}{4n^2 - 1} \cos 2\pi (2n) f_c t. \qquad (6.4\text{-}4)$$

If we substitute Eq. (6.4-4) into Eq. (6.4-3), we obtain

$$z(t) = \frac{2A_c}{\pi}[1 + ms(t)] - \frac{4A_c}{\pi}[1 + ms(t)]$$

$$\cdot \sum_{n=1}^{\infty} \frac{(-)^n}{4n^2 - 1} \cos 2\pi (2n) f_c t. \qquad (6.4\text{-}5)$$

It is clear that if $f_c > W$, an ideal low-pass filter will enable the separation of the term $2A_c[1 + ms(t)]/\pi$ from the other terms. A dc blocking capacitor will then remove the dc bias and leave us with $s(t)$. If $s(t)$ contains a dc term, that too will be removed. In TV, the loss of the dc component in $s(t)$ can be a problem, and dc restoration circuitry is required. A full-wave rectifier detector for AM signals is shown in Fig. 6.4-5.

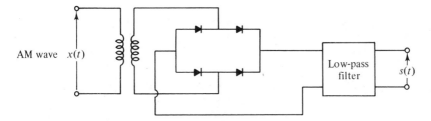

Figure 6.4-5. Detection of an AM wave with a full-wave rectifier.

The Envelope Detector

The simplest and most economical circuit for detecting AM waves is the half-wave rectifier. Despite the fact that, as we shall see in subsequent sections, both DSB and SSB have important advantages over AM, the popularity of AM over DSB and SSB is due mostly to the fact that AM waves can be detected by rectifiers, whereas both DSB and SSB require some form of synchronous detection.

An AM envelope detector with capacitive filtering is shown in Fig. 6.4-6. A thorough analysis of this circuit is given by Millman and Halkias ([6-6],

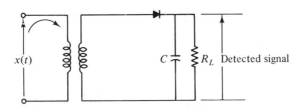

Figure 6.4-6. AM demodulation with an envelope detector.

p. 606). If the capacitor were not there, the output would simply be a half-wave-rectified version of the input (Fig. 6.4-7). However, with the capacitor filter in place, the output is essentially the input envelope with a small ripple superimposed. While the diode is conducting, the capacitor charges to the input voltage with a time constant given by the product of C and the resistance resulting from the parallel combination of R_L and the diode forward resistance R_f. Since R_f is assumed very small, for all practical purposes the input is impressed directly across the load. When the diode is reverse-biased, the capacitor discharges through R_L with a time constant equal to $R_L C$. The value of $R_L C$ should be large enough to smooth the waveform, i.e., prevent excessive ripple, but not so large as to prevent the load voltage from following the changes in the *envelope*. Figure 6.4-8 shows a closeup view of the output voltage variation in an envelope detector. The computations of the cut-in and cut-out times can be done graphically ([6-6], p. 609). Figure 6.4-8 greatly

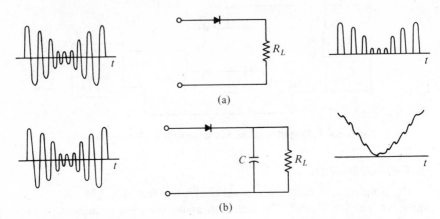

Figure 6.4-7. (a) Output without filtering; (b) output with filtering.

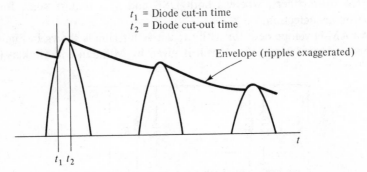

t_1 = Diode cut-in time
t_2 = Diode cut-out time

Envelope (ripples exaggerated)

Figure 6.4-8. Envelope variations in envelope detector.

exaggerates the rate of change of the envelope with respect to the carrier. At a carrier frequency of 1000 kHz and a highest audio modulating frequency of 20 kHz, there should be 50 carrier cycles for each audio cycle. Only when the audio frequency is an order (or orders) of magnitude lower in frequency than the carrier frequency does it make sense to speak of an envelope.

It should be obvious that envelope detection is fundamentally simpler and easier than homodyne detection. However, we can establish a rough equivalence between the two by looking at envelope detection from the following, approximate, point of view. The carrier waveform operates the diode as an on-off switch with switching function

$$g(t) \simeq \frac{1}{2} + \frac{2}{\pi} \sum_{n=0}^{\infty} (2n + 1)^{-1}(-1)^n \cos 2\pi(2n + 1)f_c t. \qquad (6.4-6)$$

If we ignore the loading effect of the RC low-pass filter at the output of the diode, the voltage developed across the output terminals is proportional to

$$e_0(t) = g(t)x(t). \tag{6.4-7}$$

Note that the product of $x(t)$ with the $n = 0$ component gives rise to a term that contains just $s(t)$. The addition of a low-pass filter removes all components but $s(t)$. Hence an approximate equivalent representation of the envelope detector can be constructed as in Fig. 6.4-9. When Fig. 6.4-9 is compared with Fig. 6.4-1, which shows synchronous detection, the relation between the two methods becomes more apparent.

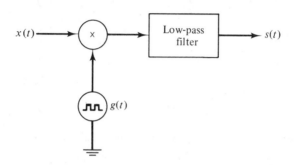

Figure 6.4-9. Equivalent representation of envelope detector which ignores loading of *RC* low-pass filter.

6.5 The Superheterodyne Receiver

There are several different types of AM receivers, but by far the most widely used is the superheterodyne (superhet) receiver shown in Fig. 6.5-1. The superheterodyne principle is used in low-cost as well as high-quality receivers.

The superhet receiver must basically carry out two functions: the amplification of the band containing the selected signal and the detection of the information $s(t)$. Superheterodyne operation refers to the frequency conversion from the variable RF (radio-frequency) to the fixed IF (intermediate-frequency) signal that is ultimately detected with an envelope detector.

In a typical situation, the signals appearing at the antenna are a superposition of a large number of radio, TV, other man-made, as well as non-man-made signals. Assume that the RF amplifier is tuned to the frequency f_c, and consider at point A, a specific AM wave of the form

$$e_A(t) = r(t) \cos 2\pi f_c t, \tag{6.5-1}$$

where

$$r(t) = a[1 + ms(t)], \quad a = \text{constant}.$$

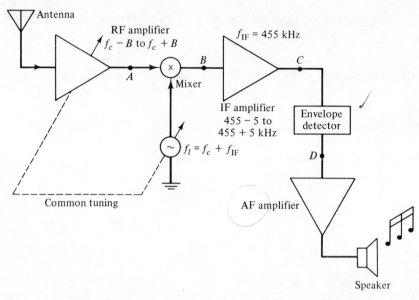

Figure 6.5-1. Principal components of the superheterodyne receiver.

At point B, the signal is $[\cos 2\pi f_{IF} t + \cos 2\pi (2f_c - f_{IF})t]$

$$e_B(t) = a_1 r(t)[\cos 2\pi f_{\mathrm{IF}} t + \cos 2\pi (2f_c + 2f_{\mathrm{IF}})t]. \qquad (6.5\text{-}2)$$

Note that the common tuning of the variable-frequency local oscillator with the RF amplifier is an ingenious way of moving the RF band past a fixed IF band-pass window. This results in much better band-pass characteristics for the IF stages, since the design of these units can be optimized for a fixed frequency. For example, the IF stage can be designed for a fixed relative bandwidth of 10/445 instead of having to handle a variable relative bandwidth, which is a significant complication.

The second term in the brackets in Eq. (6.5-2) represents the sum frequency and is rejected by the IF amplifier. At C, the signal has the form

$$e_C(t) = a_2 r(t) \cos 2\pi f_{\mathrm{IF}} t. \qquad (6.5\text{-}3)$$

The envelope detector generates the signal

$$e_D = a_3 r(t), \qquad (6.5\text{-}4)$$

and, finally, the AF (audio-frequency) amplifier amplifies $s(t)$ and drives the speaker.

It is possible to use the detected signal for automatic volume control (AVC). If the envelope variations are smoothed out completely, the output of the AVC circuit will be a dc signal proportional to the carrier amplitude. This signal can be used to control the gain of the IF stage so that if the carrier fades, the gain increases, thereby tending to keep the volume constant.

The bandwidth, $2B$, of the RF amplifier should be no less than $2W$ but may be larger since *adjacent-channel selectivity* is furnished by the IF stage. The partial selectivity of the RF stage furnishes *image-channel* rejection. This is important because otherwise image channels can be detected as well as the desired ones. To explain this phenomenon, note that without the selectivity of the RF stage, there are in fact two channels that will pass the IF stage and hence be detected. If the local oscillator frequency f_l is set at†

$$f_l = f_c + f_{IF},$$

then the RF frequency

$$f_{RF} = f_c$$

and its image

$$f_{RF}^{(i)} = f_c + 2f_{IF}$$

will fall within the band of the IF stage. The image is rejected by the RF stage—hence the name image-channel rejection.

The frequency band for standard AM broadcasting is 550–1600 kHz. The IF frequency is fixed at 455 kHz, and there is a 10-kHz channel bandwidth per station ([6-7], p. 21. 49). For perfect audible reproduction of speech and music, a frequency range of 30–15,000 kHz is desirable. Broadcast AM obviously falls far short of this, but high-quality broadcasting of AM signals provides a ± 2-dB response over the frequency range 30–4500 Hz. Figure 6.5-2 shows the locations in the spectrum of various signals.

Many new AM and AM/FM superheterodyne receivers now come with most of the active components located on a single 16-pin integrated-circuit

†The choice of $f_l = f_c + f_{IF}$ leads to a smaller tuning range for the local oscillator than $f_l = f_c - f_{IF}$. See Prob. 6-11.

Figure 6.5-2. Spectra of AM signals in the superheterodyne receiver. The adjacent station may pass the RF amplifier but will not pass the IF amplifier.

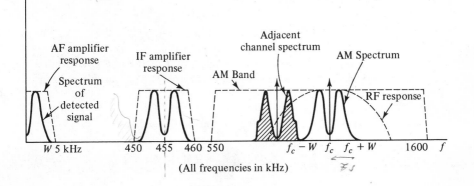

AF amplifier response

Spectrum of detected signal

IF amplifier response

AM Band

Adjacent channel spectrum

AM Spectrum

RF response

W 5 kHz   450   455   460 550 $f_c - W$ f_c $f_c + W$ 1600 f

(All frequencies in kHz)

chip. A description of such a circuit is furnished by W. Deil and R. J. McFadyen, in "Single-slice superhet", in the *IEEE Spectrum* of March 1977, vol. 14, No. 3, pp. 54–57.

*6.6 The Superheterodyne Principle in Spectrum Analysis

Practical spectrum analyzers are frequently of a type shown in Fig. 6.6-1. This configuration is closely related to the superheterodyne receiver. The amplifier/filter is centered at a *fixed* frequency f_0 and has a *fixed* bandwidth, Δ.

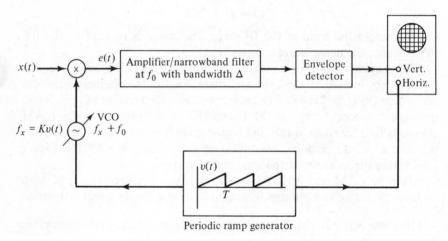

Figure 6.6-1. Scanning spectrum analyzer.

The device marked VCO is a voltage-controlled oscillator. Its frequency is proportional to the applied voltage. To understand the operation of this system we write $x(t)$ as a Fourier series:

$$x(t) = \sum_{n=1}^{\infty} 2|c_n| \cos (2\pi f_n t + \theta_n), \qquad (6.6\text{-}1)$$

where $f_n = (1/T_s)n$, T_s = period of the signal, and θ_n is the phase associated with the nth harmonic. Assume that the dc term in $x(t)$ has been filtered out. The signal $e(t)$ is given by

$$e(t) = \sum_{n=1}^{\infty} |c_n| \cos \{2\pi[(f_x - f_n)t + f_0 t] - \theta_n\}$$

$$+ \sum_{n=1}^{\infty} |c_n| \cos \{2\pi[(f_x + f_n)t + f_0 t] + \theta_n\}. \qquad (6.6\text{-}2)$$

The narrowband filter effectively passes only the component whose frequencies are in the range $[f_0 - (\Delta/2), f_0 + (\Delta/2)]$; this means that $f_n = f_x$ and that $|c_n|$ is detected and displayed on an oscilloscope whose horizontal deflection is derived from the voltage $v(t)$ which is proportional to f_x.† In the absence of an RF image rejection filter, images may be a problem and the sweep *range* must be restricted to less than $2f_0$.

If at time $t = 0$ the frequency of the VCO is f_0 and if at time $t = T$ the frequency is $f_M = f_0 + K_1 v(T)$, then the total band over which the spectrum is examined is $f_M - f_0 = K_1 v(T) = K_2 T$ (the K's are constants of proportionality). If Δ is the width of the narrowband filter, then the number of frequency resolution cells is

$$N = \frac{f_M - f_0}{\Delta}. \qquad (6.6\text{-}3)$$

A large N implies a high-resolution spectrum analyzer. During the sweep time T, the entire band $f_M - f_0$ must be examined. The time allowed per resolution cell is T/N. To get a meaningful response from the filter, its response time τ_c must satisfy

$$\tau_c \ll \frac{T}{N}.$$

At the same time we know that $\tau_c \simeq \Delta^{-1}$ (see Fig. 3.6-3 with B replaced by Δ). Hence

$$\tau_c \simeq \frac{1}{\Delta} \ll \frac{T\Delta}{f_M - f_0} \qquad (6.6\text{-}4)$$

or

$$T \gg \frac{f_M - f_0}{\Delta^2} = \frac{N^2}{f_M - f_0}. \qquad (6.6\text{-}5)$$

Thus for high resolution we require long sweep times. For $f_M - f_0 = 10{,}000$ Hz and $N = 100$, $T \gg 1$ sec. Since $T_S^{-1} > \Delta$, we must also satisfy $T \gg (f_m - f_0)T_S^2 < NT_S$. In commercial spectrum analyzers the frequency sweep and resolution adjustments are frequently mechanically coupled to implement these inequalities.

6.7 Double-Sideband (DSB)

Before considering the problem of detecting DSB we might ask why we should consider DSB in the first place if AM is so easily detected with an envelope detector. The answer lies in the fact that DSB is more efficient than AM. Since the carrier is not sent, there is considerable saving in power. To

---

†Only one harmonic component falls within the passband at any one time if $T_S^{-1} > \Delta$.

see how much power is saved, consider the time-averaged transmitted power in an AM wave, which is

$$P_{AM} = \lim_{T \to \infty} \frac{1}{T} \int_{-T/2}^{T/2} x^2(t) \, dt \equiv \langle x^2(t) \rangle \tag{6.7-1a}$$

$$= \frac{A_c^2}{2} [1 + m^2 \langle s^2(t) \rangle], \tag{6.7-1b}$$

where we have used the fact that the carrier frequency $f_c \gg$ the message bandwidth, W, and the $\langle \ \rangle$ represent the time-averaging operation indicated in Eq. (6.7-1a). The AM power can be decomposed as

$$P_{AM} = P_c + 2P_{SB}, \tag{6.7-2}$$

where $P_c = A_c^2/2$ is the carrier power and $P_{SB} = m^2 A_c^2 \langle s(t)^2 \rangle /4$ is the signal-associated power in a sideband. Since $m|s(t)| \le 1$, the maximum value of P_{SB} is $A_c^2/4$. Therefore *at least 50% of the transmitted power is in the carrier wave*, which, by itself, carries no information. Thus DSB is indeed more efficient than AM; however, DSB cannot be detected with an envelope detector, which would tend to detect $|s(t)|$ rather than $s(t)$. The detection circuitry for DSB is more complicated and requires additional electronics. As stated earlier, detection of DSB requires some form of homodyne detection. For this to work properly the local carrier must have the right frequency and phase. If the carrier has the wrong phase, then the detected signal has the form (Fig. 6.7-1) $\tilde{s}(t) = s(t) \cos \theta$. Thus for a phase error of only 45°, the detected power is reduced by half. As $\theta \to \pi/2$, the detected signal vanishes.

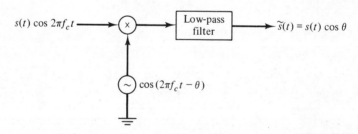

Figure 6.7-1. Detection of DSB with a carrier with the wrong phase.

The effect of demodulating with the *wrong frequency* is much worse than demodulating with the wrong phase and leads to serious distortion. To see this, consider a modulating tone of the form $s(t) = 2A_m \cos 2\pi f_m t$ that DSB-modulates a carrier $\cos 2\pi f_c t$. The DSB wave is then

$$x(t) = A_m \{\cos [2\pi(f_c + f_m)t] + \cos [2\pi(f_c - f_m)t]\}. \tag{6.7-3}$$

Now assume that the local carrier in Fig. 6.7-1 furnishes a signal $\cos [2\pi(f_c + \Delta)t]$, where Δ is the error frequency. The output of the low-pass filter is then proportional to

$$\bar{s}(t) = \cos 2\pi(f_m + \Delta)t + \cos 2\pi(f_m - \Delta)t$$
$$= 2 \cos 2\pi f_m t \cos 2\pi \Delta t. \tag{6.7-4}$$

If Δ is a very low frequency the detected signal will exhibit beats instead of a steady tone. For somewhat larger values of Δ the sum and difference frequency terms result in a completely distorted and unintelligible output.

A small amount of carrier, when added to the DSB signal, can be used to generate a carrier of the same phase and frequency at the receiver, which can then be used in homodyne detection. Adding a small amount of carrier to a DSB wave does not convert it to an AM wave, and it cannot be detected with an envelope detector. Figure 6.7-2 shows a DSB detector in which the

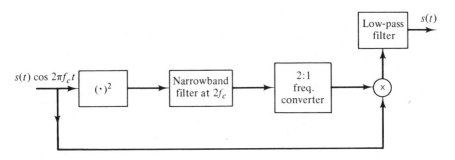

Figure 6.7-2. DSB detector using a square-law device.

carrier is generated from the sidebands with the aid of a square-law device. The narrowband filter centered at $2f_c$ produces a sine wave of frequency $2f_c$, and a 2:1 frequency converter produces the locally generated carrier frequency.

Another scheme, proposed by Costas [6-8], uses what amounts to a phase-locked loop to generate a constant-frequency carrier signal from the sidebands. The system is shown in Fig. 6.7-3. Assume that the local oscillator is at frequency $f < f_c$ and the input signal is $x(t) = s(t) \cos 2\pi f_c t$. After low-pass filtering the multiplier output signal is proportional to

$$e(t) = \frac{s^2(t)}{4} \sin [4\pi(f_c - f)t - 2\theta], \tag{6.7-5}$$

where θ is the phase difference between the local oscillator and the incoming wave when both are at the same frequency. If $f_c - f$ is small compared to frequencies at which there are significant components in $s(t)$, the amplitude variations in $e(t)$ can be filtered out, and the voltage driving the oscillator becomes essentially proportional to

$$e_{LO}(t) = K \sin [4\pi(f_c - f)t - 2\theta]. \tag{6.7-6}$$

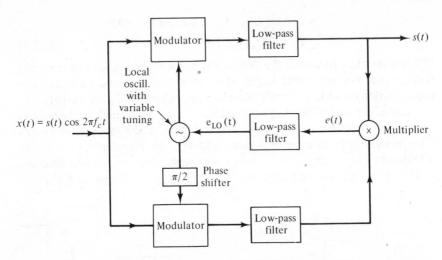

Figure 6.7-3. Costas' scheme for DSB detection.

The local oscillator will adjust its frequency until $e_{LO}(t) \rightarrow 0$. This happens not only when $f_c = f$ but when $\theta = 0$. Thus the loop drives not only to the correct frequency but the correct phase as well. Because phase-locked loops are so important in any AM-coherent detection scheme, as well as FM, we shall discuss them in greater detail in Sec. 6.10.

Demodulation with a carrier of the correct phase is essential in the case of DSB *quadrature-carrier multiplexing*. This is a bandwidth conservation scheme that enables two separate DSB signals to occupy the same band while still enabling their separation at the receiver. The central idea is that the transmitted signals are uniquely encoded by separating the phases of the two carriers (both of the same frequency) by $\pi/2$ radians. Figure 6.7-4 shows the system. The output of channel 1 is $s_1(t)$, while the output of channel 2 is $s_2(t)$. It is quite important to maintain equal gain for the two sidebands to avoid interchannel interference.

If the local carrier has the wrong phase, then we can expect interference between the channels, in addition to loss of desired signal power. Thus channel 1 would give

$$\hat{s}_1(t) = \underbrace{s_1(t) \cos \theta}_{\substack{\text{loss of} \\ \text{power}}} - \underbrace{s_2(t) \sin \theta}_{\text{interference}} \qquad (6.7\text{-}7)$$

while channel 2 would give

$$\hat{s}_2(t) = s_1(t) \sin \theta + s_2(t) \cos \theta. \qquad (6.7\text{-}8)$$

Quadrature-carrier multiplexing is used in color TV (Sec. 6.15). There,

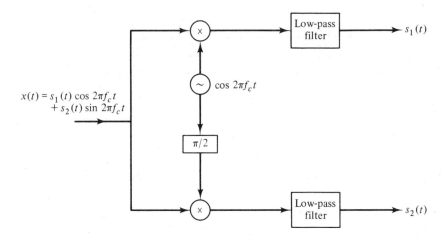

Figure 6.7-4. Detection in quadrature two-channel multiplexing.

synchronization pulses are transmitted to keep the local oscillator at the right frequency and phase.

6.8 Single-Sideband (SSB)

Since the information in DSB is in fact duplicated in the two sidebands, the transmission bandwidth of the modulated wave can be reduced by 50%, i.e., made equal to W, by sending only one sideband rather than two. In SSB systems one of the sidebands of the DSB signal is removed, usually by direct filtering, and the remaining sideband is sent by itself or with a low-level carrier. Thus SSB is more "efficient" than DSB from the point of view of bandwidth conservation.

Removing one of the sidebands by filtering is not easy to do when there are significant signal components at low frequencies. In such a case, vestigial sideband (VSB) is preferred. However, in the case of speech, articulation tests show that components below 200 Hz are not necessary for intelligibility and natural sounding speech does not require the transmission of components below 100 Hz. Even for orchestral music, reasonably good reproduction is obtained when no frequency components below 80 Hz are transmitted. An SSB generator is shown in Fig. 6.8-1. The type of signal spectrum that is conveniently filtered with a sideband filter is shown in Fig. 6.8-2. The absence of important low-frequency components in the spectrum is important because realizable sideband filters have finite transition regions in the frequency domain. Abrupt cutoff is not possible (Sec. 3.8).

An alternative to sideband filtering is suggested by the fact that the SSB signal can be expressed by

$$x(t) = [s(t) \cos 2\pi f_c t \pm \hat{s}(t) \sin 2\pi f_c t]/2, \qquad (6.8\text{-}1)$$

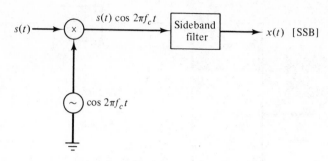

Figure 6.8-1. Generation of an SSB wave.

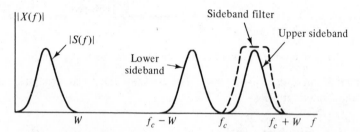

Figure 6.8-2. Sideband filtering of signals with negligible energy at low frequencies.

where $\hat{s}(t)$ is the Hilbert transform of $s(t)$. (See Sec. 2-13.) This follows from the definition of SSB according to which an upper-sideband SSB wave is given by

$$x_{SSB}(t) = \frac{1}{2} \int_{f_c}^{\infty} S(f - f_c)e^{j2\pi ft}\, df$$

$$+ \frac{1}{2} \int_{-\infty}^{-f_c} S(f + f_c)e^{j2\pi ft}\, df. \qquad (6.8\text{-}2)$$

Now let $u = f - f_c$ in the first integral and $v = f + f_c$ in the second integral. Since u and v are dummy variables, we replace them by the common dummy variable f. The result is

$$x_{SSB}(t) = \frac{1}{2} e^{j2\pi f_c t} \int_{0}^{\infty} S(f)e^{j2\pi ft}\, df + \frac{1}{2} e^{-j2\pi f_c t} \int_{-\infty}^{0} S(f)e^{j2\pi ft}\, df$$

$$= \frac{e^{j2\pi f_c t}}{4} \int_{-\infty}^{\infty} S(f)e^{j2\pi ft}\, df + \frac{e^{-j2\pi f_c t}}{4} \int_{-\infty}^{\infty} S(f)e^{j2\pi ft}\, df$$

$$+ \frac{e^{j2\pi f_c t}}{4} \int_{-\infty}^{\infty} S(f)\,\text{sgn}(f)e^{j2\pi ft}\, df + \frac{e^{-j2\pi f_c t}}{4} \int_{-\infty}^{\infty} S(f)\,\text{sgn}(-f)e^{j2\pi ft}\, df$$

$$= \frac{1}{2}[s(t)\cos 2\pi f_c t - \hat{s}(t)\sin 2\pi f_c t]. \qquad (6.8\text{-}3)$$

We use the notation introduced in Sec. 2.13,

$$\hat{s}(t) = -j \int_{-\infty}^{\infty} S(f) \, \text{sgn} \, (f) e^{j2\pi ft} \, df, \tag{6.8-4}$$

and the fact that sgn $(f) = -\text{sgn} \, (-f)$. Equation (6.8-3) is identical with Eq. (6.8-1) when the minus sign is used there.

The system suggested by this result is shown in Fig. 6.8-3. In this scheme, the central element is the quadrature filter which introduces a 90° phase shift in every frequency component in the signal band.† If we ignore the introduction of the delay elements in Fig. 6.8-3, which enable a causal system to

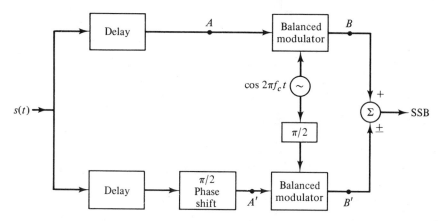

Figure 6.8-3. Generation of SSB without sideband filtering, through the use of phase shifting.

approximate a noncausal one, we see that the signal obtained at A' is the Hilbert transform of $s(t)$. Thus at A' we have

$$\hat{s}(t) = \int_{-\infty}^{\infty} H(f)S(f)e^{j2\pi ft} \, df, \tag{6.8-5}$$

where

$$H(f) = \begin{cases} -j, & f > 0 \\ 0, & f = 0 \\ j, & f < 0. \end{cases} \tag{6.8-6}$$

At B the signal is $s(t) \cos 2\pi f_c t$, and at B' it is $\hat{s}(t) \sin 2\pi f_c t$. Hence depending on the sign at the summing junction the SSB wave is described by Eq. (6.8-1). The *difference* of the signals at B and B' produces the upper-sideband (USSB) wave, and the *sum* the lower-sideband (LSSB) wave.

---

†The quadrature filter is not any less easy to build in practice than a single-sideband filter.

A technique for detecting an SSB wave is shown in Fig. 6.8-4. At A, the signal is given by [we ignore the factor of 1/2 in Eq. (6.8-1)]

$$[s(t) \cos 2\pi f_c t \pm \hat{s}(t) \sin 2\pi f_c t] \cos 2\pi f_c t = \frac{s(t)}{2}(1 + \cos 4\pi f_c t) \pm \frac{\hat{s}(t)}{2} \sin 4\pi f_c t.$$

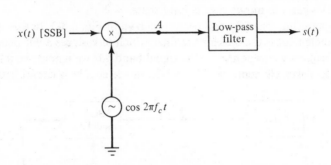

Figure 6.8-4. Homodyne detection of an SSB wave.

Hence if the low-pass filter doesn't pass the high-frequency terms, the detected signal is proportional to $s(t)$. If the carrier has the wrong phase, there will be a term proportional to $\hat{s}(t)$. For pulse-type signals, this results in serious distortion. For example, the Hilbert transform of a pulse of the form rect $[(t - T)/T]$ is shown in Fig. 2.13-5. The "horns" lead to high peak voltages. Sudden high voltages lead to flashover, energy losses by corona discharge, and destruction of insulation. For these reasons SSB would seem a doubtful choice for transmitting pulse-type signals in which abrupt signal changes occur frequently.

On the other hand, since human hearing is largely insensitive to the relative phase between the spectral components, detecting an SSB wave with a demodulating signal of the wrong phase does not generally lead to audible distortion. In fact, insofar as hearing is concerned, the demodulating signal can even have a slightly different frequency from the SSB carrier without resulting in serious audible distortion. As an illustration, if the modulating signal is a pure tone of the form $s(t) = A_m \cos 2\pi f_m t$, the upper-sideband SSB wave is proportional to

$$x(t) = A_m \cos 2\pi (f_c + f_m)t. \tag{6.8-7}$$

If this wave is now detected in a homodyne system in which the local carrier is at frequency $f_c + \Delta$, where Δ is the frequency error, the demodulated signal is proportional to

$$\hat{s}(t) = A_m \cos 2\pi (f_m - \Delta)t. \tag{6.8-8}$$

The listener will hear a tone at a slightly different frequency from the original if Δ is small. The difference may not even be discernible. Note that the "beating" effect that occurs in DSB, under the same circumstances, is absent here.

An SSB wave can be rewritten as an envelope on a carrier with a time-dependent phase $\theta(t)$ added to the linear phase $2\pi f_c t$. Thus we can write

$$x_{\text{SSB}}(t) = s(t) \cos 2\pi f_c t \pm \hat{s}(t) \sin 2\pi f_c t$$
$$= R(t) \cos [2\pi f_c t + \theta(t)], \qquad (6.8\text{-}9)$$

where

$$R(t) = \sqrt{s^2(t) + \hat{s}^2(t)}, \qquad (6.8\text{-}10\text{a})$$

$$\theta(t) = \mp\tan^{-1} \frac{\hat{s}(t)}{s(t)}. \qquad (6.8\text{-}10\text{b})$$

Equation (6.8-9) indicates that if we detect an SSB signal with an envelope detector, serious nonlinear distortion occurs. If $s(t)$ is a simple rect ($\cdot$) pulse, the detected signal will have the form shown in Fig. 6.8-5. In a real system, the peaks may be large but finite, and the signal will not extend to $t = \pm\infty$. The peaks are called the *horns*, and the rise and fall at the beginning and end of the signal are called *smears*.

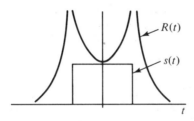

Figure 6.8-5. Result of detecting SSB with an envelope detector.

If an upper-sideband, tone-modulated SSB wave is enveloped-detected, the detected signal is proportional to

$$R(t) = \sqrt{\cos^2 2\pi f_m t + \sin^2 2\pi f_m t}$$
$$= 1.$$

Hence no tone, at any frequency, is heard. The addition of a large carrier signal, in phase with the transmitted carrier, can reduce the distortion and enable SSB detection with an envelope detector. Thus

$$x_{\text{SSB}}(t) + A \cos 2\pi f_c t = R(t) \cos [2\pi f_c t + \theta(t)], \qquad (6.8\text{-}11)$$

where $R(t)$ is now given by

$$R(t) = \sqrt{[s(t) + A]^2 + \hat{s}^2(t)} \qquad (6.8\text{-}12)$$

$$\simeq A + s(t), \qquad (6.8\text{-}13)$$

if the higher-order terms in the power series expansion of the square root are neglected. The constant bias A can be removed, and $s(t)$ can be recovered.

6.9 Vestigial Sideband (VSB)

In the case of speech and music, the generation of SSB by sideband filtering is feasible because of the absence of unimportant message components at low frequencies. Such is not the case in TV, where, as we shall see in Sec. 6.12, there are important components at very low frequencies. To maintain bandwidth conservation while eliminating the need for the critical filtering of a sideband, a compromise technique known as VSB is used. In VSB, most of one sideband is passed along with a vestige of the other sideband. VSB is widely used in TV, facsimile, and certain data-signal-transmission systems. The typical bandwidth required to transmit a VSB wave is about 1.25 that of SSB.

The VSB-modulated signal is produced by passing the DSB signal through a filter which removes part of one of the sidebands. Such a system and the spectra at various points in it are shown in Fig. 6.9-1. The filter transfer function shown in this figure removes part of the lower sideband, but the same result would be obtained if the upper sideband were partially removed. The transmission channel is not shown in Fig. 6.9-1(a), but it would come anywhere between points B and C, i.e., between the DSB modulator and the homodyne demodulator.

The main point to notice in this system is the spectrum at D, i.e., after demodulation but before the low-pass filter. Observe that the spectrum near zero frequency is the sum of the left and right partial spectra that have been shifted to $f_c = 0$ by the demodulation process. The sum spectrum should have the same shape as the original signal spectrum at point A if there is to be no distortion. Analytically we have

spectrum at A is proportional to $S(f)$,
spectrum at B is proportional to $\frac{1}{2}[S(f+f_c) + S(f-f_c)]$,
spectrum at C is proportional to $H(f)[S(f+f_c) + S(f-f_c)]$,
spectrum at D is proportional to $\frac{1}{2}H(f+f_c)[S(f+2f_c) + S(f)]$
$+ \frac{1}{2}H(f-f_c)[S(f-2f_c) + S(f)]$.

The central lobe of the spectrum at D should be $S(f)$; hence we have

$$S(f) = \frac{1}{2}[H(f+f_c)\,S(f) + H(f-f_c)S(f)]$$

or

$$H(f+f_c) + H(f-f_c) = 2. \tag{6.9-1}$$

The number 2 on the right in Eq. (6.9-1) is arbitrary and can be replaced by any constant; in fact, Eq. (6.9-1) is often written in the form

$$H(f+f_c) + H(f-f_c) = 2H(f_c). \tag{6.9-2}$$

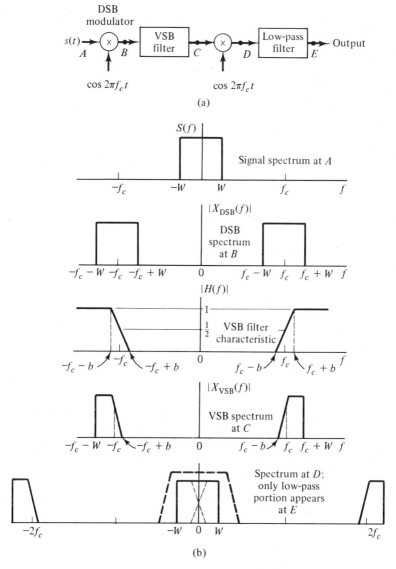

Figure 6.9-1. (a) VSB system; (b) spectra at various points in a VSB system.

The most important property of $H(f)$ is that it has odd symmetry in the neighborhood of $f = f_c$ and the 50% response level at f_c. The filter can be anywhere in the channel between the modulator and demodulator; in fact, part of it can be physically in the transmitter while the remainder is in the receiver as long as the relation in Eq. (6.9-2) is satisfied by the overall transfer function. Also, it should be noted that if the VSB filter is not exactly symmetrical as required in Eq. (6.9-2) further equalization of the spectrum is in principle possible in the low-pass filter (between *D* and *E*).

The analytical form of the VSB signal is easily obtained by inverse-Fourier-transforming the spectrum at *C*. This results in

$$x(t) = \int_{-\infty}^{\infty} H(f)[S(f+f_c) + S(f-f_c)]e^{j2\pi ft}\, df$$

$$= \int_{-\infty}^{\infty} H(f-f_c)S(f)e^{j2\pi(f-f_c)t}\, df + \int_{-\infty}^{\infty} H(f+f_c)S(f)e^{j2\pi(f+f_c)t}\, df.$$

$$(6.9\text{-}3)$$

The symmetry requirement on $H(f)$ can be expressed in the form

$$\begin{aligned} H(f-f_c) &= H(f_c) - Y(f) \\ H(f+f_c) &= H(f_c) + Y(f) \end{aligned} \quad (-W \le f \le W), \quad (6.9\text{-}4)$$

where $Y(f)$ is a function having odd symmetry about $f = 0$; i.e.,

$$Y(-f) = -Y(f) \quad (6.9\text{-}5)$$

(see Fig. 6.9-2). If we use these results in Eq. (6.9-3) and also, for convenience, let $H(f_c) = \frac{1}{2}$, we get

$$x(t) = s(t)\cos 2\pi f_c t - q(t)\sin 2\pi f_c t, \quad (6.9\text{-}6)$$

where

$$q(t) = -2j \int_{-\infty}^{\infty} Y(f)S(f)e^{j2\pi ft}\, df. \quad (6.9\text{-}7)$$

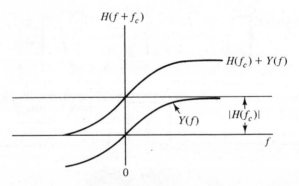

Figure 6.9-2. Transmittance of the VSB filter (upper sideband) showing the constant part $H(f_c)$ and the part having odd symmetry.

It goes without saying that the bandwidth *W* of *s*(*t*) satisfies $W < f_c$. If the function $Y(f)$ in Fig. 6.9-2 is replaced by $-Y(f)$, the lower sideband is retained; hence changing the minus sign in Eq. (6.9-6) to plus results in VSB with lower sideband retained. As the transition region becomes smaller,

$Y(f)$ approaches sgn (f), $q(t) \rightarrow \hat{s}(t)$, and the signal approaches SSB. If $Y(f) \rightarrow 0, \hat{q}(t) \rightarrow 0$, and the signal approaches DSB. For true VSB operation the frequency transition region of $H(f)$ should satisfy $0 < 2b < 2W$ (see Fig. 6.9-1). Equation (6.9-6) implies that a VSB or SSB signal can be generated from a DSB signal by injecting a quadrature component obtained from the modulating signal by a filtering operation. The quadrature component partially or totally cancels one of the sidebands. Several other ways of looking at VSB are given in [6-9].

As in the case of SSB, VSB should be detected by homodyne means. To facilitate detection at the receiver, a small pilot carrier can be transmitted along with the VSB signal to synchronize a local oscillator.† A phase-locked loop can be used for this purpose. In TV systems, the transmitted signal is not quite VSB because there is no attempt to rigidly control the transition region in the spectrum of the signal. The actual VSB shaping is done at the receiver with a VSB filter. A typical frequency response for a VSB filter for TV is shown in Fig. 6.9-3.

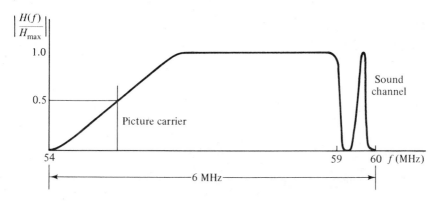

Figure 6.9-3. Frequency response of a TV receiver tuned to the 54–60-MHz channel.

*6.10 The Phase-Locked Loop

In the homodyne systems discussed in earlier sections, the local oscillator at the receiver was required to track the frequency and phase of the received carrier. This tracking can be done with a phase-locked loop (PLL) which, in simplest terms, consists of a multiplier, a linear filter/amplifier, and a voltage-controlled oscillator (VCO). Fundamentally, the phase-locked loop is an extremely narrowband filter which extracts a carrier-frequency compo-

---

†VSB can be detected by envelope detection if a large amount of carrier is added. In fact, this is how it is normally done in commercial TV.

nent from the modulated signal and removes the sideband information. For example, a PLL would extract the component $A \cos \omega_c t$ from $A[1 + ms(t)]$ $\cos \omega_c t$. Our discussion of such a loop will closely follow the analysis furnished by Viterbi [6-10], who discusses many of the sophisticated problems associated with the operation of a PLL.

Figure 6.10-1 shows the basic PLL system. We shall assume that all initial conditions are zero and that $e(t) = 0$ when $x(t) = 0$. The quiescent frequency

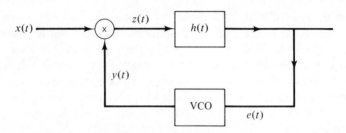

Figure 6.10-1. Phase-locked loop.

of the VCO will be denoted by ω_0 (radians per second). Use of ω_0 instead of f_0 eliminates the need for writing 2π. The signals at various points in the system are denoted

$$x(t) = \sqrt{2}\, A \sin \theta(t), \tag{6.10-1}$$

$$y(t) = \sqrt{2}\, K_1 \cos \theta'(t), \tag{6.10-2}$$

$$z(t) = x(t)y(t), \tag{6.10-3}$$

$$e(t) = \int_0^t z(\tau)h(t - \tau)\, d\tau, \tag{6.10-4}$$

and the *instantaneous* radian frequency, ω, of the VCO output signal is

$$\omega \equiv \dot{\theta}'(t) = \omega_0 + K_2 e(t). \tag{6.10-5}$$

The constant K_2 has units of radians per second per volt. The signal $z(t)$ is

$$
\begin{aligned}
z(t) &= x(t)y(t) \\
&= 2\, AK_1 \sin \theta(t) \cos \theta'(t) \\
&= K_1 A\{\sin [\theta(t) - \theta'(t)] + \sin [\theta(t) + \theta'(t)]\}, \tag{6.10-6}
\end{aligned}
$$

so that $e(t)$ is given by, from Eq. (6.10-4),

$$e(t) = \int_0^t K_1 A \sin [\theta(\tau) - \theta'(\tau)]h(t - \tau)\, d\tau$$

$$+ \int_0^t K_1 A \sin [\theta(\tau) + \theta'(\tau)]h(t - \tau)\, d\tau. \tag{6.10-7}$$

Note that $\theta(t)$ and $\theta'(t)$ are close to $\omega_0 t$. The filter $h(t)$ is essentially low-pass so that the second integral, which involves a term of frequency near

$2\omega_0$, is zero. The basic equation characterizing the PLL is obtained from Eq. (6.10-5); i.e.,

$$\theta'(t) = \omega_0 + AK_2K_1 \int_0^t \sin\left[\theta(\tau) - \theta'(\tau)\right]h(t-\tau)\, d\tau. \qquad (6.10\text{-}8)$$

To reduce Eq. (6.10-8) to a more convenient form, we make the following definitions:

$K \equiv K_1 K_2 =$ loop gain

$\phi(t) \equiv \theta(t) - \theta'(t) =$ instantaneous phase error

$\xi(t) \equiv \theta(t) - \omega_0 t =$ deviation of $\theta(t)$ about a linear phase $\omega_0 t$

$\qquad\qquad\qquad\qquad\qquad\qquad\qquad\qquad\qquad\qquad\qquad (6.10\text{-}9)$

$\xi'(t) = \theta'(t) - \omega_0 t =$ deviation of $\theta'(t)$ about a linear phase $\omega_0 t$.

Equation (6.10-8) can now be written as

$$\dot{\phi}(t) = \dot{\xi}(t) - AK \int_0^t h(t-\tau)\sin\phi(\tau)\, d\tau. \qquad (6.10\text{-}10)$$

This is a nonlinear integrodifferential equation which is quite difficult to solve. Viterbi discusses the general solution in the previously mentioned reference. We shall only discuss the noise-free linear approximation to Eq. (6.10-10).

The Linear Approximation Under Noise-Free Conditions

If the loop is near phase lock, then by definition $\phi(\tau) \simeq 0$, $\sin\phi(\tau) \simeq \phi(\tau)$, and Eq. (6.10-10) reduces to

$$\dot{\phi}(t) = \dot{\xi}(t) - AK \int_0^t h(t-\tau)\phi(\tau)\, d\tau. \qquad (6.10\text{-}11)$$

This expression approximates the PLL by a simple linear feedback loop. Such loops are studied in elementary feedback control theory and are most conveniently analyzed by use of Laplace transform methods. We assume that the reader is familiar with the general procedure. We define the Laplace transforms of $\phi(t)$, $\xi(t)$, $h(t)$, and $\xi'(t)$ by

$$\tilde{\phi}(s) = \mathcal{L}[\phi(t)] \equiv \int_0^\infty \phi(t)e^{-st}\, dt,$$

$$\tilde{\xi}(s) = \mathcal{L}[\xi(t)]$$

$$H(s) = \mathcal{L}[h(t)]$$

$$\tilde{\xi}'(s) = \mathcal{L}[\xi'(t)].$$

The Laplace transform of Eq. (6.10–11) (for $\dot{\phi}(0) = \dot{\xi}(0) = 0$) is

$$s\tilde{\phi}(s) + AKH(s)\tilde{\phi}(s) = s\tilde{\xi}(s) \qquad (6.10\text{-}12)$$

or

$$\tilde{\phi}(s) = \frac{1}{1 + AKH(s)/s}\tilde{\xi}(s) \equiv T(s)\tilde{\xi}(s). \qquad (6.10\text{-}13)$$

A block diagram of the linear model in terms of Laplace transform operators is shown in Fig. 6.10-2.

Let us now assume that $H(s)$ is the transfer function associated with an ideal integrator in parallel with a direct connection. Then $H(s) = 1 + Bs^{-1}$, where B is a gain constant. The loop will be stable if the gain constants are

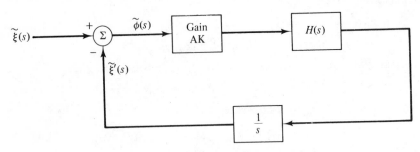

Figure 6.10-2. Block diagram of the linear model of the PLL.

adjusted so that all the poles of $T(s)$ are in the left-half s plane. Let $x(t)$ be a constant-frequency sinusoid $2\sqrt{A} \cos{(\omega t + \theta_0)}$ applied at $t = 0$. Then

$$\xi(t) = (\omega - \omega_0)t + \theta_0$$

$$\tilde{\xi}(s) = \frac{\omega - \omega_0}{s^2} + \frac{\theta_0}{s}, \tag{6.10-14}$$

and use of Eq. (6.10-13) gives

$$\tilde{\phi}(s) = \frac{s^2}{s^2 + AKs + ABK}\left(\frac{\omega - \omega_0}{s^2} + \frac{\theta_0}{s}\right). \tag{6.10-15}$$

The final value theorem then gives

$$\lim_{t \to \infty} \phi(t) = \lim_{s \to 0} s\tilde{\phi}(s) = 0. \tag{6.10-16}$$

Hence the error in tracking the instantaneous phase goes to zero, and the loop eventually locks on the correct frequency and phase.

*6.11   The Voltage-Controlled Oscillator (VCO)

We have seen that the VCO is an important component of a phase-locked loop. It has many other uses, e.g., in spectral analyzers, FM stereo demodulators, etc. A simple VCO circuit is shown in Fig. 6.11-1 ([6-3], p. 73). Let us first ignore the multiplier in the feedback path and assume a direct feedback of the signal E to the integrator. The output of the hysteric comparator is one of the two stable states E^+ or E^-. It switches to E^+ when the input exceeds V^+ and remains in that state until the input falls below V^-, whereupon it switches to E^-. It remains in that state until the input again

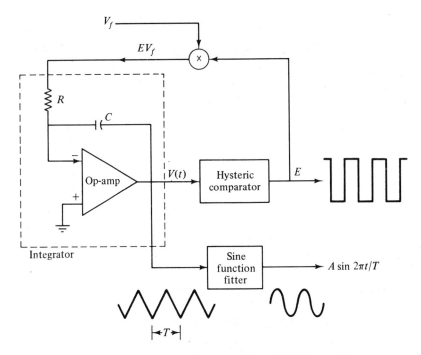

Figure 6.11-1. Simplified VCO. *Adapted from [6-3], by permission.*

exceeds V^+. Many devices, including operational amplifiers using positive feedback, can be made to develop hysteresis.†

Assume that the output has just gone into the E^+ state. The output of the integrator, starting from the time when $V = V^+$, is

$$V(t) = -\frac{1}{RC} \int_0^t E^+ \, d\tau + V^+$$

$$= V^+ - \frac{E^+}{RC}t, \qquad E^+ > 0. \tag{6.11-1}$$

When $V(t) = V^-$, E switches to E^-. The elapsed time is therefore

$$\Delta t_1 = RC \frac{V^+ - V^-}{E^+}. \tag{6.11-2}$$

Assume now that $E^- < 0$. When E switches to E^- the output of the integrator rises linearly until $V = V^+$. The elapsed time is

$$\Delta t_2 = RC \frac{V^+ - V^-}{|E^-|}. \tag{6.11-3}$$

†For example, a Schmitt trigger is a circuit that has this property ([6-11], p. 389).

The period T is $\Delta t_1 + \Delta t_2$, and the frequency is given by

$$f = \frac{1}{T} = \frac{E^+}{[1 - (E^+/E^-)](V^+ - V^-)RC}. \qquad (6.11\text{-}4)$$

If $V^+ = -V^- = V_0$ and $E^+ = -E^- = E_0$, then

$$f = \frac{E_0/V_0}{4RC}. \qquad (6.11\text{-}5)$$

Returning to Fig. 6.11-1, we see that the feedback signal is proportional to V_f, the frequency-control voltage. The frequency is then a linear function of V_f given by

$$f = \frac{V_f E_0/V_0}{4RC}. \qquad (6.11\text{-}6)$$

The output of the integrator is a symmetrical triangular wave. It can be filtered or shaped into a sine wave by a function fitter. The same can be done with the comparator output.

6.12 Television (TV)

Television is the transmission of visual images by electrical signals. The signals can be transmitted by radiation, coaxial cables, telephone wires, or esoteric means such as laser beams. The image can be defined by a brightness function, $B(x, y, t)$, which depends on three variables: the x and y coordinates of the scene and time. An electrical signal is a function of only one variable—time. For an electrical signal to absorb all the variations in $B(\cdot)$, the image must be sequentially *scanned*. The process of scanning is central in TV and requires synchronization signals to keep track of what portion of the electrical signal is associated with what portion of the image. Each image is broken down in a predetermined systematic sequence of waveforms and can be reconstructed at the receiver to produce a faithful replica of the original scene. This is in contrast to motion picture photography in which the whole scene is recorded at once as an image on photographic film. The former is an example of serial processing, while the latter might be called parallel processing. A TV system is shown in Fig. 6.12-1. The essential components are the camera; the deflection, synchronization (sync), and blanking generator; the video, sync, and blanking mixer/amplifier; and the carrier modulator. The TV camera is a device that contains optics capable of producing an image on a target consisting of a large number of photosensitive elements. An electron beam, produced by an electron gun and deflected by the signals from the deflection generator, scans the charge pattern on the photosensitive surface and produces a signal current proportional to the charge sensed at that instant; this is called the video signal. During retrace, the electron beam is

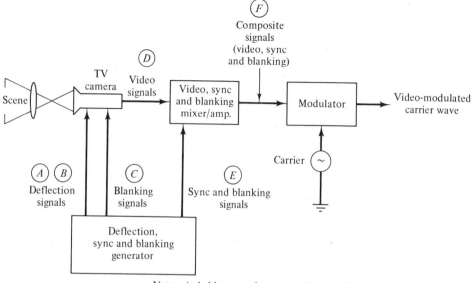

Note: circled letters refer to waveforms in Fig. 6.12-3(b)

Figure 6.12-1. Generation of TV signals.

turned off by the blanking signals. The process of scanning produces a one-to-one correspondence between the (x, y) position of the scanned image brightness and the temporal values of the video signal. There are several different types of TV camera devices, operating on variations of this principle, and they are variously called orthicon, image orthicon, vidicon, plumbicon, etc. The signal is combined with sync and blanking pulses in a mixer/amplifier and then the composite video signal is RF-modulated with a carrier in the transmitter terminal.

Camera Tubes: The Image Orthicon [6-7]

There are a number of television camera tubes in use, and they broadly fall into two classes: photoemissive and photoconductive. The image orthicon is a photoemissive tube which combines high sensitivity with wide dynamic range. Figure 6.12-2 shows the structure of the image orthicon. The principle of operation is easily understood. The optical system generates a focused image on the *photocathode*, which is photosensitive material that emits electrons from its rear surface in proportion to the amount of light impinging on its front surface. The electron image, i.e., the electron emission distribution $E(x, y)$, is proportional to the incident image brightness $B(x, y)$.

The emitted electrons are attracted to the target mosaic (TM), which is several volts positive with respect to the photocathode. The impinging primary electrons give rise to secondary emission at the TM, and these secondary

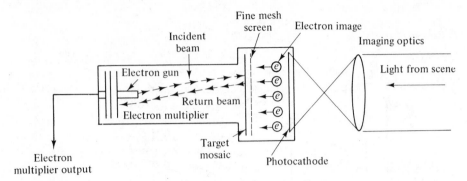

Figure 6.12-2. Image orthicon TV camera.

electrons are collected by a fine mesh screen located adjacent to the TM. A positive charge residue is left on the TM which is proportional to the brightness of the optical image. This charge configuration is essentially static during the frame interval.

The scanning electron beam furnishes the electrons to neutralize the positive charges on the rear face of the TM. If a point is highly positive (corresponding to a high-intensity image point), many electrons will be collected and the return-beam current will be low. If a point is only slightly positive, the return-beam current will be high. The variations imposed on the current constitute the *video signal*.

The electrons absorbed on the rear surface of the TM diffuse throughout the surface and neutralize the existing positive charge. The target is thus prepared for the next frame.

The return-beam current is amplified by an electron multiplier target structure surrounding the electron gun. The amplified-beam current leaves the tube through an output electrode and generates a voltage signal across a coupling resistor.

Waveforms

The scanning of the photosensitive target in the camera tube by the electron gun is, in many respects, analogous to the reading of a printed page by a human reader. The target is scanned in line-by-line fashion, from top to bottom, as shown in Fig. 6.12-3(a). The scanning pattern is commonly called a *raster* scan. A complete frame, i.e., a complete single picture of the image field, consists of a 525-line raster. However, in standard TV broadcasting practice, this is decomposed into two 262.5-line rasters that are *interlaced*. We have ignored the interlacing in Fig. 6.12-3(a). Figure 6.12-3(b) shows the important waveforms in a TV system [6-12]. The actual TV synchronization signals are considerably more complicated than the waves shown here, which have been simplified for ease of presentation. Hanson shows actual TV broadcast synchronization signals in Chapter 7 of Reference [6-13]. The first horizontal line in Fig. 6.12-3(a), i.e., from 1 to 2, is generated by a sawtooth pulse of the *X*-axis deflection waveform shown in curve *A* of Fig. 6.12-3(b). This signal, also called the fast deflection, is controlled by the timing element in the deflection, sync, and blanking generator shown in Fig. 6.12-1. The timing element produces a set of horizontal and vertical sync pulses (curve *H*) which mark the beginning of the horizontal and vertical deflection waveforms. When the sweep 1 to 2 is over, the beam is retraced to position 3. This retrace or *flyback* happens in a very brief time, corresponding to 2 to 3 in curve *A*. During the vertical (4 to 1) retrace, blanking signals (curve *C*) turn off the electron-beam signal to avoid spurious and meaningless video outputs. The combined sync/blanking waveform is shown in curve *E*. The vertical deflection signal is much slower than the horizontal deflection signal, there being 262.5 horizontal deflection pulses per vertical deflection pulse. Vertical deflection pulses are shown in curve *B*.

Assume that the camera focuses on a blackboard containing a chalk drawing of the object shown in Fig 6.12-4. The charge pattern on the photosensitive element would be pointwise proportional to the brightness of the image, and the line-by-line scan would generate a series of elemental signals as shown in curve *D* of Fig. 6.12-3(b). Each elemental signal is associated with a single horizontal sweep, and the ensemble of these signals represents, for a single picture, an orderly dissection of the image into a waveform from which we can reconstruct the scene.

Figure 6.12-3. (a) TV raster scan.

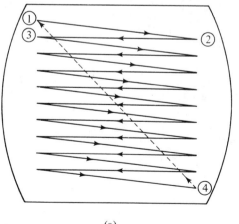

The numbers in (a) are associated with the corresponding numbers in (b).

(a)

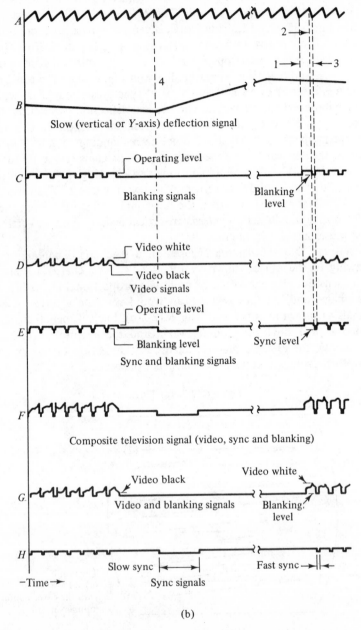

A

2 →

1 → ← 3

4

B

Slow (vertical or Y-axis) deflection signal

C

┌ Operating level

Blanking signals

Blanking level

D

┌ Video white

└ Video black

Video signals

E

┌ Operating level

└ Blanking level

Sync level

Sync and blanking signals

F

Composite television signal (video, sync and blanking)

G

Video black

Video white

Video and blanking signals

Blanking level

H

Slow sync

Fast sync →‖←

−Time →

Sync signals

(b)

Figure 6.12-3. (*Continued*) (b) TV waveforms. *Courtesy of Hughes Aircraft Co.*

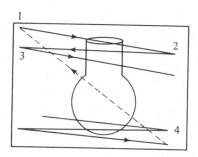

Figure 6.12-4. Scanning the charge pattern on the photosensitive surface in the camera tube.

The Receiver

A TV receiver block diagram is shown in Fig. 6.12-5. The RF signal is demodulated at the receiver terminal, and the composite signal, i.e., video, sync, and blanking signal, is transmitted to a video amplifier and sync separator. The sync signals are removed from the composite video and used to synchronize a deflection generator, while the video plus blanking signal modulates the electron beam. Without the x-y deflection signals from the deflection generator, the electron beam would produce a single flickering spot

Figure 6.12-5. Reception of TV signals.

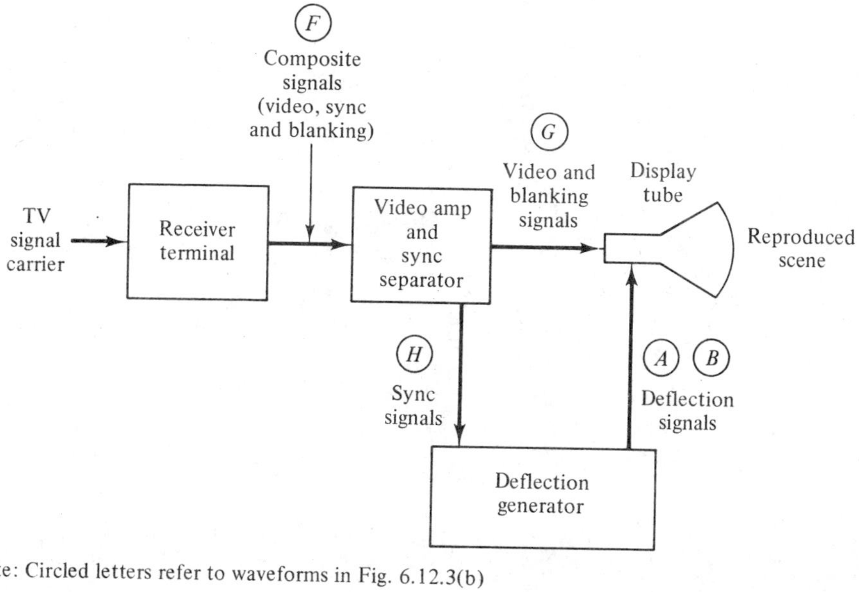

F

Composite
signals
(video, sync
and blanking)

G

Video and
blanking
signals

Display
tube

TV
signal
carrier

Receiver
terminal

Video amp
and
sync
separator

Reproduced
scene

H

Sync
signals

A B

Deflection
signals

Deflection
generator

Note: Circled letters refer to waveforms in Fig. 6.12.3(b)

of light on the phosphor screen of the display tube. The deflection signals move the electron beam in a systematic manner to reconstitute the image.

The composite video signal is shown in curve *F* in Fig. 6.12-3(b). This is the modulating signal for the RF waveform. The video and blanking signals that drive the receiver display tube are shown in curve *G*. In our discussion we have associated a large positive signal with video white and a small positive signal with video black. In actual broadcast TV, maximum signal denotes a

Figure 6.12-6. TV wave as specified by the FCC: (a) carrier wave; (b) details of composite video.

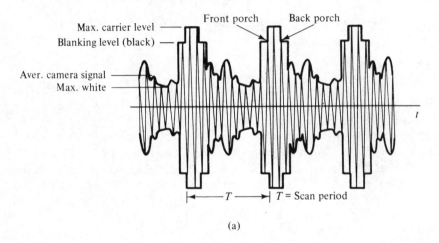

(a)

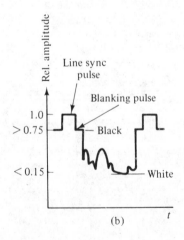

(b)

video black, while video white goes with low-level signals. The FCC standard TV wave is shown in Fig. 6.12-6.

The removal of the sync pulses from the detected video can be done with a clipping circuit, as shown in Fig. 6.12-7. The clipping should be done at the blanking level to ensure that the camera signal does not affect synchronization. The separation of the horizontal sync pulses from the vertical sync pulses cannot be done by amplitude discrimination; instead, separation by waveform content is used. The horizontal sync pulses are enhanced by a differentiating network, while the vertical sync pulses, having a much lower repetition frequency, are enhanced by an integrating circuit. A simplified block diagram of a system to separate the two sync signals is shown in Fig. 6.12-8. The actual circuits are more involved ([6-13], Chap. 7).

Figure 6.12-7. Removal of sync pulses at receiver for synchronization: (a) diode clipper; (b) comp. video input signal; (c) transfer characteristic of diode circuit; (d) separated sync pulses.

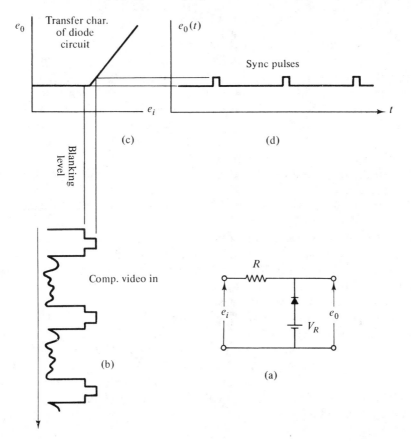

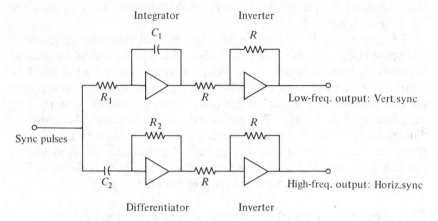

Figure 6.12-8. System for separating the vertical from the horizontal sync pulses.

6.13 Bandwidth Considerations for TV

In U.S. broadcast TV, a single TV frame consists of 525 lines of which about 490 are actually active in synthesizing the image. The inactive lines travel from the bottom to the top during the vertical retrace. The number 525 was chosen because it is composed of simple odd factors ($525 = 3 \times 5 \times 5 \times 7$) that simplify the task of frequency division. To reduce flicker in the reproduced image each full frame actually consists of two interlaced 262.5-line fields. Sixty fields are transmitted per second, two fields to a frame. The line-scanning frequency is 525×30 Hz = 15.75 kHz, and the image (frame) repetition frequency is 30 Hz. The aspect ratio (ratio of raster width to raster height) is $\frac{4}{3}$.

It is found that 60 fields (or 30 frames) per second produces a flicker-free effect on the human eye because of the phenomenon of persistence of vision. The number of picture elements scanned per second is roughly equal to $30 \times 525 \times 525 \simeq 8.3 \times 10^6$. To calculate the maximum required frequency it is customary to assume that the picture elements are arranged as alternate black and white squares along the scanning line. There are therefore two elements per cycle of the wave. The first sinusoidal harmonic of this square wave has frequency $\simeq 8.3 \times 10^6/2 = 4.15$ MHz. The actual maximum frequency in the standard video signal is around 4.2 MHz. The spectrum of the transmitted TV signal is shown in Fig. 6.13-1. One sideband is transmitted plus 25% of the other. The VSB shaping as explained in Sec. 6-9 is done at the receiver by a VSB filter. The response spectrum of the RF receiver is shown in Fig. 6.13-2.

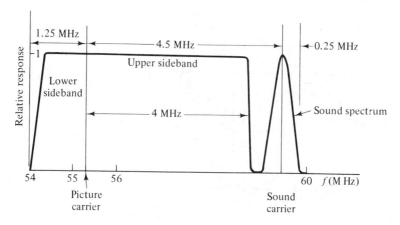

Figure 6.13-1. Spectrum of TV modulated wave (typical channel).

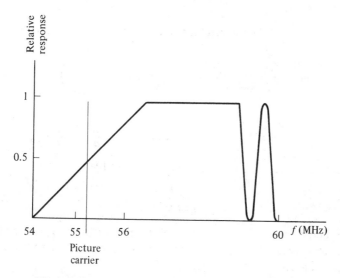

Figure 6.13-2. Frequency response of VSB filter at receiver.

6.14   Structure of the Spectrum of a TV Wave

A black/white TV frame is essentially a two-dimensional signal and can be analyzed by using a two-dimensional extension of the Fourier series. Following Bennett [6-14, p. 111], we model the scanning as a doubly infinite array of image fields. If the retrace time in a raster scan is ignored, the electrical output associated with the model (Fig. 6.14-1) is the same as in actual scanning.

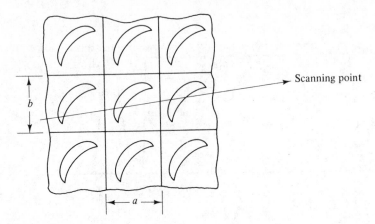

Figure 6.14-1. Model for scanning process using a doubly infinite array of image fields.

We let $B(x, y)$ denote the brightness of the still, i.e., nonchanging, image. Then

$$B(x, y) = \sum_{n=-\infty}^{\infty} \sum_{m=-\infty}^{\infty} b_{mn} \exp\left[j2\pi\left(\frac{m}{a}x + \frac{n}{b}y\right)\right], \tag{6.14-1}$$

where

$$b_{mn} = \frac{1}{ab} \int_{-a/2}^{a/2} dx \int_{-b/2}^{b/2} dy \, B(x, y) \exp\left[-j2\pi\left(\frac{m}{a}x + \frac{n}{b}y\right)\right]. \tag{6.14-2}$$

The positions x and y are related to the time variable t according to

$$\begin{aligned} x &= V_x t, \\ y &= V_y t, \end{aligned} \tag{6.14-3}$$

where V_x and V_y are the scan velocities in the x and y directions, respectively. The electrical signal $e(t)$ is proportional to $B(V_x t, V_y t)$ so that we can write (K being a constant of proportionality)

$$e(t) = K \sum_{n=-\infty}^{\infty} \sum_{m=-\infty}^{\infty} b_{mn} \exp\left[j2\pi\left(m\frac{V_x}{a}t + n\frac{V_y}{b}t\right)\right]. \tag{6.14-4}$$

The quantities a/V_x and b/V_y are recognized as the times required for a complete scan in the x and y directions, respectively. Hence V_x/a represents the line-scanning frequency, i.e., 15.75 kHz, and V_y/b represents the image repetition frequency, i.e., 30 Hz. Thus

$$e(t) = K \sum_{n=-\infty}^{\infty} \sum_{m=-\infty}^{\infty} b_{mn} \exp\left[j2\pi(m \times 15.75 \times 10^3 + n \times 30)t\right]. \tag{6.14-5}$$

The line-scanning harmonics are spaced 15.75 kHz apart, and clustered about each is an array of satellites spaced 30 Hz apart. The spectrum is shown in

Fig. 6.14-2. When there is motion in the scene being televised, there will be some smearing of the lines in the spectrum. In practice, wide gaps appear between the line-scanning harmonics. The existence of these gaps allows for the transmission of color TV signals in the same 6-MHz band allocated to black/white TV, thus enabling compatible color TV.

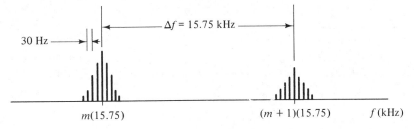

Figure 6.14-2. Spectrum of the scanned video signal.

6.15 Color Television

Compatible color TV (CCTV) is a system of color TV, developed in 1954, that fits into the existing monochrome channel assignments and permits either color or black/white reception depending on the type of receiver available.

It is well known that arbitrary color images can be reproduced by using only three primary colors: red, green, and blue. This principle is used not only in color TV but also in color photography and color printing. The three primary colors, red (R), green (G), and blue (B) when added together in equal amounts produce the colors illustrated in Fig. 6.15-1. Other colors can be produced by varying the intensities of the primary colors.

Figure 6.15-1. Additive color mixing.

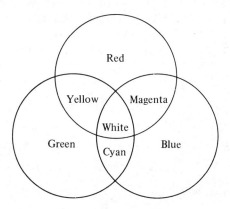

A color TV camera contains separating optics that resolve the light from the scene into the three primary colors R, G, B. A set of three camera tubes then produces three video signals, one for each primary color, $s_R(t)$, $s_G(t)$, and $s_B(t)$. These three signals could be transmitted individually and used subsequently to reconstruct a color image at the receiver. However, this method isn't used because additional bandwidth would be required, and the system would not be compatible with existing monochrome receivers.

Before discussing what is actually transmitted, we stress that any three linear, independent equations involving $s_R(t)$, $s_G(t)$, and $s_B(t)$ can be used to compute the individual signals $s_R(t)$, $s_G(t)$, and $s_B(t)$. For example, we could transmit the three independent signals $f_1(t)$, $f_2(t)$, and $f_3(t)$ given by

$$f_1(t) = \alpha_{11}s_R(t) + \alpha_{12}s_G(t) + \alpha_{13}s_B(t)$$
$$f_2(t) = \alpha_{21}s_R(t) + \alpha_{22}s_G(t) + \alpha_{23}s_B(t) \qquad (6.15\text{-}1)$$
$$f_3(t) = \alpha_{31}s_R(t) + \alpha_{32}s_G(t) + \alpha_{33}s_B(t).$$

This system of equations is conveniently written in matrix form as

$$\mathbf{f} = \mathbf{As}, \qquad (6.15\text{-}2)$$

where $\mathbf{f} = (f_1, f_2, f_3)^T$, $\mathbf{A} = [\alpha_{ij}]$ is a nonsingular 3×3 matrix, $\mathbf{s} = (s_R, s_G, s_B)^T$, and T denotes transpose. A receiver need only matrix the signals $\mathbf{f}$ according to $\mathbf{s} = \mathbf{A}^{-1}\mathbf{f}$ to recover the primary color signals. This is in fact what is actually done.

A monochrome receiver (i.e., one that shows black, white, and gray) requires a brightness or *luminance* signal, derived from the three color signals, that closely matches the luminance of a conventional, monochrome, video signal. It turns out that the required mix is given by†

$$f_Y(t) = 0.30s_R(t) + 0.59s_G(t) + 0.11s_B(t). \qquad (6.15\text{-}3)$$

The luminance produced by this signal is "compatible" with the black/white image produced by a conventional black/white TV system. Brightness levels are additive. Thus if $s_R(t) = s_G(t) = s_B(t) = 1$ (which corresponds to maximum or saturated signals), $f_Y(t) = 1$, which corresponds to the sensation of white. If blue is absent, the monochrome brightness is 0.89. However, if green is absent, the brightness falls to 0.41.

We still require two additional, independent signals for transmission of color information. These are the so-called *chrominance* signals given by

$$f_I(t) = 0.60s_R(t) - 0.28s_G(t) - 0.32s_B(t)$$
$$f_Q(t) = 0.21s_R(t) - 0.52s_G(t) + 0.31s_B(t). \qquad (6.15\text{-}4)$$

In all of the above, we have assumed that $|s_R(t)| \leq 1$, $|s_G(t)| \leq 1$, $|s_B(t)| \leq 1$.

The chrominance is that portion of the composite color signal that

---

†This is an empirical result.

represents the color information in the televised scene. The reasons the chrominance signals have the particular forms given in Eq. (6.15-4) has to do with efficient utilization of certain features of human color vision. The specifics are discussed in Reference [6-13], p. 254. However, in principle, other independent equations could serve the purpose as well.

The terms *hue* and *saturation* refer to important concepts in the accurate reconstruction of color. Hue is the attribute of colors that permits them to be classed as red, yellow, green, or blue or as intermediate between any contiguous pair of these colors. Saturation refers to chromatic purity—the freedom from dilution with white. It is a measure of the degree of difference from a gray having the same brightness. The purest color is said to be 100% saturated. A strongly saturated blue, for example, is often erroneously called a "dark" blue. A nonsaturated blue might be called light or pastel blue. However, hue and saturation are separate notions. Both hue and saturation are easily encoded by the *color vector* defined by

$$\mathbf{f}_c(t) = (f_I(t), f_Q(t)). \qquad (6.15\text{-}5)$$

When the color vector is applied to the three saturated primary colors, we obtain

$$[\mathbf{f}_c]_{\text{red}} = (0.6, 0.21) = 0.63 \angle 19°$$
$$[\mathbf{f}_c]_{\text{green}} = (-0.28, -0.52) = 0.59 \angle 242° \qquad (6.15\text{-}6)$$
$$[\mathbf{f}_c]_{\text{blue}} = (-0.32, 0.31) = 0.45 \angle 136°.$$

These are shown in Fig. 6.15-2. The magnitude of $\mathbf{f}_c(t)$, $|\mathbf{f}_c(t)|$, is proportional to the saturation, and its angle, arg $[\mathbf{f}_c(t)]$, is a measure of the hue. For example, a partially saturated magenta (red-blue) may have $s_G = 0$, $s_R = s_B = \frac{1}{2}$. The color vector is then

$$\mathbf{f}_c = (0.14, 0.26) = 0.3 \angle 62°.$$

Figure 6.15-2. $f_I\text{-}f_Q$ color plane and the three saturated primaries. Also shown is a nonsaturated magenta.

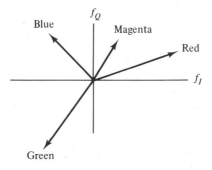

Subcarrier Determination and Frequency Interlacing

The luminance signal (also known as the Y signal) produces the monochrome image and is therefore assigned the entire 4.2-MHz bandwidth. The I-Q chrominance signals are separated from the luminance by modulating a color subcarrier whose frequency f_{cc} is halfway between the 227th and 228th harmonic of the line-scanning frequency, f_s. If f_s were 15.75 kHz, this would give

$$f_{cc} = 227.5 \times 15.75 = 3.583125 \text{ MHz.} \qquad (6.15\text{-}7)$$

However, the actual value of f_{cc} is slightly lower (3.579545 MHz to be exact) in order to avoid an objectionable beat frequency with the sound carrier which lies at 4.5 MHz above the picture carrier ([6-13], p. 246). To avoid the beat note and still maintain the relation $f_{cc} = 227.5 f_s$, the line-scanning frequency is changed to 15.73426 kHz, and the field repetition frequency is reduced to 59.94 Hz. These slight changes require no changes in existing monochrome circuitry.

The choice of $f_{cc} \simeq 3.6$ MHz represents a reasonable compromise between two factors. On the one hand, the color subcarrier should be as high as possible to minimize its effect on monochrome images. On the other hand, the satisfactory transmission of the chrominance information requires a band of at least 0.6 MHz *above* the subcarrier. The upper limit of the video passband is 4.2 MHz; hence, $4.2 - 0.6 = 3.6$ MHz $\simeq f_{cc}$.

So far we have considered the coarse determination of f_{cc}. The reason f_{cc} was chosen an odd multiple of half the line frequency will now be explained. In Sec. 6-14, it was pointed out that the spectrum of the monochrome signal consists of harmonics of f_s about which are clustered the harmonics of the frame rate (30 Hz). Motion in the televised scene slightly modifies the 30-Hz spacing, but this is of little consequence in our analysis. There are wide gaps between the harmonics of f_s where in fact additional signals can be added. The location of the color subcarrier midway between the 227th and 228th harmonics of f_s takes advantage of this fact since the line-scanning harmonics associated with the chrominance signals are located at

$$f_s\left(\frac{455}{2} \pm n\right), \qquad n = 0, 1, 2, \dots . \qquad (6.15\text{-}8)$$

The important point to note here is that these harmonics always fall in the relatively empty zones between the harmonics of the luminance signal. This is a reason compatible TV is feasible. The process of placing the chrominance carrier and its sidebands in spectral zones where there is little or no luminance information is known as frequency interlacing or *frequency interleaving*. Figure 6.15-3 illustrates the idea.

The spectral separation of chrominance and luminance information allows, at least in principle, the individual recovery of these signals. In theory, a

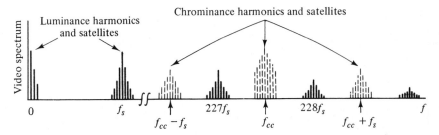

Figure 6.15-3. Interleaving of the chrominance harmonics in the luminance.

crenulated or comb-type filter could be used to admit only the chrominance and not the luminance and vice versa. However, this is not how the separation is done in practice. In fact, rather surprisingly, there is no need to separate the chrominance sidebands from the luminance signal, either in monochrome or color receivers. The effects of the chrominance sidebands on the luminance signal essentially go unnoticed by the viewer.

The meaningful luminance signal, i.e., $f_Y(t)$, and not the luminance induced by the subcarrier, dominates the chrominance because the color subcarrier frequency is exactly an odd multiple of one-half the line scanning frequency. The actual separation of chrominance from luminance is done by the time-space integration properties of the combined eye/TV-screen optical system. We shall consider the details of this phenomenon next.

Additional Properties of the Color Subcarrier

The sinusoidal variations of the subcarrier produce a flicker in the luminance which potentially could be disturbing. What are the properties of the flicker? First, since $f_{cc} \simeq 3.6$ MHz, the flickering is over a picture area no greater than a few picture resolution elements wide (there are roughly 250,000 picture resolution elements per frame—assuming 60 half-frames per second). So at worst, the sinusoidal flicker is perceived more as a kind of noise than as spatial cross talk, which would, in fact, be much more disturbing.

Second, the fact that the color subcarrier is exactly an odd multiple of one-half the line-scanning frequency produces brightness variations that are consecutively reversed both in time and space. These variations are averaged partly by the time-integration property of the phosphor on the screen, partly by the phenomenon of persistence of vision associated with the human eye/brain system, and partly by the limited resolution of human vision, which behaves, to some extent, like a *spatial* low-pass filter.

Just how does the particular choice of f_{cc} furnish out-of-phase variations? For the sake of discussion, consider a 5-line-per-field, one field-per-second, TV system. Three successive fields are shown in Fig. 6.15-4. The color subcarrier

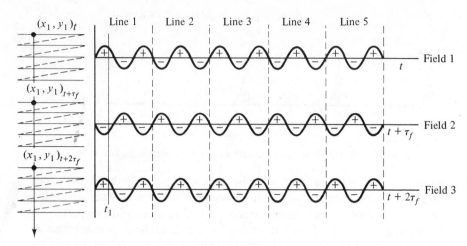

Figure 6.15-4. Effect of interlacing with a color subcarrier of frequency $f_{cc} = (2n + 1)f_s/2$, where n is an integer. Here $n = 1$.

frequency is chosen, for convenience of illustration, to be the third multiple of one-half the line-scanning frequency; i.e., $f_{cc} = 3f_s/2$. Hence there are 7.5 cycles of the subcarrier per field. The result of using such a low value of f_{cc} is to give the reader a misleading impression of the rate of averaging. In actual television, the high-frequency variations in the x direction, generated by the subcarrier and its harmonics, work to the viewer's advantage.

First, we note that the chrominance-induced brightness variations at any two adjacent points in the y direction are always out of phase and therefore tend to cancel in a *spatial* average. Second, the luminances at t and $t + \tau_f$ (τ_f is the field time) at the *same point* are always out of phase and therefore tend to cancel in a *time* average.

In Fig. 6.15-4, the start of the arrow on the extreme left indicates a particular point in the field at time t and its direction is advancing field time. At 3.6 MHz, the variations in both the time and space averages should not be noticeable to the viewer. The effective response of the eye is controlled less by the instantaneous stimulation provided by any one scan than by the integrated, *average*, stimulation furnished by several scans.

The principles behind color-subcarrier cancellations are exactly the same when $f_{cc} = 445f_s/2$. The argument also extends to the sidebands of the color subcarrier; each sideband is an odd multiple of $f_s/2$. Figure 6.15-5 illustrates the effect of the frequency interlacing technique on the net luminance signal.

In practice the cancellation of the subcarrier and its sidebands is not quite so good as the previous discussion might indicate. An important factor that contributes to less-than-perfect cancellation is the nonlinearity in the receiver circuitry and the screen. For example, positive excursions in the subcarrier

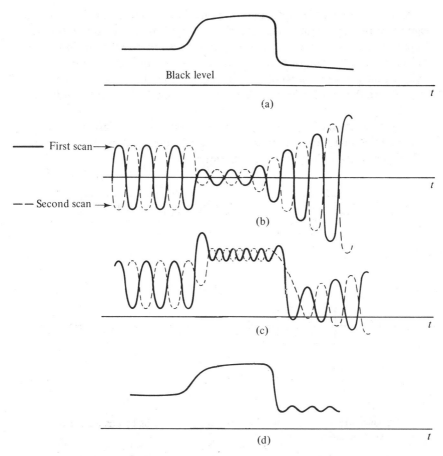

Figure 6.15-5. Effect of frequency interlacing: (a) meaningful luminance signal; (b) subcarrier signal; (c) sum of (a) + (b); (d) averaged luminance after two scans.

signal might not promote the same response as negative excursions. Another factor is that the storage of information by the eye, from one frame to the next, may not be perfect ([6-15], p. 211).

Generation of the Composite Video

The three signals f_Y, f_I, and f_Q, when suitably combined, constitute the composite color video signal. Methods for separating the three signals from each other must be found if recovery of s_R, s_G, and s_B is to be possible at the receiver. We have already discussed how the color subcarrier enables the separation of f_Y from (f_I, f_Q). We shall discuss now how the two chrominance signals are separated from each other.

Certain properties of human vision enable the bandwidths of the I and Q signals to be significantly less than the 4.2 MHz allocated to the luminance. In particular, tests show that human vision does not detect color in *very small* objects and that the color of *small* objects can be satisfactorily reproduced by utilizing only two primary colors. The result is that if the nominal bandwidths of f_I, f_Q are 0.6 and 1.6 MHz, respectively, satisfactory color reproduction is possible.

The separation of the I and Q signals is done by modulating with a variation of DSB quadrature-carrier two-channel multiplexing, discussed in Sec. 6.7. Figure 6.15-6 illustrates the technique.

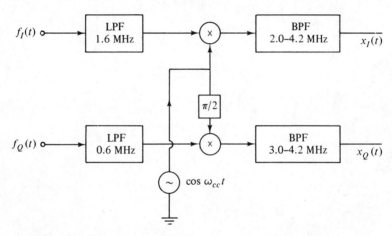

Figure 6.15-6. Separation of the chrominance signals by quadrature-carrier modulation.

The Q signal, being band-limited to 0.6 MHz, can be modulated as DSB; the I signal cannot. Hence the 0–0.6-MHz portion of the I signal is sent DSB, while the 0.6–1.6-MHz portion is sent LSSB (lower-sideband SSB). The net result is a VSB signal that can be written as

$$x_I(t) = f_I(t) \cos \omega_{cc}t + \hat{f}_{IH}(t) \sin \omega_{cc}t,$$

where $\hat{f}_{IH}(t)$ is the quadrature component needed to produce VSB modulation, analogous to the signal $q(t)$ in Eq. (6.9-6). Equivalently, for the purpose of analysis, $x_I(t)$ can be decomposed into a sum of DSB and LSSB signals as

$$x_I(t) = \underbrace{[f_I(t) - f_{IH}(t)] \cos \omega_{cc}t}_{\text{DSB}} + \underbrace{f_{IH} \cos \omega_{cc}t + \hat{f}_{IH} \sin \omega_{cc}t}_{\text{LSSB}} \qquad (6.15\text{-}9)$$

and

$$x_Q(t) = \underbrace{f_Q(t) \sin \omega_{cc}t}_{\text{DSB}}. \qquad (6.15\text{-}10)$$

Figure 6.15-7 shows the video spectra for compatible color TV. In Eq.
(6.15–9), $f_{IH}(t)$ represents the high-frequency (0.6–1.6-MHz) portion of $f_I(t)$
and $\hat{f}_{IH}$ is the Hilbert transform of f_{IH}. The reader may have noticed that
the LSSB portion of Eq. (6.15-9) has in fact twice the amplitude of the
DSB part.† This can be achieved with a band-pass filter that furnishes twice as
much gain in the 2–3.0-MHz zone than in the 3.0–4.2-MHz zone. Equivalently
a VSB filter that has a step-type characteristic can be put into the receiver
that compensates for the attenuation of one sideband in the I signal. The
filter should give a relative gain of 2 for all frequencies in the I signal between
2 and 3.0 MHz ([6–15], p. 244).

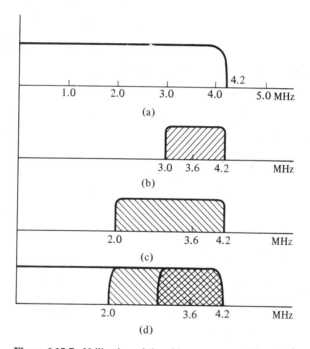

Figure 6.15-7. Utilization of the video spectrum of the color
baseband signal: (a) Y or luminance component; (b) Q
chrominance component; (c) I chrominance component;
(d) composite spectrum (amplitudes of spectra not drawn
to scale).

A simplified version of the generation of the composite baseband signal
$s_b(t)$ is shown in Fig. 6.15-8.

The color burst signal is a short sample of carrier that is added for
frequency and phase synchronization of the local oscillator at the receiver.
It is added to the trailing edge (*back porch*) of the horizontal blanking pulse.
The horizontal sync signal is derived from the subcarrier signal by countdown

†Recall that SSB is described by $x(t) = \frac{1}{2}[s(t) \cos \omega_{cc}t \pm \hat{s}(t) \sin \omega_{cc}t]$.

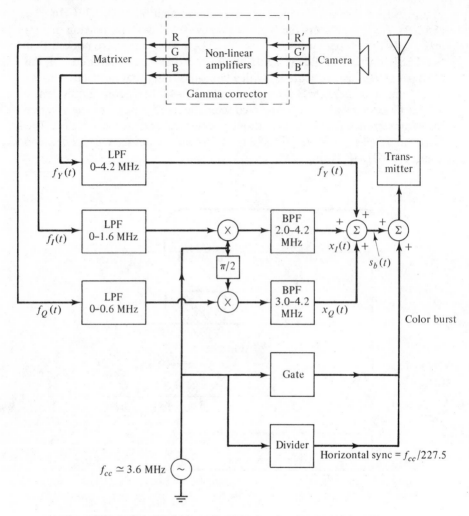

Figure 6.15-8. Simplified block diagram showing the generation of $s_b(t)$. The gamma correctors are needed to establish a linear characteristic between input and output signal.

circuitry that divides the 3.57954-MHz frequency down to the 15.734-kHz horizontal frequency. The vertical sync signal is derived from the horizontal frequency. The color burst is shown in Fig. 6.15-9.

Demodulation

Ignoring sync pulses, the baseband signal for color TV is

$$s_b(t) = f_I(t) \cos \omega_{cc} t + f_Q(t) \sin \omega_{cc} t + \hat{f}_{IH}(t) \sin \omega_{cc} t + f_Y(t). \qquad (6.15\text{-}11)$$

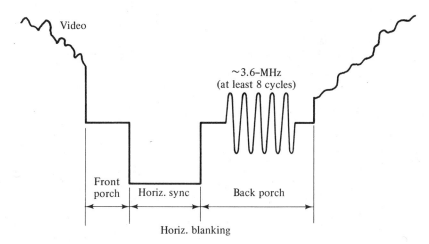

Figure 6.15-9. Horizontal blanking pulse for color TV showing the color burst.

At the receiver, demodulation of the carrier wave is done as in a black/white monochrome receiver. Demultiplexing of the baseband signal, $s_b(t)$, is done after envelope detection. The demultiplexing of $s_b(t)$ into the three primary color signals is shown in Fig. 6.15-10.

The separation of the luminance signal is done by frequency interlacing. For all practical purposes, therefore, the signal at A is $f_Y(t)$. At B' there is a signal proportional to

$$e_{B'}(t) = f_I(t) + [f_Q(t) + \hat{f}_{IH}(t)] \sin 2\omega_{cc}t + 2f_{YH}(t) \cos \omega_{cc}t$$
$$+ f_I(t) \cos 2\omega_{cc}t, \tag{6.15-12}$$

where $f_{YH}(t)$ is the high-frequency (2.0–4.2-MHz) component of the luminance signal. None of the double frequency terms pass through the low-pass filter; $f_{YH}(t) \cos \omega_{cc}t$ is centered at odd harmonics of one-half the line frequency so that it is separated, or made "invisible," by interlacing. The only effective signal at B is therefore $f_I(t)$.

At C', there is a signal proportional to

$$e_{C'}(t) = f_Q(t) - [f_Q(t) + \hat{f}_{IH}(t)] \cos 2\omega_{cc}t + f_I(t) \sin 2\omega_{cc}t$$
$$+ 2f_{YH}(t) \sin \omega_c t + \hat{f}_{IH}(t). \tag{6.15-13}$$

None of the double frequency components is passed by the low-pass filter. The terms involving $\sin \omega_c t$ are made invisible through interlacing; $\hat{f}_{IH}(t)$ has components only in 0.6–1.6 MHz. Hence it is rejected by the filter, and the only effective signal at C is proportional to $f_Q(t)$.

A conventional sync separator is used to produce the pulses necessary for control of the deflection circuitry. The sync pulses also turn on a gate circuit, which in turn admits the eight-cycle subcarrier burst. The burst is amplified

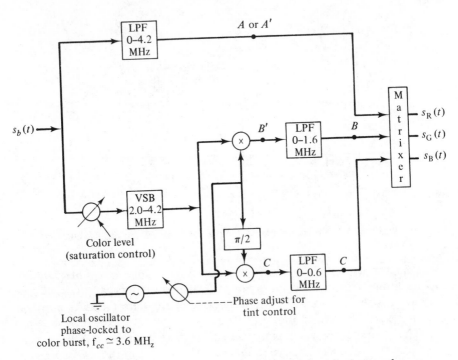

Figure 6.15-10. Demultiplexing of $s_b(t)$ and restoration of primary color signals.

and compared with the output of a local oscillator. An error voltage, proportional to the difference between burst and LO signal, is used to lock the frequency and phase of the local oscillator to the burst. Tint control is obtained by varying the phase of the LO in the color circuit. We leave it as an exercise to show that this rotates the color vector shown in Fig. 6.15-2.

Matrixing of $f_Y, f_I,$ and f_Q produces the three primary color signals:

$$s_R(t) = f_Y(t) - 0.96f_I(t) + 0.26f_Q(t)$$
$$s_G(t) = f_Y(t) - 0.28f_I(t) - 0.64f_Q(t) \qquad (6.15\text{-}14)$$
$$s_B(t) = f_Y(t) - 1.10f_I(t) + 1.7f_Q(t).$$

We leave it as an exercise (Prob. 6-31) to determine what happens when the chrominance signals are absent (i.e., a color receiver receiving monochrome images).

The TV sound signal modulates a carrier which is 4.5 MHz above the picture carrier. The sound carrier is *frequency-modulated*, a process we shall discuss in the next chapter. The complete TV channel spectrum is shown in Fig. 6.15-11.

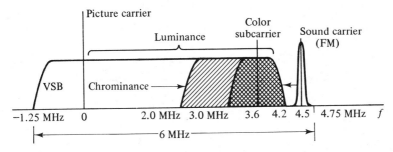

Figure 6.15-11. Standard broadcast spectrum.

6.16 Frequency-Division Multiplexing (FDM)

FDM, like TDM discussed in Sec. 4.4, is a scheme for transmitting, simultaneously, several messages over the same transmission link while enabling separation of the messages at the receiver. FDM is used in telephone systems, telemetry, stereo broadcasting, and communication networks. The Bell Telephone System multiplexes as many as 3600 4kHz separate messages on its L4 carrier system. In a sense the broadcast signals in a given geographical area form an FDM system because all of them occupy the same transmission medium at the same time and yet can be individually detected in the home receiver.

Figure 6.16-1 shows a block diagram of an FDM-AM system for several messages $\{s_i'(t), i = 1, \ldots, n\}$. The low-pass filter cuts off at W Hz, thereby ensuring that no message occupies a band wider than W Hz. The frequency-truncated message, $s_i(t)$, modulates a carrier signal at frequency f_i. The initial carrier signal is called the subcarrier and serves to uniquely encode the different messages $\{s_i(t)\}$.

For AM, the spacing between adjacent subcarriers, $\Delta f \equiv f_i - f_{i-1}$, must be at least as great as $2W$. The composite signal $s_b(t)$, consisting of the sum of all the modulated signals $\{x_i(t)\}$, is still treated as baseband because it hasn't yet modulated a high-frequency carrier wave. For AM encoding, the composite signal is

$$s_b(t) = \sum_{i=1}^{N} x_i(t) = \sum_{i=1}^{N} A_i[1 + m_i s_i(t)] \cos(2\pi f_i t + \theta_i), \qquad (6.16-1)$$

and its spectrum is shown in Fig. 6.16-2.

Modulation of a carrier by $s_b(t)$ may not be required in certain situations. Even when $s_b(t)$ modulates a carrier, the nature of the modulation may be completely different from the initial modulation of the signals $\{s_i(t)\}$. For example, the generation of $s_b(t)$ might be done with SSB encoding, as is often the case since baseband bandwidth is conserved and adjacent carrier fre-

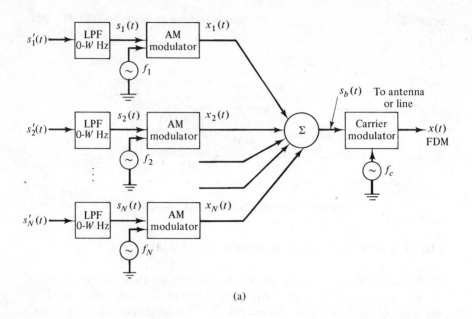

(a)

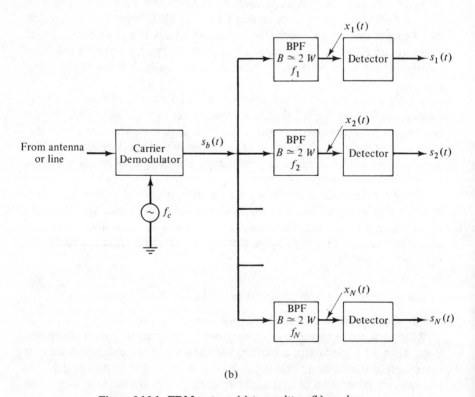

(b)

Figure 6.16-1. FDM system: (a) transmitter; (b) receiver.

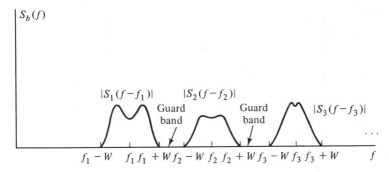

Figure 6.16-2. Spectrum of the baseband signal $s_b(t)$ in FDM.

quencies need be separated only by W Hz. The final modulation of the carrier might be by wideband frequency modulation (FM) in order to take advantage of the noise-immunity properties of this type of modulation. The designation AM/FM FDM is used to describe a frequency-division multiplexing scheme whereby $s_b(t)$ is generated by AM, while $x(t)$ is generated by FM. An example of an FM/FM FDM system for space telemetry is given in the next chapter.

In FDM systems, as well as many other communication systems, *cross talk* is an important problem and results when a signal $s_i(t)$ modulates a carrier at frequency f_j assigned to a different signal $s_j(t)$. This phenomenon results from nonlinearities in the system, and great care, including the use of feedback techniques, is taken to reduce nonlinear effects.† Another problem is the spilling over of significant spectral components from one band into a band reserved for another carrier signal. This can be caused by nonlinearities also. To reduce this kind of distortion, guard bands, as shown in Fig. 6.16-2, are used to separate adjacent-channel spectra.

Compatible Stereo

An important example of an FDM system is to be found in the generation of stereophonic signals that are compatible with monophonic receivers.

Let $s_L(t)$ and $s_R(t)$ denote the left and right message signals, respectively. The generation of the composite baseband $s_b(t)$ is shown in simplified form in Fig. 6.16-3(a).

The sum signal, $s_L(t) + s_R(t)$, can be separated by a low-pass filter and is available for monophonic reception. For stereophonic systems, $s_L(t) + s_R(t)$ is separated from the DSB signal by a low-pass filter. The DSB signal is separated from the monophonic signal by a 23–53-kHz band-pass filter and coherently detected by a locally generated (from the 19-kHz pilot) carrier at

†Another important source of cross talk is coupling between physically separated but adjacent, in the electromagnetic sense, transmission media.

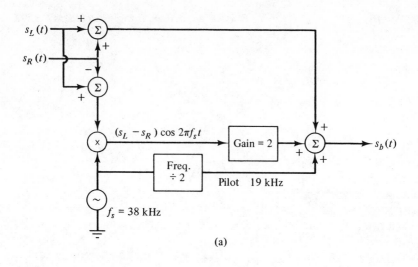

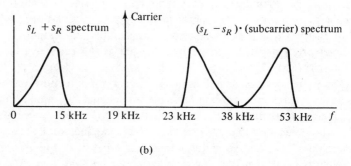

Figure 6.16-3. (a) Generation of stereo baseband signal; (b) spectrum of baseband signal.

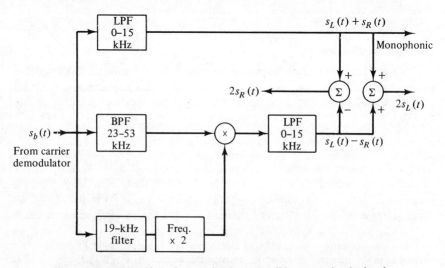

Figure 6.16-4. Detection of monophonic-compatible, stereophonic signals.

38 kHz. The two signals are then added and subtracted to furnish $s_L(t)$ and $s_R(t)$ separately. Figure 6.16-4 is a block diagram of the system.†

6.17 Summary

In this chapter we have examined the principles of amplitude modulation and their application to various systems. We saw that AM and its derivative schemes basically involve the controlled shifting of the signal spectrum to various points along the frequency scale. We saw that the fundamental operation of shifting the baseband spectrum from the origin to the frequency f_c basically requires multiplication by a carrier wave of the form $\cos 2\pi f_c t$ (or $\sin 2\pi f_c t$).

Although what is commonly called ordinary AM, i.e., DSB modulation with transmitted carrier, was seen to be inferior (with respect to power conservation and/or bandwidth requirements) to DSBSC and SSB, the ease with which ordinary AM can be detected more than makes up for its deficiencies and accounts for its widespread use in commercial broadcasting.

We investigated how TV systems manage to transmit the information in visual imagery by electrical signals. We saw that the process of scanning enables the mapping of the two-dimensional image brightness function into one-dimensional electrical signals. We showed that color TV can be made compatible with existing monochrome systems at the cost of increased complexity and the incorporation of almost every AM scheme in existence.

Finally, we studied frequency-division multiplexing (FDM) and showed how FDM can be used in stereo broadcasting that is compatible with monophonic systems.

AM systems are discussed in a number of books. Good treatments of the subject can be found in books by Carlson [6-16], Schwartz et al. [6-17], Bennett [6-14], Panter [6-18], and Rowe [6-19], but there are others.

PROBLEMS

6-1. Consider an AM wave

$$x(t) = A_c[1 + ms(t)] \cos 2\pi f_c t,$$

where $s(t)$ is a real modulating signal of bandwidth W.

a. Under what conditions does

$$x_+(t) = \frac{A_c}{2}[1 + ms(t)]e^{j2\pi f_c t}$$

represent the positive-frequency portion of $x(t)$?

---

†We shall see that in FM the detected noise spectrum increases *quadratically* with frequency. This means that, for example, in DSB-FM FDM stereo, the $s_L - s_R$ signal sits in a band 23–53 kHz where the noise is more pronounced than in the band 0–15 kHz. This accounts for the "noisyness" of stereo compared with mono reception.

b. Assuming that the conditions in part a are met, show that

$$x_+(t) = \tfrac{1}{2}[x(t) + j\hat{x}(t)],$$

where $\hat{x}(t)$ is the Hilbert transform of $x(t)$.

6-2. Assume that in the AM wave

$$x(t) = A_c[1 + ms(t)] \cos 2\pi f_c t$$

the message bandwidth $W < f_c$. Decompose $s(t)$ into $s_+(t)$ and $s_-(t)$, which represent the signals reconstructed from the positive and negative portions of the spectrum, respectively; e.g.,

$$s_+(t) = \int_0^\infty S(f) e^{j2\pi ft} \, df.$$

Sketch the spectrum of $x(t)$, and identify which of the terms $s_\pm e^{\pm j2\pi f_c t}$ produce which portions of the spectrum.

6-3. Consider an AM wave modulated by a periodic signal of the form

$$s(t) = \sum_{n=-N}^{N} c_n e^{j\omega_n t}, \qquad \omega_n = \frac{2\pi n}{T}.$$

a. What condition must be satisfied to prevent foldover distortion?
b. Show that the modulated wave can be written as

$$x(t) = A_c(1 + mc_0) \cos \omega_c t + A_c m \sum_{n=1}^{N} |c_n| \{\cos [(\omega_c + \omega_n)t + \theta_n]$$

$$+ \cos [(\omega_c - \omega_n)t - \theta_n]\}.$$

6-4. A linear half-wave rectifier is a device with transfer characteristic

$$y(t) = \begin{cases} x(t), & x(t) > 0 \\ 0, & \text{otherwise.} \end{cases}$$

Show that if $x(t)$ is a standard AM wave, the output of the rectifier contains a term proportional to $s(t)$. Design a system that will recover $s(t)$.

6-5. Consider the modulator shown in Fig. P6-5. The diode D has infinite backward resistance and forward resistance R_f. Assume piecewise linearity for the diode operating characterstic and $s(t) \ll A_c$.

Figure P6-5

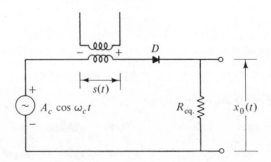

a. Compute the output $x_0(t)$ by replacing the nonlinear circuit by an equivalent linear time-varying circuit. Show that the output can be written as

$$x_0(t) = [A_c \cos \omega_c t + s(t)]g(t),$$

where $g(t)$ is an appropriate switching function. Determine $g(t)$.

b. What is the required filtering if an AM wave is desired? Assume that the signal bandwidth, W, satisfies $W < f_c$.

6-6. Because of a nonlinearity in the system, a DSB generator produces a modulated signal

$$x(t) = s(t) \cos \omega_c t + a_1[s(t) \cos \omega_c t]^2.$$

What relation between carrier frequency f_c and signal bandwidth W must be satisfied to enable the removal of the error term $[s(t) \cos \omega_c t]^2$?

6-7. A DSB wave is modulated by the following signal:

$$s(t) = \sum_{n=1}^{5} n^{-1} \cos n\omega_m t, \qquad \omega_m < \omega_c.$$

a. What constraint must be applied to prevent sideband foldover?

b. Assuming that the constraint in part a is satisfied, sketch the positive-frequency spectrum of the DSB wave.

6-8. Consider an AM wave modulated by a tone at frequency f_n:

$$x(t) = A_c[1 + m \cos \omega_n t] \cos \omega_c t.$$

Because of inadequate filtering, the upper sideband associated with $x(t)$ is totally attenuated. Assuming that $m \ll 1$, compute the resultant envelope.

6-9. Consider the analog modulator shown in Fig. P6-9. The half-square-law devices have transmittance

$$e_{\text{out}} = \begin{cases} Ke_{\text{in}}^2, & e_{\text{in}} > 0 \\ 0, & e_{\text{in}} \leq 0. \end{cases}$$

Figure P6-9

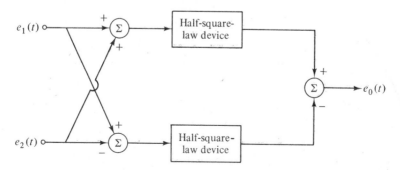

Take $e_1(t) = A_1[1 + ms(t)]$ and $e_2(t) = A_2 \cos \omega_c t$. Assume that $e_1 + e_2$ as well as $e_1 - e_2$ are greater than zero. Show that the output is an AM wave with modulation index strictly less than unity.

6-10. It is possible to increase the modulation of an AM wave before transmission by subtracting excess carrier from the modulated signal. Given an AM wave with m strictly less than 1,

$$x(t) = A[1 + ms(t)] \cos \omega_c t,$$

how much carrier must be removed before unity modulation is achieved? This technique is known as *carrier cancellation*.

6-11. Design a superheterodyne receiver for the following parameters: 50 channels, message bandwidth = 7.5 kHz, RF tuning range to begin at 600 kHz. Let $f_{IF} = 500$ kHz. What are reasonable values of the IF and AF bandwidths? What is the advantage of taking the local oscillator frequency as $f_{LO} = f_c + f_{IF}$ instead of $f_c - f_{IF}$? Sketch a block diagram of the receiver.

6-12. Show that in the regenerative frequency divider of Fig. 6.4-3 the output frequency is f_0/n if the filter rejects the frequency $(2n - 1)f_0/n$.

6-13. An equivalent circuit of the last IF stage and envelope detector in an AM demodulator is shown in Fig. P6-13.

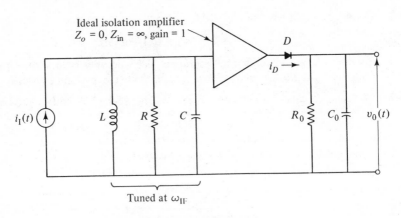

Tuned at ω_{IF}

Figure P6-13

a. Explain qualitatively how this circuit works.

b. Take $i_1(t) = I \cos \omega_{IF} t$; show that $v_0(t)$ is proportional to I. By extension, if $i_1(t) = g(t) \cos \omega_{IF} t$ with $g(t) \geq 0$ and "slowly varying," then $v_0(t)$ is proportional to $g(t)$.

6-14. (DSB and SSB) Consider a square wave of amplitude A and period T. Let $\omega_m \equiv 2\pi/T$. Assume that the square wave is made band-limited by passing it through a low-pass filter with cutoff frequency $f_{co} = Nf_m$. Call the output of the low-pass filter $s(t)$.

a. Obtain an expression for the DSBSC wave

$$x(t) = s(t) \cos \omega_c t$$

in terms of the Fourier components of $s(t)$. Assume that $\omega_c > N\omega_m$.

b. Show that the upper-sideband SSB signal can be written as

$$x(t) = \frac{A}{\pi} \sum_{n=1,3,5...}^{N} \frac{\sin (n\pi/2)}{n} \cos (\omega_c + n\omega_m)t.$$

c. Expand $x(t)$ in part b into the form

$$x(t) = \alpha(t) \cos \omega_c t - \beta(t) \sin \omega_c t,$$

and consider the SSB wave at $t = T/4$. What can be said about the peak power in $x(t)$ as $N \longrightarrow \infty$? What possible disadvantage can you foresee in the use of SSB modulation?

6-15. A method proposed by D. K. Weaver ("A Third Method of Generation and Detection of SSB Signals," *Proc. IRE*, vol. 44, pp. 1703–1705, Dec. 1956) for generating SSB is shown in Fig. P6-15. The method is applicable to

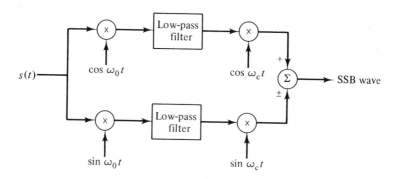

Figure P6-15

signals with finite energy gap near zero frequency. Let the signal be given by

$$s(t) = \sum_{n=1}^{N} c_n \cos (\omega_n t + \phi_n),$$

where $\omega_L \leq \omega_n \leq \omega_L + 2\pi W$ for all n in $n = 1, \ldots, N$. Let $\omega_0 = \omega_L + \pi W$ and $\omega_c \gg \omega_0 > W$.

a. Explain how this system produces an SSB wave by sketching the spectrum at each point. What is the required cutoff frequency for the LPF? Why is it necessary for the signal to have a low-frequency gap?

b. What advantage does this method have over the phase-shifting method discussed in Sec. 6.8?

6-16. Consider the modulated wave

$$x(t) = A_m s(t) \cos \omega_c t + A_m q(t) \sin \omega_c t.$$

a. What is the envelope of the resulting signal if an in-phase carrier term $A_c \cos \omega_c t$ is added to $x(t)$?

b. Let $q(t) = \hat{s}(t)$, where $\hat{s}(t)$ is the Hilbert transform of $s(t)$. Under what conditions can $x(t) + A_c \cos \omega_c t$ be reasonably detected with an envelope detector? (The resulting scheme is called compatible single-sideband or CSSB.)

c. Describe $q(t)$ when $x(t)$ represents, in turn, a (1) DSBSC, (2) SSB, and (3) VSB wave.

6-17. Consider the following expression for an AM-type modulated wave:

$$x(t) = s(t) \cos \omega_c t - q(t) \sin \omega_c t,$$

where $S(f)$ and $Q(f)$, which represent the Fourier transforms of $s(t)$ and $q(t)$, respectively, are zero for $|f| \geq f_c$. Let $x(t)$ appear as an input to a system with impulse response $h(t)$ and transmittance $H(f) = \mathcal{F}[h(t)]$. Show that the output $y(t) = x(t) * h(t)$ can be written as

$$y(t) = r(t) \cos \omega_c t - p(t) \sin \omega_c t,$$

where

$$r(t) = h_1(t) * s(t) - h_2(t) * q(t)$$

and

$$p(t) = h_2(t) * s(t) + h_1(t) * q(t).$$

The relations between $h_1(t)$, $h_2(t)$, and $h(t)$ are implied by

$$h_c(t) = h_1(t) + jh_2(t),$$

where

$$h_c(t) \equiv \mathcal{F}^{-1}[H_c(f)] = \mathcal{F}^{-1}\{1/2 \cdot [1 + \operatorname{sgn}(f + f_c)]H(f + f_c)\}.$$

For a discussion of this problem, see [6-14], p. 193.

6-18. Use the results obtained in Prob. 6-17 to design a filter with transmittance $H(f)$ that will convert

a. A DSB wave into a SSB wave.

b. A VSB wave into a SSB wave.

6-19. (Message scrambling)

a. Analyze the system shown in Fig. P6-19 and show that a reversal of the message spectrum takes place. The resulting "message" represents scrambled speech.

b. Design a system that will work as an unscrambler. *Hint*: Consider an identical system.

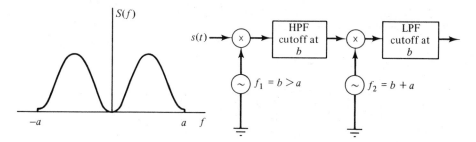

Figure P6-19

6-20. The system shown in Fig. P6-20 is proposed for securing the confidentiality of $s(t)$ in a DSB system. Let $H(f) = H_0 e^{j\phi(f)}$, where $\phi(f)$ is the phase-scrambling function and is real. Let $H(f) \simeq 0$ for $f > f_0$, where $W \leq f_0 < 2f_c - W$. Let $S(f) = 0$ for $f > W$. Explain how the system works. (In optics this procedure is known as coding through a random phase mask.)

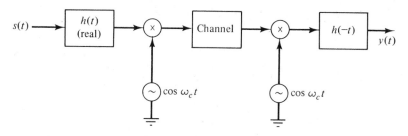

Figure P6-20

6-21. Compute the Fourier coefficients b_{mn} in Eq. (6.14-5) for a brightness distribution $B(x, y)$ as shown in Fig. P6-21.

Figure P6-21

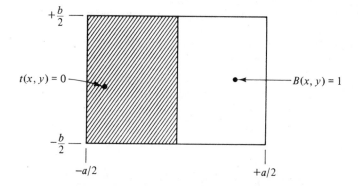

6-22. Repeat Prob. 6-21 except consider a brightness distribution consisting of a Ronchi ruling (Fig. P6-22).

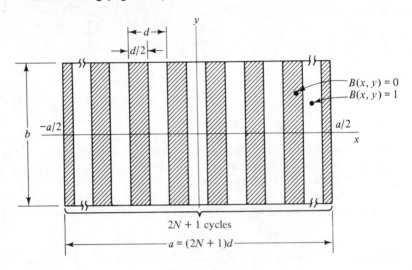

Figure P6-22

6-23. In Sec. 6.14, the scanning process was modeled as a *point* moving across the video field. In actual practice, a finite aperture must be used. Define the aperture by a transmittance function $h(\xi, \eta)$, where (ξ, η) are rectangular coordinates defined with respect to an origin at the center of the aperture and

$$h(\xi, \eta) = \frac{\text{transmitted brightness}}{\text{incident brightness}} \quad \text{at } (\xi, \eta).$$

a. Show that if the aperture is centered at (x, y) the video current is proportional to

$$I_h = \int_{-\infty}^{\infty} \int_{-\infty}^{\infty} h(\xi, \eta) B(x + \xi, y + \eta) \, d\xi \, d\eta,$$

where B is the incident brightness.

b. Write an expression for $h(\xi, \eta)$ when the aperture is a rectangular hole of width w and height h.

c. Compute the current when the aperture in part b is applied to the brightness distribution in Fig. P6-21.

6-24. Use the Fourier series expansion of $B(x, y)$ in Eq. (6.14-1) to show that the current I_h of Prob. 6-23 can be written as

$$I_h = \sum_{n=-\infty}^{\infty} \sum_{m=-\infty}^{\infty} H_{mn}^* c_{mn} \exp\left[j2\pi\left(\frac{m}{a}x + \frac{n}{b}y\right)\right],$$

where H_{mn} is given by

$$H_{mn} = \int_{-\infty}^{\infty} \int_{-\infty}^{\infty} h(\xi, \eta) \exp\left[-j2\pi\left(\frac{m}{a}\xi + \frac{n}{b}\eta\right)\right] d\xi \, d\eta$$

and $h(\xi, \eta)$ is a real function. What effect does the filter $\{H_{mn}\}$ have on the spectrum of the video signal?

6.25. The aspect ratio $\mathcal{A}$ is the ratio of relative image width W to image height L in the TV image. Show that if the resolution, in resolvable lines per unit distance, is to be the same in the horizontal and vertical directions, then

$$\mathcal{A} \equiv \frac{W}{L} = \frac{n_H}{n_V},$$

where n_V, n_H are the number of rows and columns of the finest (i.e., most detailed) checkerboard pattern that can be resolved (Fig. P6-25).

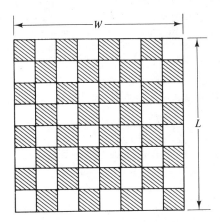

Figure P6-25

6-26. (Continuation of Prob. 6-25) If a sinusoidal signal of maximum video frequency W is applied to a TV monitor, a sequence of dark and light zones result, much as a checkerboard row. The *effective* line-scan time, T_E, is the difference between the total line-scan time, T_L, and the horizontal retrace time, T_{HR}.

a. Show that $n_H = 2WT_E = 2W(T_L - T_{HR})$.

b. The "utilization" factor or Kell factor, K, is a number that relates n_V to the useful number of scan lines per frame. If N = total number of scan lines and N_{VR} = number of scan lines blanked out during the vertical retrace, then

$$n_V = K(N - N_{VR}).$$

If the electron scanning beam were perfectly aligned with the rows of the checkerboard pattern and suffered no randomness in position, K would be 1. In practice K is less than 1 ($K \simeq 0.7$). Use this fact and the results from Prob. 6-25 and this problem to show that the bandwidth W can be written as

$$W = 0.35\mathcal{A}\frac{N}{T_L}\frac{1 - N_{VR}/N}{1 - T_{HR}/T_L}.$$

6-27. (Continuation of Prob. 6-26) The frame time T_F is the time required to produce a complete image by scanning. Show that T_F is given by

$$T_F = \frac{0.714 \alpha n_V^2}{W(1 - N_{VR}/N)(1 - T_{HR}/T_L)}.$$

6-28. A Mars-Earth facsimile transmission system has the following parameters: horizontal sweep = 300°, vertical sweep = 60°, number of horizontal lines = 9150, number of gray levels = 64, carrier frequency = 2.2 GHz (2.2 × 10⁹ Hz), and maximum signaling rate = 4000 bits/sec. Estimate the frame time T_f if both N_{VR} and T_{HR} are negligible and can be taken as zero.

6-29. In Fig. 6.15-5, explain why the average luminance signal obtained after averaging the composite signal (luminance plus subcarrier signal) after two scans is slightly different from the original luminance signal. Why is the average luminance "bumpy" in regions where the composite signal overshoots the black level?

6-30. Consider a five-line-per-field, one-field-per-second, TV system in which the color subcarrier f_{cc} is chosen to be the *first* multiple of one-half the line-scanning frequency f_s.
  a. How many cycles of subcarrier are there per field?
  b. How many cycles of subcarrier are there per line?
  c. Draw a time-phase diagram as in Fig. 6.15-4 to verify that the subcarrier-induced luminance variations tend to cancel in time and space averaging.

6-31. Explain what happens when a color TV monitor receives a black/white picture? The ability of a color monitor to reproduce a black/white image is termed *reverse compatibility*.

6-32. Design an FDM system that accommodates 10 12-kHz channels. Two of the 12 kHz should be reserved for guard bands, and the remaining spectrum should accommodate a 10-kHz signal. Choose an appropriate set of subcarriers. What kind of subcarrier modulation do you recommend? What is the bandwidth of the carrier wave if DSB modulation is chosen?

REFERENCES

[6-1] *IRE Dictionary of Electronics Terms and Symbols*, The IEEE Inc., New York, 1961, p. 92.

[6-2] B. GILBERT, "A New Wide-Band Amplifier Technique," *IEEE J. Solid State Circuits*, SC-3, pp. 353–365, Dec. 1968.

[6-3] D. H. SHEINGOLD, ed., *Non-Linear Circuits Handbook*, Analog Devices, Inc., Norwood, Mass., 1974.

[6-4] A. B. CARLSON, *Communication Systems: An Introduction to Signals and Noise in Electrical Communications*, McGraw-Hill, New York, 1968.

[6-5] V. H. GRINICH and H. G. JACKSON, *Introduction to Integrated Circuits*, McGraw-Hill, New-York, 1975.

[6-6]   J. MILLMAN and C. C. HALKIAS, *Electronic Devices and Circuits*, McGraw-Hill, New York, 1967.

[6-7]   K. HENNEY, *Radio Engineering Handbook*, McGraw-Hill, New York, 1950.

[6-8]   J. P. COSTAS, "Synchronous Communication," *Proc. IRE*, **44**, pp. 1713–1718, Dec. 1956.

[6-9]   F. S. HILL, JR., "On Time Domain Representations for Vestigial Sideband Signals," *Proc. IEEE*, **62**, pp. 1032–1033, July 1974.

[6-10] A. J. VITERBI, *Principles of Coherent Communications*, McGraw-Hill, New York, 1966.

[6-11] J. MILLMAN and H. TAUB, *Pulse, Digital and Switching Waveforms*, McGraw-Hill, New York, 1965.

[6-12] J. W. SANDBERG, "Assembling and Displaying Slow-Scan TV Pictures," Hughes Customer Application Notes 91-11-009, Hughes Aircraft Co., Ocean Side, Calif., April 1973.

[6-13] L. H. HANSEN, *Introduction to Solid-State Television Systems*, Prentice-Hall, Englewood Cliffs, N.J., 1969.

[6-14] W. R. BENNETT, *Introduction to Signal Transmission*, McGraw-Hill, New York, 1970.

[6-15] J. W. WENTWORTH, *Color Television Engineering*, McGraw-Hill, New York, 1955.

[6-16] A. B. CARLSON, *Communication Systems: An Introduction to Signals and Noise in Electrical Communications*, 2nd ed., McGraw-Hill, New York, 1975.

[6-17] M. SCHWARTZ, W. R. BENNETT, and S. STEIN, *Communication Systems and Techniques*, McGraw-Hill, New York, 1966.

[6-18] P. F. PANTER, *Modulation, Noise and Spectral Analysis*, McGraw-Hill, New York, 1965.

[6-19] H. E. ROWE, *Signals and Noise in Communication Systems*, Van Nostrand Reinhold, New York, 1965.

7

Angle Modulation

7.1 Introduction

Angle modulation encompasses *phase modulation* (PM) and *frequency modulation* (FM) and refers to the process by which the quantity $\theta(t)$ in the expression

$$x(t) = A \cos \theta(t) \tag{7.1-1}$$

is controlled by a message $s(t)$. The amplitude A is intended to be constant, and any variations in A constitute a form of noise. As always, we shall assume that the class of modulating signals $\{s(t)\}$ is band-limited to W Hz. Although both FM and PM will be discussed in this chapter, the emphasis will be on FM since it is by far the most important angle-modulation process in use.

Historically, FM was first correctly analyzed by John R. Carson [7-1], who discussed the relatively wide bandwidths required. E. H. Armstrong [7-2] was among the first to recognize the noise suppression properties of FM and designed an FM modulator based on PM. Commercial broadcast FM is radiated in the band extending from 88 to 108 MHz. A single channel is nominally 200 kHz wide.

Tests have shown that an interfering audio signal will create objectionable interference if its level is as high as 30–40 dB below the desired signal ([7-3], pp. 21–65). Thus the feasible service areas are the zones in which the desired component of the resulting audio is at least 35 dB above the interference. In the case of AM (amplitude modulation, *not* angle modulation) the power ratio of desired AM to interference must be at least 35 dB if the interfering wave is on the same carrier frequency. In the case of FM systems, however, the ratio of desired FM to interference need only be 6 dB to meet the same audio criterion. A property of FM that makes this possible is that the information in FM signals is in the zero crossings of the wave and not in its amplitude where the primary effect of the interference is manifested. The amplitude variations induced by the interference are removed in the FM receiver by a limiter circuit.

The main advantages of FM over AM are

1. Improved signal-to-noise ratio. Tests have shown as much as a 25-dB increase in this ratio over AM with respect to automobile ignition, X-ray generation, and other man-made interference.
2. A smaller geographical interference area when two nearby FM transmitters are operating simultaneously on the same frequency.
3. Less radiated power required for the same signal-to-noise ratio.
4. More efficient use of transmitting equipment.
5. The existence of uniform and well-defined service areas for a given transmitter since the FM signal-to-noise ratio remains high until the field intensity reaches a low value (threshold effect).

Against these important advantages, FM also suffers some serious drawbacks. An FM wave typically requires a large bandwidth, up to 20 times the amount required for AM. FM systems are generally more complicated than corresponding AM ones. FM modulation is also strongly nonlinear; this means that superposition doesn't hold and that the analysis of FM waves is more difficult than the analysis of AM. In fact we shall see that FM analysis uses more approximations and is less rigorous than is the analysis for AM.

7.2 Definitions

The instantaneous radian frequency is defined by

$$\omega = \lim_{\Delta t \to 0} \frac{\Delta \theta}{\Delta t} = \dot{\theta}(t) \quad \text{(radians per second)}, \qquad (7.2\text{-}1)$$

or equivalently,

$$f = \frac{1}{2\pi} \dot{\theta}(t) \quad \text{(hertz).}$$

The fact that frequency, which is a measure of the number of cycles per second, changes continuously with time should not be considered paradoxical. The same concept holds in mechanics where velocity can change continuously with time despite the fact that it is a measure of the number of meters traversed per second.

In Eq. (7.2-1), let $\theta(t) = \omega_c t + \theta_0$, where ω_c and θ_0 are constants. We then obtain

$$x(t) = A \cos(\omega_c t + \theta_0). \qquad (7.2\text{-}2)$$

This type of wave is called an unmodulated carrier and conveys no information. The instantaneous radian frequency is a constant given by

$$\dot{\theta}(t) = \frac{d}{dt}(\omega_c t + \theta_0) = \omega_c,$$

i.e., the carrier frequency. The instantaneous phase is a linear function of time and increases in direct proportion to t, with slope ω_c.

A phase-modulated wave is described by the expression

$$x_{PM}(t) = A \cos [\omega_c t + \theta_0 + K's(t)], \qquad (7.2\text{-}3)$$

where K' is a constant with units of radians per volt if $s(t)$, the modulating signal, has units of volts. If we normalize $s(t)$ to satisfy $|s(t)| \leq 1$, we find that the peak phase deviation from the unmodulated phase $\omega_c t + \theta_0$ is just K'.

Let the modulating signal be a pure tone of radian frequency ω_m; i.e.,

$$s(t) = A_m \cos \omega_m t; \qquad (7.2\text{-}4)$$

then

$$x_{PM}(t) = A \cos [\omega_c t + A_m K' \cos \omega_m t], \qquad (7.2\text{-}5)$$

where θ_0 has, without loss of generality, been set equal to zero. The constant

$$K_d \equiv A_m K' \qquad (7.2\text{-}6)$$

is called the *phase deviation* and is a measure of the maximum shift of the phase of $x_{PM}(t)$ from $\omega_c t$. The phase deviation depends on the *amplitude* of the tone.

An FM wave is described by

$$x(t) = A \cos \left[\omega_c t + \theta_0 + K'' \int_{t_0}^{t} s(\lambda)\, d\lambda \right], \qquad (7.2\text{-}7)$$

where K'' is a constant with units of radians per volt-second. By choosing t_0 appropriately we can cancel θ_0. As in the case of PM, we shall set θ_0 equal to zero.

The instantaneous phase is†

$$\theta(t) = \omega_c t + K'' \int^{t} s(\lambda)\, d\lambda, \qquad (7.2\text{-}8)$$

and the instantaneous frequency f (we omit writing $f(t)$ which could easily be misread as "function of time") is given by

$$f = \frac{\dot{\theta}(t)}{2\pi} = f_c + Ks(t), \qquad f_c \equiv \frac{\omega_c}{2\pi}, \qquad K \equiv \frac{K''}{2\pi}. \qquad (7.2\text{-}9)$$

From Eq. (7.2-8) we see that if we first integrate $s(t)$ and then allow it to phase-modulate a carrier we have achieved FM modulation indirectly. In fact this method of producing an FM wave was first used by Armstrong and is called *indirect* FM.

---

†The arbitrary nature of the lower limit is indicated by leaving it off altogether.

Let us again consider a modulating signal consisting of a pure tone, as in Eq. (7.2-4). Then the FM wave takes the form

$$x_{FM}(t) = A \cos \left(\omega_c t + \frac{A_m K''}{\omega_m} \sin \omega_m t \right). \qquad (7.2\text{-}10)$$

The maximum frequency deviation from the frequency of the unmodulated carrier is

$$\Delta\omega \equiv \left| \frac{d}{dt} \left(\frac{A_m K''}{\omega_m} \sin \omega_m t \right) \right|_{\max} = A_m K''. \qquad (7.2\text{-}11)$$

The quantity $\Delta\omega$ or $\Delta f = \Delta\omega/2\pi$ is appropriately called the *frequency deviation*. The frequency deviation divided by the modulating frequency is the maximum phase difference between the FM wave and the unmodulated carrier; it is called the *modulation index* and is frequently denoted by β; i.e.,

$$\beta \equiv \frac{\Delta\omega}{\omega_m} = \frac{\Delta f}{f_m} \qquad \text{(radians)}. \qquad (7.2\text{-}12)$$

Equations (7.2-5) and (7.2-10) are very much alike. In the steady state, for a fixed ω_m and K' and K'' adjusted so that $K' = K''/\omega_m$, there would be no way of distinguishing the PM wave from the FM wave from a single trace. Only by varying the modulating frequency ω_m, while keeping the amplitude constant, could one distinguish between the two waves. No change in peak phase deviation would be observed in the PM case, while an inverse relation between ω_m and peak phase deviation would be observed in the FM case.

Both the PM and FM waves can, under conditions of sinusoidal modulation, be described by a wave of the form

$$x(t) = A \cos [\omega_c t + \xi \sin (\omega_m t + \gamma)], \qquad (7.2\text{-}13)$$

where $\xi = \beta$ and $\gamma = 0$ for FM and $\xi = K_d$ and $\gamma = \pi/2$ for PM.

The resolution of $x(t)$ into a sinusoidal spectrum is central to the analysis of angle modulation. It is considered in the next section.

7.3 Resolution of Angle-Modulated Waves into Sinusoids

We recall the formula for the Taylor series expansion of $\exp Y$:

$$e^Y = \sum_{l=0}^{\infty} \frac{Y^l}{l!}. \qquad (7.3\text{-}1)$$

If we apply a Taylor series expansion to the function

$$g(\xi, z) \equiv e^{\xi(z-z^{-1})/2} = e^{\xi z/2} \cdot e^{-\xi z^{-1}/2},$$

we obtain the double series

$$g(\xi, z) = \sum_{i=0}^{\infty} \sum_{l=0}^{\infty} \frac{\xi^{i+l}}{2^{i+l}} \frac{z^{i-l}}{i! \, l!} (-)^l.$$

The substitution $n = i - l$ enables us to write

$$g(\xi, z) = \sum_{n=-\infty}^{\infty} z^n \sum_{l=0}^{\infty} \frac{(\xi/2)^{n+2l}(-)^l}{(n+l)!\,l!}.$$ (7.3-2)

In Eq. (7.3-2), the convergent series

$$J_n(\xi) \equiv \sum_{l=0}^{\infty} \frac{(\xi/2)^{n+2l}(-)^l}{(n+l)!\,l!}$$ (7.3-3)

is the nth-order Bessel function of the first kind and is consistent with the integral definition

$$J_n(\xi) = \frac{1}{2\pi} \int_{-\pi}^{\pi} e^{j(\xi \sin \lambda - n\lambda)} \, d\lambda.$$ (7.3-4)

To show the equivalence between Eqs. (7.3-3) and (7.3-4) requires only that we expand the exponential inside the integral into a Taylor series and carry out the integration. For example, with $n = 0$, we obtain, from Eq. (7.3-4),

$$J_0(\xi) = \frac{1}{2\pi} \int_{-\pi}^{\pi} e^{j\xi \sin \lambda} \, d\lambda$$

$$= \frac{2}{\pi} \sum_{k=0,2,4,\ldots}^{\infty} \frac{\xi^k (-)^{k/2}}{k!} \int_0^{\pi/2} \sin^k \lambda \, d\lambda.$$ (7.3-5)

If we use the formula [7-4], p. 258

$$\int_0^{\pi/2} \sin^k \lambda \, d\lambda = \frac{1 \cdot 3 \cdot 5 \cdots (k-1)}{2 \cdot 4 \cdot 6 \cdots k} \cdot \frac{\pi}{2} \qquad (k \text{ even}),$$ (7.3-6)

we obtain Eq. (7.3-3) with $n = 0$. In view of Eq. (7.3-3) we can write

$$g(\xi, z) = \sum_{n=-\infty}^{\infty} J_n(\xi) z^n.$$ (7.3-7)

If we let $z = \exp(j\omega_m t)$, we obtain

$$e^{j\xi \sin \omega_m t} = \sum_{n=-\infty}^{\infty} J_n(\xi) e^{jn\omega_m t}.$$ (7.3-8)

Equation (7.3-8) is known as the Bessel-Jacobi identity. By multiplying both sides by $\exp(j\omega_c t)$ and setting real and imaginary parts equal, we can write the important formulas

$$\cos(\omega_c t + \xi \sin \omega_m t) = \sum_{n=-\infty}^{\infty} J_n(\xi) \cos(\omega_c t + n\omega_m t),$$ (7.3-9a)

$$\sin(\omega_c t + \xi \sin \omega_m t) = \sum_{n=-\infty}^{\infty} J_n(\xi) \sin(\omega_c t + n\omega_m t).$$ (7.3-9b)

Equations (7.3-9) are the basic formulas for the sinusoidal resolution of an angle-modulated wave. They can be directly generalized to any number of modulating components, although the resulting expressions are not succinct.

For example, if the modulating signal has the form

$$s(t) = A_1 \cos \omega_1 t + A_2 \cos \omega_2 t, \tag{7.3-10}$$

then the modulation indices are $\beta_1 = A_1 K''/\omega_1$ and $\beta_2 = A_2 K''/\omega_2$ and the complex form of an FM wave can be written as

$$e^{j(\omega_c t + \beta_1 \sin \omega_1 t + \beta_2 \sin \omega_2 t)} = \sum_{n=-\infty}^{\infty} \sum_{k=-\infty}^{\infty} J_n(\beta_1) J_k(\beta_2) e^{j(\omega_c t + n\omega_1 t + k\omega_2 t)}. \tag{7.3-11}$$

Equating real and imaginary parts results in

$$\cos(\omega_c t + \beta_1 \sin \omega_1 t + \beta_2 \sin \omega_2 t)$$

$$= \sum_{n=-\infty}^{\infty} \sum_{k=-\infty}^{\infty} J_n(\beta_1) J_k(\beta_2) \cos(\omega_c + n\omega_1 + k\omega_2)t \tag{7.3-12a}$$

and

$$\sin(\omega_c t + \beta_1 \sin \beta_1 t + \beta_2 \sin \omega_2 t)$$

$$= \sum_{n=-\infty}^{\infty} \sum_{k=-\infty}^{\infty} J_n(\beta_1) J_k(\beta_2) \sin(\omega_c + n\omega_1 + k\omega_2)t. \tag{7.3-12b}$$

If $\beta_2 = 0$, all k components vanish except for $k = 0$. The result is Eqs. (7.3-9). The frequencies $\omega_c + n\omega_1, n = 0, \pm 1, \pm 2, \ldots$, are generated by the modulating tone $A_1 \cos \omega_1 t$. Similarly, when $\beta_1 = 0$, we obtain only the frequencies $\omega_c + k\omega_2$, $k = 0, \pm 1, \pm 2, \ldots$, which are generated by $A_2 \cos \omega_2 t$. However, when both modulating tones are active, we obtain all possible combinations of $\omega_c + n\omega_1 + k\omega_2$. The new frequencies $n\omega_1 + k\omega_2$ ($n, k \neq 0$) clearly show how nonlinear FM and, more generally, angle modulation are. Figure 7.3-1 illustrates the point for $f_2 > f_1$.

Properties of the Bessel Functions

The Bessel functions $J_n(\xi)$ can be defined by an integral, as in Eq. (7.3-4), or by a generating series as in Eq. (7.3-3). They are the solution of the second-order differential equation known as Bessel's equation:

$$\xi^2 \frac{d^2\eta}{d\xi^2} + \xi \frac{d\eta}{d\xi} + (\xi^2 - n^2)\eta = 0.$$

The odd-ordered Bessel functions are odd functions of the argument, while the even-ordered Bessel functions are even functions of the argument. Further,

$$J_n(\xi) = \begin{cases} J_{-n}(\xi), & n \text{ even} \\ -J_{-n}(\xi), & n \text{ odd.} \end{cases} \tag{7.3-13}$$

Figure 7.3-2 is a graph of the first few Bessel functions.

1. *Asymptotic behavior.* For ξ large and positive, we have

$$J_0(\xi) = \left(\frac{2}{\pi\xi}\right)^{1/2} \left[P_0(\xi) \cos\left(\xi - \frac{\pi}{4}\right) - Q_0(\xi) \sin\left(\xi - \frac{\pi}{4}\right) \right],$$

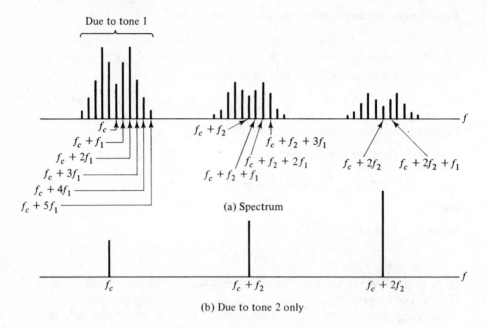

Figure 7.3-1. Spectra of an FM wave generated by two modulating tones. (a) The components due to tone 1 are clustered around f_c. (b) The components due to tone 2. All other components are beat frequencies generated by the non-linear interaction of tones 1 and 2 in the FM modulation process.

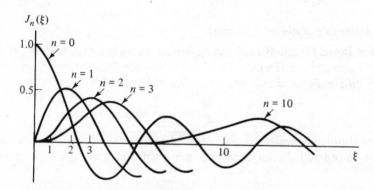

Figure 7.3-2. Variations of $J_n(\xi)$ with ξ.

where $P_0(\xi) \simeq 1$ and $Q_0(\xi) \simeq (8\xi)^{-1}$. Hence the asymptotic value of $J_0(\xi)$ is

$$J_0(\xi) = \left(\frac{2}{\pi\xi}\right)^{1/2} \cos\left(\xi - \frac{\pi}{4}\right). \qquad (7.3\text{-}14)$$

Each of the first pair of sideband frequencies has an amplitude of $J_1(\xi)$. This function, like all other $J_n(\xi)$ for $n \neq 0$, is zero at $\xi = 0$. For ξ large and positive,

$$J_1(\xi) = \left(\frac{2}{\pi\xi}\right)^{1/2}\left[P_1(\xi)\cos\left(\xi - \frac{3\pi}{4}\right) - Q_1(\xi)\sin\left(\xi - \frac{3\pi}{4}\right)\right],$$

where $P_1(\xi) \simeq 1$ and $Q_1(\xi) \simeq 3(8\xi)^{-1}$. The asymptotic value is

$$J_1(\xi) = \left(\frac{2}{\pi\xi}\right)^{1/2}\sin\left(\xi - \frac{\pi}{4}\right). \tag{7.3-15}$$

In general, we can write the asymptotic relation for all n as

$$J_n(\xi) \simeq \left(\frac{2}{\pi\xi}\right)^{1/2}\cos\left(\xi - \frac{n\pi}{2} - \frac{\pi}{4}\right), \qquad \xi \longrightarrow \infty. \tag{7.3-16}$$

2. *Zeros.* The zeros of the Bessel functions are not evenly spaced. The first few zeros for $J_0(\xi)$ and $J_1(\xi)$ are given in Table 7.3-1. For large ξ the

TABLE 7.3-1   ZEROS OF BESSEL FUNCTIONS

$J_0(\xi_n) = 0$		$J_1(\xi_n) = 0$	
Roots ξ_n	$J_1(\xi_n)$	Roots ξ_n	$J_0(\xi_n)$
2.4048	0.5191	0.0000	1.0000
5.5201	−0.3403	3.8317	−0.4028
8.6537	0.2715	7.0156	0.3001
11.7915	−0.2325	10.1735	−0.2497
14.9309	0.2065	13.3237	0.2184
18.0711	−0.1877	16.4706	−0.1965
21.2116	0.1733	19.6159	0.1801

Bessel functions behave somewhat like damped sinusoids. As n increases, $J_n(\xi)$ reaches its maximum value at a greater distance from the origin. However, a most important observation is that $J_n(\xi)$ quickly falls off in amplitude for values of $n \gg \xi$. In fact n need not really be much larger than ξ. Figure 7.3-3 is a graph of $J_n(\xi)$ vs. n with ξ an integer parameter. Non-integer-ordered Bessel functions are also in use and can be defined by Eq. (7.3-3) except that the factorial definition is extended according to

$$n! = \int_0^\infty \xi^n e^{-\xi}\, d\xi \qquad \text{for } n > 0. \tag{7.3-17}$$

The integral is the definition of the *gamma function*, $\Gamma(n + 1)$, which satisfies the recursion relation $\Gamma(n + 1) = n\Gamma(n)$. When n is an integer, it is easy to show that $\Gamma(n + 1) = n!$. For this reason the gamma function is also called the factorial function.

We return now to the analysis of FM. Equations (7.3-12) can be generalized to a signal with M Fourier components. Thus if

$$s(t) = \sum_{i=1}^{M} A_i \cos(\omega_i t + \psi_i),$$

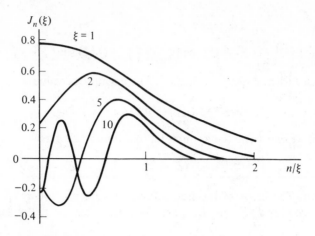

Figure 7.3-3. Variations of $J_n(\xi)$ vs. n with ξ as a parameter.

the FM wave takes the form

$$x_{\text{FM}}(t) = A_c \cos \left[\omega_c t + \sum_{i=1}^{M} \beta_i \sin (\omega_i t + \psi_i) \right], \qquad (7.3\text{-}18)$$

which can be written as

$$x_{\text{FM}}(t) = \text{Re} \left(\exp \left\{ j[\omega_c t + \sum_{i=1}^{M} \beta_i \sin (\omega_i t + \psi_i)] \right\} \right)$$

$$= \text{Re} \left(\sum_{k_1=-\infty}^{\infty} \sum_{k_2=-\infty}^{\infty} \cdots \sum_{k_M=-\infty}^{\infty} J_{k_1}(\beta_1) \cdots J_{k_M}(\beta_M) \right.$$

$$\left. \cdot \exp \left\{ j \left[\omega_c t + \sum_{i=1}^{M} k_i(\omega_i t + \psi_i) \right] \right\} \right)$$

$$= \sum_{k_1} \cdots \sum_{k_M} \left[\prod_{i=1}^{M} J_{k_i}(\beta_i) \right] \cos \left[\omega_c t + \sum_{i=1}^{M} k_i(\omega_i t + \psi_i) \right]. \qquad (7.3\text{-}19)$$

In theory, this tedious but general expression enables us to compute the bandwidth to any degree of accuracy; in practice, the bandwidth is determined by other means.

Angle-modulated waves do not have the appearance of AM waves. For example, consider a modulating signal consisting of a linear ramp; i.e., $s(t) = btu(t)$, where b is a constant and $u(t)$ is the unit step. For $t > 0$, the PM and FM waves are described by

PM: $\quad x_{\text{PM}}(t) = A \cos (\omega_c t + K'bt), \qquad$ inst. frequency $\dot{\theta} = \omega_c + K'b$

FM: $\quad x_{\text{FM}}(t) = A \cos \left(\omega_c t + \dfrac{K''b}{2} t^2 \right), \qquad \dot{\theta} = \omega_c + K''bt.$

These results enable us to draw the waveforms in Fig. 7.3-4 for the *periodic-ramp*-modulating signal.

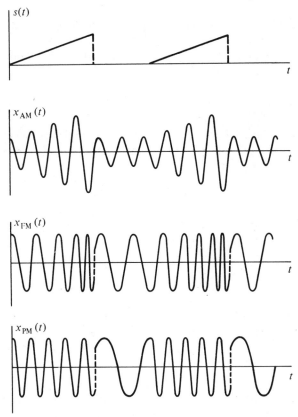

Figure 7.3-4. Carrier waveforms for AM, FM, and PM when the modulation is a periodic ramp as shown. The PM wave switches discontinuously between two frequencies separated by an amount $K'b/2\pi$ Hz.

7.4 Narrowband Angle Modulation

The analysis of angle-modulated waves is considerably simplified if the phase deviation, K_d, in PM or the modulation index, β, in FM is considerably smaller than 1 radian. Consider a PM wave modulated by a single-frequency tone at radian frequency ω_m. Then

$$
\begin{aligned}
x_{PM}(t) &= \cos(\omega_c t + K_d \cos \omega_m t) \\
&= \cos \omega_c t \cos(K_d \cos \omega_m t) - \sin \omega_c t \sin(K_d \cos \omega_m t) \\
&= \cos \omega_c t \left(1 - \frac{K_d^2 \cos^2 \omega_m t}{2!} + \frac{K_d^4 \cos^4 \omega_m t}{4!} - \cdots \right) \\
&\quad - \sin \omega_c t \left(K_d \cos \omega_m t - \frac{K_d^3 \cos^3 \omega_m t}{3!} + \cdots \right), \quad (7.4\text{-}1)
\end{aligned}
$$

where we have replaced $\cos(K_d \cos \omega_m t)$ and $\sin(K_d \cos \omega_m t)$ by their Maclaurin series expansion.

If $K_d < 0.2$ radian, then, to a good approximation, we can omit all powers above the first and obtain

$$x_{PM}(t) = \cos \omega_c t - K_d \sin \omega_c t \cos \omega_m t$$

$$= \cos \omega_c t - \tfrac{1}{2} K_d [\sin(\omega_c + \omega_m)t + \sin(\omega_c - \omega_m)t]. \quad (7.4\text{-}2)$$

The spectrum of such a wave is shown in Fig. 7.4-1 and superficially resembles AM in that the only significant frequencies are a pair of symmetrical sidebands at $f = f_c - f_m$ and $f = f_c + f_m$.

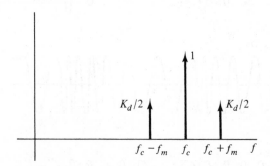

Figure 7.4-1. Amplitude spectrum of narrowband PM.

However, the sidebands of narrowband PM sum to a component in quadrature with the carrier, while in AM the sidebands sum to a component in phase with the carrier. This is easily seen from a phasor diagram.† The complex form of Eq. (7.4-2) is

$$\tilde{x}_{PM}(t) = e^{j\omega_c t}\left[1 + \frac{K_d}{2}e^{j(\omega_m t + \pi/2)} + \frac{K_d}{2}e^{j(-\omega_m t + \pi/2)}\right]$$

$$= e^{j\omega_c t} X_{PM}(t), \quad (7.4\text{-}3)$$

where the $X_{PM}(t)$ are the three terms in square brackets. The real wave $x_{PM}(t)$ is reconstructed from

$$x_{PM}(t) = \mathrm{Re}\,[e^{j\omega_c t} X_{PM}(t)]. \quad (7.4\text{-}4)$$

The factor $\exp(j\omega_c t)$ is the carrier phasor and rotates in the counterclockwise direction at an angular velocity ω_c. Without loss of generality, we can suppress its effect by considering all rotations relative to it. The important

†For the reader not familiar with this notion, the phasor is basically a vector representation of the complex or "analytic" form of a signal $x(t)$. For example, $x(t) = A \cos(\omega_c t + \theta)$ can be represented by the complex signal $\tilde{x}(t) = A \exp[+j(\omega_c t + \theta)]$. The phasor would then be interpreted as having length A and angle $\omega_c t + \theta$ from the real axis. It would be drawn as a vector whose projection on the real axis is $x(t)$.

quantity is then the resultant $X_{PM}(t)$. It is shown in Fig. 7.4-2(a). The phasor diagram for AM is shown in Fig. 7.4-2(b). Note that in PM the amplitude of the wave hardly changes but the phase does, while in AM there is no phase change but a considerable amplitude change.

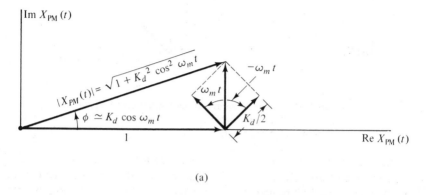

(a)

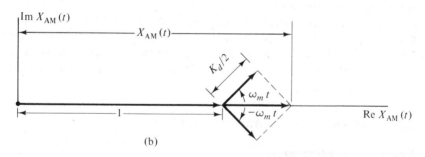

(b)

Figure 7.4-2. Phasors for narrowband PM and AM: (a) PM; (b) AM.

The same results can be obtained for narrowband FM (NBFM). In this case

$$x_{FM}(t) = \cos(\omega_c t + \beta \sin \omega_m t),$$

which, for $\beta \ll 1$ radian (i.e., $\beta \leq 0.2$ radian), gives

$$x_{FM}(t) \simeq \cos \omega_c t - \frac{\beta}{2}[\cos(\omega_c - \omega_m)t - \cos(\omega_c + \omega_m)t]. \quad (7.4\text{-}5)$$

Equation (7.4-5) was derived using techniques similar to those leading up to Eq. (7.4-2). The phasor diagram is obtained from the complex wave

$$\tilde{x}_{FM}(t) = e^{j\omega_c t}X_{FM}(t),$$

where

$$X_{FM}(t) = 1 + \frac{\beta}{2}e^{-j(\omega_m t - \pi)} + \frac{\beta}{2}e^{j\omega_m t}. \quad (7.4\text{-}6)$$

Figure 7.4-3 shows how the sidebands sum to a component in quadrature with the carrier.

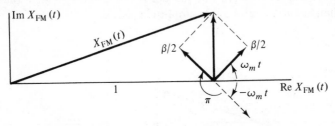

Figure 7.4-3. Phasor diagram for narrowband FM. As in the case of PM, the sidebands sum to a component in quadrature with the carrier.

The spectrum of narrowband angle modulation is thus similar to AM except for the phase quadrature of the sidebands with respect to the carrier. This suggests a method of generating narrowband angle modulation in terms of AM circuitry with which we are already familiar. We recall that to generate AM we could add a carrier to a DSB output produced by a balanced modulator [Fig. 7.4-4(a)]. If we phase-shift the carrier by 90° before addition, we produce a PM wave [Fig. 7.4-4(b)]. If $s(t)$ is integrated before modulation, we produce an FM wave [Fig. 7.4-4(c)]. These systems apply to narrowband angle modulation only.

7.5 Bandwidth Considerations for FM

Single-Tone Modulation

Nowhere in the study of FM does there seem to be more "rule-of-thumb" computations and less rigor than in the determination of the required transmission bandwidth. First, we have to decide what we mean by bandwidth. We cannot use the "strict" definition of bandwidth that requires all components of the spectrum to be zero outside the band because, according to that criterion, even a tone-modulated FM wave would produce an infinite bandwidth. The actual bandwidth is of course determined by the spread of *significant* components in the spectrum. Estimates of the spread of these significant components can be obtained from the following considerations.

From Eq. (7.3-9a) we have

$$x_{FM}(t) = \cos(\omega_c t + \beta \sin \omega_m t) = \sum_{n=-\infty}^{\infty} J_n(\beta) \cos(\omega_c + n\omega_m t). \qquad (7.5-1)$$

Figure 7.3-3 shows that for $n > \beta$ the value of $J_n(\beta)$ falls off very quickly, especially when β is large. So, to a first approximation, we have that the bandwidth is determined by the 2β nonnegligible components above and

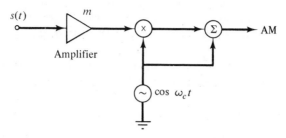

Stark and Tuteur 7.4.4a

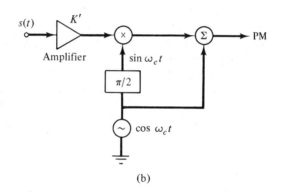

(b)

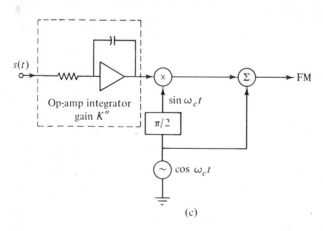

(c)

Figure 7.4-4. (a) Generation of AM; (b) generation of NBPM; (c) generation of NBFM.

below f_c (i.e., β above, β below). Since adjacent components are separated by f_m Hz, the total bandwidth is $2\beta f_m$ = twice the frequency deviation Δf. However, a shortcoming of this result is that if $\beta \ll 1$, as is the case in NBFM, Δf may be less than f_m, and we would have no modulation at all if we assign a bandwidth $2\Delta f$. To retain some modulation we must keep at least the first pair of sidebands at $f_c \pm f_m$. Thus, to ensure that we have modulation by the tone for any β we could require that the bandwidth be given by

$$B_{CR} = 2\Delta f + 2f_m = 2f_m(1 + \beta) \qquad (7.5\text{-}2)$$

$$\simeq 2f_m\beta = 2\Delta f, \qquad \beta \gg 1$$
$$\simeq 2f_m, \qquad\qquad \beta \ll 1 \;\; (\text{NBFM}). \qquad (7.5\text{-}3)$$

Equation (7.5-2) is a handy rule of thumb and a special case of a bandwidth formula called *Carson's rule* (the subscript CR is for Carson's rule).

For high-quality FM transmission or for modulation with high-frequency signals the "$n > \beta$" rule for determining significant components may not suffice and we then need to be more specific about what we mean by significant Fourier components. Suppose that we define as significant all components for which $|J_n(\beta)| \geq \epsilon$. This criterion requires that significant sidebands be at least $100\epsilon\%$ of the amplitude of the unmodulated carrier. Let L be the largest integer, for β fixed, for which this criterion is met; i.e.,

$$|J_L(\beta)| \geq \epsilon, \qquad |J_{L+1}(\beta)| < \epsilon.$$

Then L depends on β and ϵ, and the required bandwidth is

$$B_\epsilon = 2L(\beta, \epsilon)f_m. \qquad (7.5\text{-}4)$$

Figure 7.5-1 is a plot of $L(\beta, \epsilon)$ vs. β for the 1% ($\epsilon = 0.01$) and 10% sideband ($\epsilon = 0.1$) criteria. The 10% criterion may lead to slight distortion, while 1% is overly conservative. The dashed line is a reasonable compromise ([7-5], p. 240).

To obtain an idea of the relation among bandwidth, number of significant components, and β, let us consider commercial FM where the FCC requires that Δf be no greater than 75 kHz and the modulating tones range from, say, 30 Hz to 15 kHz. The computation of β, B_{CR}, $L(\beta, \epsilon)$, and B_ϵ for $\epsilon = 0.01$ and 0.1 is summarized in Table 7.5-1. This table demonstrates several interest-

TABLE 7.5-1   BANDWIDTHS FOR VARIOUS TONES, $\Delta f = 75$ KHZ

f_m (kHz)	β	$L(\beta, 0.01)$	$L(\beta, 0.1)$	$B_{0.01}$ (kHz)	$B_{0.1}$ (kHz)	B_{CR} (kHz)
15	5	8	6	240	180	180
10	7.5	11	8	220	160	170
3.75	20	22	20	165	150	157
1	75	75	75	150	150	152
0.05	1500	1500	1500	150	150	150

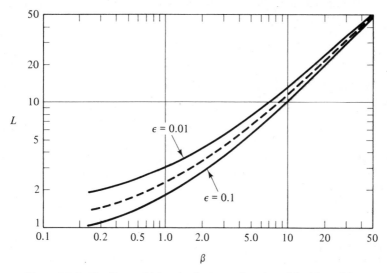

Figure 7.5-1. Significant sideband pairs L as a function of β. *Adapted from Carlson ([7-5], p. 240), by permission.*

ing facts concerning FM bandwidth requirements. First, by assuming that all tones produce the maximum frequency deviation, we see that the largest modulating frequency requires the largest bandwidth and that the largest β's are associated with the smallest bandwidths. Second, for large values of β, the bandwidth is simply twice the frequency deviation, regardless of which criterion is chosen. Third, Carson's rule for estimating bandwidth corresponds closely to the 10% sideband bandwidth, although it is somewhat more conservative than this criterion. The FCC channel-width allocation for FM stations is 200 kHz. Although this bandwidth pertains to modulation by a signal $s(t)$ which is not a pure tone, we shall see that its determination is based on the same principles as in the case of tone modulation.

We have assumed so far that all tones produce the maximum frequency deviation of 75 kHz. But if the various tone amplitudes do not produce the same frequency deviation, then it is not simply the highest frequency that requires the greatest bandwidth. For example, let a 15-kHz tone have an amplitude that is only 40% of the maximum amplitude. Then the frequency deviation is 30 kHz and $\beta = 2$. If we use the 1.0% criterion, then $L(2, 0.01) = 6$, giving $B_{0.01} = 150$ kHz. Likewise, the 10% criterion gives $L(2, 0.1) = 3$, which furnishes $B_{0.1} = 90$ kHz. These bandwidths are much lower than the required bandwidths for full-amplitude tone modulation. Thus the bandwidth is seen to depend on frequency deviation *and* modulation frequency. This being the case, we might ask, Given an ensemble of tones, $\{s_i(t)\}$, which single modulating tone will require the greatest transmission bandwidth? An approximate analytical approach to this problem is furnished by Carson's formula [Eq. (7.5-2)]

$$B_{CR} = 2f_m(1 + \beta)$$

$$= 2(KA_m + f_m).$$ (7.5-5)

Clearly, the greatest bandwidth is determined by the tone $s^*(t)$, with amplitude A_m^* and frequency f_m^*, which satisfies

$$\max_{s_i} 2(KA_{m_i} + f_{m_i}) = 2(KA_m^* + f_m^*),$$ (7.5-6)

where A_{m_i} and f_{m_i} are the amplitude and frequency, respectively, of $s_i(t)$. In the *worst-case* analysis, we let $A_m = 1$ and $f_m = W$. The resulting bandwidth, B_{max}, is given by

$$B_{max} = 2([\Delta f]_{max} + W),$$ (7.5-7)

where $[\Delta f]_{max}$ is the maximum allowed frequency deviation and W is the highest frequency in the baseband.

The use of Carson's rule to compute B_{max} is somewhat arbitrary. For example, Carlson ([7-5], p. 241) uses the relation

$$L(\beta, \bar{\epsilon}) \simeq \beta + C,$$ (7.5-8)

which is an approximate fit to the dashed line in Fig. 7.5-1. $\bar{\epsilon}$ is a compromise between the 0.1 and 0.01 sideband criteria (the dashed line in Fig. 7.5-1), and C is a constant representing the extrapolated ordinate when $\beta = 0$. The constant C lies between 1 and 2, but setting $C = 2$ furnishes a formula which estimates the required bandwidth somewhat more conservatively than Carson's rule of thumb. Use of Eq. (7.5-8) in Eq. (7.5-4) furnishes the formula

$$B_{\bar{\epsilon}} = 2(\Delta f + 2f_m).$$ (7.5-9)

The worst-case bandwidth involving a signal of maximum amplitude at the highest baseband frequency is

$$[B_{\bar{\epsilon}}]_{max} = 2[(\Delta f)_{max} + 2W].$$ (7.5-10)

For $(\Delta f)_{max} = 75$ kHz and $W = 15$ kHz, this formula gives a bandwidth of 210 kHz, which is quite close to the 200 kHz assigned to commercial FM channels in the United States.

It is interesting to examine the structure of the FM spectrum under conditions of changing β when (1) the modulating frequency f_m is fixed but the frequency deviation Δf is allowed to increase and (2) the modulating frequency f_m is decreasing but the frequency deviation is held constant. Consider (1) first. When $\beta \ll 1$, the bandwidth is just $2f_m$, and only a single pair of sidebands appears. As β increases, the bandwidth increases in direct proportion to the number of significant sidebands that appear. The spacing between adjacent orders is fixed at f_m. For $\beta \gg 1$, the bandwidth is $2\Delta f$. The result is shown in Fig. 7.5-2(a).

Now consider case (2). For small β, the bandwidth is just $2f_m$, as before. As β increases due to decreasing f_m, the bandwidth approaches $2\Delta f$. The number of components may be very large, but the frequency spacing between them will be very small. The result is shown in Fig. 7.5-2(b). Increasing β still further does not affect the bandwidth, which can, however, be increased (decreased) by increasing (decreasing) the amplitude of the modulating tone.

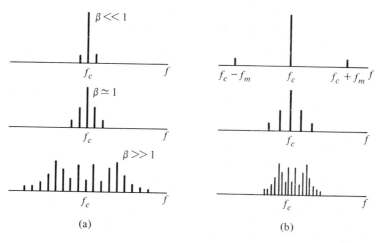

Figure 7.5-2. Spectrum of an FM wave under conditions of changing β: (a) β increasing due to Δf increasing; (b) β increasing due to f_m decreasing.

Modulation by a Composite Signal

The exact determination of the bandwidth for an arbitrary modulating signal is difficult to do without resorting to a numerical evaluation of an expression of the type given in Eq. (7.3-19). However, for some signals, particularly those that have no significant components above some highest frequency, say W, such as speech or music, the required transmission bandwidth can be estimated from other considerations. In particular the bandwidth formulas developed earlier in this section can be extended to furnish useful estimates of bandwidth, although the strong nonlinearity of FM does not permit the results to be as rigorous as we might like. The bandwidth estimation formulas are sometimes empirically adjusted to furnish estimates that are more in line with the bandwidths that are required in practice.

To help us formulate a reasonable estimate of the required bandwidth, we recall that when a carrier is modulated by two tones f_1 and f_2, sidebands occur at frequencies $f_c + kf_2 + nf_1$, where n and k are positive and negative integers. For $f_2 > f_1$, the beat-frequency sidebands show up as satellites around the frequencies $f_c + kf_2$, and the amplitudes of these components are proportional to the product $J_n(\beta_1)J_k(\beta_2)$, which becomes negligible when $J_k(\beta_2)$ becomes negligible. To a first approximation, then, we can say that in the case of a composite signal the bandwidth is determined by a full-amplitude tone (i.e., one that produces maximum frequency deviation) at the highest significant frequency, W, in the baseband. If β^* denotes the ratio of maximum frequency deviation to W, then the bandwidth can be estimated from Eq. (7.5-4), which furnishes

$$B_\epsilon = 2L(\beta^*, \epsilon)W. \tag{7.5-11}$$

Carson's rule is more convenient to use and furnishes

$$B_{CR} = 2W(1 + \beta^*), \tag{7.5-12}$$

and the "conservative" linear approximation, i.e., Eq. (7.5-10), furnishes

$$[B_\epsilon]_{max} = 2W(2 + \beta^*). \tag{7.5-13}$$

EXAMPLE

In commercial FM, the FCC allows $[\Delta f]_{max} = 75$ kHz. Let the bandwidth of $s(t)$ be $W = 15$ kHz. Then $\beta^* = 5$. Carson's rule gives $B_{CR} = 180$ kHz, while Eq. (7.5-13) gives 210 kHz. The latter figure is closer, on the conservative side, to the 200 kHz nominally assigned.

Variations from the above formulas are also in use. For example, a formula that is used in FM telephony is

$$B = 2W(1 + \alpha\beta^*),$$

where α is a constant that depends on the quality of transmission. For commercial telephony $\alpha = 1$, which, for $[\Delta f]_{max} = 15$ kHz and $W = 3$ kHz (this bandwidth enables voice identification), furnishes $\beta^* = 5$. The bandwidth is then 36 kHz. For higher-fidelity communication larger values of α are necessary.

7.6 Indirect Generation of FM: The Armstrong Method

E. H. Armstrong was the first to demonstrate the feasibility and merits of FM by using an indirect technique to generate FM. The so-called indirect FM (IFM) method consists of three important steps:

1. Integrate the modulating signal $s(t)$ to produce the signal

$$z(t) = \int^t s(\lambda) \, d\lambda.$$

2. Phase-modulate a carrier with $z(t)$ to produce NBFM.
3. Use a system of frequency multipliers to convert NBFM to wideband FM.

A block diagram of an IFM system is shown in Fig. 7.6-1. At point A in Fig. 7.6-1, the output is

$$\xi(t) = K_1 \int^t s(\lambda)\, d\lambda = K_1 z(t), \qquad (7.6\text{-}1)$$

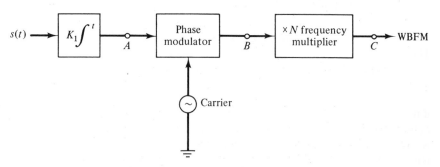

Figure 7.6-1. Method of generating indirect FM.

and at point B the output is proportional to

$$x_1(t) = \cos\left[\omega_{c_1}t + K_1'' \int^t s(\lambda)\, d\lambda\right]$$

$$= \cos\left(\omega_{c_1}t + \beta_1 \sin \omega_m t\right) \qquad (7.6\text{-}2)$$

if sine-wave modulation is assumed. The phase deviation $\beta_1 \equiv (\Delta f)_1/f_m$ is typically much less than 0.5 radian so that $x_1(t)$ can be considered, within reason, an NBFM wave. The frequency multiplier produces a signal with frequency ω given by

$$\omega = N\dot\theta_1(t), \qquad (7.6\text{-}3)$$

where $\theta_1(t) \equiv \omega_{c_1}t + \beta_1 \sin \omega_m t$. Hence

$$\omega = N\omega_{c_1} + N\beta_1\omega_m \cos \omega_m t.$$

The new phase is proportional to the integral of $N\dot\theta_1(t)$ so that at C we have

$$x(t) = \cos(N\omega_{c_1}t + N\beta_1 \sin \omega_m t). \qquad (7.6\text{-}4)$$

By writing $\omega_c = N\omega_{c_1}$ and $\beta = N\beta_1$, we can rewrite Eq. (7.6-4) in standard form; i.e.,

$$x(t) = \cos(\omega_c t + \beta \sin \omega_m t),$$

where β is now a value typical of wideband FM.

EXAMPLE

Let f_m range from 100 to 10,000 Hz, and let the maximum frequency deviation, Δf, at the output be 50 kHz. Then

$$\beta_{\min} = \frac{50 \times 10^3}{10^4} = 5$$

$$\beta_{\max} = \frac{50 \times 10^3}{10^2} = 500.$$

If $[\beta_1]_{\max} = 0.5$, then the required frequency multiplication is

$$N = \frac{\beta_{\max}}{[\beta_1]_{\max}} = \frac{500}{0.5} = 1000.$$

The maximum allowed frequency deviation at the *input* is

$$\Delta f_1 = \frac{50 \times 10^3}{1000} = 50 \text{ Hz}$$

if the maximum specified Δf at the output is 50 kHz. If the initial carrier frequency f_{c_1} were, say, 200 kHz, then the final frequency would be $f_c = 200$ MHz. This figure is too high for standard FM broadcasting, and frequency converters are used to reduce f_c to the desired band. For example, if we heterodyne $x(t)$ in Eq. (7.6-4) with a carrier wave of frequency, f_{LO}, the modulation index $N\beta_1$ remains unaffected but the wave is shifted to the new carrier frequency, $Nf_c - f_{\text{LO}}$.

A block diagram of an Armstrong-type indirect transmitter is shown in Fig. 7.6-2. The largest modulation index furnished by the NBFM system is determined by the lowest modulating frequency ($\simeq 50$ Hz), and this determines the required frequency multiplication. For example, if $\beta_1 \simeq 0.5$ for a 50-Hz tone, then the 75-kHz frequency deviation requires a $\beta = 1500$. The required multiplication is $1500/0.5 = 3000$. The largest tone frequency determines the bandwidth. The value of β_1 for the 15-kHz signal is approximately 1.7×10^{-3}. The extremely small initial values of β_1 are required to prevent distortion due to amplitude-modulation effects that can occur in the generation of NBFM by the method shown in Fig. 7.4-4.

The indirect method of producing FM enjoys good frequency stability, which constitutes a distinct advantage over the direct method (Sec. 7.7). However, the repeated stages of frequency multiplication as well as the extra heterodyning require considerable circuit complexity, which must be judged a disadvantage. The signal integrator must operate over a frequency range on the order of 1000 to 1. If the integration is done with RC elements in an operational amplifier configuration, the integrating capacitor must have high Q and small leakage over the frequency range of the signal.

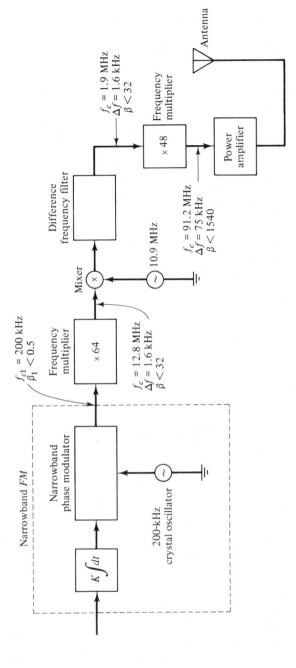

Figure 7.6-2. Armstrong-type indirect FM transmitter. *Adapted from M. Schwartz, Information Transmission, Modulation and Noise, McGraw-Hill Book Company, New York, 1959, p. 134, by permission.*

7.7 Generation of FM by Direct Methods

In the direct generation of FM, all that is essentially required is a device whose output frequency varies linearly with the level of the applied signal. The voltage-controlled oscillator is such a device, and a particular design of a VCO was discussed in the previous chapter. Another technique for achieving the same result is to use a tuned circuit oscillator in which one of the resonant circuit elements is a variable reactance. A useful device in this respect is the varactor, which is a diode whose barrier capacitance C depends on the applied voltage. The capacitance depends inversely on the width of the space charge layer, which in turn depends on the externally applied voltage. The capacitance vs. applied voltage relation for varactor diodes can be approximated, over a limited region of operation, by a linear function. A description of varactors (also called varicaps or voltacaps) is furnished in [7-6], p. 137. Figure 7.7-1 shows typical capacitance vs. applied voltage variations for varactor diodes.

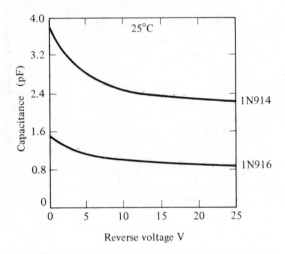

Figure 7.7-1. Capacitance variation with applied voltage for silicon diodes. *Courtesy of Fairchild Semiconductor Corporation.*

Figure 7.7-2 shows, in simplified form, a tuned circuit oscillator in which the resonant frequency is controlled by a varactor diode.

The resistors R_1, R_2, and R_e furnish the quiescent self-bias to start the oscillations and the dynamic self-bias is obtained from the $R_2 C''$ combination due to the flow of base current. The polarities of the primary and secondary windings are reversed, which furnishes a 180° phase shift, in the collector-to-base feedback; the additional 180° phase shift between base and

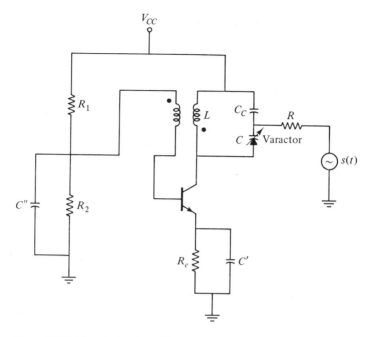

Figure 7.7-2. Tuned circuit oscillator with a voltage-controlled varactor in the tuned circuit.

collector gives a net loop phase shift of zero. The capacitance C_C represents a high impedance to $s(t)$, but if $C_C \gg C$, the instantaneous resonant frequency is unaffected by C_C and is given by (winding losses are disregarded)

$$\omega = [LC]^{-1/2}. \qquad (7.7\text{-}1)$$

Now assume that when a signal $s(t)$ is applied the capacitance can be described by

$$C = C_0\left[1 - \frac{\Delta C}{C_0}s(t)\right], \qquad |s(t)| < 1, \qquad (7.7\text{-}2)$$

where C_0 is the capacitance when $s(t) = 0$ and ΔC represents the maximum deviation in capacitance from C_0. If we assume that $\Delta C/C_0 \ll 1$, then the instantaneous oscillator frequency may be written as

$$\omega \simeq \omega_c\left[1 + \frac{\Delta C}{2C_0}s(t)\right], \qquad (7.7\text{-}3)$$

where $\omega_c = [LC_0]^{1/2}$. The output signal can be written as

$$x(t) = \cos\left[\omega_c t + K'' \int^t s(\lambda)\, d\lambda\right], \qquad (7.7\text{-}4)$$

which is the standard form for an FM wave, with $K'' \equiv (\omega_c \Delta C)/2C_0$. We can obtain some idea of how large ΔC must be from Eq. (7.7-3). For example, if

$f_c = 30$ MHz and the maximum frequency deviation is $\Delta f = 25$ kHz, then

$$\frac{\Delta C}{2C_0} = \frac{\Delta f}{f_c} = \frac{25}{30} \times 10^{-3} = 0.833 \times 10^{-3}.$$

A block diagram for an illustrative direct FM system is shown in Fig. 7.7-3. The main advantages of direct FM is the reduced need for frequency multiplication and heterodyne frequency conversion. On the minus side, the high carrier frequencies must be stabilized against even small drifts since, in the example considered, a drift of only 0.08 % corresponds to the maximum

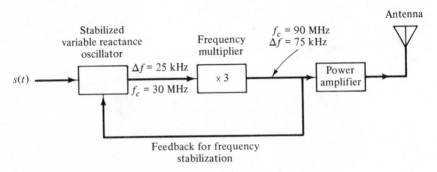

Figure 7.7-3. Simplified diagram of a direct FM system.

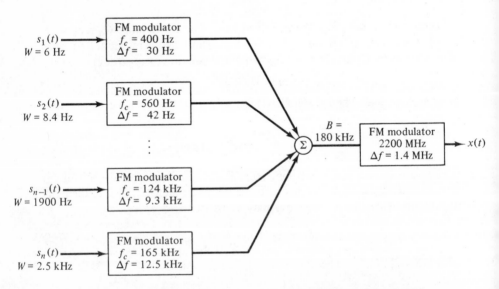

Figure 7.7-4. FM/FM system for space telemetry (proportional-bandwidth subcarrier channels). For each channel $\Delta f/W \simeq 5$ and $\Delta f/f_c = 7.5\%$. The nth carrier is given by $f_n = f_{n-1}(W_n/W_{n-1})$. *From M. Schwartz ([7-7], p. 256), by permission.*

allowed frequency deviation. For this reason, feedback is used to stabilize the oscillator frequency. A frequency drift error signal can be detected with an appropriate discriminator and used to control the oscillator.

Some typical direct FM systems for use in space telemetry are given in [7-7] pp. 255–256. An example of such a system, that uses multiplexing of several channels, is shown in Fig. 7.7-4.

7.8 FM Signals in Linear Networks

Before discussing the problem of FM demodulation, it is worthwhile to consider the problem of transmitting an FM wave through a linear network. Formally at least, the solution is straightforward. If we denote the network transmittance by $|H(\omega)| \exp[-j\gamma(\omega)]$, where $|H(\omega)|$ is the magnitude of the transmittance and $\gamma(\omega)$ is the phase, the response of the network to

$$x(t) = \cos(\omega_c t + \beta \sin \omega_m t) \tag{7.8-1}$$

is

$$y(t) = \sum_{n=-\infty}^{\infty} |H(\omega_c + n\omega_m)| J_n(\beta) \cos[(\omega_c + n\omega_m)t - \gamma(\omega_c + n\omega_m)], \tag{7.8-2}$$

where we have used Eq. (7.3-9a); i.e.,

$$x(t) = \sum_{n=-\infty}^{\infty} J_n(\beta) \cos(\omega_c + n\omega_m)t. \tag{7.8-3}$$

Equation (7.8-2) gives the solution that is associated with an FM wave modulated by a single tone. Even in this most elementary case the expression is very complicated and furnishes little insight. For arbitrary modulating signals, an expression such as Eq. (7.3-19) would be needed, and the meaning would be even less clear.

The Quasi-Static Method (QSM)

An alternative to the direct but possibly tedious FM-response computations described by the previous equations is the so-called quasi-static method. The underlying assumption in the QSM is that the steady-state solution evaluated at the instantaneous frequency is a good first approximation which can be systematically improved by the addition of higher-order terms. Unfortunately convergence of the resultant series is not guaranteed, and estimation of the remaining error at any particular stage of the calculation is not easy ([7-8], p. 229). However, we shall not be concerned with convergence problems in the discussion that follows.

We write

$$x(t) = \cos\left[\omega_c t + K'' \int^t s(\lambda)\, d\lambda\right] = \text{Re } z(t), \tag{7.8-4}$$

where

$$z(t) = e^{j[\omega_c t + \mu(t)]} \qquad (7.8\text{-}5)$$

and

$$\mu(t) \equiv K'' \int^t s(\lambda)\, d\lambda. \qquad (7.8\text{-}6)$$

Let the transmittance of the network (Fig. 7.8-1) be $H(\omega)$. The response $y(t)$ is given by

$$y(t) = \mathrm{Re}\, \eta(t) = \mathrm{Re} \int_{-\infty}^{\infty} Z(\omega)H(\omega)e^{j\omega t}\,\frac{d\omega}{2\pi}, \qquad (7.8\text{-}7)$$

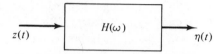

Figure 7.8-1. Transmission of an FM signal through a linear network.

where the capital letters indicate Fourier transforms of the lowercase functions; i.e., $Z(\omega) = \mathfrak{F}[z(t)]$, etc. We describe the transmittance by

$$H(\omega) = A(\omega)e^{-j\gamma(\omega)}, \qquad (7.8\text{-}8)$$

where $A(\omega)$ and $\gamma(\omega)$ are real. Now assume that $H(\omega)$ can be expanded in a Taylor series about $\omega = \omega_c$. Then

$$H(\omega) = H(\omega_c) + H'(\omega_c)(\omega - \omega_c) + H''(\omega_c)\frac{(\omega - \omega_c)^2}{2!}$$

$$+ \cdots + H^{(n)}(\omega_c)\frac{(\omega - \omega_c)^n}{n!} + \cdots, \qquad (7.8\text{-}9)$$

where

$$H^{(i)}(\omega) \equiv \frac{d^i}{d\omega^i}[H(\omega)]_{\omega = \omega_c}. \qquad (7.8\text{-}10)$$

If we substitute Eq. (7.8-9) into Eq. (7.8-7) and consider $\eta(t)$ rather than $\mathrm{Re}\,\eta(t)$, we obtain

$$\eta(t) = \sum_{n=0}^{\infty} \frac{1}{n!} H^{(n)}(\omega_c) \int_{-\infty}^{\infty} Z(\omega)(\omega - \omega_c)^n e^{j\omega t}\,\frac{d\omega}{2\pi} \qquad (7.8\text{-}11)$$

$$= \sum_{n=0}^{\infty} \frac{1}{n!} H^{(n)}(\omega_c)e^{j\omega_c t}$$

$$\cdot \int_{-\infty}^{\infty} Z(\omega)(\omega - \omega_c)^n e^{j(\omega - \omega_c)t}\,\frac{d\omega}{2\pi}. \qquad (7.8\text{-}12)$$

Observe that

$$\frac{1}{j}\frac{d}{dt}\left[\int_{-\infty}^{\infty} Z(\omega)e^{j(\omega-\omega_c)t}\frac{d\omega}{2\pi}\right] = \int_{-\infty}^{\infty} Z(\omega)(\omega - \omega_c)e^{j(\omega-\omega_c)t}\frac{d\omega}{2\pi}$$

(7.8-13)

and, more generally, that

$$\frac{1}{j^n}\frac{d^n}{dt^n}\int_{-\infty}^{\infty} Z(\omega)e^{j(\omega-\omega_c)t}\frac{d\omega}{2\pi} = \int_{-\infty}^{\infty} Z(\omega)(\omega - \omega_c)^n e^{j(\omega-\omega_c)t}\frac{d\omega}{2\pi}.$$

(7.8-14)

Since

$$z(t) = \int_{-\infty}^{\infty} Z(\omega)e^{j\omega t}\frac{d\omega}{2\pi},$$

it follows that

$$z(t)e^{-j\omega_c t} = \int_{-\infty}^{\infty} Z(\omega)e^{j(\omega-\omega_c)t}\frac{d\omega}{2\pi},$$

so that, from Eq. (7.8-14),

$$\frac{1}{j^n}\frac{d^n}{dt^n}[e^{-j\omega_c t}z(t)] = \int_{-\infty}^{\infty} Z(\omega)(\omega - \omega_c)^n e^{j(\omega-\omega_c)t}\frac{d\omega}{2\pi}.$$ (7.8-15)

If we use Eqs. (7.8-15) and (7.8-5) in Eq. (7.8-12), we obtain

$$\eta(t) = \sum_{n=0}^{\infty} \frac{1}{n!}H^{(n)}(\omega_c)e^{j\omega_c t}\frac{1}{j^n}\frac{d^n}{dt^n}[e^{-j\omega_c t}z(t)]$$

$$= z(t)\{H(\omega_c) + \dot{\mu}(t)H'(\omega_c) + \tfrac{1}{2}H''(\omega_c)$$

$$\cdot[\dot{\mu}(t)]^2 - j\ddot{\mu}(t)] + \cdots\}.$$ (7.8-16)

Now consider the special case where $\ddot{\mu}(t)$ and all higher time derivatives of the modulating signal can be ignored, as well as the higher-frequency derivatives (say above the second) of the transmittance $H(\omega)$. Then Eq. (7.8-16) can be approximated by

$$\eta(t) \simeq z(t)\{H(\omega_c) + \dot{\mu}(t)H'(\omega_c) + \tfrac{1}{2}H''(\omega_c)[\dot{\mu}(t)]^2\}.$$ (7.8-17)

The instantaneous frequency is $\omega_c + \dot{\mu}(t)$. An expansion for $H[\omega_c + \dot{\mu}(t)]$ about the carrier frequency ω_c gives

$$H[\omega_c + \dot{\mu}(t)] = H(\omega_c) + H'(\omega_c)\dot{\mu}(t)$$

$$+ \frac{H''(\omega_c)}{2}[\dot{\mu}(t)]^2 + \cdots.$$ (7.8-18)

But Eq. (7.8-18), with higher terms omitted, is precisely the term in brackets in Eq. (7.8-17). Hence for "slowly" varying $\mu(t)$ (with time) and slowly varying $H(\omega)$ (with frequency) we can write

$$\eta(t) \simeq z(t)H[\omega_c + \dot{\mu}(t)]. \qquad (7.8\text{-}19)$$

The interpretation of Eq. (7.8-19) is that, to a first approximation and under the conditions given above, the complex response is simply the product of the complex input and the transmittance evaluated at the instantaneous frequency $\omega_c + \dot{\mu}(t)$.

Bennett et al. ([7-8], p. 232) give numerous references on the QSM. When the modulating signal changes abruptly, as in square-wave modulation, or the transmittance $H(\omega)$ contains many wiggles over the band of significant instantaneous frequencies, the QSM is generally not useful.

7.9 Demodulation of FM Signals

Slope Detection

Demodulation of an FM wave requires a device that produces an output signal whose amplitude is a linear function of the frequency of the input signal. The single-tuned circuit shown in Fig. 7.9-1 can, with the proper choice of parameters and proper choice of operating region, furnish the required transfer characteristic.

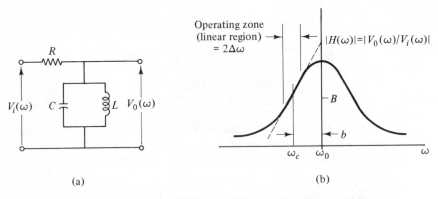

(a)   (b)

Figure 7.9-1. Use of a single-tuned circuit to demodulate an FM wave: (a) the circuit; (b) the amplitude of the transfer function.

If the linear region of the amplitude response can be described by an equation of the form

$$|H(\omega)| = A(\omega - \omega_c) + B, \qquad (7.9\text{-}1)$$

then use of Eq. (7.8-19) furnishes

$$\eta(t) = [A\dot{\mu}(t) + B] \exp\left[j\{\omega_c t + \mu(t) - \gamma[\omega_c + \dot{\mu}(t)]\}\right], \qquad (7.9\text{-}2)$$

where $\dot{\mu}(t) = Ks(t)$ and $\gamma(\omega)$ is the phase of $H(\omega)$.

The response $y(t)$ is simply the real part of $\eta(t)$. Thus

$$y(t) = [A\dot{\mu}(t) + B] \cos \{\omega_c t + \mu(t) - \gamma[\omega_c + \dot{\mu}(t)]\}. \tag{7.9-3}$$

If $y(t)$ is injected into an envelope detector and the dc term B is removed by a blocking capacitor, we recover the signal $s(t)$. Detection of FM waves in which the slope of a tuned circuit is used in the manner described above is quite appropriately called *slope detection*. It is interesting to investigate the conditions required for slope detection with single-tuned circuits. To this end, consider Fig. 7.9-1, where $\omega_0 = (LC)^{-1/2}$, ω_c is the carrier frequency, $\omega_0 - \omega_c \equiv b$, and $\Delta\omega$ denotes the frequency deviation of the input signal frequency from the carrier frequency. The magnitude of the transfer function $H(\omega)$ is given by

$$|H(\omega)| = \frac{1}{\sqrt{1 + C^2 R^2 [(\omega^2 - \omega_0^2)/\omega]^2}}. \tag{7.9-4}$$

If we substitute $\omega = \omega_c + \Delta\omega$, $\omega_0 = \omega_c + b$ and assume that

$$\begin{align} &\text{(i)} \quad \Delta\omega \ll b \ll \omega_c \\ &\text{(ii)} \quad b^2 \ll 2\omega_c \Delta\omega, \tag{7.9-5} \\ &\text{(iii)} \quad b \ll [2CR]^{-1}, \end{align}$$

then Eq. (7.9-4) can be written in the form

$$|H(\omega)| = A(\omega - \omega_c) + B, \tag{7.9-6}$$

where $B \equiv 1 - 2[CR]^2 b^2$ and $A \equiv [2CR]^2 b$. We leave the details as an exercise for the reader.

A much better system than a single-tuned circuit is the balanced discriminator shown in Fig. 7.9-2. The transfer function is shown in Fig. 7.9-3. One resonant circuit is tuned to $\omega_c + b$, while the other is tuned to $\omega_c - b$.

Figure 7.9-2. Balanced discriminator.

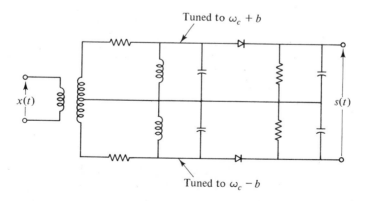

Tuned to $\omega_c + b$

$x(t)$

$s(t)$

Tuned to $\omega_c - b$

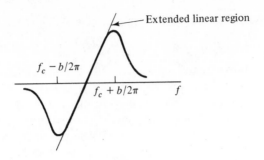

Figure 7.9-3. Equivalent transfer function of balanced discriminator.

The total transfer characteristic over the linear region is described by

$$H_T(\omega) = [1 - 2(CR)^2 b^2] + (2CR)^2 b(\omega - \omega_c)$$
$$- \{[1 - 2(CR)^2 b^2] - (2CR)^2 b(\omega - \omega_c)\}$$
$$= A(\omega - \omega_c), \tag{7.9-7}$$

where $A \equiv 2(2CR)^2 b$. The assumption of a linear region is only an approximation valid in the region around f_c. The response, from Eq. (7.9-2), is just

$$\eta(t) = A\dot{\mu}(t) \exp [j\{\omega_c t + \mu(t) - \gamma[\omega_c + \dot{\mu}(t)]\}], \tag{7.9-8}$$

and the envelope detector furnishes, without the use of a blocking capacitor, the signal $s(t)$.

Phase-Locked Loop (PLL)

The PLL was discussed in the previous chapter in connection with synchronous detection. In that application, the PLL was used as an extremely narrowband filter to remove the carrier component of the AM signal, thereby enabling homodyne detection through mixing. The PLL can also be used to demodulate FM signals. The operation of the PLL in this connection is easily explained by considering the circuit in Fig. 7.9-4. In Sec. 6-10, we showed that when the loop was near or at phase lock then $\theta(t) \simeq \theta'(t)$. If

Figure 7.9-4. Use of a phase-locked loop to demodulate FM.

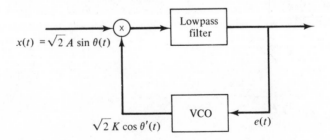

$x(t)$ is an FM wave and the loop is at or near phase lock, then

$$\theta'(t) \simeq \theta(t) = \omega_c t + K'' \int^t s(\lambda)\, d\lambda. \qquad (7.9\text{-}9)$$

The VCO frequency, $\dot{\theta}'(t)$, is given by

$$\dot{\theta}'(t) = \omega_0 + K_v e(t), \qquad (7.9\text{-}10)$$

where ω_0 is the frequency of the VCO when $e(t) = 0$ and K_v is a constant. The instantaneous input frequency is

$$\dot{\theta}(t) = \omega_c + K'' s(t). \qquad (7.9\text{-}11)$$

Upon comparing Eq. (7.9-10) with Eq. (7.9-11) we see that, when $\dot{\theta}'(t) = \dot{\theta}(t)$,

$$e(t) = K_v^{-1}(\omega_c - \omega_0) + K_v^{-1} K'' s(t). \qquad (7.9\text{-}12)$$

Hence, except for a removable dc term, we see that $e(t)$ is proportional to $s(t)$ and that demodulation has been achieved. The low-pass filter must reject the sum frequency signal produced by the multiplier but admit all significant components in the baseband signal $s(t)$. The PLL usually follows the IF stage in the receiver. The spurious amplitude variations in the FM signal should be removed before reaching the PLL to avoid distortion in the output. A limiter circuit is therefore often included in the IF stage to remove these variations.

Phase-Shift Discriminator (PSD)

The PSD is widely used for FM demodulation and exists in several configurations. Figure 7.9-5(a) shows a PSD using a doubly tuned circuit in which there is an ac connection between the primary and secondary windings of the transformer. A simplified equivalent circuit for the ac is shown in Fig. 7.9-5(b). The capacitors labeled C_b are ac bypass capacitors. The resistor R_e, shown in dashed lines, is sometimes added to equalize the Q's of the two circuits; however, it is omitted in our discussion. How the PSD works can best be understood by considering the phase of E_2 with respect to E_1. The transformer together with the capacitor consists of a tuned circuit. At the resonant frequency ω_0 (which depends on the Thevenin's equivalent inductance) the circuit is resistive, and the current I is in phase with E_1. Hence E_2 lags E_1 by 90°. If $\omega > \omega_0$, the circuit is inductive, and the current I lags E_1 by some angle. The voltage E_2 always lags I by 90°; hence E_2 lags E_1 by an angle greater than 90°. When $\omega < \omega_0$, the circuit is capacitive, and the current leads E_1. Hence E_2 lags E_1 by an angle less than 90°. With these observations in mind we can now return to Fig. 7.9-5(b) and draw the phasor relationships for the voltages across D_1 and D_2. These are shown in Fig. 7.9-6. Under normal operation both tuned circuits are tuned to the carrier frequency ω_c; hence $\omega_0 = \omega_c$.

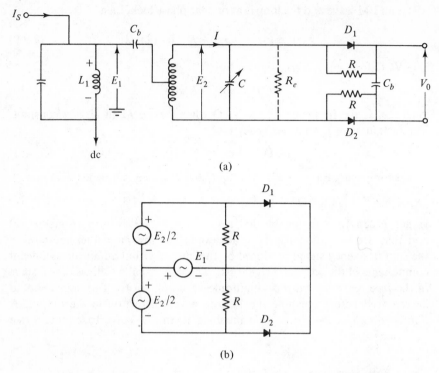

Figure 7.9-5. (a) Phase-shift discriminator; (b) simplified ac equivalent circuit.

The rectified output voltage, V_0, is proportional to the difference in the amplitudes (envelopes) of the ac, i.e.,

$$|E_{D_1}| - |E_{D_2}|,$$

and therefore is zero in the absence of modulation. The circuit parameters are chosen to force the output voltage to be a linear function of the deviation from ω_c. When tuning the PSD, the primary is adjusted for symmetry of response on either side of ω_c, and the secondary is adjusted to obtain zero response at ω_c ([7-3], p. 12.35).

Ratio Detector

The ratio detector performs the double function of demodulation and limiting in a single unit and hence is widely used in less costly FM receivers. A circuit realization of a ratio detector is shown in Fig. 7.9-7. Other ratio-detector configurations are to be found in [7-3], p. 19.82 and 12.36. The major difference between this circuit and the PSD is that one of the rectifying diodes is reversed and a large capacitor, C_e, is placed across the load resistors

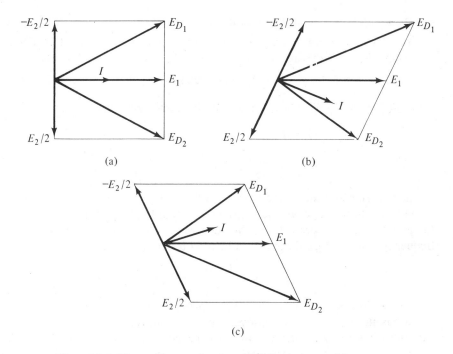

Figure 7.9-6. Phasor diagrams for phase-shift discriminator: (a) resonance, $\omega = \omega_0$; (b) $\omega > \omega_0$; (c) $\omega < \omega_0$.

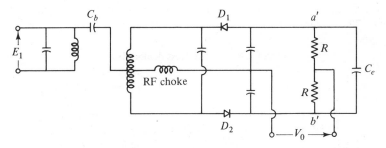

Figure 7.9-7. Ratio detector.

R. The effect of the large capacitor is to maintain a constant voltage across the secondary, which results in an output signal substantially unaffected by rapid changes in the input level. The capacitor C_e can, in effect, be replaced by a battery whose value is the peak voltage that appears across $a'b'$. The time constant associated with the load resistors R and C_e must allow for V_0 to follow variations in $s(t)$, although variations in the amplitude A of $x(t)$ are typically smoothed out.

FM Demodulation by Direct Differentiation

Any device that can differentiate a signal can, potentially, be used for FM detection. Consider the signal

$$x(t) = A \cos \theta(t), \tag{7.9-13}$$

where $\theta(t) = \omega_c t + K'' \int^t s(\lambda) \, d\lambda$. Direct differentiation gives

$$\dot{x}(t) = -A\dot{\theta}(t) \sin \theta(t)$$

$$= -A\omega_c \left[1 + \frac{s(t)}{\omega_c} K'' \right] \sin \left[\omega_c t + K'' \int^t s(\lambda) \, d\lambda \right]. \tag{7.9-14}$$

We have been assuming all along that the frequency deviation is small compared to ω_c; hence the envelope $A\omega_c[1 + K''s(t)/\omega_c]$ never goes negative, and Eq. (7.9-14) has the form of an AM wave with perturbations in the carrier frequency. An envelope detector can be used to recover the term

$$1 + \frac{K''s(t)}{\omega_c},$$

and removal of the dc yields a signal proportional to $s(t)$. As usual, envelope variations in $x(t)$ should be removed by some kind of limiting, since such variations constitute a form of noise if detected. The process of limiting is essentially nonlinear, and bandpass filtering is used to extract a constant-amplitude FM wave at frequency f_c from the various harmonics that are generated. A balanced differentiator is shown in Fig. 7.9-8. It is known as the Clarke-Hess frequency demodulator and is discussed in detail in [7-9], p. 586. The differentiation is done by the capacitors C, which produce a

Figure 7.9-8. Clarke-Hess demodulator.

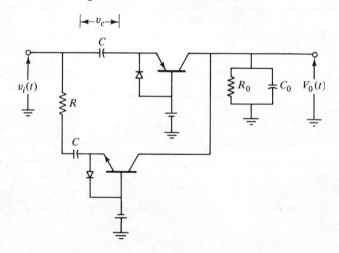

current proportional to $\dot{v}_c(t)$. The R_0C_0 filter at the output admits only the baseband component of the rectified voltage which is proportional to $s(t)$.

By choosing $RC = \omega_c^{-1}$, the output voltage is zero in the absence of modulation. The circuit has several advantages, including relative insensitivity to low-level amplitude-variation noise.

Quadrature Detector

The quadrature detector is shown in Fig. 7.9-9.

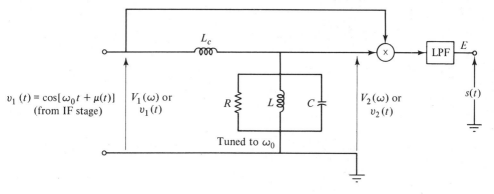

Figure 7.9-9. Quadrature detector.

The input to the demodulator is typically a 10.7-MHz IF signal; the output, at E, is the detected signal. The resonant circuit is normally tuned to the carrier frequency of the IF signal, ω_0, and the Q of the circuit is chosen to satisfy

$$2Q \ll \frac{\omega_0}{\Delta\omega}, \qquad \Delta\omega \equiv \omega - \omega_0, \qquad (7.9\text{-}15)$$

where $\Delta\omega$ is the frequency deviation. To understand how the circuit works, we first compute the transfer function $V_2(\omega)/V_1(\omega)$, where $V_2(\omega) = \mathfrak{F}[v_2(t)]$ and $V_1(\omega) = \mathfrak{F}[v_1(t)]$. This is

$$\frac{V_2(\omega)}{V_1(\omega)} = H(\omega) = \frac{1}{1 + (L_c/L)[1 - (\omega/\omega_0)^2] + j\omega(L_c/R)}. \quad (7.9\text{-}16)$$

If we recall that the Q of a parallel RLC circuit is given by

$$Q = \frac{R}{\omega_0 L}, \qquad (7.9\text{-}17)$$

then use of Eq. (7.9-15) enables us to write, to a very good approximation,

$$H(\omega) \simeq A \exp\left\{ j\left[\frac{\pi}{2} - 2Q\left(\frac{\omega - \omega_0}{\omega_0}\right) \right] \right\}, \qquad (7.9\text{-}18)$$

where A is $R/L_c\omega_0$. We are assuming that we are operating near resonance and that the reactance of L_c is large†. We now apply the quasi-static method and write the input signal as

$$z(t) = e^{j[\omega_0 t + \mu(t)]}, \qquad [v_1(t) = \text{Re } z(t)] \qquad (7.9-19)$$

where

$$\mu(t) = K'' \int^t s(\lambda) \, d\lambda. \qquad (7.9-20)$$

The complex output signal, from Eqs. (7.8-19) and (7.9-18), is

$$\eta(t) = z(t) \exp\left[j\left(\frac{\pi}{2} - 2Q\frac{\dot{\mu}}{\omega_0}\right) \right]$$

$$= \exp\left\{ j\left[\frac{\pi}{2} + \mu(t) + \omega_0 t - 2Q\frac{\dot{\mu}}{\omega_0}\right] \right\}, \qquad (7.9-21)$$

where gain constants have been ignored. The output voltage is

$$v_2(t) = \text{Re } [\eta(t)]$$

$$= -\sin\left[\omega_0 t + \mu(t) - 2Q\frac{\dot{\mu}}{\omega_0} \right]. \qquad (7.9-22)$$

The signal obtained at point E after the balanced modulator is proportional to

$$v_1(t)v_2(t) = \sin\left(2Q\frac{\dot{\mu}}{\omega_0} \right)$$

$$\simeq 2Q\left(\frac{K''}{\omega_0}\right)s(t) \qquad \text{(sum frequency term assumed filtered out).}$$

$$(7.9-23)$$

The last line follows from Eq. (7.9-15) and the facts that $\dot{\mu}(t) = K''s(t)$ and $|s(t)| \leq 1$. From Eq. (7.9-23) we see that the signal $s(t)$ has been recovered from the modulated carrier. The two key steps in this scheme are the $\pi/2$ phase shifting of the input signal and the taking of the product of the phase-shifted signal with the original. The circuit is basically a variation of the slope circuit except that it is the linear region of the *phase* characteristic, around resonance, that is used.

Zero-Crossing Detector

Another type of FM detector takes advantage of the fact that a measure of the instantaneous frequency is the number of zero crossings per unit time. It is in fact not difficult to show that the message can be reconstructed from knowledge of the zero crossings alone. Consider the FM wave

$$x(t) = A \cos\left[\omega_c t + K'' \int^t s(\lambda) \, d\lambda \right] = A \cos \theta(t). \qquad (7.9-24)$$

†More specifically, we are assuming that $R/(L_c\omega_0) \ll 1$.

Let t_1 and t_2 be the times associated with two adjacent zero crossings. Then

$$\theta(t_2) - \theta(t_1) = \pi = \omega_c(t_2 - t_1) + K'' \int_{t_1}^{t_2} s(\lambda)\, d\lambda. \qquad (7.9\text{-}25)$$

The bandwidth, W, of the message $s(t)$ is assumed much less than the bandwidth, B, of the modulated wave. Hence $s(t)$ is essentially constant over the interval $[t_1, t_2]$, and $s(\lambda)$ can be taken outside the integral and given any argument t in $[t_1, t_2]$. Under these circumstances we obtain

$$[\omega_c + K''s(t)](t_2 - t_1) = \pi. \qquad (7.9\text{-}26)$$

The term $\omega_c + K''s(t)$ is simply the derivative of the instantaneous phase. Hence it represents the instantaneous frequency ω_i or f_i, which is

$$\omega_i = \omega_c + K''s(t) \simeq \frac{\pi}{t_2 - t_1} \qquad (7.9\text{-}27)$$

or

$$f_i = f_c + Ks(t) = \frac{1}{2(t_2 - t_1)}, \qquad K \equiv \frac{K''}{2\pi}. \qquad (7.9\text{-}28)$$

If we count the number of zero crossings in an interval T which is large compared to f_c^{-1} but small compared to B^{-1}, we can assume that $s(t)$ is reasonably constant over T while ensuring that we have a reasonable number of zero crossings. If n_T denotes the number of zero crossings in T, then the spacings between adjacent zero crossings will not depart significantly from $t_2 - t_1$, and

$$n_T \simeq \frac{T}{t_2 - t_1} \qquad (7.9\text{-}29)$$

so that

$$f_i = f_c + Ks(t) \simeq \frac{n_T}{2T}. \qquad (7.9\text{-}30)$$

Hence n_T can be used to recover $s(t)$. A number of practical systems are available that work on this principle ([7-9], p. 618). For example, one system uses a monostable multivibrator that is triggered on the positive-sloping edge of a hard-limited FM wave and produces a pulse of short duration every time it is triggered. A subsequent low-pass filter with time constant T serves as an integrator and does the averaging indicated in Eq. (7.9-30). A balancing branch can be used to eliminate the f_c term in Eq. (7.9-30), allowing for $s(t)$ to be obtained directly.

Summary of Detectors

It might be useful at this point to briefly point out some of the merits of the various FM demodulators. The single-tuned circuit shown in Fig. 7.9-1 suffers from the fact that the range of its linear region may be too small to handle the full frequency deviation of the signal. Excursions away from the linear region generate terms involving $H''(\omega)$ and higher derivatives in Eq. (7.8-18) which could produce significant distortion. Also a dc signal is produced in the absence of modulation which requires a dc blocking capacitor, which, in turn, decreases the low-frequency response of the demodulator.

The balanced slope detector shown in Fig. 7.9-2 is better than the single-tuned circuit but may be hard to tune to give a good linear response. Also a limiter is still required.

The PLL is generally designed for bandwidths that are narrower than the 15 kHz required for demodulation of broadcast FM. In FM receivers the PLL is generally used as a low-pass filter for extracting a pilot tone for demodulation of the stereo signal (see Fig. 6.16-4). The PSD and ratio detector furnish good linearity and are widely used. The ratio detector also functions as a limiter (i.e., the capacitor C_e in Fig. 7.9-7) and is, therefore, an economical circuit that appears in less costly receivers.

The balanced Clarke-Hess demodulator shown in Fig. 7.9-8 gives good linearity and also functions partly as a limiter. It also makes no use of inductances, which is an advantage in integrated-circuit technology. It should, however, be used with germanium transistors ([7-9], p. 586). Also it is well to remember that high-frequency noise is always a potential problem in any device that uses a differentiator.

The quadrature detector offers very high linearity, especially if instead of the simple resonant circuit shown in Fig. 7.9-9 a doubly tuned resonant circuit is used. It is commonly used in high-quality receivers. Except for the tuned circuit, the rest of the detector, including the IF amplifier, is usually mounted on a single chip (e.g., RCA CA 3089E).

The zero-crossing detector is capable of better than 0.1% linearity over frequency deviations approaching the carrier frequency. Commercial units are available that handle carrier frequencies as high as 100 MHz, although above 10 MHz, divide-by-two and divide-by-ten counters are used to relax rather strenuous circuit requirements (e.g., recovery time of the pulse generator must be less than 10 nanoseconds). In general, this type of detector is best suited when exceptional linearity over very large frequency deviations is required. It is less useful when the frequency deviation is a small fraction of the carrier frequency.

*7.10   Interference in FM

When an interfering wave $y(t) = B \cos \omega_i t$ is superimposed on an FM wave, the instantaneous phase, as well as the amplitude, is disturbed. Presence of a signal exerts a masking effect on the interference which could lead to less severe requirements on the interference rejection ability of the system. For this reason, the study of the effect of interference on an unmodulated carrier is important. Let $x(t) = A \cos \omega_c t$ represent the unmodulated carrier. The total signal, i.e., interference plus carrier, is then

$$z(t) = A \cos \omega_c t + B \cos \omega_i t. \qquad (7.10\text{-}1)$$

If $\cos \omega_i t$ is written as $\cos (\omega_i - \omega_c + \omega_c)t$ and expanded according to the formula

$$\cos (a + b) = \cos a \cos b - \sin a \sin b,$$

we obtain a useful representation for $z(t)$ in terms of envelope and phase:

$$z(t) = R(t) \cos [\underbrace{\omega_c t + \phi(t)}_{\theta(t)}], \qquad (7.10\text{-}2)$$

where

$$R(t) = [(A + B \cos \omega_D t)^2 + B^2 \sin^2 \omega_D t]^{1/2}, \qquad (7.10\text{-}3)$$

$$\phi(t) = \tan^{-1} \frac{B \sin \omega_D t}{A + B \cos \omega_D t}, \qquad (7.10\text{-}4)$$

and

$$\omega_D \equiv \omega_i - \omega_c. \qquad (7.10\text{-}5)$$

The quantity ω_D is the difference between the carrier frequency and the frequency of the interfering wave.†

The phasor diagram for the resultant wave is shown in Fig. 7.10-1.

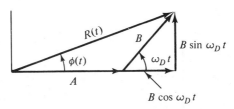

Figure 7.10-1. Phasor diagram showing the effect of interference.

The most interesting case in practice is when $B/A < 1$. With $r \equiv B/A$, we can rewrite Eqs. (7.10-3) and (7.10-4) as

$$R(t) = A(1 + r^2 + 2r \cos \omega_D t)^{1/2}, \qquad (7.10\text{-}6)$$

$$\phi(t) = \tan^{-1} \frac{r \sin \omega_D t}{1 + r \cos \omega_D t}. \qquad (7.10\text{-}7)$$

For the case $r \ll 1$, $R(t)$ and $\phi(t)$ can be simplified to

$$R(t) \simeq 1 + r \cos \omega_D t, \qquad \phi(t) \simeq r \sin \omega_D t$$

†The interfering wave can also be regarded as a component at frequency ω_i from a noise signal.

so that

$$z(t) \simeq A(1 + r \cos \omega_D t) \cos (\omega_c t + r \sin \omega_D t). \qquad (7.10\text{-}8)$$

To a first approximation, then, the effect of a small-amplitude interference wave is to amplitude- *and* phase-modulate the carrier such that the resulting modulation index is r.

An ideal PM detector is insensitive to $R(t)$ and furnishes an output proportional to $\phi(t)$. Likewise, an ideal FM detector delivers an output proportional to $\dot{\phi}(t)$. The computation of $\dot{\phi}(t)$ is straightforward if we recall that

$$\frac{d}{dt} (\tan^{-1} u) = \frac{1}{1 + u^2} \frac{du}{dt}.$$

The result is

$$\dot{\phi}(t) = r\omega_D \frac{r + \cos \omega_D t}{1 + 2r \cos \omega_D t + r^2} \equiv \omega_e \qquad (7.10\text{-}9a)$$

and

$$\dot{\theta}(t) = \omega_c + r\omega_D \frac{r + \cos \omega_D t}{1 + 2r \cos \omega_D t + r^2} \qquad [\theta = \omega_c t + \phi(t)]. \qquad (7.10\text{-}9b)$$

The quantity $\dot{\phi}(t)$ represents the frequency modulation induced by the interfering wave and represents the (error) departure of the instantaneous frequency from ω_c. Since $\dot{\phi}(t)$ is an error term, we denote it by ω_e and refer to it as *error frequency*.

An examination of Eq. (7.10-9a) shows that

$$\omega_e = \begin{cases} \dfrac{-r}{1 - r}\omega_D, & \omega_D t = (2n + 1)\pi \\[2mm] \dfrac{r^2}{1 + r^2}\omega_D, & \omega_D t = (2n + 1)\dfrac{\pi}{2} \\[2mm] \dfrac{r}{1 + r}\omega_D, & \omega_D t = 2n\pi, \end{cases} \qquad (7.10\text{-}10)$$

where $n = 0, \pm 1, \ldots$. Because of interference, one obtains the signal shown in Fig. 7.10-2 instead of detecting a null which an ideal detector would furnish in the absence of a modulating signal $s(t)$.

The analysis of more complicated cases such as interference between FM waves in common or adjacent channels follows the same principles, although the algebra gets more involved. In common channel interference, two messages from different sources appear in the demodulated signal. This phenomenon is known as *cross talk* and is generally very disturbing to the listener. Corrington [7-10] and Panter [7-1] give extensive discussions of various types of interference and show how the distortion in the detected signal varies with the parameter r.

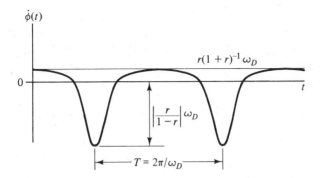

Figure 7.10-2. Detected instantaneous frequency associated with the interference of two carrier-type signals.

7.11 Preemphasis and Deemphasis Filtering (PDF)

An examination of the results obtained in the previous section shows that, other things being equal, the interference increases with increasing value of ω_D [see, for example, Eq. (7.10-9a) or (7.10-10)]. For many signals of common interest, such as speech and music, most of the energy is generally located in the lower frequency range. The small-amplitude high-frequency components of the baseband do not produce the full frequency deviation that is allowed. Hence the assigned bandwidth is not uniformly filled with signal energy, there being less at high frequencies. Unfortunately, as we shall see in Chapter 10, the receiver noise density increases as the square of frequency. The net result can be an intolerably low ratio of signal-to-noise power at the higher signal frequencies. To offset this undesirable phenomenon, an ingenious scheme known as preemphasis and deemphasis filtering is employed. The central idea is easily explained. At the transmitter, where signal levels are high and noise is, to some extent, under control, the modulating signal, $s(t)$, is purposely altered with the aim of enhancing the higher frequencies relative to the lower ones. This tailoring of the signal produces distortion, and the receiver must therefore invert the process to furnish an undistorted version of $s(t)$. The initial alteration of the signal that produces high-frequency enhancement is known as preemphasis filtering; the inverse process is deemphasis filtering. The deemphasis filter enhances the low frequencies relative to the higher ones. The net gain is that the noise now adds to a *strong* high-frequency signal and is suppressed in the deemphasis filtering. By the simple expedient of boosting the most noise-susceptible portion of the signal *before* noise becomes a problem and then inverting the process, an increase in signal-to-noise ratio is manifest at the receiver.

The technique of PDF is not confined to FM systems. However, it is widely used in FM because it is effective and the PDF circuitry is relatively simple. The deemphasis filter should be located in the receiver *after* the discriminator. Preemphasis filtering is done at the transmitter before modulation. A typical preemphasis network is shown in Fig. 7.11-1. The transfer function for the PE filter is

$$H_{PE}(\omega) = \hat{K}\frac{1 + j\omega\tau_1}{1 + j\omega\tau_2} \simeq \hat{K}(1 + j\omega\tau_1) \qquad \text{for } \omega < \omega_2 \equiv \frac{1}{\tau_2}, \qquad (7.11\text{-}1)$$

where $\hat{K} \equiv R_2(R_1 + R_2)^{-1}, \tau_1 = CR_1$, and $\tau_2 = CR_1R_2(R_1 + R_2)^{-1} \simeq CR_2$.

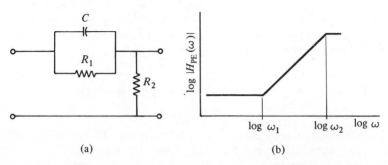

(a)   (b)

Figure 7.11-1. (a) Preemphasis filter; (b) asymptotic frequency response.

A typical value for τ_1 is 75 μsec. This means that components above 2.1 kHz are "emphasized." The value of τ_2 is relatively unimportant provided only that $(2\pi\tau_2)^{-1}$ is at least as great as the highest audio frequency (in hertz) for which preemphasis is desired; for quality reception this might be around 15 kHz.

The deemphasis filter must perform the inverse of the preemphasis filter to avoid a net distortion of $s(t)$. A deemphasis network is shown in Fig. 7.11-2. The transfer function for the DE network is

$$H_{DE}(\omega) = \frac{1}{1 + j\omega\tau_1}, \qquad (7.11\text{-}2)$$

where $\tau_1 = R_1C$. The break frequency is 2.1 kHz when $\tau_1 = 75$ μsec.

We note that

$$H_{PE}(\omega)H_{DE}(\omega) \simeq \hat{K} \qquad \text{for } \omega < \omega_2, \qquad (7.11\text{-}3)$$

which is the requirement for no distortion.

An interesting observation can be made with reference to Eq. (7.11-1). In the "active" region where $\omega_1 < \omega < \omega_2$, the transfer function of the preemphasis filter is approximately

$$H_{PE}(\omega) \simeq j\hat{K}\omega\tau_1. \qquad (7.11\text{-}4)$$

Thus if $S(\omega)$ represents the message spectrum, the response of the filter is seen to be the derivative of the signal. But frequency-modulating a carrier with

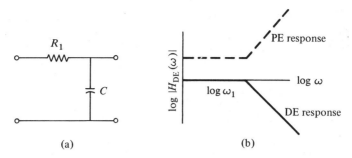

Figure 7.11-2. (a) Deemphasis filter; (b) asymptotic frequency response.

the *derivative* of a signal is equivalent to *phase-modulating* with the signal. Hence preemphasized FM is a mixture of FM and PM. In this connection it is interesting to compare the interference susceptibility of FM with that of PM. Equation (7.10-7) states that a PM detector produces an error signal $\phi_e(t)$ whose peak value is independent of the interfering frequency ω_D. For $r \ll 1$, the peak detected error is

$$\phi_e(t) = r, \qquad (7.11\text{-}5)$$

while in the case of FM, the peak detected error is proportional to

$$\dot{\phi}_e = \omega_D r.$$

Hence we deduce that FM is superior to PM at low values of ω_D and that PM is superior to FM at large values of ω_D. Suitably designed FM with PDF incorporates the best features of both FM and PM. At low values of ω_D, the PDF should be inactive so that the detected interference is as in FM. At higher values of ω_D, the response of the preemphasis filter is essentially as in PM. The net result is shown in Fig. 7.11-3.

Figure 7.11-3. Detected interference for FM, PM, and FM with preemphasis.

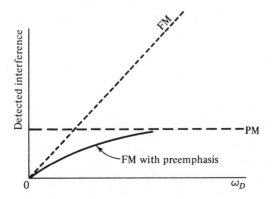

7.12 The FM Receiver

A block diagram of an FM receiver is shown in Fig. 7.12-1.

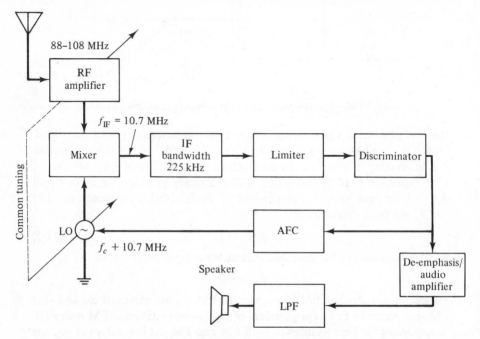

Fignre 7.12-1. Block diagram of an FM receiver.

Most FM receivers are of the superheterodyne type. The tuning controls of the local oscillator and RF amplifier are mechanically coupled so that the ouput of the mixer is a constant-carrier-frequency (10.7 MHz) FM signal. The frequency of 10.7 MHz is slightly larger than one-half the FM broadcasting band of 20 MHz. This means that all image frequencies lie outside the broadcasting band. For example, if the local oscillator frequency is set at $f_{LO} = f_c + 10.7$ MHz, then the FM signals at f_c and $f_c + 2(10.7)$ MHz will be admitted by the IF stage. The frequency $f_c + 2(10.7)$ MHz is the image frequency. However, *there is no signal* in the FM broadcast band at $f_c + 2(10.7)$ MHz, as is clear from the following: When f_c has its lowest value of 88 MHz, the image frequency is 109.4 MHz, which is outside the broadcast band. The possibility still remains, however, that image interference may be encountered from other services outside the 88–108-MHz band. One such service is the airport-aircraft communication channel which operates above the FM band. The ability to reject image frequencies is a criterion of quality for an FM receiver, as is its ability to reject adjacent channels in a given geographical area (these are usually spaced no less than 400 kHz above and

356

below the station). The latter is described by the *selectivity* of the receiver. Selectivity can be tested by tuning to a given station and injecting a second signal into an adjacent channel. The strength of the second signal is then raised until it becomes audible; the stronger the required signal, the more selective the receiver.

Other important criteria are harmonic distortion (HD) and intermodulation distortion (IMD). HD refers to the levels of undesired signals, produced at harmonics of signal frequencies, that are generated by nonlinearities in the system. HD depends on power level and generally increases as the audio-power level increases. IMD refers to the levels of undesired signals, produced at frequencies which are not harmonics of the signal frequencies. A power series expansion of a nonlinear characteristic would show that undesired harmonic and *inharmonic* (discordant) frequencies are produced which are likely to fall in the middle or upper portion of the audio range. This makes excessive IMD very objectionable, and considerable efforts are made by manufacturers to keep IMD small in quality receivers. Many other figures of merit are used in describing the quality of FM receivers. Descriptions of these are given in the literature [7-12].

7.13 Summary

We began the chapter by giving a mathematical description of angle modulation and deriving a Fourier resolution for FM and PM signals modulated by a single, constant-frequency, tone. We then used these mathematical tools to derive approximate bandwidths for FM waveforms modulated by arbitrary signals.

The quasi-static method for dealing with the problem of transmission of FM signals through linear networks was then discussed and applied to the slope detector.

Several types of FM detectors were discussed and analyzed, including some suitable for manufacture as integrated circuits because they lack inductances.

We ended the chapter by considering FM interference and preemphasis and deemphasis filtering. References [7-5], [7-7], and [7-8] are useful for further reading, as are References [7-10] and [7-11], in which some of the more complicated mathematical and statistical problems associated with FM interference are discussed. We shall pick up on signal-to-noise considerations in Chapter 10 of this book.

PROBLEMS

7-1. A real wave with arbitrary angle modulation $\phi(t)$ can be written as

$$x(t) = A \cos [\omega_c t + \phi(t)].$$

a. Show that $x(t)$ may be written as

$$x(t) = \text{Re } z(t),$$

where $z(t) = u(t)e^{j\omega_c t}$ and $u(t) = Ae^{j\phi(t)}$.

b. Explain why $z(t)$ cannot be regarded as a complex analytic signal even if $\phi(t)$ is band-limited.

c. It is known from practice that for phase and frequency modulation the spectrum of $u(t)$ ultimately drops off rapidly with increasing $|\omega|$. Explain therefore why, if ω_c is large enough, we can write

$$z(t) \simeq \xi(t) \equiv x(t) + j\hat{x}(t),$$

where $\hat{x}(t)$ is the Hilbert transform of $x(t)$.

7-2. Define the average power in $x(t)$ over an interval T to be

$$P_{\text{ave}} = \frac{1}{T} \int_{-T/2}^{T/2} x^2(t)\, dt.$$

What should be the size relationships between T, ω_c, and the bandwidth of $\exp[j2\phi(t)]$ in order for the average power in $x(t) = A \cos[\omega_c t + \phi(t)]$ to register $A^2/2$?

7-3. At first glance it might seem simpler to define FM by adding a signal-dependent frequency to ω_c; i.e.,

$$x(t) = \cos \underbrace{[\omega_c + \omega(t)]}_{\theta(t)} t.$$

a. Compute the instantaneous frequency $\omega_i \equiv \dot{\theta}(t)$ for the general case and when $\omega(t)$ is sinusoidal.

b. Demonstrate the inadequacy of this definition by considering what happens when $t \longrightarrow \infty$ and $\omega(t)$ is sinusoidal.

7-4. Compute the instantaneous phase and frequency for both PM and FM when the modulating signal, $s(t)$, is as shown in Fig. P7-4.

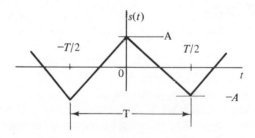

Figure P7-4

7-5. The spectra of AM and narrowband PM, although similar, enjoy important differences. Use phasor diagrams as in Fig. 7.4-2 to compute the maximum variations in the amplitudes of the waves in the two cases. Which is greater?

7-6. Repeat Prob. 7-5 when FM is compared with AM instead of PM.

7-7. Verify that the three systems shown in Fig. 7.4-4 do in fact generate AM, NBPM, and NBFM when $s(t)$ is a sinusoidal modulating signal. What polarities should be shown at the summation points?

7-8. A 50-MHz carrier is to be frequency-modulated by a 20-kHz sine wave.
   a. The frequency deviation is 20 Hz. What is a minimum appropriate bandwidth?
   b. The frequency deviation is increased to 1.0 MHz. What is the appropriate bandwidth? Why is the approximation $B_{CR} \simeq 2\Delta f$ inadequate for part a?

7-9. Consider an FM wave

$$x(t) = \text{Re } x_c(t)$$
$$= \text{Re } A_c \exp\{j[\omega_c t + \phi(t)]\},$$

where $x_c(t)$ is the complex form of the wave and Re means "real part of." Let the modulating signal as well as $\phi(t)$ be periodic with period $T_m = 2\pi/\omega_m$. Show that $x_c(t)$ may be written as

$$x_c(t) = A_c \sum_{n=-\infty}^{\infty} c_n e^{j(\omega_c + n\omega_m)t},$$

where

$$c_n = \frac{1}{T_m} \int_{-T_m/2}^{T_m/2} e^{j\phi(t)} e^{-jn\omega_m t} \, dt.$$

7-10. Let the instantaneous frequency $\omega_i = \omega_c + \dot{\phi}$ be as shown in Fig. P7-10. Use the results of Prob. 7-9 to show that

$$x(t) = A_c \sum_{n=-\infty}^{\infty} \frac{2\beta}{\pi(\beta^2 - n^2)} \sin\left[(\beta - n)\frac{\pi}{2}\right] \cos(\omega_c + n\omega_m)t,$$

where $\beta \equiv \Delta\omega/\omega_m$.

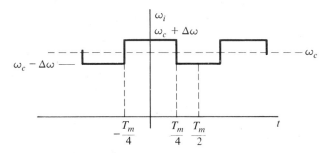

Figure P7-10

7-11. Assume that there exists an FM broadcasting system for which $\Delta f = 50$ kHz. Compute a table similar to Table 7.5-1 by considering sinusoidal modulating frequencies that range from 20 to 0.05 kHz. Which tones require the largest bandwidths?

7-12. Assume that for a particular FM system a sinusoidal tone at frequency f_m generates a frequency deviation given by

$$\Delta f = 75 e^{-[(f_m - 500)/1000]^2} \quad \text{(kHz)} \quad (f_m \text{ in Hz}).$$

Plot the required bandwidth, as determined by Carson's rule, for sinusoidal tones with frequencies varying over the audio range.

7-13. Plot the amplitude-frequency spectra of an FM wave for $\beta = 0.1, 1, 5, 10$. Make two plots in which (a) β increases because Δf increases but f_m is held fixed and (b) β increases because f_m decreases but Δf is held fixed. Verify that as β becomes large the bandwidth approaches $2\Delta f$. Choose any reasonable increments.

7-14. Consider the phase-modulated wave

$$x(t) = \cos[\omega_c t + K' s(t)], \qquad |s(t)| \leq 1.$$

a. Explain why K' must be restricted to $K' < \pi$.
b. Discuss the required bandwidth for PM with a sinusoidal modulating signal. What happens when the phase deviation K_d changes? What happens when the modulating frequency f_m changes?

7-15. It is desired to design an Armstrong-type indirect FM transmitter. Let the modulating frequencies, f_m, range from $f_m = 50$ to $f_m = 5000$ Hz. The maximum frequency deviation at the output is to be no greater than 75 kHz. The initial NBFM $[\beta_1]_{\max}$ is restricted to 0.2. Compute the required frequency multiplication and the largest frequency deviation at the input.

7-16. An FM signal is to be demodulated with a slope detector followed by an envelope detector. Assume that the (one-sided) transfer function associated with the slope detector is described by

$$H(\omega) = k_1(\omega - \omega_c)^2 + k_2(\omega - \omega_c) + k_3,$$

where, for simplicity the $\{k_i\}$ are real constants in appropriate units.
a. Compute the complex response, $\eta(t)$, to the complex signal

$$z(t) = \exp\{j[\omega_c t + \mu(t)]\}.$$

b. Let the modulation be a periodic ramp, i.e., $s(t) = At$, $0 \leq t \leq T$, etc. Compute the per-cycle power ratio of signal to distortion in the envelope-detected output.

7-17. Design a single-tuned RLC network (Fig. 7.9-1) to slope-detect an FM signal with carrier frequency of 50 MHz and $\Delta f = 75$ kHz.

7-18. It was shown in Sec. 7.9 that demodulation of an FM wave is possible by direct differentiation followed by envelope detection and dc blocking. Explain how delay lines might be used to approximate the derivative of the wave by realizing the quantity.

$$\dot{x}(t) \simeq \frac{x(t) - x(t - \tau)}{\tau}.$$

Draw the system, and suggest how small τ must be in order for the right-hand side to be a good approximation of the derivative.

7-19. Consider an FM wave $x(t) = A \cos \theta(t)$, where $\theta(t) = \omega_c t + K'' \int^t s(\lambda)\, d\lambda$.

Let t_1, t_2 $(t_2 > t_1)$ denote the time of two adjacent zeros of $x(t)$. Assume that

$$\int_{t_1}^{t_2} s(\lambda)\, d\lambda \simeq s(t)(t_2 - t_1), \qquad \text{where } t_1 \le t \le t_2.$$

Show that

$$K'' s(t) = \frac{\pi}{T_{AC}} - \omega_c,$$

where $T_{AC} = t_2 - t_1$. This result implies that $s(t)$ can be detected by counting the zero crossings in $x(t)$. What property of FM waves account for this?

7-20. (Generalization of Prob. 7-19) Let n_z denote the number of zero crossings in time T_z. Show that if T_z satisfies

$$f_c^{-1} < T_z \ll W^{-1},$$

where W is the bandwidth of $s(t)$, then

$$Ks(t) \simeq \frac{n_z}{2T_z} - f_c.$$

Hence the signal can be detected by using a zero-crossing counter. Why must the inequalities on T_z be satisfied for this method to work?

7-21. (Multipath transmission in FM) Consider two FM waves coming from the same source but reaching the receiver by different paths. Let the direct wave be given by

$$x_1(t) = \cos \underbrace{[\omega_c(t - t_1) + \beta \sin \omega_m(t - t_1)]}_{\psi_1(t)},$$

and let the reflected (indirect) wave be given by

$$x_2(t) = \rho \cos \underbrace{[\omega_c(t - t_2) + \beta \sin \omega_m(t - t_2)]}_{\psi_2(t)},$$

where $\rho \le 1$ and $t_1(t_2)$ is the time for the direct (reflected) wave to reach the receiver. Draw a phasor diagram for this situation and show that the resultant wave can be written as

$$x_R(t) = A(t) \cos \phi(t),$$

where

$$A \equiv \sqrt{1 + \rho^2 + 2\rho \cos \psi(t)}$$

$$\psi(t) \equiv \psi_2(t) - \psi_1(t)$$

$$\phi(t) \equiv \psi_1(t) + \tan^{-1} \frac{\rho \sin \psi(t)}{1 + \rho \cos \psi(t)}.$$

7-22. (Problem 7-21 continued) Consider the resultant wave $x_R(t)$ in Prob. 7-21. What is the detected signal if $x_R(t)$ is sent to a limiter followed by a balanced discriminator? (A balanced discriminator removes the dc term ω_c.)

REFERENCES

[7-1] J. R. CARSON, "Notes on the Theory of Modulation," *Proc. IRE*, **10**, pp. 57–64, Feb. 1922.

[7-2] E. H. ARMSTRONG, "Method of Reducing the Effect of Atmospheric Disturbances," *Proc. IRE*, **16**, pp. 15–26, Jan. 1928.

[7-3] K. HENNEY, ed., *Radio Engineering Handbook*, 5th ed., McGraw-Hill, New York, 1959.

[7-4] M. ABRAMOWITZ and I. A. STEGUN, *Handbook of Mathematical Functions*, Dover, New York, 1965.

[7-5] A. B. CARLSON, *Communication Systems: An Introduction to Signals and Noise in Electrical Communications*, McGraw-Hill, New York, 1968.

[7-6] J. MILLMAN and C. C. HALKIAS, *Electronic Devices and Circuits*, McGraw-Hill, New York, 1967.

[7-7] M. SCHWARTZ, *Information Transmission, Modulation and Noise*, 2nd ed., McGraw-Hill, New York, 1970.

[7-8] M. SCHWARTZ, W. R. BENNETT, and S. STEIN, *Communication Systems and Techniques*, McGraw-Hill, New York, 1966.

[7-9] K. K. CLARKE and D. T. HESS, *Communication Circuits: Analysis and Design*, Addison-Wesley, Reading, Mass., 1971.

[7-10] M. S. CORRINGTON, "Frequency Modulation Caused by Common and Adjacent Interference," *RCA Review*, **7**, pp. 552–560, Dec. 1946.

[7-11] P. F. PANTER, *Modulation, Noise and Spectral Analysis*, McGraw-Hill, New York, 1965.

[7-12] *Understanding High Fidelity*, 2nd ed., Pioneer Electronic Corporation, Tokyo, 1975.

8

Elements
of Probability

8.1 Introduction

In this chapter we present some of the basic ideas of probability theory. We assume that the reader has had some prior exposure to set theory and is somewhat familiar with the notion of sets, set operations, etc. The material discussed here is fairly standard and is covered in much greater depth in such excellent books as those by Parzen [8-1] and Papoulis [8-2].

Probability as a Measure of Frequency of Occurrence

One approach to defining the probability of an event E is to perform an experiment n times. The number of times that E appears is denoted by n_E. Then the probability of E occurring is defined according to

$$P(E) = \lim_{n \to \infty} \frac{n_E}{n}. \qquad (8.1\text{-}1)$$

Quite clearly, since $n_E \le n$, we must have $0 \le P(E) \le 1$. One difficulty with this approach is that we can never perform the experiment an infinite number of times so that we can only estimate $P(E)$ from a finite number of trials. Second, we *postulate* that n_E/n approaches a limit in some sense as n goes to infinity. But consider flipping a fair coin 1000 times. The likelihood of getting exactly 500 heads is very small; in fact if we flipped the coin 10,000 times, the likelihood of getting exactly 5000 heads is even smaller. As $n \to \infty$, the event of observing exactly $n/2$ heads becomes vanishingly small. Yet our intuition demands that $P(\text{head}) = \frac{1}{2}$ for a fair coin. Suppose we choose a $\delta > 0$; then we shall find experimentally that if the coin is truly fair, the number of times that

$$\left| \frac{n_E}{n} - \frac{1}{2} \right| > \delta \qquad (8.1\text{-}2)$$

as n becomes large becomes very small. Thus although it is very unlikely that at any stage of this experiment, especially when n is large, n_E/n is exactly $\frac{1}{2}$, this ratio will nevertheless hover around $\frac{1}{2}$, and the number of times it will make significant excursion away from the vicinity of $\frac{1}{2}$ in the sense of Eq. (8.1-2) becomes very small indeed.

Despite these problems with the frequency definition of probability, the relative frequency concept is essential in applying probability theory to the physical world.

Probability as the Ratio
of Favorable to Total Outcomes

In this approach, which is not experimental, the probability of an event is computed *a priori*† by counting the number of ways N_E that E can occur and forming the ratio N_E/N, where N is the number of all possible outcomes, i.e., the number of all alternatives to E plus N_E. An important notion here is that all outcomes are equally likely. Since "equally likely" is really a way of saying equally probable, the reasoning is somewhat circular. Suppose we throw a pair of unbiased dice and ask what is the probability of getting a seven? We partition the outcome into 36 equally likely outcomes as shown in Table 8.1-1, where each entry is a possible outcome. The total number of

TABLE 8.1-1

First die

		1	2	3	4	5	6
	1	2	3	4	5	6	7
Second die	2	3	4	5	6	7	8
	3	4	5	6	7	8	9
	4	5	6	7	8	9	10
	5	6	7	8	9	10	11
	6	7	8	9	10	11	12

outcomes is 36 if we keep the dice distinct. The number of ways of getting a seven is $N_7 = 6$. Hence

$$P(\text{getting a seven}) = \tfrac{6}{36} = \tfrac{1}{6}.$$

---

†*A priori* means relating to reasoning from self-evident propositions or presupposed by experience. *A posteriori* means relating to reasoning from observed facts.

EXAMPLE

Throw a fair coin twice (note that since no physical experimentation is involved, there is no problem in postulating an ideal "fair coin"). The outcomes are *HH, HT, TH, TT*. The probability of getting at least one tail is computed as follows: With E denoting the event of getting at least one tail, the event E is the set of outcomes

$$E = \{HT, TH, TT\}.$$

The number of elements in E is $N_E = 3$; the number of all outcomes, N, is 4. Hence

$$P(\text{at least one } T) = \frac{N_E}{N} = \frac{3}{4}.$$

Axiomatic Approach

This is the approach followed in most modern textbooks on the subject. To develop it we first define certain notions.

The *sample space* or *sample description space* S of a random event or experiment with random outcomes is the space of descriptions of all possible elementary outcomes of the experiment.

EXAMPLES

1. The experiment consists of throwing a coin twice. Then $S = \{HT, TH, HH, TT\}$.

2. The experiment consists of choosing a person at random and counting the hair on his (or her) head. Then $S = \{0, 1, \ldots\}$, i.e., the set of nonnegative integers. An event E is a subset of S written $E \subset S$. Since $S \subset S$, S is also an event and is called the *certain event*. Hence S is both the sample space and the certain event.

3. The experiment consists of throwing a coin twice. Then $S = \{HH, HT, TH, TT\}$. The event, E, of getting at least one tail is

$$E = \{HT, TH, TT\} \subset S.$$

The empty set, i.e., the set containing no descriptions in S, is the impossible event, denoted by $\varnothing$. If S consists of n discrete elementary events, i.e., basic descriptions $s_i, i = 1, \ldots, n$, then there are 2^n distinct events that can be defined over S. For example, with $S = \{H, T\}$ (one throw of a coin), the events are

$$\{H\}, \quad \{T\}, \quad \{S\}, \quad \{\varnothing\}.$$

Exercise: Show that $m = 2^n$ events can be defined over S.

Before giving the axiomatic definition of probability we remind the reader that the union (sum) of two sets is the set of all elements that are in at least one of the two sets. The intersection (product) of two sets is the set of all elements that appear only in both sets. A symbolic representation of union, $\cup$, and intersection,† $\cap$, is obtained with the use of *Venn diagrams*. In Fig.

†The intersection of two sets, $A \cap B$, will be written as AB.

8.1-1(a) the union of the sets E and F is indicated by the shaded area. In Fig. 8.1-1(b), the intersection of E and F is shown by the shaded area. Another concept that we need is that of a *Borel field* $\mathfrak{B}$. In general a *field* $\mathfrak{M}$ is a class of sets such that

1. If $A \in \mathfrak{M}$, then $A^c \in \mathfrak{M}$ (A^c is the complement of A; it is the set of all elements not in A).
2. If $A \in \mathfrak{M}$ and $B \in \mathfrak{M}$, then $A \cup B \in \mathfrak{M}$, $AB \in \mathfrak{M}$, and $\varnothing \in \mathfrak{M}$.
3. Generalization: If $A_1, \ldots, A_n$ belong to $\mathfrak{M}$, so do the union $A_1 \cup A_2 \cup \cdots \cup A_n$ and the intersection $A_1 A_2 \cdots A_n$.

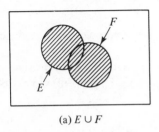

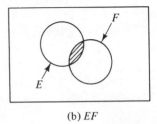

(a) $E \cup F$ (b) EF

Figure 8.1-1

A *Borel* field $\mathfrak{B}$ is a field that is closed under any countable set of unions, intersections, and combinations thereof. Thus if $A_1, \ldots, A_n, \ldots$ belong to $\mathfrak{B}$, so do

$$\bigcup_{i=1}^{\infty} A_i \quad \text{and} \quad \prod_{i=1}^{\infty} A_i.$$

If S has a countable number of elements, then every subset of S is an event, and therefore the set of all subsets of S forms a Borel field. However, when S is not countable, i.e., say when $S = R =$ the real line, then not every subset of S can be constructed from countable unions and intersections. Such subsets are said to be nonprobabilizable and cannot be considered as events. These sets cannot be assigned a probability measure without violating the axioms in Sec. 8.2. However, such sets do not normally occur in problems of engineering interest.

8.2 Axiomatic Definition of Probability

Probability is a function $P(\cdot)$ that assigns to every $E \subset S$ a number $P(E)$ called the probability of E such that

(i) $P(E) \geq 0,$ (8.2-1)

(ii) $P(S) = 1,$ (8.2-2)

(iii) $P(E \cup F) = P(E) + P(F)$ if $EF = \varnothing.$ (8.2-3)

Note that E and F must be *events*; that is, they must be subsets of S in $\mathfrak{B}$. A *probability space* $\mathfrak{IC}$ refers to the triplet

$$(S, \mathfrak{B}, P), \tag{8.2-4}$$

where S is the sample description space of a random phenomenon, $\mathfrak{B}$ is the Borel field of events, and P is the probability measure whose domain is S and whose range is $[0, 1]$.

Exercises:

1. Show that $P(\varnothing) = 0$.
2. Show that with $E \in \mathfrak{B}$, $F \in \mathfrak{B}$,

$$P(EF^c) = P(E) - P(EF).$$

3. Show that $P(E) = 1 - P(E^c)$.
4. Show that $P(E \cup F) = P(E) + P(F) - P(EF)$.

EXAMPLES

1. The experiment consists of throwing a coin once. Hence

$$S = \{H, T\}.$$

The Borel field of events consists of the following sets: $\{H\},\{T\},\{S\},\{\varnothing\}$. With $P(H)$ assumed $\frac{1}{2} = P(T)$ we have

$$P(H) = P(T) = \tfrac{1}{2}, \qquad P(S) = 1, \quad P(\varnothing) = 0.$$

2. The experiment consists of throwing a die once. The outcome is the number of dots n_i appearing on the upface of the die. The set S is given by $S = \{1, 2, 3, 4, 5, 6\}$. The Borel field of events consists of 2^6 elements. Some are

$$\{\varnothing\}, \quad \{S\}, \quad \{1\}, \quad \{1, 2\}, \quad \{1, 2, 3\}, \quad \{1, 4, 6\}, \quad \{1, 2, 4, 5\}.$$

We assign

$$P(i) = \tfrac{1}{6}, \qquad i = 1, \dots, 6 \quad \text{(an assumption)}.$$

All probabilities can now be computed from the basic axioms and the assumed probabilities for the elementary events. Thus with $A = \{1\}$ and $B = \{2, 3\}$ we obtain $P(A) = \frac{1}{6}$. Also $P(A \cup B) = P(A) + P(B)$, since $AB = \varnothing$. Further, $P(B) = P(2) + P(3) = \frac{2}{6}$ so that

$$P(A \cup B) = \tfrac{1}{6} + \tfrac{2}{6} = \tfrac{1}{2}.$$

3. The experiment consists of picking at random a numbered ball from 12 balls numbered 1–12 from an urn:

$$S = \{1, \dots, 12\}.$$

Let

$$A = \{1, \dots, 6\}, \qquad B = \{3, \dots, 9\},$$

$$A \cup B = \{1, \dots, 9\}, \qquad AB = \{3, 4, 5, 6\}, \qquad AB^c = \{1, 2\},$$

$$B^c = \{1, 2, 10, 11, 12\}, \qquad A^c = \{7, \dots, 12\}, \qquad A^c B^c = \{10, 11, 12\},$$

$$(AB)^c = \{1, 2, 7, 8, 9, 10, 11, 12\}.$$

Hence
$$P(A) = P(1) + P(2) + \cdots + P(6)$$
$$P(B) = P(3) + \cdots + P(9)$$
$$P(AB) = P(3) + \cdots + P(6).$$
If $P(1) = \cdots = P(12) = \frac{1}{12}$, then $P(A) = \frac{1}{2}$, $P(B) = \frac{7}{12}$, $P(AB) = \frac{4}{12}$, etc.

8.3 Joint and Conditional Probabilities; Independence

Consider a probability space $(S, \mathfrak{B}, P)$. Let A and B be two events defined over this space. Then AB is an event. The joint probability of the event A and B is $P(AB)$. The frequency interpretation is as follows:

$$P(A) \simeq \frac{n_A}{n}$$

$$P(B) \simeq \frac{n_B}{n} \qquad \text{for } n \text{ large} \qquad (8.3\text{-}1)$$

$$P(AB) \simeq \frac{n_{AB}}{n},$$

where n_{AB} is the number of times *both A and B* occur. The conditional probability of A given B is defined by

$$P(A|B) = \frac{P(AB)}{P(B)}, \qquad P(B) > 0. \qquad (8.3\text{-}2)$$

Frequency interpretation: From Eq. (8.3-1) we have

$$P(A|B) \simeq \frac{n_{AB}/n}{n_B/n} = \frac{n_{AB}}{n_B}, \qquad n_B > 0, \qquad (8.3\text{-}3)$$

i.e., the ratio of the number of times that A and B both occur to the number of times at least B occurs.

From Eq. (8.3-2) we have $P(AB) = P(A|B)P(B)$. Two events A, B are said to be independent if

$$P(AB) = P(A)P(B). \qquad (8.3\text{-}4)$$

Interpretation: If $P(A|B) = P(A)$, then observing B has no effect on the probability of A. Hence we say A and B are independent. In general we say that $A_1, \ldots, A_n$ are *independent* if

$$P(A_1 \ldots A_n) = \prod_{i=1}^{n} P(A_i). \qquad (8.3\text{-}5)$$

8.4 Total Probability and Bayes' Theorem

Let $A_1, \ldots, A_n$ be n mutually exclusive events. Let $\bigcup_{i=1}^{n} A_i = S$. With B denoting any event defined over the same probability space we have

$$P(B) = P(B|A_1)P(A_1) + \cdots + P(B|A_n)P(A_n). \qquad (8.4\text{-}1)$$

Proof: Since $A_i A_j = \varnothing$, $i \neq j$, we have

$$BS = B = B(A_1 \cup A_2 \cup \cdots \cup A_n) = \bigcup_{i=1}^{n} BA_i. \tag{8.4-2}$$

But $(BA_i)(BA_j) = \varnothing$ since $A_i A_j = \varnothing$. Hence

$$P(BA_1 \cup \cdots \cup BA_n) = P(BA_1) + \cdots + P(BA_n).$$

Using

$$P(BA_i) = P(B \mid A_i)P(A_i),$$

we obtain

$$P(B) = P\left[\bigcup_{i=1}^{n} (BA_i)\right] = \sum_{i=1}^{n} P(B \mid A_i)P(A_i). \tag{8.4-3}$$

$P(B)$ in Eq. (8.4-3) is sometimes called the total (unconditional) probability of the event B.

Bayes' Theorem

A formula that is widely used in statistical communication theory, pattern recognition, and statistical inference is Bayes' theorem, which can easily be developed from the results of the above discussion. With $P(A_i) > 0$, $i = 1, \ldots, n$, $P(B) > 0$, $\bigcup_{i=1}^{n} A_i = S$, and $A_i A_j = \varnothing$ for $i \neq j$, we obtain

$$P(A_j \mid B) = \frac{P(A_j B)}{P(B)} = \frac{P(B \mid A_j)P(A_j)}{\sum_{i=1}^{n} P(B \mid A_i)P(A_i)}. \tag{8.4-4}$$

The probability $P(A_j \mid B)$ is sometimes called the *a posteriori* probability of A_j given B. Equation (8.4-4) enables the computation of the *a posteriori* probability in terms of the *a priori* conditional probabilities $P(B \mid A_i)$.

EXAMPLES

1. In a communication system a zero or one is transmitted with $P(X = 0) \equiv P_0$, $P(X = 1) \equiv P_1 = 1 - P_0$, respectively. Due to noise in the channel, a zero can be received as a one with probability β, and a one can be received as a zero also with probability β. A one is observed. What is the probability that a one was transmitted?

Solution: The structure of the channel is shown in Fig. 8.4-1. Hence

$$P(X = 1 \mid Y = 1) = \frac{P(X = 1, \ Y = 1)}{P(Y = 1)}$$

$$= \frac{P(Y = 1 \mid X = 1)P(X = 1)}{P(Y = 1 \mid X = 1)P(X = 1) + P(Y = 1 \mid X = 0)P(X = 0)}$$

$$= \frac{P_1(1 - \beta)}{P_1(1 - \beta) + P_0\beta}.$$

If $P_0 = P_1 = \frac{1}{2}$, the *a posteriori* probability $P(X = 1 \mid Y = 1)$ depends on β, as shown in Fig. 8.4-2. The channel is said to be noiseless if $\beta = 1$ or $\beta = 0$.

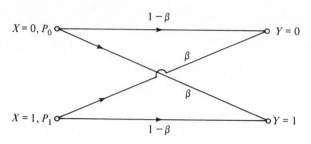

Figure 8.4-1

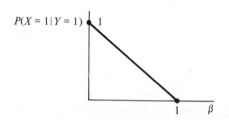

Figure 8.4-2

2. ([8-1], p. 119) Suppose there exists a (fictitious) test for cancer with the following properties. Let

A = event that the test states that tested person has cancer

B = event that person has cancer

A^c = event that test states person is free from cancer

B^c = event that person is free from cancer.

It is known that $P(A|B) = P(A^c|B^c) = 0.95$ and $P(B) = 0.005$. Is the test a good test?

Solution: To answer this question we should like to know the likelihood that a person actually has cancer if the test so states, i.e., $P(B|A)$. Hence

$$P(B|A) = \frac{P(B)P(A|B)}{P(A|B)P(B) + P(A|B^c)P(B^c)} = \frac{(0.005)(0.95)}{(0.95)(0.005) + (0.05)(0.995)}$$

$$= 0.087.$$

Hence in only 8.7% of the cases tested will a positive test reaction actually indicate that the person has cancer. This test has a very high false-positive rate and in this sense cannot be regarded as a good test.

8.5 Bernoulli Trials

Consider the very simple experiment that consists of a single trial with a binary outcome X: a success $\{X = s\}$ with probability p or a failure $\{X = f\}$

with probability $q = 1 - p$. Thus $P(s) = p, P(f) = q$, and the sample description space is $S = \{s, f\}$.

Suppose we do the experiment twice. The new sample description space S_2, written $S_2 = S \times S$, is the set of all ordered 2-tuples

$$S_2 = \{ss, sf, fs, ff\}.$$

The product $S \times S$ is called the Cartesian product. If we do n independent trials, the sample space is

$$S_n = \underbrace{S \times S \times \cdots \times S}_{n \text{ times}}$$

and contains 2^n elementary outcomes each of which is an ordered n-tuple. Thus

$$S_n = \{a_1, \ldots, a_M\}, \qquad \text{where } M = 2^n,$$

and $a_i = z_1 z_2 \cdots z_n$, where $z_j = s$ or f. Since each outcome z_j is independent of any other outcome, the joint probability $P(z_1, \ldots, z_n) = P(z_1)P(z_2) \cdots P(z_n)$. Thus the probability of a given ordered set of k successes and $n - k$ failures is simply $p^k q^{n-k}$. For example, suppose we throw a coin three times with $p = P(H)$ and $q = P(T)$. The probability of the event $\{HTH\}$ is $pqp = p^2 q$. The probability of the event $\{THH\}$ is also $p^2 q$. The different events leading to two heads and one tail are

$$E_1 = \{HHT\}$$
$$E_2 = \{HTH\}$$
$$E_3 = \{THH\}.$$

If F denotes the event of getting two heads and one tail, then $F = E_1 \cup E_2 \cup E_3$. Since $E_i E_j = \varnothing, i \neq j$, we obtain $P(F) = P(E_1) + P(E_2) + P(E_3) = 3p^2 q$.

Very frequently we are interested in this kind of result, namely, the probability of getting k successes in n tries without specification of order. The probability law that applies to this case is the *binomial law* $b(k; n, p)$, which is the probability of getting k successes in n *independent* tries with individual Bernoulli-trial success probability p. The probability law is given by

$$b(k; n, p) = \binom{n}{k} p^k q^{n-k}, \tag{8.5-1}$$

where

$$\binom{n}{k} \equiv \frac{n!}{k!(n-k)!} = \text{binomial coefficient.} \tag{8.5-2}$$

The coefficient $\binom{n}{k}$ is the number of ways of getting unordered subsets of size k that may be formed from the members of a set of size n. How do we obtain this coefficient? Consider an urn containing n distinguishable balls. Suppose we draw k balls without replacement. From these k balls we can form $k!$ distinguishable (ordered) subsets. The total number of ordered subsets must equal $k!x_k$, where x_k is the different subsets of size k we can form from n. But $k!x_k$ must equal $n(n-1)\cdots(n-k+1)$ since this is the number of distinguishable subsets we can draw from the urn directly. Hence

$$x_k k! = n(n-1)\cdots(n-k+1) = \frac{n!}{(n-k)!} \qquad (8.5\text{-}3)$$

or

$$x_k = \frac{n!}{k!(n-k)!} = \binom{n}{k}. \qquad (8.5\text{-}4)$$

EXAMPLES

1. Suppose $n = 4$; i.e., there are four balls numbered 1–4 in the urn. The number of distinguishable, ordered, samples of size 2 that can be drawn without replacement is 12, i.e., $\{1, 2\}, \{1, 3\}, \{1, 4\}, \{2, 1\}, \{2, 3\}, \{2, 4\}, \{3, 1\}, \{3, 2\}, \{3, 4\}, \{4, 1\}, \{4, 2\}, \{4, 3\}$. The number of distinguishable unordered sets is 6, i.e.,

$$\{1, 2\} \quad \{1, 3\} \quad \{1, 4\} \quad \{2, 4\} \quad \{3, 4\} \quad \{2, 3\}$$
$$\{2, 1\} \quad \{3, 1\} \quad \{4, 1\} \quad \{4, 2\} \quad \{4, 3\} \quad \{3, 2\}$$

From the formula we obtain this result directly; i.e.,

$$\binom{n}{k} = \frac{4!}{2!2!} = 6.$$

2. Ten independent, binary pulses per second arrive at a receiver. The error probability (i.e., a zero received as a one or vice versa) is 0.001. What is the probability of at least one error per second?

$$P(\text{at least one error/sec}) = 1 - P(\text{no errors/sec})$$

$$= 1 - \binom{10}{0}(0.001)^0(0.999)^{10} = 1 - (0.999)^{10} \simeq 0.01.$$

Observation:

$$P(S) = \sum_{k=0}^{n} b(k; n, p) = 1. \qquad (8.5\text{-}5)$$

Why?

8.6 Further Discussion of the Binomial Law

We shall write down some self-evident formulas for further use. The probability $B(k; n, p)$ of at least k successes in n tries is given by

$$B(k; n, p) = \sum_{i=k}^{n} b(i; n, p) = \sum_{i=k}^{n} \binom{n}{i} p^i q^{n-i}. \qquad (8.6\text{-}1)$$

The probability of fewer than k successes in n tries is

$$\sum_{i=0}^{k-1} b(i; n, p) = 1 - B(k; n, p). \qquad (8.6\text{-}2)$$

The probability of more than k successes but no more than j successes is

$$\sum_{i=k+1}^{j} b(i; n, p). \qquad (8.6\text{-}3)$$

8.7   The Poisson Law

Suppose that $n \gg 1$, $p \ll 1$ but that np remains finite, say $np = a$. Recall that $q = 1 - p$. Hence

$$\binom{n}{k} p^k (1 - p)^{n-k} \simeq \frac{1}{k!} a^k \left(1 - \frac{a}{n}\right)^{n-k},$$

where $n(n - 1) \cdots (n - k + 1) \simeq n^k$ because the binomial probability will have significant values around $k = np \ll n$. Hence in the limit as $n \to \infty$, $p \to 0$, $np = a$, $k \ll n$, we obtain

$$b(k; n, p) \simeq \frac{1}{k!} a^k \left(1 - \frac{a}{n}\right)^{n-k} = \frac{a^k}{k!} e^{-a} \qquad (\text{as } n \to \infty). \qquad (8.7\text{-}1)$$

EXAMPLES

1. A computer contains 10,000 components. Each component fails independently from the others, and the yearly failure probability per component is 10^{-4}. What is the probability that the computer will be working at the end of the year? Assume that the computer fails if one or more components fail.

Solution:

$$p = 10^{-4}, \qquad n = 10,000, \qquad k = 0, \qquad np = 1.$$

Hence

$$b(0; 10,000, 10^{-4}) = \frac{1^0}{0!} e^{-1} = \frac{1}{e} = 0.368.$$

2. Suppose that n independent points are placed at random in an interval $(0, T)$. Let $0 < t_1 < t_2 < T$ and $t_2 - t_1 = \Delta t$. Let $\Delta t/T \ll 1$ and $n \gg 1$. What is the probability of observing exactly k points in Δt?

Solution: Consider a single point occurring in $[0, T]$. The probability of the point appearing in Δt is $\Delta t/T$. Let $p = \Delta t/T$. Every other point has the same probability of being in Δt. Hence the probability of finding k points in Δt is the binomial law

$$P(k \text{ points in } \Delta t) = \binom{n}{k} p^k q^{n-k}. \qquad (8.7\text{-}2)$$

With $n \gg 1$ and $\Delta t/T \ll 1$, we use the approximation in Eq. (8.7-1) to give

$$b(k; n, p) \simeq \left(\frac{n\Delta t}{T}\right)^k \frac{e^{-n\Delta t/T}}{k!} \tag{8.7-3}$$

$$= (\lambda \Delta t)^k \frac{e^{-\lambda \Delta t}}{k!}, \tag{8.7-4}$$

where $\lambda \equiv n/T$ is the "average" number of points per unit interval.

Equations (8.7-1) and (8.7-4) are examples of the *Poisson probability law*. The Poisson law with parameter a ($a > 0$) is defined by[†]

$$P(k \text{ events}) = e^{-a}\frac{a^k}{k!}, \tag{8.7-5}$$

where $k = 0, 1, 2, \ldots$. With $a = \lambda \Delta t$, where λ is the average number of events per unit time, and Δt the interval $(t, t + \Delta t)$, the probability of k events in Δt is

$$P[k \text{ events in } (t, t + \Delta t)] = e^{-\lambda \Delta t}\frac{(\lambda \Delta t)^k}{k!}. \tag{8.7-6}$$

In Eq. (8.7-6) we assumed that λ was independent of t. If λ depends on t, the probability of k events in $(t, t + \Delta t)$ is

$$\exp\left[-\int_t^{t+\Delta t} \lambda(\xi)\, d\xi\right]\frac{1}{k!}\left[\int_t^{t+\Delta t} \lambda(\xi)\, d\xi\right]^k. \tag{8.7-7}$$

8.8 Conclusion

The reader who has had no prior exposure to the subject matter of this chapter may want to consult additional material on probability theory. Such a reader will benefit from reading Chapters 1–3 in Parzen [8-1] or the introductory chapters in any of the several good textbooks devoted to probability theory and random processes: Papoulis [8-2], Cramer [8-3], Feller [8-4], and Drake [8-5].

In the next chapter we shall discuss the theory of random variables in greater detail than we have discussed the elements of probability theory here. Random variables have great application in all of the sciences. Not only are they useful in modeling numerical-valued random phenomena, but, suitably interpreted, they can be used to analyze *random waveforms* which occur often in electrical communications.

[†]We use the term *event* here rather than "success."

PROBLEMS

8-1. An urn contains three balls numbered 1, 2, 3. The experiment consists of drawing a ball at random, recording the number, and replacing the ball before the next ball is drawn. This is called sampling with replacement. What is the probability of drawing the same ball twice?

8-2. An experiment consists of drawing two balls *without* replacement from an urn containing six balls numbered 1–6. Describe the sample description space S. What is S if the first ball is replaced before the second is drawn?

8-3. The experiment consists of measuring the heights of each partner of a randomly chosen married couple. Describe S in convenient notation.

8-4. In Prob. 8-3, let E be the event that the man is shorter than the woman. Describe E in convenient notation.

8-5. An urn contains 10 balls numbered 1–10. Let E be the event of drawing a ball numbered no greater than 5. Let F be the event of drawing a ball numbered greater than 3 but less than 9. Evaluate E^c, F^c, EF, $E \cup F$, EF^c, E^cF, $E^c \cup F^c$, $EF^c \cup E^cF$, $EF \cup E^cF^c$, $(E \cup F)^c$, and $(EF)^c$. Express these events in words.

8-6. An experiment consists of drawing two balls at random, with replacement, from an urn containing five balls numbered 1–5. Three students "Dim," "Dense," and "Smart" were asked to compute the probability p that the sum of numbers appearing on the two draws equals 5. Dim computed $p = \frac{2}{15}$, arguing that there are 15 distinguishable unordered pairs and only 2 are favorable, i.e., (1, 4) and (2, 3). Dense computed $p = \frac{1}{9}$, arguing that there are 9 distinguishable sums (2–10) of which only 1 was favorable. Smart computed $p = \frac{4}{25}$, arguing that there were 25 distinguishable ordered outcomes of which 4 were favorable, i.e., (4, 1), (3, 2), (2, 3), and (1, 4). Why is $p = \frac{4}{25}$ the correct answer? Explain what is wrong with the reasoning of Dense and Dim.

8-7. Use the axioms given in Eqs. (8.2-1)–(8.2-3) to show the following ($E \in \mathcal{B}, F \in \mathcal{B}$):
a. $P(\varnothing) = 0$.
b. $P(EF^c) = P(E) - P(EF)$.
c. $P(E) = 1 - P(E^c)$.
d. $P(E \cup F) = P(E) + P(F) - P(EF)$.

8-8. Use the appropriate axiom and the results of Prob. 8-7 to show that the probability of $\{E \text{ or } F\}$ is given by

$$P(E \text{ or } F) = P(E) + P(F) - 2P(EF).$$

Hint: Write the event $\{E \text{ or } F\}$ as a union of two disjoint sets.

8-9. A fair die is tossed twice (a die is said to be fair if all outcomes $1, \ldots, 6$ are equally likely). Given that a three appears on the first toss, what is the probability of obtaining the sum seven in two tosses?

8-10. A random-number generator generates integers from 1 to 9 (inclusive). All outcomes are equally likely; each integer is generated independently of any previous integer. Let Σ denote the sum of two consecutively generated integers; i.e., $\Sigma = N_1 + N_2$. Given that Σ is odd, what is the conditional probability that Σ is 7? Given that $\Sigma > 10$, what is the conditional probability that at least one of the integers is > 7? Given that $N_1 > 8$, what is the conditional probability that Σ will be odd?

8-11. In the trinary communication channel shown in Fig. P8-11 a 3 is sent three times more frequently than a 1, and a 2 is sent two times more frequently than a 1. A 1 is observed; what is the conditional probability that a 1 was sent?

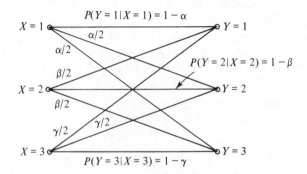

Figure P8-11.

8-12. Derive the result in Eq. (8.5-5), namely that

$$P(S) = \sum_{k=0}^{n} b(k; n, p) = 1,$$

where S is the certain event and $b(\cdot)$ is the binomial probability law.

8-13. War-game strategists make a living by solving problems of the following type. There are 6 incoming ballistic missiles (BM's) against which are fired 12 antimissile missiles (AMM's). The AMM's are fired so that two AMM's are directed against each BM. The single-shot-kill probability (SSKP) of an AMM is 0.8. The SSKP is simply the probability that an AMM destroys a BM. Assume that the AMM's don't interfere with each other and that an AMM can, at most, destroy only the BM against which it is fired. Compute the probability that (a) all BM's are destroyed, (b) at least one BM gets through to destroy the target, and (c) exactly one BM gets through.

8-14. Assume in Prob. 8-13 that the target was destroyed by the BM's. What is the conditional probability that only one BM got through?

8-15. An odd number of people want to play a game which requires two teams made up of even numbers of players. To decide who shall be left out to act as umpire, each of the N persons tosses a fair coin with the following stipulation: If there is one person whose outcome (be it heads or tails) is different from the rest of the group, that person will be the umpire. Assume that there are 11 players. What is the probability that a player will be "odd man out" on the first play?

8-16. In Prob. 8-15, derive a formula for the probability that the "odd man out" will occur on the nth play. *Hint*: Consider each play as an independent Bernoulli trial with success if an odd man out occurs and failure otherwise.

8-17. A smuggler, trying to pass himself off as a glass-bead importer, attempts to smuggle diamonds by mixing diamond beads among glass beads in the proportion of 1 diamond bead per 1000 glass beads. A harried customs inspector examines a sample of 100 beads. What is the probability that the smuggler will be caught?

8-18. Assume that a faulty receiver produces audible clicks to the great annoyance of the listener. The average number of clicks per second depends on the receiver temperature and is given by $\lambda(\tau) = 1 - e^{-\tau/10}$, where τ is time from turn-on. Derive a formula for the probability of $0, 1, 2, \ldots$ clicks during the first 10 sec of operation after turn-on.

REFERENCES

[8-1] E. PARZEN, *Modern Probability Theory and Its Applications*, Wiley, New York, 1960.

[8-2] A. PAPOULIS, *Probability, Random Variables, and Stochastic Processes*, McGraw-Hill, New York, 1965.

[8-3] H. CRAMER, *Mathematical Methods of Statistics*, Princeton University Press, Princeton, N.J., 1946.

[8-4] W. FELLER, *An Introduction to Probability Theory and Its Applications*, Wiley, New York, 1957.

[8-5] A. W. DRAKE, *Fundamentals of Applied Probability Theory*, McGraw-Hill, New York, 1967.

9

Random Variables

9.1 Introduction

Many random phenomena have sample description spaces which are sets of real numbers: the voltage $v_n(t)$, at time t, across a noisy resistor; the arrival time of the next customer at a movie theatre; the number of photons in a light pulse; the brightness level at a particular point on the TV screen; the number of times a light bulb will switch on before failing; the lifetime of a given living person; the number of hairs on the head of a person picked at random in New York City, etc. In all these cases the sample description space can be conveniently chosen as R, the real line. The set R is convenient even when the sample description space of a particular random phenomenon can be written more "compactly," as in the case of the number of hairs on a person's head where a compact sample space might be the set of integers $S = \{0, 1, \ldots\}$. By letting R be the sample description space we simply add a lot of descriptions whose probability is zero. Although the set R may contain subsets of probability zero, the advantage of considering R is that we can generate a unified theory that will cover all random phenomena with real-valued numerical outcomes.

When R is the sample description space, every event is a subset of R. However, not every subset of R is necessarily an event. There exist certain sets that are not countable unions or intersections of intervals or points in R, and these sets are said to be nonprobabilizable. However, as pointed out in Chapter 8, such sets are of no importance in engineering, and we shall forget about them. All sets of engineering importance can be generated from countable unions and intersections of intervals of the form (a, b), $(a, b]$, [*a, b*), and [*a, b*],† where *a* and *b* can be finite or infinite numbers. Such sets are called Borel sets (see Chapter 8), and the family of such sets is called the Borel field of events.

When the sample description space *S* is not numerical, we might want to generate a sample space from *S* which is numerical. For example, we might want to encode written text into signals for purposes of transmission or storage in a computer. Also, in many problems in communication theory we shall want to operate on random phenomena to generate new random phenomena by means of mathematical operations. To achieve these ends, we introduce and apply the notion of a random variable.

Let *S* be a sample description space (numerical or otherwise), and let its elements be {*s*}. A random variable $X(\cdot)$ is a *function* that assigns to every $s \in S$ a *number* $X(s)$. We also require that for every Borel set of numbers B the set $\{s: X(s) \in B\}$ is an event, i.e., is in the domain of the probability function $P(\cdot)$.‡ The domain of $X(\cdot)$ is *S*, and its image is *R* (see Fig. 9.1-1.)

In communication theory we are rarely concerned with the functional form of *X*, and we rarely specify the set *S*, although it is always implied. Our interest is not in generating the numbers $\{X(s)\}$ from the elements of *S*. Primarily we want to compute the probability that *X* will take values in a certain prespecified set. We give some examples below.§

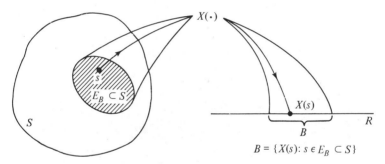

$$B = \{X(s): s \in E_B \subset S\}$$

Figure 9.1-1. Symbolic representation of the function $X(\cdot)$.

EXAMPLES

1. A bus arrives at random in [0, *T*]; let *t* denote the time of arrival. The sample description space *S* is $S = \{t: t \in [0, T]\}$. An r.v. *X* is defined by

$$X(t) = \begin{cases} 1, & t \in \left[\dfrac{T}{4}, \dfrac{T}{2}\right] \\ 0, & \text{otherwise.} \end{cases}$$

†$(a, b) = (a < x < b)$, $[a, b) = (a \leq x < b)$, etc.

‡A mathematical definition of an r.v. $X(\cdot)$ is that *X* be a function on *S* to *R* such that the *inverse images* under *X* of all Borel sets in *R* are events.

§The notation r.v. will frequently be used for "random variable."

Assume that the arrival time is uniform over $[0, T]$. We can now compute the probability $P[X(t) = 1]$ or $P[X(t) = 0]$ or $P[X(t) \leq 5]$.

2. An urn contains three colored balls. The balls are colored white, black, and red. The experiment consists of choosing a ball at random from the urn. The sample description space is $S = \{W, B, R\}$. The random variable X is defined by

$$X(s) = \begin{cases} \pi, & s = W \text{ or } B \\ 0, & s = R. \end{cases}$$

We can compute the probability $P(X \leq x_1)$, where x_1 is any number.

9.2 Probability Distribution Function

In the first example we gave in the previous section, the basic set of events are $[\{R\}, \{0\}, \{1\}, \{\varnothing\}]$ for which the probabilities are $P(R) = 1$, $P(X = 0) = \frac{3}{4}$, $P(X = 1) = \frac{1}{4}$, and $P(\varnothing) = 0$. From these probabilities, we can infer any other probabilities, such as, for example, $P(X \leq 0.5)$. In many cases it is awkward to write down $P(\cdot)$ for every event. For this reason we introduce the notion of the probability distribution function (PDF). The PDF† is a pointwise function of x which contains all the information necessary to compute $P(E)$ for any E in the Borel set of events. The PDF $F_X(x)$ is defined by

$$F_X(x) = P(X \leq x). \tag{9.2-1}$$

For the present we shall denote random variables by capital letters, e.g., X, Y, and Z, and the values they can take by lowercase letters, x, y, and z.

Properties of $F_X(x)$

1. $F_X(\infty) = 1$, $F_X(-\infty) = 0$.
2. $F_X(x_1) \leq F_X(x_2) \rightarrow x_1 \leq x_2$; i.e., $F_X(x)$ is a nondecreasing function of x.
3. $F_X(x)$ is continuous from the right, i.e.,

$$F_X(x) = \lim_{\epsilon \to 0} F_X(x + \epsilon), \qquad \epsilon > 0.$$

Proof of Property 2

Consider the event $\{x_1 < X \leq x_2\}$ with $x_2 > x_1$. It is nonempty and in $\mathcal{B}$.‡ Hence

$$0 \leq P(x_1 < X \leq x_2) \leq 1.$$

But $\{X \leq x_2\} = \{X \leq x_1\} \cup \{x_1 < X \leq x_2\}$, and

$$\{X \leq x_1\}\{x_1 < X \leq x_2\} = \varnothing.$$

---

†PDF should not be confused with pdf, which will stand for probability density function.

‡We continue to use $\mathcal{B}$ as a symbol for the Borel field of events except that events in this chapter are specifically sets of numbers on the real line.

Hence

$$F_X(x_2) = F_X(x_1) + P(x_1 < X \le x_2)$$

or

$$P(x_1 < X \le x_2) = F_X(x_2) - F_X(x_1) \ge 0 \quad \text{for } x_2 > x_1. \quad (9.2\text{-}2)$$

Exercise: Compute the probabilities of the events $\{X \le a\}$, $\{a \le X < b\}$, $\{a \le X \le b\}$, $\{a < X \le b\}$, and $\{a < X < b\}$ in terms of $F_X(x)$ and $P(X = x)$.

EXAMPLE

Toss a coin once; $S = \{H, T\}$ with $P(H) = p$ and $P(T) = q$; define $X(\cdot)$ by $X(H) = \pi$, $X(T) = 0$.

Computation of $F_X(x)$

1. $x < 0$: The event $\{X \le x\} = \varnothing$, and $F_X(x) = 0$.
2. $0 \le x < \pi$: The event $\{X \le x\}$ is equivalent to the event $\{T\}$ since

$$X(H) = \pi > x$$
$$X(T) = 0 \le x.$$

Hence $F_X(x) = q$.
3. $x \ge \pi$: The event $\{X \le x\}$ is the certain event since

$$X(H) = \pi \le x$$
$$X(T) = 0 \le x.$$

The solution is drawn in Fig. 9.2-1.

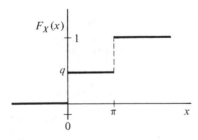

Figure 9.2-1. Probability distribution associated with the coin-tossing experiment.

9.3 Probability Density Function (pdf)

The pdf, if it exists, is given by

$$f(x) = \frac{dF(x)}{dx}, \quad (9.3\text{-}1)$$

where $F(x) \equiv F_X(x)$† since we are dealing only with a single random variable.

Properties

If $f(x)$ exists, then

(i) $\displaystyle \int_{-\infty}^{\infty} f(\xi)\, d\xi = F(\infty) - F(-\infty) = 1,$ \hfill (9.3-2)

(ii) $\displaystyle F(x) = \int_{-\infty}^{x} f(\xi)\, d\xi = P(X \leq x),$ \hfill (9.3-3)

(iii) $\displaystyle F(x_2) - F(x_1) = \int_{-\infty}^{x_2} f(\xi)\, d\xi - \int_{-\infty}^{x_1} f(\xi)\, d\xi$

$$= \int_{x_1}^{x_2} f(\xi)\, d\xi. \tag{9.3-4}$$

Interpretation of f(x)

$$P(x < X \leq x + \Delta x) = F(x + \Delta x) - F(x).$$

If $F(x)$ is continuous in its first derivative, then, for sufficiently small Δx,

$$F(x + \Delta x) - F(x) = \int_{x}^{x+\Delta x} f(\xi)\, d\xi \simeq f(x)\,\Delta x.$$

Hence for small Δx

$$P(x < X \leq x + \Delta x) \simeq f(x)\,\Delta x. \tag{9.3-5}$$

EXAMPLE

The univariate normal (Gaussian) pdf is

$$f(x) = \frac{1}{\sqrt{2\pi\sigma^2}} \exp\left[-\frac{1}{2}\left(\frac{x - \mu}{\sigma}\right)^2\right]. \tag{9.3-6}$$

There are two independent parameters: $\sigma \equiv$ standard deviation ($\sigma^2 =$ variance) and $\mu \equiv$ the mean (Fig. 9.3-1). Consider a normal random variable, X, with mean

†We shall dispense with the subscript X on $F(\cdot)$ unless we deal with more than one r.v.

Figure 9.3-1. Univariate normal pdf.

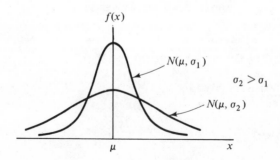

μ and standard deviation σ; we usually denote this by $X: N(\mu, \sigma)$. We have

$$P(a < X \le b) = \frac{1}{\sqrt{2\pi\sigma^2}} \int_a^b \exp\left[-\left(\frac{1}{2}\right)\left(\frac{x - \mu}{\sigma}\right)^2\right] dx. \qquad (9.3\text{-}7)$$

With $\beta = (x - \mu)/\sigma$, $d\beta = (1/\sigma)\, dx$, $b' = (b - \mu)/\sigma$, and $a' = (a - \mu)/\sigma$, we obtain

$$P(a < X \le b) = \frac{1}{\sqrt{2\pi}} \int_{a'}^{b'} \exp\left(-\frac{1}{2}\beta^2\right) d\beta$$

$$= \frac{1}{\sqrt{2\pi}} \int_0^{b'} \exp\left(-\frac{1}{2}\beta^2\right) d\beta - \frac{1}{\sqrt{2\pi}} \int_0^{a'} \exp\left(-\frac{1}{2}\beta^2\right) d\beta$$

$$\equiv \Phi(b') - \Phi(a'). \qquad (9.3\text{-}8)$$

The function $\Phi(x)$ is tabulated. It is sometimes referred to as the error function†
of x. Hence if X is normal,

$$P(a < X \le b) = \Phi\left(\frac{b - \mu}{\sigma}\right) - \Phi\left(\frac{a - \mu}{\sigma}\right). \qquad (9.3\text{-}9)$$

9.4   Continuous, Discrete, and Mixed Random Variables

If $F(x)$ is continuous for every x and its derivative exists everywhere except
at a countable set of points, then we say that X is a continuous random
variable. At points x where $F'(x)$ exists, the pdf is $f(x) = F'(x)$. At points
where $F'(x)$ doesn't exist, we can assign any positive number to $f(x)$; $f(x)$
will then be defined for every x, and we are free to use the following important formulas:

$$F(x) = \int_{-\infty}^{x} f(\xi)\, d\xi, \qquad (9.4\text{-}1)$$

$$P(x_1 < X \le x_2) = \int_{x_1}^{x_2} f(\xi)\, d\xi, \qquad (9.4\text{-}2)$$

and

$$P(B) = \int_{\text{all } \xi: \xi \in B} f(\xi)\, d\xi, \qquad (9.4\text{-}3)$$

where, in Eq. (9.4-3), B is any event (i.e., any Borel set of real numbers).
Equation (9.3-6) is an example of the pdf for a continuous r.v.

A *discrete random variable* has a staircase type of distribution function
(Fig. 9.4-1). A probability measure for discrete r.v.'s is the probability *frequency* function, which is also known as the probability mass function

---

†The function erf (x) is more usually defined by $(2/\sqrt{\pi}) \int_0^x e^{-t^2}\, dt$; hence $\Phi(x) =$
$\frac{1}{2}$ erf $(x/\sqrt{2})$.

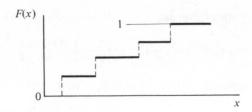

Figure 9.4-1. Probability distribution function
for a discrete r.v.

$P(x_i)$. It is defined by

$$P(x_i) \equiv P(X = x_i) = F(x_i) - F(x_i^-), \qquad (9.4\text{-}4)$$

where x_i^- is the point taken on the left of the jump at x_i.

Frequency Interpretation

For large n, $P(x_i) \simeq n_i/n$. The pdf for this case consists of impulse functions; i.e.,

$$f(x) = \frac{dF(x)}{dx} = \sum_i P(x_i)\delta(x - x_i) \qquad (9.4\text{-}5)$$

and

$$P(x_1 < X \le x_2) = \int_{x_1^-}^{x_2^+} f(\xi)\, d\xi. \qquad (9.4\text{-}6)$$

The upper limit in the integration includes the impulse at $x = x_2$, while the lower limit avoids the integration of the impulse at $x = x_1$.

A mixed r.v. has a discontinuous probability distribution function but not of the staircase type (Fig. 9.4-2).

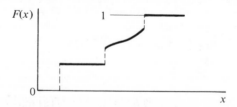

Figure 9.4-2. Probability distribution function
for a mixed r.v.

9.5 Conditional and Joint Distributions

Consider the event C consisting of all outcomes $s \in S$ such that $X(s) \le x$ and $s \in B \subset S$. The event C is then the set product of the two events $\{s: X(s) \le x\}$ and $\{s: s \in B\}$. We define the *conditional distribution function of*

X given the event B as

$$F(x \mid B) = \frac{P(C)}{P(B)} = \frac{P(X \leq x, B)}{P(B)}, \qquad (9.5\text{-}1)$$

where $P(X \leq x, B)$ is the probability of the joint event $\{X \leq x\} \cap B$ with $P(B) \neq 0$. It is not difficult to prove ([9-1], p. 105) that $F(x \mid B)$ has all the properties of an ordinary distribution; i.e., $F(x_1 \mid B) \leq F(x_2 \mid B) \rightarrow x_1 \leq x_2$, $F(-\infty \mid B) = 0$, $F(\infty \mid B) = 1$, etc. The conditional pdf is simply

$$f(x \mid B) = \frac{d}{dx} [F(x \mid B)]. \qquad (9.5\text{-}2)$$

Generally the event B will be expressed in terms of X.

EXAMPLE

Let $B = \{X \leq 10\}$. We wish to compute $F(x \mid B)$.

1. For $x \geq 10$, the event $\{X \leq 10\}$ is a subset of the event $\{X \leq x\}$. Hence $P(X \leq 10, X \leq x) = P(X \leq 10)$, and use of Eq. (9.5-1) gives

$$F(x \mid B) = \frac{P(X \leq x, X \leq 10)}{P(X \leq 10)} = 1.$$

2. For $x \leq 10$, the event $\{X \leq x\}$ is a subset of the event $\{X \leq 10\}$. Hence $P(X \leq 10, X \leq x) = P(X \leq x)$ and

$$F(x \mid B) = \frac{P(X \leq x)}{P(X \leq 10)}.$$

The result is shown in Fig. 9.5-1. We leave it as an exercise for the reader (Prob. 9-2) to compute $F(x \mid B)$ when $B = \{b < X \leq a\}$.

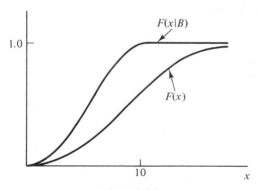

Figure 9.5-1

Joint Distribution

Suppose we are given two random variables X, Y defined on the same probability space $(S, \mathcal{B}, P)$. The event $\{X \leq x, Y \leq y\} = \{X \leq x\} \cap \{Y \leq y\}$

consists of all outcomes $s \in S$ such that $X(s) \le x$ *and* $Y(s) \le y$. The *joint distribution function* of X and Y is defined by

$$F_{XY}(x, y) = P(X \le x, Y \le y). \tag{9.5-3}$$

Since $\{X \le \infty\}$ and $\{Y \le \infty\}$ are certain events, we obtain

$$\{X \le x, Y \le \infty\} = \{X \le x\}, \tag{9.5-4}$$

$$\{X \le \infty, Y \le y\} = \{Y \le y\}, \tag{9.5-5}$$

so that

$$F_{XY}(x, \infty) = F_X(x), \tag{9.5-6}$$

$$F_{XY}(\infty, y) = F_Y(y). \tag{9.5-7}$$

The joint pdf, if it exists, is given by

$$f_{XY}(x, y) = \frac{\partial^2}{\partial x \, \partial y} [F_{XY}(x, y)]. \tag{9.5-8}$$

By twice integrating Eq. (9.5-8), we obtain

$$F_{XY}(x, y) = \int_{-\infty}^{x} d\xi \int_{-\infty}^{y} d\eta \, f_{XY}(\xi, \eta). \tag{9.5-9}$$

The functions $F_X(x)$ and $F_Y(y)$ are called *marginal* distributions if they are derived from a multivariate distribution as in Eqs. (9.5-6) and (9.5-7). Thus

$$F_X(x) = F_{XY}(x, \infty) = \int_{-\infty}^{x} d\xi \int_{-\infty}^{\infty} dy \, f_{XY}(\xi, y), \tag{9.5-10}$$

$$F_Y(y) = F_{XY}(\infty, y) = \int_{-\infty}^{y} d\eta \int_{-\infty}^{\infty} dx \, f_{XY}(x, \eta). \tag{9.5-11}$$

Since the marginal densities are given by

$$f_X(x) = F_X'(x)$$
$$f_Y(y) = F_Y'(y),$$

we obtain, by differentiating Eqs. (9.5-10) and (9.5-11),

$$f_X(x) = \int_{-\infty}^{\infty} f_{XY}(x, y) \, dy, \tag{9.5-12}$$

$$f_Y(y) = \int_{-\infty}^{\infty} f_{XY}(x, y) \, dx. \tag{9.5-13}$$

For discrete random variables we obtain equivalent results. Given the joint probability mass function $P_{XY}(X = x_i, Y = y_k)$ for all x_i, y_k, we compute the marginal mass function from

$$P_X(X = x_i) = \sum_{\text{all } y_k} P_{XY}(X = x_i, Y = y_k), \tag{9.5-14}$$

$$P_Y(Y = y_k) = \sum_{\text{all } x_i} P_{XY}(X = x_i, Y = y_k). \tag{9.5-15}$$

If X, Y are independent r.v.'s, then

$$F_{XY}(x, y) = F_X(x)F_Y(y), \tag{9.5-16}$$

$$f_{XY}(x, y) = f_X(x)f_Y(y), \tag{9.5-17}$$

$$P_{XY}(X = x_k, Y = y_j) = P_X(X = x_k)P_Y(Y = y_j). \tag{9.5-18}$$

EXAMPLE

$$f_{XY}(x, y) = \frac{1}{2\pi\sigma^2} \exp\left[-\frac{1}{2\sigma^2}(x^2 + y^2)\right]$$

$$= \frac{1}{\sqrt{2\pi\sigma^2}} \exp\left(-\frac{1}{2}\frac{x^2}{\sigma^2}\right) \frac{1}{\sqrt{2\pi\sigma^2}} \exp\left(-\frac{1}{2}\frac{y^2}{\sigma^2}\right). \tag{9.5-19}$$

Therefore X and Y are independent.

9.6 Functions of Random Variables

In later work we shall have to deal with random variables that have been acted upon by operators and shall have to know how to compute the statistics of the output. We shall give some examples.

EXAMPLES

1. Let $Z = \max(X, Y)$, X, Y independent. What is $F_Z(z)$?

$$F_Z(z) = P(Z \leq z) = P[\max(X, Y) \leq z].$$

But

$$P[\max(X, Y) \leq z] = P(X \leq z, Y \leq z)$$

$$= P(X \leq z)P(Y \leq z) = F_X(z)F_Y(z).$$

2. The square-law device (Fig. 9.6-1):

$$F_Y(y) = P(Y \leq y) = P(X^2 \leq y) = P(-\sqrt{y} \leq X \leq \sqrt{y})$$

$$= F_X(\sqrt{y}) - F_X(-\sqrt{y}) + P(X = -\sqrt{y}).$$

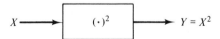

$$X \longrightarrow \boxed{(\cdot)^2} \longrightarrow Y = X^2$$

Figure 9.6-1

But $P(X = -\sqrt{y}) = 0$ if X is a continuous r.v. Hence

$$f_Y(y) = \frac{d}{dy}F_Y(y) = \begin{cases} \dfrac{1}{2\sqrt{y}}f_X(\sqrt{y}) + \dfrac{1}{2\sqrt{y}}f_X(-\sqrt{y}), & y > 0 \\ 0, & y < 0. \end{cases} \tag{9.6-1}$$

3. Let the phase X of a sine wave be uniformly distributed in $[-\pi, \pi]$; i.e.,

$$f_X(x) = \begin{cases} \dfrac{1}{2\pi}, & |x| < \pi \\ 0, & \text{otherwise.} \end{cases}$$

Let $Y = \sin X$; what is $f_Y(y)$? To compute $F_Y(y)$ we observe from Fig. 9.6-2 that for

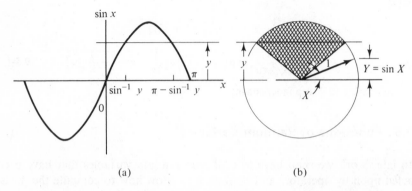

(a)   (b)

Figure 9.6-2. (a) The function $y = \sin x$; (b) the event $\{\sin X \le y\}$ is viewed as a unit vector that can rotate to any angle in the clear zone only.

$1 \ge y \ge 0$ the event $\{\sin X \le y\}$ is given by $\{\pi - \sin^{-1} y < X \le \pi\} \cup \{-\pi < X \le \sin^{-1} y\}$. Since these are disjoint, we have

$$P(\sin X \le y) = P(-\pi < X < \sin^{-1} y) + P(\pi - \sin^{-1} y < X \le \pi)$$
$$= F_X(\pi) - F_X(\pi - \sin^{-1} y) + F_X(\sin^{-1} y) - F_X(-\pi) = F_Y(y)$$

so that

$$f_Y(y) = \frac{dF_Y(y)}{dy} = f_X(\pi - \sin^{-1} y)\frac{1}{\sqrt{1 - y^2}} + f_X(\sin^{-1} y)\frac{1}{\sqrt{1 - y^2}}$$

$$= \frac{1}{\pi}\frac{1}{\sqrt{1 - y^2}}, \quad 0 \le y \le 1. \tag{9.6-2}$$

Exercise: Repeat the calculation for $y < 0$, and show that the complete solution is

$$f_Y(y) = \begin{cases} \dfrac{1}{\pi}\dfrac{1}{\sqrt{1 - y^2}}, & |y| < 1 \\ 0, & \text{otherwise.} \end{cases} \tag{9.6-3}$$

4. Given $Z = X + Y$, X, Y independent r.v.'s, what is $f_Z(z)$?

Comment: This situation arises frequently in communication systems where, say, X represents a stochastic signal and Y represents noise at a particular instant of time:

$$F_Z(z) = P(X + Y \le z) = \iint_{x+y\le z} f_X(x)f_Y(y) \, dx \, dy$$

$$= \int_{-\infty}^{\infty} dx \, f_X(x) \int_{-\infty}^{z-x} f_Y(y) \, dy$$

$$= \int_{-\infty}^{\infty} F_Y(z-x)f_X(x) \, dx. \qquad (9.6\text{-}4)$$

Hence

$$f_Z(z) = \frac{d}{dz}F_Z(z) = \int_{-\infty}^{\infty} f_Y(z-x)f_X(x) \, dx$$

$$= f_Y(z) * f_X(z). \qquad (9.6\text{-}5)$$

Comment: When X, Y are independent, the pdf of the sum is the *convolution* of the pdf's. When X, Y are not independent, the solution is

$$f_Z(z) = \int_{-\infty}^{\infty} f_{XY}(x, z-x) \, dx. \qquad (9.6\text{-}6)$$

The general expression for evaluating the probability of any event E when X, Y are continuous r.v.'s is

$$P(E) = \iint_{\text{all } x,y \in E} f_{XY}(x, y) \, dx \, dy. \qquad (9.6\text{-}7)$$

5. $X + Y$ injected into a square-law device; i.e., $W = Z^2 = (X + Y)^2$ (Fig. 9.6-3). Let X and Y be independent, identically distributed r.v.'s, uniformly dis-

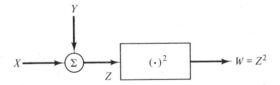

Figure 9.6-3

tributed in $(-\frac{1}{2}, \frac{1}{2})$. From Eq. (9.6-5) we compute $f_Z(z) = (1 - |z|) \, \text{rect} \, (z/2)$. Now we use the result derived in Eq. (9.6-1) to obtain

$$f_W(w) = \begin{cases} \dfrac{1}{\sqrt{w}}(1 - \sqrt{w}), & 0 \le w \le 1 \\ 0, & \text{otherwise.} \end{cases} \qquad (9.6\text{-}8)$$

9.7 A General Formula for Determining the pdf of a Function of a Single Random Variable

Suppose we are given a continuous r.v. X with its pdf $f_X(x)$. Consider the continuous transformation $y = g(x)$. What is the pdf of Y when X is applied to the system with transmittance $g(\cdot)$?

Solution: The event $\{y < Y \leq y + dy\}$ can be written as a union of disjoint elementary events on the probability space of X. If the equation $y = g(x)$ has n real roots $x_1, \ldots, x_n$, then the disjoint events have the form $E_i = \{x_i - |dx_i| < X < x_i\}$ if $g'(x_i)$ is negative or $E_i = \{x_i < X < x_i + |dx_i|\}$ if $g'(x_i)$ is positive. (See Fig. 9.7-1.) In either case, it follows from the definition of the pdf that $P(E_i) = f(x_i)|dx_i|$. Hence

$$P(y < Y < y + dy) = f_Y(y)|dy|$$

$$= \sum_{i=1}^{n} f_X(x_i)|dx_i|, \qquad (9.7\text{-}1)$$

or, equivalently, if we divide through by $|dy|$,

$$f_Y(y) = \sum_{i=1}^{n} f_X(x_i)\left|\frac{dx_i}{dy}\right|.$$

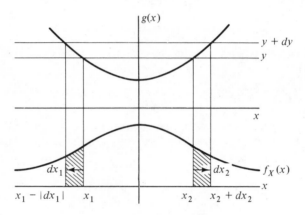

Figure 9.7-1

Since $y = g(x)$ and $dy/dx_i = g'(x_i)$, we obtain the important formula

$$f_Y(y) = \sum_{i=1}^{n} \frac{f_X(x_i)}{|g'(x_i)|} \qquad x_i = x_i(y). \qquad (9.7\text{-}2)$$

Equation (9.7-2) is a fundamental equation that is very useful in solving problems where the transmittance $g(x)$ has several roots. If, for a given y, the equation $y - g(x)$ has no real roots, then $f_Y(y) = 0$. Figure 9.7-1 illustrates the case when $n = 2$. From Fig. 9.7-1 we see that the event $\{y < Y \leq y + dy\}$ is identical to the event $\{x_1 - |dx_1| < X \leq x_1\} \cup \{x_2 < X \leq x_2 + |dx_2|\}$, and Eq. (9.7-1) follows with $n = 2$ in this case.

EXAMPLE

1. Let $g(\cdot)$ be the square-law device discussed earlier. Then $y - x^2 = 0$ has two roots, i.e., $x_1 = -\sqrt{y}$, $x_2 = +\sqrt{y}$ for $y > 0$. For $y < 0$, $y - x^2$ has no real roots; hence $f_Y(y) = 0$. Suppose X is a normal r.v. with mean zero and variance

unity; then from Eq. (9.7-2),

$$f_Y(y) = \frac{1}{\sqrt{2\pi}} \frac{\exp[-\frac{1}{2}(\sqrt{y})^2]}{2\sqrt{y}} + \frac{1}{\sqrt{2\pi}} \frac{\exp[-\frac{1}{2}(-\sqrt{y})^2]}{2\sqrt{y}}, \quad y > 0$$

$$= \begin{cases} \frac{1}{\sqrt{2\pi y}} \exp(-\frac{1}{2}y), & y > 0 \\ 0, & \text{otherwise.} \end{cases} \tag{9.7-3}$$

2. Let $g(x_1, x_2) = x_1^2 + x_2^2$. With $Z_i = X_i^2$, $i = 1, 2$, we compute $f_{Z_i}(z)$ from Eq. (9.7-2) and compute the pdf of $W = Z_1 + Z_2$ from Eq. (9.6-5).

Comment: When X_1 and X_2 are normal, independent random variables, the computation of $Y = X_1^2 + X_2^2$ is facilitated by a direct approach. Thus

$$F_Y(y) = P(X_1^2 + X_2^2 \le y) = \iint\limits_{x_1^2 + x_2^2 \le y} f_{X_1 X_2}(x_1, x_2)\, dx_1\, dx_2. \tag{9.7-4}$$

Note that the surface of integration is a circle about the origin (Fig. 9.7-2). We must compute

$$F_Y(y) = \iint\limits_{x_1^2 + x_2^2 \le y} \frac{1}{2\pi\sigma^2} \exp\left[-\frac{1}{2}\left(\frac{x_1^2 + x_2^2}{\sigma^2}\right)\right] dx_1\, dx_2.$$

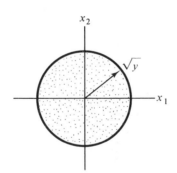

Figure 9.7-2

Let $x_1 = r\cos\theta$, $x_2 = r\sin\theta$; then $r^2 = x_1^2 + x_2^2$ and $\theta = \tan^{-1} x_2/x_1$. Hence

$$F_Y(y) = \int_{\theta=0}^{2\pi} \int_0^{\sqrt{y}} \frac{r}{2\pi\sigma^2} \exp\left(-\frac{1}{2}\frac{r^2}{\sigma^2}\right) dr\, d\theta = 1 - e^{-y/2\sigma^2}, \quad y > 0$$

$$= 0, \, y < 0 \text{ (note that } Y \text{ cannot take on negative values).}$$

The result may be conveniently written as

$$F_Y(y) = (1 - e^{-y/2\sigma^2})u(y), \tag{9.7-5}$$

where $u(y) = 1$ for $y > 0$ and $u(y) = 0$ for $y < 0$. The pdf is thus

$$f_Y(y) = \frac{dF_Y(y)}{dy} = \frac{1}{2\sigma^2} e^{-y/2\sigma^2} u(y), \tag{9.7-6}$$

which is known as the exponential probability law. Had we considered instead the pdf of $Y = \sqrt{X_1^2 + X_2^2}$, the surface of integration would have been a circular disk of radius y. For this case

$$F_Y(y) = \int_{\theta=0}^{2\pi} \int_0^y \frac{r}{2\pi\sigma^2} \exp\left(-\frac{1}{2}\frac{r^2}{\sigma^2}\right) dr \, d\theta$$

$$= \begin{cases} 1 - e^{-y^2/2\sigma^2}, & y > 0 \\ 0, & y < 0. \end{cases} \tag{9.7-7}$$

The pdf in this case is

$$f_Y(y) = \frac{y}{\sigma^2} e^{-y^2/2\sigma^2} u(y), \tag{9.7-8}$$

which is the *Rayleigh* density function. It is also known as the χ ("chi") distribution with two degrees of freedom ([9-2], p. 181).

9.8 Averages

The average value of X, also called the expected or mean value, is written $E(X)$ (read "the operator E acting on X") and is defined for a discrete r.v. by

$$\mu \equiv E(X) = \sum_i x_i P(X = x_i), \tag{9.8-1}$$

where E is the expectation operator and the summation over i really means "sum over all x_i such that $P(X = x_i) > 0$." For a continuous r.v. we have

$$\mu \equiv E(X) = \int_{-\infty}^{\infty} x f_X(x) \, dx. \tag{9.8-2}$$

In more advanced work† it is shown that $E(X)$ can also be calculated according to

$$E(X) = \lim_{\substack{\Delta x \to 0 \\ k \to \infty}} \sum_k x_k P(x_k < X \le x_k + \Delta x)$$

$$= \lim_{\substack{\Delta x \to 0 \\ k \to \infty}} \sum_k x_k [F(x_k + \Delta x) - F(x_k)]$$

$$= \int_R x \, dF. \tag{9.8-3}$$

The definitions in Eqs. (9.8-1) and (9.8-2) extend to more general cases. Thus with $g(X)$ denoting any function of X we have

$$E[g(X)] = \int_{-\infty}^{\infty} g(x) f_X(x) \, dx, \quad X \text{ continuous}, \tag{9.8-4}$$

$$E[g(X)] = \sum_i g(x_i) P(X = x_i), \quad X \text{ discrete}. \tag{9.8-5}$$

†See, for example, [9-3], p. 166.

The conditional expectation of X given $Y = y$, written $E_Y(X)$ or $E(X \mid Y = y)$, is defined by

$$E_Y(X) = \int_{-\infty}^{\infty} x f_{X \mid Y}(x \mid y) \, dx, \qquad X \text{ continuous}, \qquad (9.8\text{-}6)$$

$$E_Y(X) = \sum_i x_i P[X = x_i \mid Y = y_j], \quad X, Y \text{ discrete}. \qquad (9.8\text{-}7)$$

Note that the conditional expectations are functions of the particular value of y that the random variable Y has taken on.

For several variables,

$$E[g(X, Y, Z)] = \int_{-\infty}^{\infty} \int_{-\infty}^{\infty} \int_{-\infty}^{\infty} g(x, y, z) f_{XYZ}(x, y, z) \, dx \, dy \, dz, \qquad (9.8\text{-}8)$$

and

$$E(X + Y + Z) = \int_{-\infty}^{\infty} \int_{-\infty}^{\infty} \int_{-\infty}^{\infty} (x + y + z) f_{XYZ}(x, y, z) \, dx \, dy \, dz$$

$$= E(X) + E(Y) + E(Z). \qquad (9.8\text{-}9)$$

The proof of Eq. (9.8-9) is left as an exercise. This result can be generalized to

$$E\left[\sum_{i=1}^{N} X_i \right] = \sum_{i=1}^{N} E[X_i].$$

Note that *independence is not required*. The notation $E(X) = \bar{X}$ is also frequently used. We use it in what follows.

9.9 Moments

We consider functions $g(\cdot)$ of X of the form X^r. We define the *rth moment* of X by

$$\zeta_r \equiv \overline{X^r} = \int_{-\infty}^{\infty} x^r f_X(x) \, dx. \qquad (9.9\text{-}1)$$

The extension to the discrete case is straightforward. We omit writing expressions for the discrete case unless the extension from the continuous case is not obvious. We see from Eq. (9.9-1) that $\zeta_0 = 1$, $\zeta_1 = \mu$. The *rth central moment* is more widely used. It is defined by

$$m_r \equiv E[(X - \bar{X})^r]$$

$$= \int_{-\infty}^{\infty} (x - \mu)^r f_X(x) \, dx. \qquad (9.9\text{-}2)$$

The most widely used moments are the first and second. The second central moment m_2 is also called the *variance* or dispersion. The symbol σ^2 is frequently used in place of m_2.

Exercise: Show that if $X: N(\mu, \sigma)$, the expected value of X is μ and the variance $m_2 = \sigma^2$. Also show that $m_3 = 0$ and $m_4 = 3\sigma^4$. Hence the symbols μ, σ^2 in the Gaussian pdf are aptly chosen.

Joint Moments

The ijth joint central moment of X and Y is defined by

$$m_{ij} = E[(X - \bar{X})^i(Y - \bar{Y})^j]$$

$$= \int_{-\infty}^{\infty} \int_{-\infty}^{\infty} (x - \bar{X})^i(y - \bar{Y})^j f_{XY}(x, y)\, dx\, dy. \qquad (9.9\text{-}3)$$

The most widely used joint central moment is m_{11}, which is called the covariance. The correlation coefficient is defined by

$$\rho \equiv \frac{m_{11}}{\sqrt{m_{20}m_{02}}}. \qquad (9.9\text{-}4)$$

Exercise: Show by direct substitution into Eq. (9.9-3) that $m_{11} = \rho\sigma_x\sigma_y$, where $f_{XY}(x, y)$ is given by

$$f_{XY}(x, y) = \frac{1}{2\pi\sigma_x\sigma_y\sqrt{1 - \rho^2}}$$

$$\cdot \exp\left\{-\frac{1}{2(1 - \rho^2)}\left[\frac{(x - \bar{X})^2}{\sigma_x^2} - \frac{2\rho(x - \bar{X})(y - \bar{Y})}{\sigma_x\sigma_y} + \frac{(y - \bar{Y})^2}{\sigma_y^2}\right]\right\}$$

$$(9.9\text{-}5)$$

and $\sigma_x^2 = m_{20}$ and $\sigma_y^2 = m_{02}$. Equation (9.9-5) is the formula for two jointly normal random variables X, Y.

Exercise: Show that if X, Y are jointly Gaussian but uncorrelated random variables, i.e., $\rho = 0$, they are also independent.

Exercise: Show that any two statistically independent random variables are always uncorrelated. *Warning:* The converse is not generally true.

9.10 Moment-Generating and Characteristic Functions

In the previous section we have seen that the computation of moments requires a summation or integration for every moment that we want calculated. We shall now present a method for computing the moments from a single function which produces all moments by routine differentiation.

Definition: The moment-generating function $\theta(t)$ of a random variable X is given by

$$\theta(t) = E(e^{tX}). \qquad (9.10\text{-}1)$$

When X is continuous we obtain

$$\theta(t) = \int_{-\infty}^{\infty} e^{tx} f_X(x)\, dx, \qquad (9.10\text{-}2)$$

and when X is discrete we obtain

$$\theta(t) = \sum_i e^{tx_i} P(X = x_i). \qquad (9.10\text{-}3)$$

Assume that X is continuous; then

$$e^{tX} = 1 + tX + \frac{(tX)^2}{2!} + \cdots + \frac{(tX)^n}{n!} + \cdots, \qquad (9.10\text{-}4)$$

from which we obtain, assuming that $\zeta_r = \overline{X^r}$ exists,

$$E(e^{tX}) = 1 + t\mu + \frac{t^2}{2!}\zeta_2 + \cdots + \frac{t_n}{n!}\zeta_n + \cdots, \qquad (9.10\text{-}5)$$

which is a linear combination of all the moments of X. To obtain any single moment, observe that (the differentiation is with respect to t)

$$\theta'(t)\Big|_{t=0} = E(Xe^{tX})_{t=0} = \mu, \qquad (9.10\text{-}6)$$

$$\theta''(t)\Big|_{t=0} = E(X^2e^{tX})_{t=0} = \zeta_2, \qquad (9.10\text{-}7)$$

$$\theta'''(t)\Big|_{t=0} = E(X^3e^{tX})_{t=0} = \zeta_3, \quad \text{etc}, \qquad (9.10\text{-}8)$$

Hence the rth moment of X is obtained from

$$\zeta_r = \frac{d^r}{dt^r}[\theta(t)]\Big|_{t=0}. \qquad (9.10\text{-}9)$$

EXAMPLES

1. Let $X = N(\mu, \sigma)$; then by direct evaluation of the integral

$$\theta(t) = \frac{1}{\sqrt{2\pi\sigma^2}} \int_{-\infty}^{\infty} \exp\left[-\frac{1}{2}\left(\frac{x-\mu}{\sigma}\right)^2\right] e^{tx}\, dx$$

we obtain

$$\theta(t) = \exp\left(t\mu + \tfrac{1}{2}t^2\sigma^2\right) \equiv e^{f(t)}, \qquad (9.10\text{-}10)$$

where $f(t) = t\mu + \tfrac{1}{2}t^2\sigma^2$.

The first two moments are calculated to be

$$\theta'(t)\Big|_{t=0} = (\mu + t\sigma^2)e^{f(t)}\Big|_{t=0} = \zeta_1 = \mu, \qquad (9.10\text{-}11)$$

$$\theta''(t)\Big|_{t=0} = [\sigma^2 + (\mu + t\sigma^2)^2]e^{f(t)}\Big|_{t=0} = \zeta_2 = \sigma^2 + \mu^2. \qquad (9.10\text{-}12)$$

2. Let X be a discrete r.v. with a binomial distribution. The moment-generating function is

$$\theta(t) = \sum_{k=0}^{n} e^{tk}\binom{n}{k}p^k q^{n-k}$$

$$= \sum_{k=0}^{n} \binom{n}{k}(e^t p)^k q^{n-k}, \quad p+q=1. \qquad (9.10\text{-}13)$$

By the binomial expansion theorem, this is simply $(pe^t + q)^n$. Hence for the binomial random variable

$$\theta(t) = (pe^t + q)^n, \qquad (9.10\text{-}14)$$

$$\theta'(t) = npe^t(pe^t + q)^{n-1} = np \quad \text{at } t = 0, \qquad (9.10\text{-}15)$$

so that $\mu = np$. Similarly $\theta''(t)$ is given by

$$\theta''(t) = \frac{d}{dt}\theta'(t) = npe^t(pe^t + q)^{n-1}$$

$$+ n(n-1)p^2e^{2t}(pe^t + q)^{n-2}, \qquad (9.10\text{-}16)$$

so that

$$\zeta_2 = npq + n^2p^2. \qquad (9.10\text{-}17)$$

Note that the second central moment is $m_2 = \zeta_2 - \mu^2 = npq$. One could similarly define a *joint moment-generating* function according to

$$\theta_{X_1X_2\cdots X_n}(t_1, \ldots, t_n) = E\left[\exp\left(\sum_{i=1}^n t_iX_i\right)\right]$$

$$= \sum_{k_1=0}^\infty \sum_{k_2=0}^\infty \cdots \sum_{k_n=0}^\infty \frac{t_1^{k_1}}{k_1!}\frac{t_2^{k_2}}{k_2!}\cdots\frac{t_n^{k_n}}{k_n!}\overline{X_1^{k_1}\cdots X_n^{k_n}}. \quad (9.10\text{-}18)$$

For example, with $n = 2$ we obtain

$$\theta_{X_1X_2}(t_1, t_2) = E[e^{t_1X_1+t_2X_2}], \qquad (9.10\text{-}19)$$

so that the first and second moments can be obtained from

$$\overline{X_1} = \frac{\partial}{\partial t_1}\theta_{X_1X_2}(0, 0) \equiv \mu_1, \qquad (9.10\text{-}20)$$

$$\overline{X_1^2} = \frac{\partial^2}{\partial t_1^2}\theta_{X_1X_2}(0, 0) = \mu_1^2 + m_{20}, \qquad (9.10\text{-}21)$$

$$\overline{X_1X_2} = \frac{\partial^2}{\partial t_1\,\partial t_2}\theta_{X_1X_2}(0, 0) = m_{11} + \mu_1\mu_2, \qquad (9.10\text{-}22)$$

$$\overline{X_2^2} = \frac{\partial^2}{\partial t_2^2}\theta_{X_1X_2}(0, 0) = \mu_2^2 + m_{02}, \qquad (9.10\text{-}23)$$

$$\overline{X_2} = \frac{\partial}{\partial t_2}\theta_{X_1X_2}(0, 0) \equiv \mu_2, \qquad (9.10\text{-}24)$$

where the arguments $(0, 0)$ mean that all derivatives are evaluated at $t_1 = t_2 = 0$. Note that the variances and covariances can be expressed in terms of the above according to

$$m_{20} = \frac{\partial^2}{\partial t_1^2}[\theta_{X_1X_2}(0, 0)] - \left|\frac{\partial}{\partial t_1}[\theta_{X_1X_2}(0, 0)]\right|^2, \qquad (9.10\text{-}25)$$

$$m_{11} = \frac{\partial^2}{\partial t_1\,\partial t_2}[\theta_{X_1X_2}(0, 0)] - \frac{\partial}{\partial t_1}[\theta_{X_1X_2}(0, 0)]\frac{\partial}{\partial t_2}[\theta_{X_1X_2}(0, 0)]. \quad (9.10\text{-}26)$$

We have seen that the pdf of the sum of independent random variables involves the convolution of their pdf's. Thus if $Z = X_1 + \cdots + X_n$, where

$X_i, i = 1, \ldots, n$, are independent random variables, the pdf of Z is furnished by

$$f_Z(z) = f_{X_1}(z) * f_{X_2}(z) * \cdots * f_{X_n}(z), \tag{9.10-27}$$

i.e., the repeated convolution product.

The actual evaluation of Eq. (9.10-27) can be very tedious. However, we know from our studies of Fourier transforms that the Fourier transform of a convolution product is the product of the individual transforms. This property of the Fourier transform is also used conveniently in probability theory by introducing a function closely related to the moment-generating function called the *characteristic function*. The characteristic function $\Phi_X(\omega)$ for a continuous r.v. is defined conventionally by

$$\Phi_X(\omega) = E(e^{j\omega X}) = \theta_X(j\omega) = \int_{-\infty}^{\infty} e^{j\omega x} f_X(x)\,dx, \tag{9.10-28}$$

which is seen to be the complex conjugate of the Fourier transform of $f_X(x)$. All the moments of a random variable *that exist* can be obtained from

$$E(X^k) = \frac{1}{j^k} \frac{d^k}{d\omega^k} \Phi_X(\omega)\bigg|_{\omega=0}. \tag{9.10-29}$$

Equation (9.10-29) is obtained by differentiating Eq. (9.10-28) k times and multiplying by $(-j)^k$ at $\omega = 0$.

Consider the pdf of $Z = X_1 + X_2$, where X_1 and X_2 are independent r.v.'s with pdf's $f_{X_1}(x)$ and $f_{X_2}(x)$. Let us compute the characteristic function of Z. By definition

$$\Phi_Z(\omega) = E(e^{j\omega Z}) = E[e^{j\omega(X_1+X_2)}] = E[e^{j\omega X_1}]E[e^{j\omega X_2}]$$
$$= \Phi_{X_1}(\omega)\Phi_{X_2}(\omega). \tag{9.10-30}$$

The same result is obviously obtained if $\Phi_Z(\omega)$ is computed directly from $f_Z(z)$:

$$\Phi_Z(\omega) = \int_{-\infty}^{\infty} f_Z(z)e^{j\omega z}\,dz, \tag{9.10-31}$$

$$= \int_{-\infty}^{\infty} dz\,e^{j\omega z} \int_{-\infty}^{\infty} f_{X_1}(z - x)f_{X_2}(x)\,dx$$

$$= \int_{-\infty}^{\infty} dx\,f_{X_1}(x)e^{j\omega x} \int_{-\infty}^{\infty} du\,f_{X_2}(u)e^{j\omega u}$$

$$= \Phi_{X_1}(\omega)\Phi_{X_2}(\omega). \tag{9.10-32}$$

Thus the characteristic function of the sum of independent r.v.'s is the product of the characteristic functions. Note that by the Fourier inversion theorem the pdf of Z can be obtained directly from

$$f_Z(z) = \int_{-\infty}^{\infty} \Phi_Z(\omega)e^{-j\omega z} \frac{d\omega}{2\pi}. \tag{9.10-33}$$

For a discrete random variable X, the characteristic function is defined by

$$\Phi_X(\omega) = \sum_i e^{j\omega x_i} P(X = x_i).$$ (9.10-34)

Even though X is discrete, $\Phi_X(\omega)$ is a continuous function of ω. Further, the magnitude of $\Phi_X(\omega)$ is bounded from above by unity since

$$|\Phi_X(\omega)| = |\sum_i e^{j\omega x_i} P(X = x_i)| \leq \sum_i |e^{j\omega x_i} P(X = x_i)|$$

$$= \sum_i P(X = x_i) = 1.$$ (9.10-35)

Hence $|\Phi_X(\omega)| \leq 1$. A similar result holds when X is continuous. Equation (9.10-29) can be used to compute the moments whether X is discrete or continuous.

EXAMPLES

1. Let X_1 be $N(\mu_1, \sigma_1)$ and X_2 be $N(\mu_2, \sigma_2)$. Compute the pdf of $Z = X_1 + X_2$.

Solution: We can obtain $\Phi_{X_i}(\omega)$, $i = 1, 2$, by direct integration. However, since $\Phi_{X_i}(\omega) = \theta_{X_i}(j\omega)$, we use

$$\theta_{X_i}(t) = \exp\left(t\mu_i + \tfrac{1}{2}t^2\sigma_i^2\right),$$

from which we obtain

$$\Phi_{X_1}(\omega) = \exp\left(j\omega\mu_1 - \tfrac{1}{2}\omega^2\sigma_1^2\right),$$
$$\Phi_{X_2}(\omega) = \exp\left(j\omega\mu_2 - \tfrac{1}{2}\omega^2\sigma_2^2\right).$$ (9.10-36)

Hence

$$\Phi_Z(\omega) = \Phi_{X_1}(\omega)\Phi_{X_2}(\omega)$$
$$= \exp\left(j\omega\mu - \tfrac{1}{2}\omega^2\sigma^2\right),$$ (9.10-37)

where $\mu \equiv \mu_1 + \mu_2$ and $\sigma^2 \equiv \sigma_1^2 + \sigma_2^2$. The pdf of Z is

$$f_Z(z) = \int_{-\infty}^{\infty} \Phi_Z(\omega) e^{-j\omega z} \frac{d\omega}{2\pi}$$

$$= \frac{1}{\sqrt{2\pi\sigma^2}} \exp\left[-\frac{1}{2}\left(\frac{z - \mu}{\sigma}\right)^2\right].$$ (9.10-38)

Thus the sum of two normal, independent r.v.'s is a normal r.v. whose mean is the sum of the means and whose variance is the sum of the variances. A similar result holds true for the sum of n normal, independent, random variables. The pdf of the sum is Gaussian with mean $\mu = \sum_{i=1}^{n} \mu_i$ and variance $\sigma^2 = \sum_{i=1}^{n} \sigma_i^2$.

2. Compute the first few moments of $Y = \sin\theta$ if θ is uniformly distributed in $[0, 2\pi]$.

Solution: We use the result in Eq. (9.8-4); i.e., if $Y = g(X)$, then $\bar{Y} = \int_{-\infty}^{\infty} y f_Y(y)\, dy = \int_{-\infty}^{\infty} g(x) f_X(x)\, dx$. Hence

$$E(e^{j\omega Y}) = \int_{-\infty}^{\infty} e^{j\omega y} f_Y\, dy$$

$$= \int_{-\infty}^{\infty} e^{j\omega \sin\theta} f_\Theta(\theta)\, d\theta$$

$$= \frac{1}{2\pi} \int_{-\infty}^{\infty} e^{j\omega \sin\theta}\, d\theta$$

$$= J_0\,(\omega), \tag{9.10-39}$$

where $J_0(\omega)$ is the Bessel function of the first kind of order zero. A power series expansion of $J_0(\omega)$ gives

$$J_0(\omega) = 1 - \left(\frac{\omega}{2}\right)^2 + \frac{1}{2!2!}\left(\frac{\omega}{2}\right)^4 - \cdots. \tag{9.10-40}$$

Hence all the odd-order moments are zero. From Eq. (9.10-29) we compute

$$\overline{Y^2} = (-1)\frac{d^2}{d\omega^2}[J_0(\omega)]\Big|_{\omega=0} = \frac{1}{2}, \tag{9.10-41}$$

$$\overline{Y^4} = (+1)\frac{d^4}{d\omega^4}[J_0(\omega)]\Big|_{\omega=0} = \frac{3}{8}. \tag{9.10-42}$$

As in the case of joint moment-generating functions, we can define the joint characteristic function by

$$\Phi_{X_1 \cdots X_n}(\omega_1, \ldots, \omega_n) = E[\exp\,(j\sum_{i=1}^{n} \omega_i X_i)]. \tag{9.10-43}$$

By the Fourier inversion property, the joint pdf is the inverse Fourier transform of $\Phi_{X_1 \cdots X_n}(\omega_1, \ldots, \omega_n)$. Thus

$$f_{X_1 \cdots X_n}(x_1, \ldots, x_n) = \frac{1}{(2\pi)^n} \int_{-\infty}^{\infty} \cdots \int_{-\infty}^{\infty} \Phi_{X_1 \cdots X_n}(\omega_1, \ldots, \omega_n) \exp\,(-j\sum_{i=1}^{n} \omega_i x_i)$$

$$\cdot\, d\omega_1 d\omega_2 \cdots d\omega_n. \tag{9.10-44}$$

The rkth joint moment can be obtained by differentiation. Thus with X, Y denoting any two random variables, we have

$$\zeta_{rk} \equiv E(X^r Y^k) = (-j)^{r+k}\frac{\partial^{r+k}\Phi(\omega_1, \omega_2)}{\partial\omega_1^r\, \partial\omega_2^k}\Big|_{\omega_1=\omega_2=0} \tag{9.10-45}$$

Exercise: Show that if X is $N(0, \sigma)$, then

$$\zeta_n = m_n = \begin{cases} 1\cdot 3 \,\cdots\, (n-1)\sigma^n, & n \text{ even} \\ 0, & n \text{ odd}. \end{cases} \tag{9.10-46}$$

Show also that $E|X|^n$ is given by

$$E|X|^n = \begin{cases} m_n, & n \text{ even} \\ \dfrac{\sqrt{2}}{\pi} 2^{(n-1)/2} \left(\dfrac{n-1}{2}\right)! \sigma^n, & n \text{ odd} \end{cases} \qquad (9.10\text{-}47)$$

9.11 Two Functions of Two Random Variables

Suppose we are given two random variables X, Y and two real functions $w = g(x, y)$, $z = h(x, y)$. We form the two random variables $W = g(X, Y)$, $Z = h(X, Y)$ and consider the computation of $f_{WZ}(w, z)$.

This problem is a direct extension of the problem considered in Sec. 9.7, and the solution has the same form as Eq. (9.7-2) except extended to two dimensions. We denote by (x_i, y_i), $i = 1, \ldots, n$, the n real solutions to the equations

$$g(x_i, y_i) = w, h(x_i, y_i) = z, \qquad i = 1, \ldots, n. \qquad (9.11\text{-}1)$$

At each root we evaluate a normalizing factor called the Jacobian, denoted J_i at the ith root, and defined by

$$J_i = \left| \frac{\partial w}{\partial x} \frac{\partial z}{\partial y} - \frac{\partial z}{\partial x} \frac{\partial w}{\partial y} \right|_{x=x_i, y=y_i}. \qquad (9.11\text{-}2)$$

The joint pdf of W and Z is given by

$$f_{WZ}(w, z) = \sum_{i=1}^{n} \frac{f_{XY}(x_i, y_i)}{J_i}, \qquad (9.11\text{-}3)$$

where $x_i = x_i(w, z)$ and $y_i = y_i(w, z)$. The proof is given in several places, for example, in [9-1], p. 201. Note that J_i plays the same roles as $|g'(x_i)|$ in Eq. (9.7-2), which considered the single-variable transformation. We illustrate the application of Eq. (9.11-3) with an important example.

EXAMPLE

Let $W = X$ and $Z = X/Y$. Compute $f_{WZ}(w, z)$ and $f_Z(z)$ in terms of $f_{XY}(x, y)$.

Solution: The only solution to the set of equations

$$w = x$$

$$z = \frac{x}{y}$$

is $x = w$, $y = w/z$. Hence

$$J_1 = \left| \frac{x}{y^2} \right| = \frac{z^2}{|w|}$$

and

$$f_{WZ}(w, z) = f_{XY}\left(w, \frac{w}{z}\right) \frac{|w|}{z^2}. \qquad (9.11\text{-}4)$$

Suppose $f_{XY}(x, y)$ is given by

$$f_{XY}(x, y) = \frac{1}{2\pi\sigma_1\sigma_2\sqrt{1 - \rho^2}} \exp\left\{-\left[\frac{1}{2(1 - \rho^2)}\left(\frac{x^2}{\sigma_1^2} - \frac{2\rho xy}{\sigma_1\sigma_2} + \frac{y^2}{\sigma_2^2}\right)\right]\right\}; \quad (9.11\text{-}5)$$

then, from Eq. (9.11-4), $f_{WZ}(w, z)$ is given by

$$f_{WZ}(w, z) = \frac{|w|}{2\pi\sigma_1\sigma_2 z^2\sqrt{1 - \rho^2}} \exp\left\{-\left[\frac{w^2}{2(1 - \rho^2)}\left(\frac{1}{\sigma_1^2} - \frac{2\rho}{z\sigma_1\sigma_2} + \frac{1}{z^2\sigma_2^2}\right)\right]\right\}.$$

$$(9.11\text{-}6)$$

To obtain $f_Z(z)$, we integrate $f_{WZ}(w, z)\, dw$ over all values of w. Thus

$$f_Z(z) = \int_{-\infty}^{\infty} f_{WZ}(w, z)\, dw. \quad (9.11\text{-}7)$$

Despite the formidable appearance of $f_{WZ}(w, z)$, this is actually an easy integration. The result is

$$f_Z(z) = \frac{\sqrt{1 - \rho^2}\,\sigma_1\sigma_2/\pi}{\sigma_2^2(z - \rho\sigma_1/\sigma_2)^2 + \sigma_1^2(1 - \rho^2)}. \quad (9.11\text{-}8)$$

The distribution function of Z is given by

$$F_Z(z) = \int_{-\infty}^{z} f_Z(u)\, du. \quad (9.11\text{-}9)$$

Here, too, despite the formidable appearance of $f_Z(z)$, the integration is easy. We leave it as an exercise for the reader to show that†

$$F_Z(z) = \frac{1}{2} + \frac{1}{\pi}\tan^{-1}\frac{\sigma_2 z - \rho\sigma_1}{\sigma_1\sqrt{1 - \rho^2}}. \quad (9.11\text{-}10)$$

In the special case where X and Y are independent r.v.'s with common variance σ^2, then

$$f_Z(z) = \frac{1}{\pi}\frac{1}{z^2 + 1} \quad (9.11\text{-}11)$$

and

$$F_Z(z) = \frac{1}{2} + \frac{1}{\pi}\tan^{-1} z. \quad (9.11\text{-}12)$$

Suppose we are interested in the probability that $X/Y < 0$. Then with the aid of Eq. (9.11-10) we obtain

$$P\left(\frac{X}{Y} < 0\right) = P(Z < 0) = F_Z(0) = \frac{1}{2} - \frac{1}{\pi}\tan^{-1}\frac{\rho}{\sqrt{1 - \rho^2}}. \quad (9.11\text{-}13)$$

Let

$$\alpha \equiv \tan^{-1}\frac{\rho}{\sqrt{1 - \rho^2}}$$

†Alternatively, the integral can be looked up in tables of integrals such as those by B.O. Peirce and R.M. Foster, *A Short Table of Integrals*, Ginn, Boston 1956, Eq. (70).

(see Fig. 9.11-1). Then

$$P\left(\frac{X}{Y} < 0\right) = \frac{1}{2} - \frac{1}{\pi}\alpha \qquad (9.11\text{-}14)$$

$$= \frac{\beta}{\pi}, \qquad (9.11\text{-}15)$$

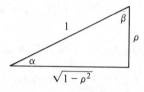

Figure 9.11-1

where $\beta = (\pi/2) - \alpha = \cos^{-1} \rho$. Considering the complexity of the joint Gaussian pdf, it is somewhat surprising that we end up with such simple expressions. The probability that $X/Y < 0$ will come up when we discuss Van Vleck's theorem in Sec. 10.10.

9.12 Conclusion

The material in this chapter is standard and is discussed in depth in the two excellent books by Papoulis [9-1] and Parzen [9-2], to which the reader is referred for more details. For a deeper discussion of probability theory and stochastic phenomena, the reader should consult the work by Doob [9-4] or the treatise by Loève [9-3]. The latter gives a thorough discussion of the measure theoretic foundations of probability theory.

PROBLEMS

9-1. In a restaurant known for its unusual service, the time X, in minutes, that a customer has to wait before he captures the attention of a waiter is specified by the following distribution function:

$$F_X(x) = \left(\frac{x}{2}\right)^2 \qquad \text{for } 0 \le x \le 1$$

$$= \frac{x}{4} \qquad \text{for } 1 \le x \le 2$$

$$= \frac{1}{2} \qquad \text{for } 2 \le x \le 10$$

$$= \frac{x}{20} \qquad \text{for } 10 \le x \le 20$$

$$= 1 \qquad \text{for } x \ge 20.$$

(a) Sketch $F_X(x)$. (b) Compute and sketch the pdf $f_X(x)$. Verify that the area under the pdf is indeed unity. (c) What is the probability that the customer will have to wait (1) at least 10 minutes, (2) less than 5 minutes, (3) between 5 and 10 minutes, (4) exactly 1 minute?

9-2. Show that the conditioned distribution of X given the event $A = \{b < X \le a\}$ is

$$F_X(x \mid A) = \begin{cases} 0, & x < b \\ \dfrac{F_X(x) - F_X(b)}{F_X(a) - F_X(b)}, & b \le x < a \\ 1, & x > a. \end{cases}$$

9-3. In the following pdf's, compute the constant B required for proper normalization:

a. Cauchy $(\alpha < \infty, \beta > 0)$:

$$f(x) = \frac{B}{1 + [(x - \alpha)/\beta]^2}, \qquad -\infty < x < \infty.$$

b. Maxwell $(\alpha > 0)$:

$$f(x) = \begin{cases} Bx^2 e^{-x^2/\alpha^2}, & x > 0 \\ 0, & \text{otherwise.} \end{cases}$$

c. Beta $(b > -1, c > -1)$:

$$f(x) = \begin{cases} Bx^b(1 - x)^c, & 0 \le x \le 1 \\ 0, & \text{otherwise.} \end{cases}$$

(See formula 6.2.2 on p. 258 of [7-4].)

d. Chi-square $(\sigma > 0, n = 1, 2, \ldots)$:

$$f(x) = \begin{cases} Bx^{(n/2)-1} \exp(-x/2\sigma^2), & x > 0 \\ 0, & \text{otherwise.} \end{cases}$$

9-4. A noisy resistor produces a voltage $v_n(t)$. At $t = t_1$, the noise level $X \equiv v_n(t_1)$ is known to be a Gaussian r.v. with pdf

$$f_X(x) = \frac{1}{\sqrt{2\pi\sigma^2}} \exp\left[-\frac{1}{2}\left(\frac{x}{\sigma}\right)^2\right].$$

Compute and plot the probability that $|X| > k\sigma$ for $k = 1, 2, \ldots$.

9-5. A noisy waveform is sampled at instants $t_1, t_2, \ldots, t_N$. The samples form a set of N random variables with joint pdf given by

$$f_{X_1 \cdots X_N}(x_1, \ldots, x_N) = \frac{1}{(2\pi\sigma^2)^{N/2}} \exp\left(-\frac{1}{2\sigma^2} \sum_{i=1}^{N} x_i^2\right)$$

Are the $\{X_i\}$ independent of each other? Compute the probability that at least one sample exceeds the threshold 3σ.

9-6. (The general joint Gaussian pdf for two r.v.'s) Two random variables X_1, X_2 are said to be jointly Gaussian if their joint pdf is given by

$$f_{X_1 X_2}(x_1, x_2) = \frac{1}{2\pi\sigma_1\sigma_2\sqrt{1 - \rho^2}}$$

$$\exp\left\{-\frac{1}{2(1 - \rho^2)}\left[\frac{(x_1 - \mu_1)^2}{\sigma_1^2} - \frac{2\rho(x_1 - \mu_1)(x_2 - \mu_2)}{\sigma_1\sigma_2} + \frac{(x_2 - \mu_2)^2}{\sigma_2^2}\right]\right\}.$$

   a. How many independent parameters characterize this pdf?
   b. Show that if $\rho = 0$, X_1, X_2 are independent.
   c. Show that the marginal pdf's are Gaussian.
   d. Show that the loci of constant probability are ellipses.
   e. What is true about the principal axes of the ellipses of constant probability when $\rho = 0$?
   f. Assume that $\rho = 0$; under what circumstances do the ellipses degenerate into circles?

9-7. Let X be an r.v. with probability distribution function $F_X(x)$ and pdf $f_X(x)$. A new r.v. is generated by the transform $Y = aX$. Compute $F_Y(y)$ when (a) $a > 0$ and (b) $a < 0$.

9-8. Let X be an r.v. with probability distribution function $F_X(x)$ and pdf $f_X(x)$. What is $F_Y(y)$ when Y is the output of an ideal half-wave rectifier; i.e.,

$$Y = \begin{cases} X, & X = 0 \\ 0, & \text{otherwise?} \end{cases}$$

9-9. Let X be an r.v. with probability distribution $F_X(x)$. Assume that $F_X(\cdot)$ is invertible; i.e., $F_X^{-1}(x)$ exists and is well defined for each x. Show that the transformation $Y = F_X(X)$ produces a uniformly distributed r.v.

9-10. Let X be a discrete r.v. with binomial probability law; i.e.,

$$P(X = k) = \binom{n}{k}p^k q^{n-k}, \quad k = 0, 1, \ldots, \quad p + q = 1$$

Compute $E(X)$, σ_X^2.

9-11. Let X be a discrete r.v. with Poisson probability law with parameter $\lambda > 0$; i.e.,

$$P(X = k) = e^{-\lambda}\frac{\lambda^k}{k!}, \quad k = 0, 1, \ldots.$$

Compute $E(X)$, σ_X^2.

9-12. (Tchebycheff inequality) Let X be an r.v. with pdf $f(x)$ with mean and variance given by μ, σ^2, respectively.
   a. Show that

$$\sigma^2 \geq \int_{|x-\mu|>k\sigma} (x - \mu)^2 f(x)\, dx,$$

   where k is an arbitrary constant, $k > 0$.
   b. Show that

$$\int_{|x-\mu|>k\sigma} (x - \mu)^2 f(x)\, dx \geq k^2\sigma^2 \int_{|x-\mu|>k\sigma} f(x)\, dx.$$

c. Use parts a and b and the fact that

$$P(|X - \mu| \geq k\sigma) = \int_{|x - \mu| > k\sigma} f(x)\, dx$$

to show that

$$P(|X - \mu| \geq k\sigma) \leq \frac{1}{k^2}.$$

This important result is known as the *Tchebycheff inequality*.

9-13. Let X be a uniform r.v. with pdf

$$f_X(x) = \begin{cases} (b - a)^{-1}, & a < x < b \\ 0, & \text{otherwise.} \end{cases}$$

Compute the mean and variance of X.

9-14. Compute the mean and variance of X when X is an r.v. that has as pdf the Rayleigh law:

$$f_X(x) = \begin{cases} \dfrac{x}{\alpha^2} \exp\left[-\dfrac{1}{2}\left(\dfrac{x}{\alpha}\right)^2\right], & x > 0,\ \alpha > 0 \\ 0, & \text{otherwise.} \end{cases}$$

9-15. Show that the rth central moment of X

$$m_r = \int_{-\infty}^{\infty} (x - \mu)^r f_X(x)\, dx$$

can be expanded into a series

$$m_r = \sum_{i=0}^{r} (-1)^i \binom{r}{i} \mu^i \zeta_{r-i},$$

where ζ_{r-i} is defined in Eq. (9.9-1).

9-16. Let X, Y be two random variables. Show that $E(X + Y) = E(X) + E(Y)$. Extend this result to $E(Z)$ where $Z = \sum_{i=1}^{N} X_i$ and $X_i, i = 1, \ldots, N$, are r.v.'s.

9-17. a. Show that if $h_1(x) \leq h_2(x)$ for all x, then

$$E[h_1(X)] \leq E[h_2(X)].$$

b. Show that for any $h(\cdot)$

$$|E[h(X)]| \leq E|h(X)|.$$

9-18. Use the result of Prob. 9-15 or use a direct calculation to show that the variance of an r.v. X is given by

$$\sigma^2 = \zeta_2 - \mu^2.$$

9-19. The expectation of X is frequently estimated from the sample mean $\hat{\mu}$:

$$\hat{\mu} = \frac{1}{N} \sum_{i=1}^{N} X_i,$$

where the $\{X_i\}$ are N independent outcomes of X. Compute the mean and variance of $\hat{\mu}$. How fast does the variance of $\hat{\mu}$ go to zero with N?

9-20. (Regression) Two random variables X, Y are suspected of having a strong linear dependence on the basis of a scatter diagram. The scatter diagram is a plot of actual observations (x_i, y_i) on X, Y (Fig. P9-20). Assume that the dependence is modeled as $Y_p = \alpha + \beta X$.

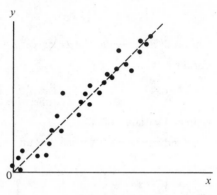

Figure P9-20

a. Compute the mean-square error $\overline{\epsilon^2}$ between the predicted and actual value of Y; i.e.,

$$\overline{\epsilon^2} = E[Y - Y_p]^2.$$

b. Compute

$$\frac{\partial \overline{\epsilon^2}}{\partial \alpha}, \quad \frac{\partial \overline{\epsilon^2}}{\partial \beta}$$

to find the "best" α, β in the sense of minimizing $\overline{\epsilon^2}$.

c. Show that the best predictor passes through $E(X)$, $E(Y)$.

9-21. Let X_1 and X_2 be independent, normal r.v.'s with mean and variance given by μ_1, σ_1^2 and μ_2, σ_1^2, respectively. Show that the mean and variance (μ, σ^2) of $X_1 + X_2$ is $\mu = \mu_1 + \mu_2$ and $\sigma^2 = \sigma_1^2 + \sigma_2^2$.

9-22. Let X be an r.v. that takes on values 0 and 1 with probability

$$P(X = 1) = p, \quad P(X = 0) = q = 1 - p.$$

a. Compute the moment-generating function.

b. Compute the first few moments of X.

c. Compute the mean and variance of an r.v. X using the moment-generating technique if $f_X(k) = e^{-\lambda}\lambda^k/k!$

9-23. (Log-normal distribution) The r.v. X is said to be *log-normal* if $Y = \log X$ is normal. Compute the mean and variance of X using moment-generating functions. *Hint*: $E(X) = E[e^{tY}]_{t=1}$. Similarly, $E(X^2) = E[e^{tY}]_{t=2}$, etc.

9-24. Let Θ be uniformly distributed in $[-\pi/2, \pi/2]$. Compute the pdf of $Y = \sin \Theta$ using the method of characteristic functions. *Hint*: Use the approach that leads up to Eq. (9.10-39); i.e., consider the transformation $y = \sin\theta$ in

$$\Phi_Y(\omega) = \int_{-\pi/2}^{\pi/2} e^{j\omega\sin\theta}\frac{d\theta}{\pi}$$

and compare with

$$\Phi_Y(\omega) = \int_{-\infty}^{\infty} e^{j\omega y} f_Y(y)\, dy.$$

9-25. Use the method of characteristic functions to compute the first four moments of X if the pdf of X is the exponential law; i.e.,

$$f_X(x) = \begin{cases} \lambda e^{-\lambda x}, & x > 0, \lambda > 0 \\ 0, & \text{otherwise.} \end{cases}$$

9-26. Let X be a uniform r.v. over the interval $[-a, a]$. Let Y be a uniform r.v. over the interval $[0, 2a]$. X and Y are independent. Compute the characteristic function of $Z = X + Y$. Use this result to compute

$$\bar{Z} \quad \text{and} \quad \overline{[Z - \bar{Z}]^2}.$$

9-27. Consider the system in Fig. P9-27. Let X and Y be correlated, jointly normal r.v.'s with pdf's as given in Eq. (9.9-5). Use the method of-joint characteristic functions to compute $\bar{W}$.

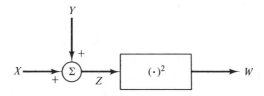

Figure P9-27

9-28. In communication theory, waveforms of the type

$$w(t) = x(t)\cos\omega t - y(t)\sin\omega t$$

appear quite often. At a particular instant of time, say $t = t_1$, $X \equiv x(t_1)$ and $Y \equiv y(t_1)$ are known to be Gaussian, uncorrelated r.v.'s. Compute the joint pdf of the *envelope* $Z \equiv (X^2 + Y^2)^{1/2}$ and *phase* $\phi \equiv \tan^{-1} Y/X$ of $w(t_1)$.

REFERENCES

[9-1] A. PAPOULIS, *Probability, Random Variables, and Stochastic Processes*, McGraw-Hill, New York, 1965.

[9-2] E. PARZEN, *Modern Probability Theory and Its Applications*, Wiley, New York, 1960.

[9-3] M. LOÈVE, *Probability Theory*, Van Nostrand Reinhold, New York, 1962.

[9-4] J. L. DOOB, *Stochastic Processes*, Wiley, New York, 1953.

10

Random Processes

10.1 Introduction

In communication systems, one frequently encounters waveforms that display irregular and unpredictable fluctuations in some characteristic of the wave. The waveforms associated with ordinary speech and music, the noise voltage across a resistor, electromagnetic emission from the sun and stars, the envelope of a TV signal, the instantaneous electron emission current in a tube, and telegraph and telephone signals all exhibit some inherent randomness which makes them "interesting." The study of waveforms of this type brings us to the subject of this chapter, which is *random* or *stochastic* processes.† A rigorous study of random processes is a formidable undertaking and well beyond our intention. Our aim is to present enough of the theory to enable the reader to make input-output computations for typical communication systems, when the signals are random. An important calculation of this type is the computation of the *signal-to-noise* $(\mathcal{S}/\mathcal{N})$ ratio for a communication system; communication engineers should understand how the parameters (i.e., gain, bandwidth, etc.) of the system influence the $\mathcal{S}/\mathcal{N}$ ratio.

Why study random processes? For one thing, because they include the most important waveforms in engineering, the ones that really contain information. If a waveform is completely predictable, if no characteristic of it (amplitude, phase, start time, spectrum, etc.) is unknown, then there is no point in sending it—it contains no information. A monochromatic sine wave of known amplitude, frequency, and phase is, despite its esthetic quality, extremely dull; its future values are as predictable as yesterday's weather.

†The word stochastic is derived from the Greek word "stochastiko," which means ". . . skillful in aiming or guessing at . . . ."

Whereas the unpredictability of some waveforms reflects their ability to impart information, the unpredictability of other waveforms is regarded only as noise and represents both an irritation and a challenge to designers. The thermal noise voltage observed across a resistor is an example of this type of waveform. This signal is generated by the thermal agitation of the electrons in the resistor; if we were interested in the kinetics of electron motion in that resistor, then, conceivably, the variations in the wave might give us useful information. However, in a communication-systems setting, the resistor noise voltage is a source of interference. Precisely how it limits our capability to enjoy, say, listening to music in an AM receiver is the concern of communication engineers.

The thermal noise voltage is an example of a signal with irregular fluctuations. Knowledge of the waveform for past values of time does not enable us to predict future values. However, not all random waveforms have irregular fluctuations. A sine wave with random phase remains a sine wave for all time, and knowledge of it for (say) $t < t_0$ specifies its values for $t > t_0$. Nevertheless, such a waveform is an example of a random process.

10.2 Definition of a Random Process

As we shall see, a random process is essentially nothing more than a collection of random variables related through a suitable indexing set, say T. If T is the set of integers, the random process is said to be of a discrete type. Otherwise, the random process is said to be continuous. However, this definition does not make clear how random variables are related to random waveforms. Suppose that we are performing an experiment with random outcomes whose sample description space is S. We already know that a random variable X is a function on S, which means that for every $s \in S$, $X(\cdot)$ assigns a number $X(s)$. Now let us assign to every $s \in S$ a function of two parameters, $x(t, s)$, where t denotes time and is an element of the indexing set T.† The family of functions $\{x(t, s), s \in S \; t \in T\}$ then furnishes, depending on what is held fixed, four objects:

	$x(t, s)$	
	s fixed	*s* variable
t fixed	$x(t, s)$ is a number	$x(t, s)$ is a single random variable
t variable	$x(t, s)$ is a single function of time, also called a sample function	$x(t, s)$ is a family of sample functions or equivalently a collection of random variables (a random process)

†Most random signals of interest are one-dimensional time variations. However, if the indexing set is $R \times R$ with elements (x, y), we would have a *spatial* random process. Such processes occur in optics.

This is the point of view taken by Papoulis [10-1] and clearly shows how random processes are related to random variables. If t is held fixed, say at t_1, but s is variable, then $x(t_1, s) \equiv X_1$ is a random variable. For $t_2 \equiv t_1 + \Delta$, we have $x(t_2, s) \equiv X_2$, a new random variable. Our process is generated by stepping the index parameters t. Note that both t and s must be varied; keeping t fixed by letting $x(t, \cdot)$ roam over S generates a random variable; varying t generates the random process.

As a somewhat more concrete example of a random process, consider the noise voltage $v(t)$ across a thermally agitated resistor. The value of this voltage at a particular instant of time, say $t = t_1$, constitutes a random variable $X_1 \equiv v(t_1)$. To obtain the statistics of X_1 we could take a very large number $(n \longrightarrow \infty)$ of identical resistors and record their waveforms. These might look as in Fig. 10.2-1.

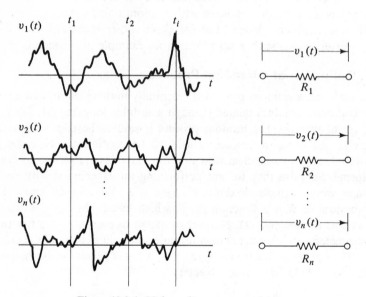

Figure 10.2-1. Noise voltage across resistors.

At $t = t_1$, the totality of observed levels represents the values that X_1 can take on. In particular, at $t = t_1$, we can obtain the relative frequency of occurrence of the event $\{x_1 < X_1 \le x_1 + \Delta_1\}$ and estimate the pdf of X_1. Of course this can be repeated for X_2 at t_2, X_3 at t_3, etc. The *joint pdf* of X_1 and X_2 can similarly be obtained by considering the relative frequency of the event $\{x_1 < X_1 \le x_1 + \Delta_1, x_2 < X_2 \le x_2 + \Delta_2\}$. Proceeding in this way, we can obtain the nth-order joint pdf of $X_1, X_2, \ldots, X_n$.

In this example, the underlying experiment can be thought of as randomly picking a resistor. The sample description space S of the underlying experiment is the set of all n resistors. Once we have picked a particular resistor, say the ith, the resultant waveform $v_i(t) \equiv v(t, s_i)$ is a single function of time. On the other hand, the noise voltage at $t = t_i$ is a random variable whose sample description space is the totality of amplitudes recorded at $t = t_i$ for our ensemble of resistors.

Notation

In denoting a random process, the functional dependence of waveforms and random variables on the outcomes $s \in S$ of the underlying experiment is usually suppressed. In general the underlying experiment is not made explicit although it is implied. The dependence on time, however, is made explicit. Some authors use bold or capital letters to differentiate a random process from an ordinary function of time. Others use lowercase symbols such as $x(t)$ to denote the sample functions of the process and reserve x_t for the random variable generated by fixing t [10-2]. Our own preference is to reserve the use of capital letters such as X, Y, Z for Fourier transforms of x, y, z. Hence, in general, we shall use a lowercase symbol $x(t)$ to denote the random variable at t or a sample function of the process. The specific interpretation of $x(t)$ will be inferred from the context. The use of lowercase letters is in distinct contrast to the notation used in Chapter 9, but the adoption of the former will enable us to dispense with introducing cumbersome or exotic symbols. On some occasions, specifically when we deal with vectors of random variables such as in the Gaussian process, we shall revert to capital notation to indicate that we are dealing with vectors instead of scalars. Also we shall occasionally use the notation $\{x(t)\}$ to indicate an ensemble of sample functions when it is necessary to do so. Finally, because of our desire to reserve capital letters primarily for Fourier transforms, we shall generally omit the use of capital subscripts on probability distribution or density functions.

EXAMPLE

Consider the experiment of throwing a fair die. The sample description space S is $S = \{1, 2, 3, 4, 5, 6\}$. Let the random process be defined by

$$x(t) = kt \qquad \text{if } s = k, k = 1, \ldots, 6. \tag{10.2-1}$$

The sample functions are shown in Fig. 10.2-2. At $t = 1$, $x(1)$ is an r.v. that takes on values $k = 1, \ldots, 6$ with probability $P[x(1) = k] = \frac{1}{6}$. All other values of $x(1)$ are associated with the impossible event. At $t = 2$, $x(2)$ is an r.v. that takes on values $2k$ for $k = 1, \ldots, 6$ with probability $P[x(2) = 2k] = \frac{1}{6}$.

The average value of $x(1)$ is

$$\overline{x(1)} = \tfrac{1}{6} \sum_{k=1}^{6} k = 3.5,$$

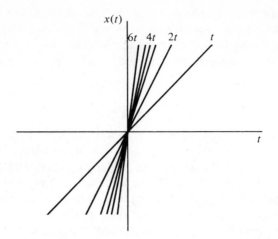

Figure 10.2-2. Sample functions of the process $x(t) = kt$.

while

$$\overline{x(2)} = \tfrac{1}{6} \sum_{k=1}^{6} 2k = 7.$$

We see that the mean value of $x(t)$ depends on t. Consequently this is an example of a *nonstationary* random process. Processes for which the mean and higher moments are independent of t are said to be *stationary*. We shall say more about such processes later.

In the example considered, the sample functions were very regular, and knowing $x(t)$ for $t < t_1$ completely specifies $x(t)$ for all t. Regularity of sample functions is not the usual case, however, and most of our concern will be with functions that are highly irregular, i.e., noise-like. Knowledge of such functions for $t < t_1$ is insufficient to predict future values exactly.

10.3 Statistics of a Random Process

Consider the random variable $x(t_1)$, where t_1 is a fixed value of t. With x_1 denoting a number, the probability of the event $\{x(t_1) \leq x_1\}$ is the probability distribution function (PDF) of $x(t_1)$; i.e.,

$$F(x_1, t_1) = P[x(t_1) \leq x_1]. \tag{10.3-1}$$

The pdf† is obtained by differentiating Eq. (10.3-1). Thus

$$f(x_1, t_1) = \frac{\partial F}{\partial x_1}. \tag{10.3-2}$$

---

†Recall that pdf stands for the probability density function.

$F(x_1, t_1)$ is called the first-order distribution of the process $x(t)$. The event $\{x(t_1) \leq x_1\}$ consists of all $s \in S$ such that $x(t_1, s) \leq x_1$. Similarly, given t_1 and t_2, $x(t_1)$ and $x(t_2)$ represent two random variables. Their joint distribution is the second-order distribution of $x(t)$. It is given by

$$F(x_1, x_2; t_1, t_2) = P[x(t_1) \leq x_1, x(t_2) \leq x_2]. \qquad (10.3\text{-}3)$$

The associated second-order pdf is

$$f(x_1, x_2; t_1, t_2) = \frac{\partial^2 F}{\partial x_1 \, \partial x_2}. \qquad (10.3\text{-}4)$$

The nth-order joint distribution function $F(x_1, \ldots, x_n; t_1, \ldots, t_n)$ and the nth order pdf are direct extensions of the above. Thus

$$F(x_1, \ldots, x_n; t_1, \ldots, t_n = P[x(t_1) \leq x_1, \ldots, x(t_n) \leq x_n], \qquad (10.3\text{-}5)$$

$$f(x_1, \ldots, x_n; t_1, \ldots, t_n) = \frac{\partial^n F}{\partial x_1 \cdots \partial x_n}. \qquad (10.3\text{-}6)$$

A random process is *completely characterized* if its nth-order statistics are known for all n. In other words, it is required that

$$F(x_1, \ldots, x_n; t_1, \ldots, t_n)$$

is known for every n and for every set and combination of the arguments $(x_1, \ldots, x_n)$ and $(t_1, \ldots, t_n)$. Fortunately, most problems of engineering importance require knowledge of only the first two orders.

Stationary Random Processes

A random process is said to be strict-sense stationary (sss) if

$$P[x(t_1) \leq x_1, \ldots, x(t_n) \leq x_n]$$
$$= P[x(t_1 + \tau) \leq x_1, \ldots, x(t_n + \tau) \leq x_n] \qquad (10.3\text{-}7)$$

for every combination of $\mathbf{t} = (t_1, \ldots, t_n)$, $\mathbf{x} = (x_1, \ldots x_n)$, τ, and n. The pdf of an sss process is independent of time shifts; this implies that sss waveforms cannot be switched on and off; i.e., they must be on for all time.

Averages of a Random Process

The average, expected value, or mean $\mu(t)$, of a random process is defined by

$$\mu(t) = E[x(t)] = \int_{-\infty}^{\infty} x f(x, t) \, dx. \qquad (10.3\text{-}8)$$

For a stationary process, $\mu(t)$ is independent of time. The proof is easy:

$$\mu(t + \tau) = E[x(t + \tau)] = \int_{-\infty}^{\infty} x f(x, t + \tau) \, dx$$
$$= \int_{-\infty}^{\infty} x f(x, t) \, dx = \mu(t).$$

The *autocorrelation* function (also called the correlation or self-correlation) is given by $R(t_1, t_2) = E[x(t_1)x(t_2)] = \int_{-\infty}^{\infty} \int_{-\infty}^{\infty} x_1 x_2 f(x_1, x_2; t_1, t_2)\, dx_1\, dx_2$. For a stationary process, this depends only on $\tau = t_2 - t_1$ (why?) and is written $R(\tau)$. The autocovariance or, simply, covariance is given by

$$r(t_1, t_2) = E\{[x(t_1) - \mu(t_1)][x(t_2) - \mu(t_2)]\}$$
$$= R(t_1, t_2) - \mu(t_2)\,\mu(t_1). \tag{10.3-9}$$

For an sss process, the autocovariance, like the autocorrelation, depends only on $t_2 - t_1 = \tau$. We write it as $r(\tau)$. The nth joint moment of a random process is

$$E[x(t_1)x(t_2) \cdots x(t_n)]$$
$$= \int_{-\infty}^{\infty} \int_{-\infty}^{\infty} x_1 x_2 \cdots x_n f(x_1, x_2, \ldots, x_n; t_1, t_2, \ldots, t_n)\, dx_1 \cdots dx_n. \tag{10.3-10}$$

Definition: A random process is said to be *wide-sense or weakly stationary* (wss) if

(i) $\mu(t) = E[x(t)]$ is independent of t,   (10.3-11a)

(ii) $R(t_1, t_2) = R(t_2 - t_1) = R(\tau)$,   (10.3-11b)

(iii) $R(0) = E[x^2(t)] < \infty$.   (10.3-11c)

Such processes are very important in practice. Unless otherwise stated we shall assume that a process is at least wide-sense stationary. Note that all sss processes are wss. The reverse is not true.

Properties of the Correlation Function

It is easily shown that $R(\tau)$ has the following properties:

(i) $R(\tau) = R(-\tau)$

(ii) $R(0) \geq R(\tau)$ (10.3-12)

(iii) $R(0) = \sigma^2 + \mu^2$,

where $\sigma^2 \equiv r(0)$ is the variance of the wss random process $x(t)$. Properties (i) and (iii) follow from the definition of $R(\tau)$ and property (ii) from the fact that

$$E[x(t_2) - x(t_1)]^2 \geq 0.$$

The correlation function is a measure of the statistical dependence of two random variables separated by a distance τ in time. If the process $x(t)$ contains no periodic or dc components, we generally expect $x(t_1)$ and $x(t_2)$ to be strongly dependent if $t_2 - t_1$ is small and essentially independent if $t_2 - t_1$ is large. For example, the outdoor temperature an hour from now is much more dependent on the present temperature than will be the temperature

365 days later. Such processes are said to satisfy a *strong mixing condition* [10-3]. For such processes $\lim_{\tau \to \infty} E[x(t)x(t + \tau)] \to E[x(t)]E[x(t + \tau)] = \mu^2$. Most processes of engineering interest that do not have periodic components other than a dc component behave in this way.

10.4 Examples of Correlation Function Computations

The Full-Random Binary Waveform: The "Telegraph" Signal

A typical sample function of this process is shown in Fig. 10.4-1. The average number of polarity switches (zero crossings) per unit time is λ. The probability of getting exactly k crossings in time τ is given by the Poisson law:

$$p(k) = e^{-\lambda\tau} \frac{(\lambda\tau)^k}{k!}. \tag{10.4-1}$$

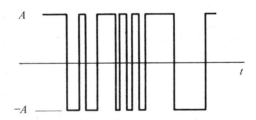

Figure 10.4-1. Sample function of the telegraph signal.

Note that the events $\{k$ points in $(0, t)\}$, $k = 0, 1, \ldots$, are mutually exclusive. Let $x(t)$ denote the value of the waveform at t. The correlation function is given by

$$\begin{aligned}
R(t_1, t_2) &= E[x(t_2)x(t_1)] \\
&= A^2 P[x(t_1) = A, x(t_2) = A] \\
&\quad + A(-A)P[x(t_1) = A, x(t_2) = -A] \\
&\quad + (-A)(A)P[x(t_1) = -A, x(t_2) = A] \\
&\quad + (-A)^2 P[x(t_1) = -A, x(t_2) = -A]. \tag{10.4-2}
\end{aligned}$$

We assume that for *any* t_1, t_2 the events $\{x(t_1) = A, x(t_2) = A\}$ and $\{x(t_1) = -A, x(t_2) = -A\}$ are equally likely. The same assumption† holds for the events $\{x(t_1) = -A, x(t_2) = A\}$ and $\{x(t_1) = A, x(t_2) = -A\}$. We also

†What would happen if we went a step further and assumed that for any t_1, t_2 the four events $\{x(t_1) = \pm A, x(t_2) = \pm A\}$ are equally likely?

assume that $P[x(t_1) = A] = P[x(t_2) = A] = \frac{1}{2}$. These assumptions make the waveform statistically symmetric above and below the axes. Hence

$$R(t_1, t_2) = 2A^2\{P[(x(t_1) = A, x(t_2) = A] - P[x(t_1) = -A, x(t_2) = A]\}$$
$$= A^2\{P[x(t_2) = A \,|\, x(t_1) = A] - P[x(t_2) = -A \,|\, x(t_1) = A]\}.$$

$$(10.4\text{-}3)$$

With $\tau \equiv t_2 - t_1$, we obtain, with the help of Eq. (10.4-1),

$$P[x(t_2) = A \,|\, x(t_1) = A] = \sum_{k \text{ even}} e^{-\lambda\tau} \frac{(\lambda\tau)^k}{k!} \qquad (10.4\text{-}4a)$$

and

$$P[x(t_2) = -A \,|\, x(t_1) = A] = \sum_{k \text{ odd}} e^{-\lambda\tau} \frac{(\lambda\tau)^k}{k!}. \qquad (10.4\text{-}4b)$$

A direct substitution of Eqs. (10.4-4) into Eq. (10.4-3) yields

$$R(t_1, t_2) = R(\tau) = A^2 e^{-2\lambda\tau}, \qquad \tau > 0. \qquad (10.4\text{-}5)$$

To obtain this result we used the fact that

$$\sum_{k \text{ even}} e^{-\lambda\tau} \frac{(\lambda\tau)^k}{k!} - \sum_{k \text{ odd}} e^{-\lambda\tau} \frac{(\lambda\tau)^k}{k!} = e^{-\lambda\tau} \sum_{k=0}^{\infty} \frac{(\lambda\tau)^k}{k!}(-)^k$$
$$= e^{-2\lambda\tau}.$$

The complete solution that includes $\tau < 0$ (i.e., if $t_1 > t_2$) is

$$R(\tau) = A^2 e^{-2\lambda|\tau|}. \qquad (10.4\text{-}6)$$

This function is shown in Fig. 10.4-2.

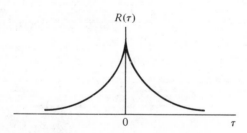

$R(\tau)$

0 $\qquad\qquad$ τ

Figure 10.4-2. Correlation function of the telegraph signal.

Sine Wave with Random Phase

The random process $x(t)$ has sample functions $\{\cos(\omega_0 t + \theta)\}$, where θ denotes the values that a random variable Θ can take on and ω_0 is a constant. A sample function is shown in Fig. 10.4-3. If Θ is uniformly distributed

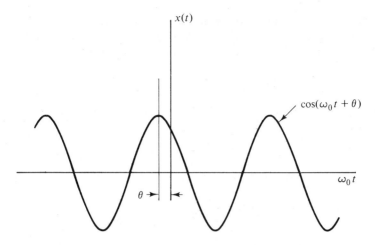

Figure 10.4-3. Sample function of the sine wave with random phase process.

in $[-\pi, \pi]$, the correlation function is given by

$$E[x(t_1)x(t_2)] = \frac{1}{2\pi} \int_{-\pi}^{\pi} \cos(\omega_0 t_1 + \theta) \cos(\omega_0 t_2 + \theta) \, d\theta$$

$$= \frac{1}{2} \cos \omega_0(t_2 - t_1). \tag{10.4-7}$$

The process is wss, and the correlation function can be written in terms of $\tau \equiv t_2 - t_1$ as

$$R(\tau) = \tfrac{1}{2} \cos \omega_0 \tau. \tag{10.4-8}$$

Notice that the sample functions in this case are extremely regular; for a given element s of the sample space S, knowledge of $x(t)$ for $t < t_1$ specifies all future values.

Integrals of Random Processes

Given a random process $x(t)$, what meaning shall we give to the integral

$$\xi = \int_a^b x(t) \, dt? \tag{10.4-9}$$

Recall that a random process can be viewed as an ensemble of sample functions. If the integral in Eq. (10.4-9) converges, in the Riemann sense, for almost† every sample function $x(t, s)$, $s \in S$, then the numbers $\xi(s)$ define a

---

†Except, conceivably, for a "small" set of probability zero. If the integral does not converge for some $s \in S$, it is still possible to define $\xi(s)$ in terms of the limit of a sum of random variables, i.e., $\xi = \sum_{i=1}^{n} x(t_i) \, \Delta t_i$, which converges in the mean-square sense. We shall use this viewpoint on occasion.

random variable with sample description space $\{\xi(s) : s \in S\}$. Integrals of random processes are important in the practical estimation of the low-order moments of certain types of processes called ergodic processes.

Time Averages and Ergodicity

Consider the limit, frequently called a *time average* of $x(t)$,

$$\hat{\mu} = \lim_{T \to \infty} \frac{1}{2T} \int_{-T}^{T} x(t) \, dt, \qquad (10.4.10)$$

where $x(t)$ denotes a stationary random process. Clearly if $\mu = E[x(t)]$, then

$$E[\hat{\mu}] = \lim_{T \to \infty} \frac{1}{2T} \int_{-T}^{T} E[x(t)] \, dt$$

$$= \mu. \qquad (10.4-11)$$

Now in order to argue that $\hat{\mu}$ (an r.v.) is equal to μ, we must show that the variance of $\hat{\mu}$, written $\sigma_{\hat{\mu}}^2$, tends to zero as $T \to \infty$. If $E[\hat{\mu}] = \mu$ and $\sigma_{\hat{\mu}}^2 = 0$, then we can say that the time average equals the ensemble average and $x(t)$ has the property of *ergodicity of the mean*. The notion of ergodicity is extremely important in engineering because we don't usually have available a large number of sample functions from which to do an ensemble average. The notion of a random process consisting of a large ensemble of waveforms is useful as a model but does not reflect real-life situations too well. Usually we have to estimate the statistics of a random process from an observation on a single function. Hence the question arises, To what extent does a single sample function represent the entire ensemble? This is something we cannot know with certainty. But if we can reasonably argue that the process is *ergodic*, then a single function is all we need.

A process $x(t)$ is said to be ergodic if all of its statistics can be obtained by observing a single function $x(t, s)$ (s fixed) of the process. For an ergodic process all the moments can be determined from time averages performed on $x(t, s)$. In practice, however, most of the time we only require ergodicity of the first two moments; i.e.,

$$\hat{\mu} = \lim_{T \to \infty} \frac{1}{2T} \int_{-T}^{T} x(t) \, dt = \mu, \qquad (10.4-12)$$

$$\hat{R}(\tau) = \lim_{T \to \infty} \frac{1}{2T} \int_{-T}^{T} x(t) x(t + \tau) \, dt = R(\tau). \qquad (10.4-13)$$

We reiterate that in order for Eqs. (10.4-12) and (10.4-13) to have any meaning, the variances of $\hat{\mu}$ and $\hat{R}(\tau)$ must be zero as $T \to \infty$. Ergodic processes must be stationary, but the converse is not necessarily true (Probs. 10-5 and 10-6).

10.5 The Power Spectrum

Although the sample functions of many random processes have irregular shapes and cannot easily (if at all) be described by an equation, it still seems reasonable to talk about frequency content, bandwidth, average power, etc. For example, the "grassy"-looking waveform shown in Fig. 10.5-1(a) is expected to have a higher bandwidth than the smoother-looking waveform shown in Fig. 10.5-1(b).

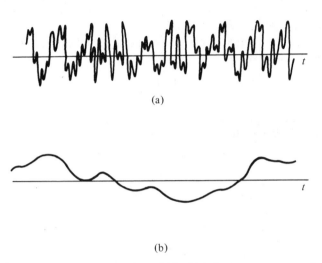

(a)

(b)

Figure 10.5-1. Sample function of (a) a high-frequency process and (b) a low-frequency process.

In earlier chapters we found that the analysis of nonperiodic, finite-energy signals is conveniently done with the use of the Fourier transform. However, when $x(t)$ is a random process, the condition

$$\int_{-\infty}^{\infty} |x(t)| \, dt < \infty \tag{10.5-1}$$

is generally not satisfied,† and therefore the Fourier transform may not exist. Even integrals of the form

$$X_T(\omega) = \int_{-T}^{T} x(t) e^{-j\omega t} \, dt, \tag{10.5-2}$$

where T is finite, are not a useful frequency decomposition of a random process because $X_T(\omega)$, for fixed ω, is a random variable whose statistics are

†For a wss process $x(t)$ is never "switched off," and therefore the integral of $|x(t)|$ grows without limit. It would be more precise to say that the r.v. $y = \int_{-T}^{T} |x(t)| \, dt$ has unbounded variance as $T \longrightarrow \infty$. This is true even when $x(t)$, and not just its magnitude, is considered.

generally difficult to compute. In Sec. 10-4 we had suggested that a way to view the integral of a random process was as a limit of a sum of many random variables. Hence computing the pdf of $X_T(\omega)$ would, typically, involve many repeated superposition-type integrations leading to extremely involved expressions.

For the reasons given above, a sinusoidal resolution of a random process is generally not practical. On the other hand, a sinusoidal resolution of the *average power* in $x(t)$ is not only frequently possible but is extremely useful from an engineering point of view. Such a resolution is called the *power spectrum* or the *power spectral density*.

We already know that for a deterministic signal $x(t)$ the time-averaged power over an interval $(-T, T)$ is given by

$$P_T = \frac{1}{2T} \int_{-T}^{T} x^2(t)\, dt. \tag{10.5-3}$$

If $x(t)$ is a random process, however, the object, P_T, as given in Eq. (10.5-3) is a random variable. However, the quantity

$$E(P_T) = \frac{1}{2T} \int_{-T}^{T} \overline{x^2(t)}\, dt \tag{10.5-4}$$

is a number and plays the same role for a random process as P_T in Eq. (10.5-3) plays for a deterministic function of time. If $x(t)$ is a stationary random process, we might be interested in the average power over all time, which is

$$\lim_{T \to \infty} E(P_T) \equiv \bar{P} = \lim_{T \to \infty} \frac{1}{2T} \int_{-T}^{T} \overline{x^2(t)}\, dt. \tag{10.5-5}$$

This definition is only useful if the limit exists.

Equation (10.5-5) is a direct extension of familiar results. To determine the distribution of power in *frequency*, we proceed as follows: Define $x_T(t)$ by

$$x_T(t) = \begin{cases} x(t), & |t| < T \\ 0, & |t| > T, \end{cases}$$

and let $X_T(f) = \mathfrak{F}[x_T(t)]$. By Parseval's theorem

$$E(P_T) = \int_{-\infty}^{\infty} \frac{\overline{x_T^2(t)}}{2T}\, dt = \int_{-\infty}^{\infty} \frac{\overline{|X_T(f)|^2}}{2T}\, df, \tag{10.5-6}$$

and, by Eq. (10.5-5), we obtain

$$\bar{P} = \int_{-\infty}^{\infty} \lim_{T \to \infty} \frac{\overline{|X_T(f)|^2}}{2T}\, df, \tag{10.5-7}$$

where we have interchanged the order of integration and limiting. The quantity

$$W(f) \equiv \lim_{T \to \infty} \frac{\overline{|X_T(f)|^2}}{2T} \tag{10.5-8}$$

is commonly taken as the definition of the power spectrum. Note that the *expectation* must be taken *before* the limiting operation to ensure convergence of $W(f)$. In terms of $W(f)$, the average power can be written as

$$\bar{P} = \int_{-\infty}^{\infty} W(f)\, df. \tag{10.5-9}$$

The relation between $W(f)$ and the correlation function $R(\tau)$ is furnished by the *Wiener-Khintchine theorem*. We shall consider this next.

Theorem: If, for a stationary process,

$$\int_{-\infty}^{\infty} |\tau R(\tau)|\, d\tau < \infty, \tag{10.5-10}$$

then

$$W(f) = \int_{-\infty}^{\infty} R(\tau) e^{-j2\pi f \tau}\, d\tau. \tag{10.5-11}$$

In words: If Eq. (10.5-10) is satisfied, then the power spectrum is the Fourier transform of the autocorrelation function. The Wiener-Khintchine theorem is of fundamental importance in the theory of stationary random processes.

Proof of the Wiener-Khintchine Theorem

From the definition of $X_T(f)$ we have

$$\overline{|X_T(f)|^2} = \int_{-T}^{T} ds \int_{-T}^{T} dt\, \overline{x(t)x(s)}\, e^{-j2\pi f(s-t)}$$

$$= \int_{-T}^{T} ds \int_{-T}^{T} dt\, R(s - t) e^{-j2\pi f(s-t)}. \tag{10.5-12}$$

We have used the fact that $\overline{x(t)x(s)} \equiv R(s - t)$ and $\overline{|X_T(f)|^2} = \overline{X_T(f)X_T^*(f)}$. Note that the region of integration is the inside of a square of side T centered at the origin of the st plane. The evaluation of Eq. (10.5-12) can be simplified if we use the coordinate transformation $\alpha \equiv s - t$, $\beta \equiv s + t$. Since $\alpha = 0$ implies $s = t$ and $\beta = 0$ implies $s = -t$, the new region of integration is the inside of a rotated square or "diamond," shown in Fig. 10.5-2. For $\alpha < 0$, the integration with respect to β goes from $-(2T + \alpha)$ to $2T + \alpha$. However, for $\alpha > 0$, the integration on β goes from $-(2T - \alpha)$ to $2T - \alpha$. Hence the integration is broken up into two integrals as in

$$\overline{|X_T(f)|^2} = \tfrac{1}{2}\left\{ \int_{-2T}^{0} d\alpha\, R(\alpha)e^{-j2\pi f\alpha} \int_{-2T-\alpha}^{2T+\alpha} d\beta \right.$$

$$\left. + \int_{0}^{2T} d\alpha\, R(\alpha)e^{-j2\pi f\alpha} \int_{-2T+\alpha}^{2T-\alpha} d\beta \right\}. \tag{10.5-13}$$

The factor of $\tfrac{1}{2}$ results from the fact that the Jacobian of the transformation is 2. The integration with respect to β gives

$$\overline{|X_T(f)|^2} = \int_{-2T}^{2T} R(\alpha)e^{-j2\pi f\alpha}\{2T - |\alpha|\}\, d\alpha. \tag{10.5-14}$$

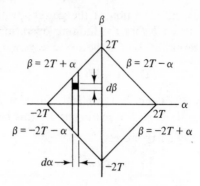

Figure 10.5-2. Region of integration for proof of Wiener-Khintchine theorem.

Hence

$$\lim_{T\to\infty} \frac{\overline{|X_T(f)|^2}}{2T} = \lim_{T\to\infty} \int_{-2T}^{2T} R(\alpha)e^{-j2\pi f\alpha}\,d\alpha$$

$$- \lim_{T\to\infty} \int_{-2T}^{2T} \frac{|\alpha|R(\alpha)}{2T}e^{-j2\pi f\alpha}\,d\alpha. \qquad (10.5\text{-}15)$$

Because

$$\int_{-\infty}^{\infty} |\alpha R(\alpha)|\,d\alpha$$

is bounded, the limit of the second integral in Eq. (10.5-15) is zero. Thus

$$\lim_{T\to\infty} \frac{\overline{|X_T(f)|^2}}{2T} = \int_{-\infty}^{\infty} R(\alpha)e^{-j2\pi f\alpha}\,d\alpha$$

$$= W(f), \qquad (10.5\text{-}16)$$

which proves Eq. (10.5-11).

Properties of W(f)

From the definition of $W(f)$, we have that

$$W(f) \geq 0. \qquad (10.5\text{-}17)$$

Also, since $R(\tau)$ is an even function of τ, we obtain

$$W(f) = \int_{-\infty}^{\infty} R(\tau)\cos 2\pi f\tau\,d\tau, \qquad (10.5\text{-}18)$$

which shows that $W(f)$ is an *even* function of f. By the Fourier inversion property it follows that

$$R(\tau) = \int_{-\infty}^{\infty} W(f)e^{j2\pi f\tau}\,df.$$

In particular, $R(\tau)$ evaluated at $\tau = 0$ gives the total power in $x(t)$:

$$R(0) = \int_{-\infty}^{\infty} W(f)\, df = \sigma^2 + \mu^2. \qquad (10.5\text{-}19)$$

In words: The integral of $W(f)$ over $(-\infty, \infty)$ is the sum of the dc and ac power in the process $x(t)$. Finally, setting $f = 0$ in Eq. (10.5-18) shows that $W(0)$ is the area under the curve $R(\tau)$; i.e.,

$$W(0) = \int_{-\infty}^{\infty} R(\tau)\, d\tau. \qquad (10.5\text{-}20)$$

Asymptotic Behavior of W(f)

The asymptotic behavior of $W(f)$ for large $|f|$ can be determined from the properties of $R(\tau)$ in the vicinity of $\tau = 0$. Formally we obtain

$$\left. \frac{dR(\tau)}{d\tau} \right|_{\tau=0} \equiv R'(0) = \int_{-\infty}^{\infty} j2\pi f W(f)\, df, \qquad (10.5\text{-}21)$$

$$R''(0) = -\int_{-\infty}^{\infty} (2\pi f)^2 W(f)\, df. \qquad (10.5\text{-}22)$$

If $R'(\tau)$ is discontinuous at the origin, $R''(0)$ doesn't exist. Hence

$$\int_{-\infty}^{\infty} f^2 W(f)\, df$$

doesn't converge, which indicates that the asymptotic behavior of $W(f)$ is such that $W(f)$ decreases no more slowly than f^{-3}; i.e.,

$$W(f) \geq \frac{K}{f^3} \qquad (10.5\text{-}23)$$

as $f \to \infty$ (K is a constant). On the other hand, if $R'(\tau)$ is continuous at the origin,

$$\int_{-\infty}^{\infty} f^2 W(f)\, df \qquad (10.5\text{-}24)$$

exists, indicating that

$$W(f) < \frac{k}{f^3} \qquad \text{as } f \to \infty. \qquad (10.5\text{-}25)$$

(See Prob. 10-14.)

*Estimating W(f) from Measurements

The power spectrum is one of the most important quantities in electrical as well as other branches of engineering. It is important in determining the bandwidth of random signals and in the subsequent design of filters to pass or reject such signals. It is useful in distinguishing between an earthquake and a man-made explosion. It can be used to differentiate between normal and abnormal brain waves or normal and abnormal heart signals. It is even useful in economics where it has been used to analyze the behavior of stock-market fluctuations, etc. For all these reasons, the measurement of $W(f)$ has been given prominent consideration in the literature, and entire books have been written on the subject. The trouble is that $W(f)$ is a somewhat complicated *average*, and, in general, it cannot be measured directly and certainly not exactly. It can only be estimated, albeit with a high degree of precision. We shall briefly consider below some techniques for estimating $W(f)$. Let $x(t)$ be a stationary random process; then

$$W(f) = \lim_{T \to \infty} \frac{\overline{|X_T(f)|^2}}{2T} \qquad (10.5\text{-}26)$$

is its power spectrum. The quantity

$$\mathcal{W}_T(f) = \frac{1}{2T} \left| \int_{-T}^{T} x(t) e^{-j2\pi ft} \, dt \right|^2 \qquad (10.5\text{-}27)$$

would seem to be a reasonable estimate of $W(f)$. Unfortunately for many processes, including the all-important Gaussian process, the quantity $\mathcal{W}_T(f)$ doesn't converge in a statistical sense (i.e., in the mean) to $W(f)$ even when $T \to \infty$ ([10-2], p. 107).

If $\mathcal{W}(f)$ is to be a "good" estimate of $W(f)$, it should, on the average, closely approximate $W(f)$ and at the same time be *stable*; i.e., its variance should be low. The degree to which $\mathcal{W}(f)$ approximates $W(f)$ is an indication of *fidelity* and is measured by a quantity called the *bias*, given by

$$\beta(f) = \overline{|\mathcal{W}(f) - W(f)|}. \qquad (10.5\text{-}28)$$

A low value of $\beta(f)$ is clearly desirable. However, $\beta(f)$ doesn't tell the whole story since it is a function of average values only. Equally important therefore is the stability of $\mathcal{W}(f)$. The stability of $\mathcal{W}(f)$ is a measure of its repeatability. A stable estimate is repeatable from record to record and is free from noise-like fluctuations unrelated to $W(f)$. The stability of $\mathcal{W}(f)$ is commonly given by the ratio

$$\frac{\text{var}\,[\mathcal{W}(f)]}{[\overline{\mathcal{W}(f)}]^2}, \qquad (10.5\text{-}29)$$

where var $[\mathcal{W}(f)]$ is shorthand for $\overline{|\mathcal{W}(f) - \overline{\mathcal{W}}(f)|^2}$. If the ratio is close to unity, the estimate is very unstable.

There are several techniques available for extracting faithful and stable estimates from observations on $x(t)$. One approach is to generate $\mathcal{W}_T(f)$ as in Eq. (10.5-27) and then convolve it with a smoothing filter, $B(f)$. The smoothing filter $B(f)$ is sometimes called a *spectral window*, and the smoothing operation is described by

$$\mathcal{W}_B(f) = \int_{-\infty}^{\infty} \mathcal{W}_T(f - f')(B(f') \, df'. \qquad (10.5\text{-}30)$$

A direct computation can be done to show that in the Gaussian case (and even somewhat more generally)

$$\frac{\text{var}\,[\mathcal{W}_T(f)]}{[\mathcal{W}_T(f)]^2} \simeq 1, \qquad T \text{ large.} \tag{10.5-31}$$

However, when spectral smoothing is done,

$$\frac{\text{var}\,[\mathcal{W}_B(f)]}{[\mathcal{W}_B(f)]^2} \simeq \frac{E}{T}, \qquad T \text{ large,} \tag{10.5-32}$$

where E is the energy in the window; i.e.,

$$E \equiv \int_{-\infty}^{\infty} |B(f)|^2 \, df. \tag{10.5-33}$$

Clearly, the number E/T can be made as small as desirable by choice of $B(f)$, but the increased stability is usually achieved at the expense of lowered fidelity. A very stable estimate may grossly fail to give a faithful rendition of $W(f)$ and hence be useless. In general a compromise must be reached between fidelity and stability. The choice of $B(f)$, or, equivalently, its inverse Fourier transform $b(\tau)$ (called a *lag* or *data* window), is determined from the criterion of goodness set up for the estimator.

An example of smoothing by the method of Eq. (10.5-30), i.e., direct realization of the convolution, is shown in Fig. 10.5-3. The example is taken from optics, and the random process is the overlapping circular-grain model, which is discussed in Chapter 13 (Fig. 13.10-2). Figure 10.5-3(a) shows a computer printout of the unsmoothed power spectrum estimate, $\mathcal{W}_T(f)$. What should be clear from this figure is the rapid fluctuation in intensity for small changes in frequency. This is symptomatic of an unstable spectrum.

Figure 10.5-3(b) shows $\mathcal{W}_B(f)$, computed as in Eq. (10.5-30) with the convolution done by digital computer. The particular spectral window, $B(f)$, that was used is given by

$$B(f) = \begin{cases} 3K(1 - \frac{3}{2}\pi^2 K f^2), & |f| < f_0 \\ 0, & |f| > f_0, \end{cases}$$

where K, a normalizing constant, is related to f_0 according to

$$K \equiv \frac{2}{3\pi^2 f_0^2}.$$

This window has certain optimal properties when used with two-dimensional processes. Specifically, it furnishes a minimum-bias spectral estimate for a fixed f_0 [10-4]. Problem 10-15 deals with spectral smoothing.

The smoothed spectrum is seen to be free of meaningless amplitude fluctuations; the inherent shape of the spectrum, however, has been preserved for further analysis.

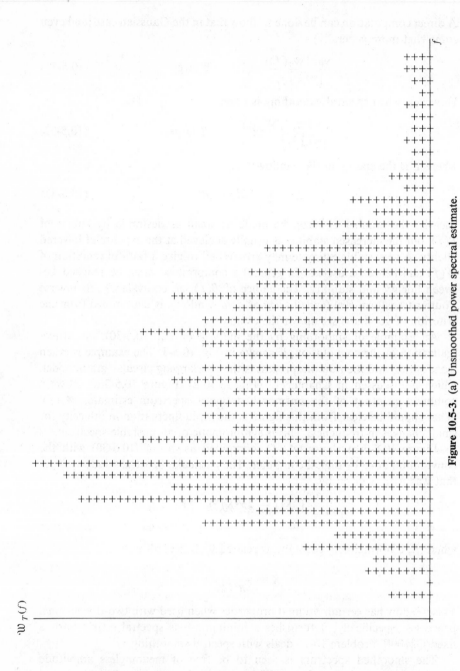

$\mathscr{W}_T(f)$

f

Figure 10.5-3. (a) Unsmoothed power spectral estimate.

$\mathcal{W}_B(f)$

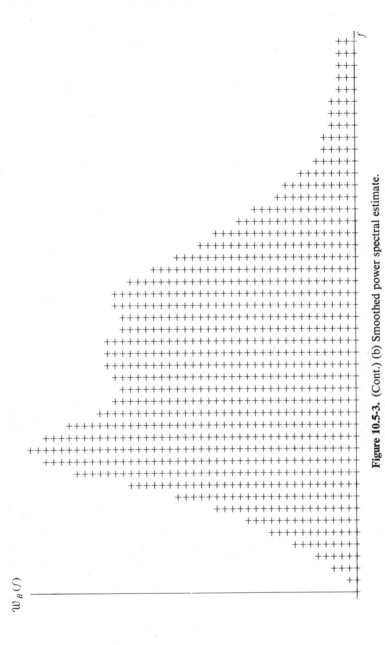

Figure 10.5-3. (Cont.) (b) Smoothed power spectral estimate.

f

An alternative to Eq. (10.5-30) is to start with the *sample autocorrelation* function

$$\mathcal{R}_T(\tau) = \begin{cases} \dfrac{1}{2T - |\tau|} \displaystyle\int_{-T+|\tau|/2}^{T-|\tau|/2} x\left(t - \dfrac{|\tau|}{2}\right) x\left(t + \dfrac{|\tau|}{2}\right) dt, & |\tau| < 2T \\ 0, & |\tau| > 2T. \end{cases} \quad (10.5\text{-}34)$$

The peculiar limits on the integral are not so peculiar after a moment's reflection: Recall that the sample function is observed only over $[-T, T]$. We now form the product

$$\mathcal{R}_T(\tau) b(\tau)$$

and compute the inverse Fourier transform,

$$\mathcal{W}_B(f) = \int_{-\infty}^{\infty} \mathcal{R}_T(\tau) b(\tau) e^{-j2\pi f\tau} \, d\tau. \quad (10.5\text{-}35)$$

Equation (10.5-35) is equivalent to Eq. (10.5-30) if

$$\mathcal{R}_T(\tau) = \mathcal{F}^{-1}[\mathcal{W}_T(f)], \quad (10.5\text{-}36)$$

$$b(\tau) = \mathcal{F}^{-1}[B(f)]. \quad (10.5\text{-}37)$$

A third method avoids convolution† and takes advantage of the existence of the fast Fourier transform algorithms (FFT) discussed in Chapter 4. First, we compute $\mathcal{W}_T(f)$ as in Eq. (10.5-27). Second, we use the FFT to compute $\mathcal{R}_T(\tau)$ and form the product $\mathcal{R}_T(\tau) b(\tau)$. Finally we use the FFT to compute $\mathcal{W}_B(f)$ from Eq. (10.5-35).

Since the available record is observed only over $[-T, T]$, the sample correlation function does not contain all possible lags τ. For this reason $b(\tau)$ is usually of finite support (i.e., it is zero for τ outside some interval). Hence we write

$$b(\tau) = 0, \quad |\tau| > \tau_b.$$

Because $\mathcal{R}_T(\tau)$ is even, $b(\tau)$ is even. Also, to avoid scale changes in the estimated spectrum, $b(\tau)$ is normalized so that $b(0) = 1$. Specification of either the lag window *or* spectral window uniquely defines the other because of the Fourier transform relation that exists between them. The choice of $b(\tau)$ [or $B(f)$] is determined by the criterion of fidelity and/or stability specified by the user. The reader wishing more information on the theory of spectral estimation should consult [10-4]–[10-7] and the relevant portions of Chapter 5, especially Sec. 5-5.

---

†It is not always desirable to avoid convolution. For example, the discrete Hamming window [Eq. (5.5-4)] involves little "number crunching" on a computer.

10.6 Input-Output Relations for Random Processes in Linear Systems

The *black-box* representation of a linear, time-invariant system is shown in Fig. 10.6-1. The response of the system to the excitation $x(t)$ is given by

$$y(t) = \int_{-\infty}^{\infty} h(u)\, x(t - u)\, du. \qquad (10.6\text{-}1)$$

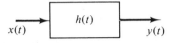

Figure 10.6-1. Representation of a linear system excited by the process $x(t)$.

We denote the correlation function associated with $y(t)$ as $R_y(\tau)$. Similarly, the correlation function associated with $x(t)$ will be written as $R_x(\tau)$. For $R_y(\tau)$, we compute

$$R_y(\tau) = \overline{y(t)y(t + \tau)} = E\left[\int_{-\infty}^{\infty} h(u)x(t - u)\, du \int_{-\infty}^{\infty} h(v)x(t + \tau - v)\, dv \right]$$

$$= \int_{-\infty}^{\infty} \int_{-\infty}^{\infty} h(u)h(v)\overline{x(t - u)x(t + \tau - v)}\, du\, dv$$

$$= \int_{-\infty}^{\infty} \int_{-\infty}^{\infty} h(u)h(v)R_x(\tau + u - v)\, du\, dv. \qquad (10.6\text{-}2)$$

If we let $z = v - u$, Eq. (10.6-2) can be written as

$$R_y(\tau) = \int_{-\infty}^{\infty} dz\, R_x(\tau - z) \int_{-\infty}^{\infty} du\, h(u)h(z + u). \qquad (10.6\text{-}3)$$

Equation (10.6-3) is a basic result and can be concisely written as

$$R_y(\tau) = \int_{-\infty}^{\infty} R_x(\tau - z)g(z)\, dz, \qquad (10.6\text{-}4)$$

where

$$g(z) \equiv \int_{-\infty}^{\infty} h(z + u)h(u)\, du. \qquad (10.6\text{-}5)$$

The function $g(z)$ is known as the filter autocorrelation function. The type of integral in Eq. (10.6-5) is sometimes called a correlation integral. Equation (10.6-4) says that the output correlation function of the process $y(t)$ is the convolution of the input-process correlation function and the filter auto-correlation. $g(z)$ is uniquely determined by $h(z)$, but the converse is not true. The output correlation function is seen to be a weighted average of the input correlation function. Equations (10.6-3)–(10.6-5) are the basic input-output relations for the second moments in the time (actually *time lag*) domain. Frequently it is simpler and more instructive to relate the input and output *power spectra.*

The Fourier transform of Eq. (10.6-4) furnishes

$$W_y(f) = W_x(f)G(f), \tag{10.6-6}$$

where $W_y(f) = \mathcal{F}[R_y(\tau)]$, $W_x(f) = \mathcal{F}[R_x(\tau)]$, and $G(f) = \mathcal{F}[g(\tau)]$. $G(f)$ is computed from

$$
\begin{aligned}
G(f) &= \int_{-\infty}^{\infty} d\tau\, e^{-j2\pi f\tau} \int_{-\infty}^{\infty} du\, h(u)h(u+\tau) \\
&= \int_{-\infty}^{\infty} du\, h(u) \int_{-\infty}^{\infty} d\tau\, h(u+\tau)e^{-j2\pi f\tau} \\
&= \int_{-\infty}^{\infty} h(u)e^{j2\pi fu}\, du \int_{-\infty}^{\infty} h(\alpha)e^{-j2\pi f\alpha}\, d\alpha \\
&= |H(f)|^2. \tag{10.6-7}
\end{aligned}
$$

Equations (10.6-6) and (10.6-7) are among the most important results in linear systems theory. They indicate that the output spectral density is simply the input spectral density multiplied by the squared magnitude of the transfer function of the system. Equation (10.6-6) is of course completely equivalent to Eq. (10.6-4). However, the fact that $g(z)$ is not uniquely related to $h(z)$ is more easily discernible from Eq. (10.6-7). Two transfer functions with the same magnitudes but different phases give rise to the same $G(f)$ and hence the same $g(z)$.

Sometimes one is interested in the correlation of two different processes. Such a quantity is called the *cross correlation* and, for real processes, is given by

$$R_{xy}(\tau) = \overline{x(t)y(t+\tau)}. \tag{10.6-8}$$

The cross-correlation function does not generally satisfy the conditions listed in Eq. (10.3-12). For the linear system with input response $h(t)$, the cross correlation for a delay τ between input, $x(t)$, and output, $y(t+\tau)$, is

$$
\begin{aligned}
R_{xy}(\tau) &= \int_{-\infty}^{\infty} \overline{x(t)x(u)}h(t+\tau-u)\, du \\
&= \int_{-\infty}^{\infty} R_x(\alpha)h(\tau+\alpha)\, d\alpha, \tag{10.6-9a} \\
&= \int_{-\infty}^{\infty} R_x(\alpha)h(\tau-\alpha)\, d\alpha. \tag{10.6-9b}
\end{aligned}
$$

In Eq. (10.6-9a) we let $\alpha = t - u$; in Eq. (10.6-9b) we let $\alpha = u - t$. The cross correlation between two outputs $y(t)$ and $z(t)$ stimulated by the same input (Fig. 10.6-2) is similarly computed to be

$$R_{yz}(\tau) = \int_{-\infty}^{\infty} R_x(\tau - z)g_{12}(z)\, dz, \qquad (10.6\text{-}10)$$

where

$$g_{12}(z) = \int_{-\infty}^{\infty} h_1(u)h_2(u + z)\, du. \qquad (10.6\text{-}11)$$

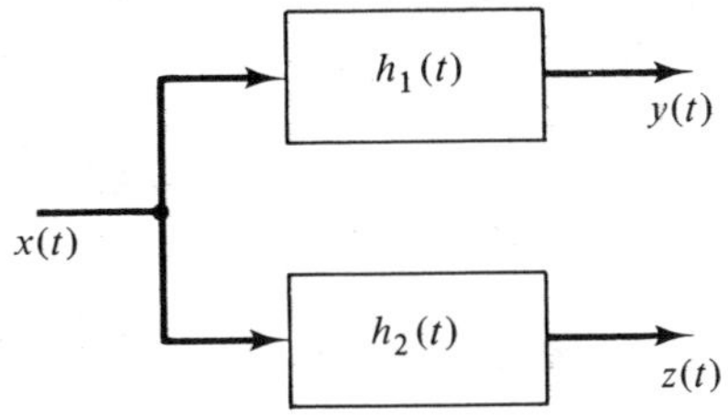

Figure 10.6-2. Two outputs produced by the same input.

The quantity $g_{12}(z)$ is known as the filter cross-correlation function, and its Fourier transform, $G_{12}(f)$, relates the *cross-spectral density* $W_{yz}(f)$ to $W_x(f)$ according to

$$W_{yz}(f) = G_{12}(f)W_x(f). \qquad (10.6\text{-}12)$$

From Eq. (10.6-11), $G_{12}(f) = H_1^*(f)H_2(f)$. In the special but important case of the cross-spectral density between the input and output of a system with transfer function $H(f)$, we obtain, from Eq. (10.6-9b),

$$W_{xy}(f) = H(f)W_x(f). \qquad (10.6\text{-}13)$$

Table 10.6-1 summarizes the most important results.

TABLE 10.6-1   BASIC INPUT-OUTPUT RELATIONS FOR RANDOM PROCESSES
IN LINEAR SYSTEMS

Relations	Time domain	Frequency domain
Input-output signals	$y(t) = \int_{-\infty}^{\infty} x(\tau)h(t - \tau)\, d\tau$ $= \int_{-\infty}^{\infty} x(t - \tau)h(\tau)\, d\tau$	$Y(f) = H(f)X(f)$
Input-output second-order moments and spectra	$R_y(\tau) = \int_{-\infty}^{\infty} R_x(u)g(\tau - u)\, du$ $g(\tau) = \int_{-\infty}^{\infty} h(u)h(u + \tau)\, du$	$W_y(f) = G(f)W_x(f)$ $G(f) = |H(f)|^2$
Cross-correlation function and cross spectra	$R_{xy}(\tau) = \int_{-\infty}^{\infty} R_x(u)h(\tau - u)\, du$	$W_{xy}(f) = H(f)W_x(f)$

10.7 The Gaussian Random Process

Consider a random process $x(t)$. Suppose we choose any k instants $t_1, \ldots,$ t_k. The corresponding k random variables $x(t_1), \ldots, x(t_k)$ constitute a random vector $\mathbf{X}$ given by

$$\mathbf{X} = [x(t_1), \ldots, x(t_k)]^T, \qquad (10.7\text{-}1)$$

where T here denotes transpose. Let $\mathbf{x}$ be a vector of k numbers in the range of $\mathbf{X}$ so that the event $\{x(t_1) \leq x_1, x(t_2) \leq x_2, \ldots, x(t_k) \leq x_k\}$ is written $\{\mathbf{X} \leq \mathbf{x}\}$. Then $x(t)$ is a Gaussian process if $\mathbf{X}$ has a jointly (multivariate) Gaussian pdf for every finite set of $\{t_i\}$ and every k.

The multivariate Gaussian pdf is given by

$$p(\mathbf{x}) = \frac{1}{(2\pi)^{k/2} |\det \mathbf{K}|^{1/2}} \exp\left[-\frac{1}{2} (\mathbf{x} - \boldsymbol{\mu})^T \mathbf{K}^{-1} (\mathbf{x} - \boldsymbol{\mu}) \right], \quad (10.7\text{-}2)$$

where $\mathbf{K}$ is the covariance matrix and $\boldsymbol{\mu}$ is the vector of means. Thus

$$\boldsymbol{\mu} = E[\mathbf{X}] = [\mu_1, \ldots, \mu_k]^T$$

$$\mathbf{K} = \begin{bmatrix} r_{11} & r_{12} & \cdots & r_{1k} \\ r_{12} & r_{22} & \cdots & \\ \cdot & & & \\ \cdot & & & \\ \cdot & & & \\ r_{1k} & & & \end{bmatrix}, \qquad (10.7\text{-}3)$$

where $r_{ij} \equiv r(t_i, t_j) = E\{[x(t_i) - \mu_i][x(t_j) - \mu_j]\}$.

Observation: Suppose the k variates are uncorrelated; i.e.,

$$\overline{[x(t_i) - \mu_i][x(t_j) - \mu_j]} = 0 \qquad \text{for } i \neq j.$$

Then

$$\mathbf{K} = \begin{bmatrix} \sigma_1^2 & & & 0 \\ & \sigma_2^2 & & \\ & & \cdot & \\ & & & \cdot \\ 0 & & & \sigma_k^2 \end{bmatrix} \equiv \text{diag}\,(\sigma_1^2, \ldots, \sigma_k^2), \qquad (10.7\text{-}4)$$

where $\sigma_i^2 = \overline{[x(t_i) - \mu_i]^2}$ and all off-diagonal entries are zero. In this case, the joint pdf $p(\mathbf{x})$ can be written as

$$p(\mathbf{x}) = \frac{1}{(2\pi)^{k/2}} \frac{1}{[\sigma_1^2 \cdots \sigma_k^2]^{1/2}} \exp\left[-\frac{1}{2} \sum_{i=1}^{k} \frac{x_i^2}{\sigma_i^2} \right]$$

$$= \frac{1}{(2\pi\sigma_1^2)^{1/2}} \exp\left[-\frac{1}{2} \frac{x_1^2}{\sigma_1^2} \right] \cdots \frac{1}{(2\pi\sigma_k^2)^{1/2}} \exp\left[-\frac{1}{2} \frac{x_k^2}{\sigma_k^2} \right]$$

$$= p(x_1)\,p(x_2) \cdots p(x_k). \qquad (10.7\text{-}5)$$

Thus in the Gaussian case, when random variables are uncorrelated they are also independent. This is in general not true for other probability laws.

Some well-known properties of the Gaussian process are

1. $p(\mathbf{x})$ depends only on $\boldsymbol{\mu}$, $\mathbf{K}$.
2. If $x(t_i)$, $i = 1, \ldots, k$, are jointly Gaussian, then each $x(t_i)$ is individually Gaussian.
3. If $\mathbf{K}$ is diag $(\sigma_1^2, \ldots, \sigma_k^2)$, then the $x(t_i)$, $i = 1, \ldots, k$, are independent.
4. Linear transformations on Gaussian r.v.'s yield Gaussian r.v.'s.
5. A wide-sense stationary Gaussian process is always strict-sense stationary.

Property 4 is a very important result. It says that if a Gaussian process is acted upon by a linear system, the output will be Gaussian. We demonstrate this in the following example.

EXAMPLE

Show that if $\mathbf{X}$ is Gaussian, then for nonsingular $\mathbf{A}$ the random vector $\mathbf{Y} = \mathbf{A}\mathbf{X} + \mathbf{b}$ is Gaussian.

Let $\mathbf{X}$ be any k r.v.'s from the process $x(t)$. Let $\mathbf{Y}$ be any k r.v.'s from the process $y(t)$. Then $\mathbf{X} = [X_1, \ldots, X_k]^T$, $\mathbf{Y} = [Y_1, \ldots, Y_k]^T$. Let $F_{\mathbf{Y}}(\mathbf{y}) \equiv P(\mathbf{Y} \leq \mathbf{y})$. Then

$$F_{\mathbf{Y}}(\mathbf{y}) = P[\mathbf{X} \leq \mathbf{A}^{-1}(\mathbf{y} - \mathbf{b})]$$
$$= F_{\mathbf{X}}[\mathbf{A}^{-1}(\mathbf{y} - \mathbf{b})],$$

where $F_{\mathbf{X}}(\mathbf{x}) = P(\mathbf{X} \leq \mathbf{x})$. $F_{\mathbf{Y}}(\mathbf{y})$ has the same form as $F_{\mathbf{X}}(\mathbf{x})$ except that the quadratic form is now

$$[\mathbf{A}^{-1}(\mathbf{y} - \mathbf{b}) - \boldsymbol{\mu}]^T \mathbf{K}^{-1}[\mathbf{A}^{-1}(\mathbf{y} - \mathbf{b}) - \boldsymbol{\mu}].$$

If we let $\boldsymbol{\alpha} \equiv \mathbf{A}^{-1}\mathbf{b} + \boldsymbol{\mu}$, the above has the form

$$(\mathbf{y}^T\mathbf{A}^{-1^T} - \boldsymbol{\alpha}^T)\mathbf{K}^{-1}(\mathbf{A}^{-1}\mathbf{y} - \boldsymbol{\alpha}) = [\mathbf{y}^T - \boldsymbol{\alpha}^T\mathbf{A}^T]\mathbf{A}^{-1^T}\mathbf{K}^{-1}\mathbf{A}^{-1}(\mathbf{y} - \mathbf{A}\boldsymbol{\alpha})$$
$$= (\mathbf{y} - \mathbf{A}\boldsymbol{\alpha})^T(\mathbf{A}^{-1^T}\mathbf{K}^{-1}\mathbf{A}^{-1})(\mathbf{y} - \mathbf{A}\boldsymbol{\alpha})$$
$$= (\mathbf{y} - \boldsymbol{\mu}')^T\mathbf{K}'^{-1}(\mathbf{y} - \boldsymbol{\mu}'). \tag{10.7-6}$$

Hence the new mean is $\boldsymbol{\mu} = \mathbf{b} + \mathbf{A}\boldsymbol{\mu}$, and the new covariance matrix is $\mathbf{K}' = \mathbf{A}\mathbf{K}\mathbf{A}^T$. Note that for a stationary Gaussian process $\boldsymbol{\mu}(t) = \boldsymbol{\mu}$ and $\mathbf{K}(t) = \mathbf{K}$; i.e., these quantities are not functions of time.

10.8   The Narrowband Gaussian Process (NGP)

When a sample function of the NGP is viewed on an oscilloscope, what appears is a waveform that resembles a sine wave with slowly varying amplitude and phase. For this reason, the NGP can be conveniently described† by

$$x(t) = V(t) \cos [2\pi f_c t + \phi(t)], \tag{10.8-1}$$

---

†This description is actually quite general, but if we want to associate $V(t)$ and $\phi(t)$ with a slowly varying envelope and phase, respectively, of a high-frequency carrier, then $x(t)$ should be a narrowband process.

where $V(t)$ is the slowly varying envelope and $\phi(t)$ is the slowly varying phase. A way to generate an NGP process is shown in Fig. 10.8-1.

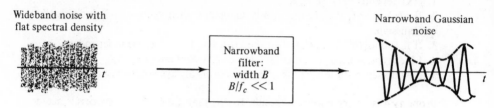

Wideband noise with flat spectral density

Narrowband filter: width B
$B/f_c \ll 1$

Narrowband Gaussian noise

Figure 10.8-1. Generation of the NGP.

Equation (10.8-1) can be written as

$$x(t) = x_c(t) \cos 2\pi f_c t - x_s(t) \sin 2\pi f_c t, \qquad (10.8\text{-}2)$$

where $\qquad x_c(t) = V(t) \cos \phi(t)$ (in-phase component),

$\qquad\qquad x_s(t) = V(t) \sin \phi(t)$ (quadrature component).

The zero-mean process $x(t)$ is band-pass with bandwidth B and centered at center frequency f_c. The modulating functions $x_c(t)$ and $x_s(t)$ are low-pass with bandwidth $B/2$. The process $x(t)$ can be viewed as the difference of two DSB waves in quadrature.

There are several ways to show that if $\overline{x^2(t)} \equiv \sigma^2$, then $\overline{x_c^2(t)} = \overline{x_s^2(t)} = \sigma^2$ and $\overline{x_c(t)x_s(t)} = 0$. One way is through the classical Rice approach described in [10-2], p. 158, in which $x(t)$ is written as a Fourier series over $[-T/2, T/2]$ and advantage is taken of the fact that the Fourier coefficients are Gaussian r.v.'s that become uncorrelated as $T \to \infty$.

Another approach is to assume the existence of fixed time T such that $1/f_c \ll T \ll 1/B$. This is not a rash assumption since in ordinary broadcast AM and FM there are, roughly, two orders of magnitude between the IF center frequency and the filter bandwidth. Under this assumption, we can compute $x_c(t)$ and $x_s(t)$ to a good approximation by treating them as Fourier coefficients and computing them likewise; i.e., to compute $x_c(t)$, multiply both sides of Eq. (10.8-2) by $\cos 2\pi f_c t$ and integrate over $[t - (T/2), t + (T/2)]$; to compute $x_s(t)$, multiply both sides of Eq. (10.8-2) by $\sin 2\pi f_c t$ and integrate over $[t - (T/2), t + (T/2)]$.† Thus

$$x_c(t) \simeq \frac{2}{T} \int_{t-T/2}^{t+T/2} x(u) \cos 2\pi f_c u \, du \qquad (10.8\text{-}3a)$$

and

$$x_s(t) \simeq -\frac{2}{T} \int_{t-T/2}^{t+T/2} x(u) \sin 2\pi f_c u \, du. \qquad (10.8\text{-}3b)$$

†In carrying out this procedure, the integrals which are discarded can be shown to be very small in comparison, in the mean-square sense, to the integral which is retained.

Now we can make several interesting observations. First, by taking the expectation of both sides in Eqs. (10.8-3) we obtain $\overline{x_c(t)} = \overline{x_s(t)} = 0$. Second, since $x_c(t)$ and $x_s(t)$ are obtained by *linear* operations on a Gaussian process, they are themselves Gaussian. Finally by using Eqs. (10.8-3a) and (10.8-3b) to directly compute $\overline{x_c^2(t)}$, $\overline{x_s^2(t)}$, and $\overline{x_c(t)x_s(t)}$, we obtain

$$\overline{x_c^2(t)} = \overline{x_s^2(t)} = \overline{x(t)^2} = \sigma^2 \tag{10.8-4}$$

and

$$\overline{x_c(t)x_s(t)} = 0. \tag{10.8-5}$$

(See Prob. 10-20.) We shall omit the details since the evaluation is a straightforward albeit tedious exercise in integration; the hardy reader is urged to check these results on his or her own. Putting all our results together, including the fact that $x_c(t)$ and $x_s(t)$ are zero-mean Gaussian processes, we find that their joint pdf is

$$f(x_{ct}, x_{st}) = \frac{1}{2\pi\sigma^2} \exp\left[-\left(\frac{x_{ct}^2 + x_{st}^2}{2\sigma^2}\right)\right], \tag{10.8-6}$$

where x_{ct} and x_{st} are numbers in the range of $x_c(t)$ and $x_s(t)$, respectively. The joint distribution function of $V(t)$ and $\phi(t)$ is obtained from

$$F(v, \phi) = P[V(t) \le v, \phi(t) \le \phi] = \iint\limits_{D} f(x_{ct}, x_{st})\, dx_{ct}\, dx_{st}, \tag{10.8-7}$$

where D is the shaded region in Fig. 10.8-2. The result is

$$F(v, \phi) = \int_0^{\phi} d\phi' \int_0^{v} \frac{1}{2\pi\sigma^2} e^{-\xi^2/2\sigma^2} \xi\, d\xi$$

$$= \begin{cases} \left(\dfrac{\phi}{2\pi}\right)(1 - e^{-v^2/2\sigma^2}), & v \ge 0, \quad 0 < \phi \le 2\pi \\ 0, & \text{otherwise.} \end{cases} \tag{10.8-8}$$

Figure 10.8-2. Region D required to compute $F(v, \phi)$.

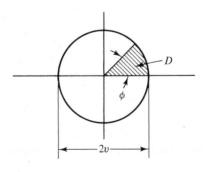

To derive Eq. (10.8-8) we used the transformation

$$\xi = (x_{ct}^2 + x_{st}^2)^{1/2} \quad \text{and} \quad \phi' = \tan^{-1} \frac{x_{st}}{x_{ct}}.$$

Equation (10.8-8) clearly shows that $\phi(t)$ and $V(t)$, for fixed t, are independent r.v.'s. The joint pdf is

$$f(v, \phi) = \frac{\partial^2 F(v, \phi)}{\partial v \, \partial \phi} = g(\phi) f(v) = \left(\frac{1}{2\pi}\right) \frac{v}{\sigma^2} e^{-v^2/2\sigma^2} u(v) \cdot [u(\phi) - u(\phi - 2\pi)],$$

$$(10.8\text{-}9)$$

where, for any ξ, $u(\xi)$ is the unit step starting at $\xi = 0$. We conclude that, for fixed but arbitrary t, $\phi(t)$ is an r.v. uniformly distributed in $[0, 2\pi]$ and $V(t)$ is an r.v. that obeys the Rayleigh probability law [Eq. (9.7-8), Chapter 9]. Figure 10.8-3 shows the pdf's.

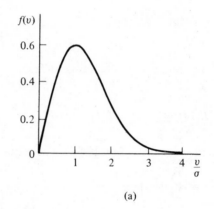

(a)

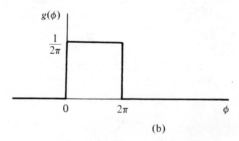

(b)

Figure 10.8-3. Probability density functions of (a) the envelope and (b) the phase of the NGP.

For a discussion and derivation of the joint densities of $V(t)$ and $V(t + \tau)$ and of $\phi(t)$ and $\phi(t + \tau)$, the reader is urged to consult Davenport and Root ([10-2], p. 161). The solution of this more difficult problem shows that although for fixed t, $V(t)$ and $\phi(t)$ are independent random variables, the joint pdf of $V(t)$, $V(t + \tau)$, $\phi(t)$, and $\phi(t + \tau)$ *cannot* be factored into pdf's depending only on the V's and ϕ's individually. Hence $V(t)$ and $\phi(t)$ are *not independent processes*.

10.9 Gaussian White Noise

So-called *white noise* is an unrealizable process, full of inherent contradictions, yet widely used in communication theory. The power spectrum of white noise is constant, i.e.,

$$W(f) = N_0, \qquad -\infty < f < \infty, \tag{10.9-1}$$

and its autocorrelation function is

$$R(\tau) = N_0 \delta(\tau). \tag{10.9-2}$$

For a stationary, zero-mean process the total power is given by

$$\sigma^2 = \int_{-\infty}^{\infty} W(f)\, df.$$

Hence white noise has infinite power.† Further, if $x(t)$ is a white-noise process, $x(t)$ and $x(t + \epsilon)$, for any fixed t and $\epsilon > 0$, are always independent r.v.'s no matter how small ϵ is. This means that any finite interval of a white-noise process contains an infinite number of independent random variables. From this observation, all types of bizarre implications follow upon which we shall not dwell. Nevertheless, despite all the inadequacies of this model, white noise is an extremely useful concept. A wideband Gaussian process that has uniform spectral density over the transmittance window of a filter can be considered a white-noise process for the sake of computations.

Many naturally occurring processes are modeled as white noise. Shot noise, semiconductor noise, and the spatial grain noise of high-resolution photographic film can all be modeled as white noise in many instances. However, in communication systems, it is thermal noise that is most often modeled as white noise. In fact thermal noise and white noise are frequently used interchangeably. The presence of thermal noise imposes fundamental limits on the performance of a communication system. For this reason we shall discuss it briefly below.

Thermal Noise

Thermal noise is generated by thermally induced interactions between charges flowing in conducting media. The most common situation is that of electrons in random motion in a resistor. J. B. Johnson and H. Nyquist

---

†The infinite power resulting from the postulation of white noise is called the *ultraviolet catastrophe*.

[10-8] studied thermal noise and showed, both from experimental and theoretical considerations, that the mean-squared noise voltage across a resistor of resistance R is given by

$$\overline{v^2(t)} = 4kTRB, \tag{10.9-3}$$

where T is the temperature in degrees Kelvin of the resistor, k is the Boltzmann constant (1.38×10^{-23} joule/$°K$), and B is any arbitrary bandwidth. The spectral density is then

$$W_0(f) = \frac{\overline{v^2(t)}}{2B} = 2kTR, \tag{10.9-4}$$

which is seen to be flat.

A more careful calculation that includes quantum-mechanical effects shows that the spectral density of the thermal voltage is ([10-9], p. 551)

$$W(f) = 2\left(\frac{hf}{2} + \frac{hf}{e^{hf/kT} - 1}\right)R, \tag{10.9-5}$$

where f is frequency and h is Planck's constant, $h = 6.6257 \times 10^{-34}$ joule/sec. The first term is negligible at frequencies $f \ll kT/h \simeq 10^{13}$ Hz. For $f \ll 10^{13}$ Hz, $\exp[hf/kT] \simeq 1 + hf/kT$ and $W(f)$ can be approximated by

$$W(f) = 2kTR. \tag{10.9-6}$$

This is the same result furnished by Eq. (10.9-4). Thus, for frequencies in use in normal communications, but not including optical or laser frequencies, the white-noise approximation is excellent. Because thermal noise is the result of a very large number of essentially independent interactions, its statistics tend to be Gaussian. This is assured by the *central-limit theorem* of statistics ([10-2], p. 81).

10.10   The Bilateral Clipper; Van Vleck's Theorem

Van Vleck's theorem states that the correlation function (and therefore the power spectrum) of a Gaussian process can be determined from the second-order statistics of the zero crossings of the process [10-10]. The result has significant implications for practical power spectra computations and can save much computer time. The computing burden can be reduced by a factor of from 5 to 7 [10-11]. To develop the main theorem, we first develop some preliminary results.

Let $x(t)$ be a zero-mean Gaussian random process. If $X \equiv x(t)$ and $Y \equiv x(t + \tau)$, then the correlation coefficient, or normalized covariance, ρ, is given by

$$\rho = \frac{\overline{XY}}{(\overline{X^2}\,\overline{Y^2})^{1/2}} = \frac{R(\tau)}{R(0)}. \tag{10.10-1}$$

From Fig. 10.10-1, we see that an *odd* number of zeros in the interval $(t, t + \tau)$ is simply the event $XY < 0$. Hence the probability $P_o(\tau)$, of an odd number of zeros in the interval $(t, t + \tau)$ is simply $P(XY < 0)$. The prob-

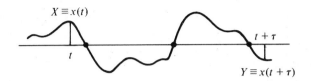

Figure 10.10-1. Zero crossings of a random process.

ability of the event $XY < 0$ or, equivalently, $X/Y < 0$ for X and Y Gaussian variates was treated in Sec. 9.11. Use of Eq. (9.11-15) gives

$$P_o(\tau) = \frac{1}{\pi} \cos^{-1} \rho$$

$$= \frac{1}{\pi} \cos^{-1} \frac{R(\tau)}{R(0)}. \qquad (10.10\text{-}2)$$

The probability of an *even* number of zeros is

$$P_e(\tau) = 1 - P_o(\tau) = 1 - \frac{1}{\pi} \cos^{-1} \frac{R(\tau)}{R(0)}. \qquad (10.10\text{-}3)$$

Equations (10.10-2) and (10.10-3) will be useful in what follows.

To remove random amplitude variations in waveforms in which information is stored in the zero crossings, a hard clipper can be used to generate square pulses which can subsequently be filtered to produce a uniform amplitude wave. A hard clipper is a nonlinear device whose action is shown in Fig. 10.10-2. Its transfer characteristic is described by

$$y = \begin{cases} 1, & x \geq 0 \\ -1, & x < 0. \end{cases} \qquad (10.10\text{-}4)$$

Let the input be a zero-mean Gaussian random process $x(t)$, and let $y(t)$ denote the output. Also let $R_y(\tau)$ and $R_x(\tau)$ denote the correlation functions

Figure 10.10-2. Bilateral hard clipper.

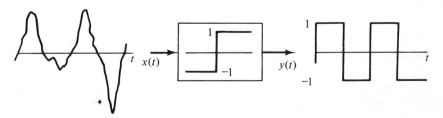

of $y(t)$ and $x(t)$, respectively. Let $Y_1 \equiv y(t)$ and $Y_2 \equiv y(t + \tau)$. Then

$$R_y(\tau) = P(Y_1 = 1, Y_2 = 1) + P(Y_1 = -1, Y_2 = -1)$$
$$-P(Y_1 = 1, Y_2 = -1) - P(Y_1 = -1, Y_2 = 1).$$

(10.10-5)

Now $P(Y_1 = 1, Y_2 = 1) + P(Y_1 = -1, Y_2 = -1)$ is simply the probability $P_e(\tau)$ that $x(t)$ makes an even number of zero crossings in τ. Similarly, $P(Y_1 = -1, Y_2 = 1) + P(Y_1 = 1, Y_2 = -1)$ is the probability that $x(t)$ makes an odd number of zero crossings in τ. Hence from Eqs. (10.10-2) and (10.10-3) we obtain

$$R_y(\tau) = \left[1 - \frac{1}{\pi} \cos^{-1} \frac{R_x(\tau)}{R_x(0)} \right] - \frac{1}{\pi} \cos^{-1} \frac{R_x(\tau)}{R_x(0)}$$

$$= 1 - \frac{2}{\pi} \cos^{-1} \frac{R_x(\tau)}{R_x(0)}. \tag{10.10-6}$$

With the help of a little diagram such as in Fig. 9.11-1, Eq. (10.10-6) can be written as

$$R_y(\tau) = \frac{2}{\pi} \sin^{-1} \left[\frac{R_x(\tau)}{\sigma^2} \right], \tag{10.10-7}$$

where $\sigma^2 \equiv R_x(0)$. Hence

$$R_x(\tau) = \sigma^2 \sin \left[\frac{\pi}{2} R_y(\tau) \right]. \tag{10.10-8}$$

Equation (10.10-8) is the key result of the theorem; it states that the correlation function $R_x(\tau)$ of a Gaussian process can be computed directly from the correlation function of the hard-clipped waveform $y(t)$.

10.11   Noise in AM and Derived Systems

In this and subsequent sections we shall compute the signal-to-noise ratios for several different modulation and detection schemes. It will be convenient to consider the information-bearing signal, $s(t)$, as an ergodic process so that the time average of $s^2(t)$, denoted by $\langle s^2(t) \rangle$, is equal to $\overline{s^2(t)}$. The average power associated with a signal such as $s(t) \cos 2\pi f_c t$ is then

$$\lim_{T \to \infty} \frac{1}{2T} \int_{-T}^{T} s^2(t) \cos^2 2\pi f_c t \, dt = \frac{\overline{s^2(t)}}{2}.$$

Although the input noise to the receiver need be neither Gaussian nor band-limited (in fact before the RF stage it typically is very broadband, and the signal-to-noise ratio there is poorly defined), the post-IF predetection noise is normally assumed narrowband and Gaussian. Depending on whether the modulation is AM (including DSB) or SSB, the effective IF bandwidth,

B_T, is $2W$ or W Hz, respectively. Since the carrier frequency is $f_c \gg B_T$ in either case, the narrowband condition holds. In what follows we shall ignore the RF stage since its primary function is the rejection of image frequencies. The two important signal-to-noise ratios are those that are computed (1) after the IF stage but before detection and (2) after detection. The former will be called the input signal-to-noise ratio and the latter the output or postdetection signal-to-noise ratio.

Synchronous Detection of DSBSC

The detection of DSBSC is shown in Fig. 10.11-1.

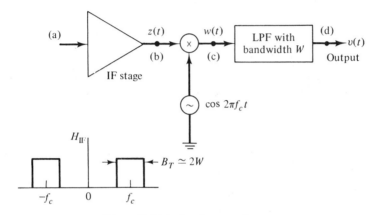

Figure 10.11-1. Synchronous detector.

The reader will recall from Chapter 6 that the spectrum of a DSBSC signal is essentially that of AM and consists of the two sidebands centered around the carrier frequency f_c [Fig. 6.2-2(b) with the carrier component removed]. For simplicity we assume that the filter has the ideal band-pass characteristic shown in Fig. 10.11-1; hence the spectral density of the noise inside the passband is constant. We let N_0 denote the two-sided noise spectral density, and assume that the bandwidth, B_T, of the IF stage is approximately $B_T \simeq 2W$. At point (b), the signal is a mixture of modulated wave and band-limited noise, $n(t)$. The signal $z(t)$ is given by

$$z(t) = A_c s(t) \cos 2\pi f_c t + n(t). \qquad (10.11\text{-}1)$$

As already stated, because $f_c/B_T \gg 1$, $n(t)$ can be considered to be an NGP. The noise power is given by†

$$\overline{n^2(t)} = \int_{-\infty}^{\infty} |H_{\text{IF}}(f)|^2 N_0 \, df = 2B_T N_0 = 4N_0 W, \qquad (10.11\text{-}2)$$

†Gain constants are arbitrarily set equal to unity because we are interested in *ratios* of signal power to noise power.

and the input signal-to-noise ratio is

$$\left(\frac{S}{\mathfrak{N}}\right)_i = \frac{P_s}{4N_0 W}, \tag{10.11-3}$$

where P_s, the power associated with the carrier-modulated signal, is given by

$$P_s = A_c^2 \overline{s^2(t)}/2 \quad \text{(DSB)}. \tag{10.11-4}$$

At (c) the signal is given by

$$w(t) = \frac{A_c s(t)}{2}(1 + \cos 4\pi f_c t) + n(t)\cos 2\pi f_c t. \tag{10.11-5}$$

However, in Sec. 10.8 we found that the NGP could be written as

$$n(t) = n_c(t)\cos 2\pi f_c t - n_s(t)\sin 2\pi f_c t. \tag{10.11-6}$$

Hence $w(t)$ can be written as

$$w(t) = \left[\frac{A_c s(t)}{2} + \frac{n_c(t)}{2}\right](1 + \cos 4\pi f_c t) - \frac{n_s(t)}{2}\sin 4\pi f_c t. \tag{10.11-7}$$

After the low-pass filter, the signal at (d) is

$$v(t) = \frac{A_c s(t)}{2} + \frac{n_c(t)}{2}, \tag{10.11-8}$$

so that only the in-phase noise component figures in the signal-to-noise ratio. The output, or postdetection, signal-to-noise ratio is

$$\left(\frac{S}{\mathfrak{N}}\right)_o = \frac{A_c^2 \overline{s^2(t)}}{\overline{n_c^2}} = \frac{P_s}{2N_0 W} = 2\left(\frac{S}{\mathfrak{N}}\right)_i, \tag{10.11-9}$$

where $\overline{n_c^2} = \overline{n_s^2} = \overline{n^2}$. Hence in the case of DSBSC modulation, the output signal-to-noise ratio is *twice* that of the input. Although DSB demodulation is simply a shifting of the sidebands to the origin of the frequency axis, the signal components add as amplitudes, while the noise adds as power. We say that the signals add *coherently*, while the noise adds *incoherently*. This is what accounts for the 3.0-dB increase in $(S/\mathfrak{N})$ ratio.

Synchronous Detection of SSB, VSB

The synchronous detector for SSB or VSB is the same as for DSBSC and is shown in Fig. 10.11-1. If the SSB signal is generated by admitting the upper sideband of the DSB signal $x(t) = A_c s(t)\cos 2\pi f_c t$ (the upper sideband is chosen for specificity) then the modulated SSB signal is

$$x(t) = \frac{A_c}{2}[s(t)\cos 2\pi f_c t - \hat{s}(t)\sin 2\pi f_c t], \tag{10.11-10}$$

where $\hat{s}(t)$ is the Hilbert transform of $s(t)$; i.e.,

$$\hat{s}(t) = \frac{1}{\pi}\int_{-\infty}^{\infty}\frac{s(\lambda)}{t - \lambda}\,d\lambda. \tag{10.11-11}$$

For simplicity, we again assume that the transmittance window of the IF stage is rectangular, as shown in Fig. 10.11-2. Referring now to Fig. 10.11-1, we find that the signal at (b) is

$$z(t) = \frac{A_c}{2}[s(t) \cos \omega_c t - \hat{s}(t) \sin \omega_c t]$$

$$+ n_c(t) \cos \omega_0 t - n_s(t) \sin \omega_0 t, \qquad (10.11\text{-}12)$$

where the switch to radian frequencies was made for brevity and where

$$\omega_0 = \omega_c + \pi W. \qquad (10.11\text{-}13)$$

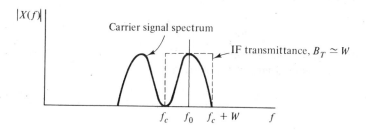

Figure 10.11-2. Transmittance of IF stage for SSB demodulation.

At point (c) we have

$$w(t) = \frac{A_c}{4} s(t)(1 + \cos 2\omega_c t) - \frac{A_c}{4} \hat{s}(t) \sin 2\omega_c t$$

$$+ \frac{n_c(t)}{2}[\cos (\omega_0 + \omega_c)t + \cos (\omega_0 - \omega_c)t]$$

$$- \frac{n_s(t)}{2}[\sin (\omega_0 + \omega_c)t + \sin (\omega_0 - \omega_c)t]. \qquad (10.11\text{-}14)$$

The low-pass filter rejects all components at frequencies greater than W Hz. Hence at (d) we obtain

$$v(t) = \frac{A_c}{4} s(t) + \frac{n_c(t)}{2} \cos (\omega_0 - \omega_c)t - \frac{n_s(t)}{2} \sin (\omega_0 - \omega_c)t. \qquad (10.11\text{-}15)$$

Since $\overline{s(t)n_c(t)} = \overline{s(t)n_s(t)} = \overline{n_c(t)n_s(t)} = 0$, the mean-square value of $v(t)$ is

$$\overline{v^2(t)} = \frac{A_c^2 \overline{s^2(t)}}{16} + \frac{\overline{n_c^2(t)}}{4}, \qquad (10.11\text{-}16)$$

where we have used the fact that $\overline{n_c^2(t)} = \overline{n_s^2(t)}$. For SSB, $\overline{n_c^2(t)} = 2B_T N_0$, so that the output signal-to-noise ratio is

$$\left(\frac{S}{\mathfrak{N}}\right)_0 = \frac{A_c^2 \overline{s^2(t)}/16}{2B_T N_0/4} \simeq \frac{P_s}{2WN_0} \qquad \text{(SSB)}, \qquad (10.11\text{-}17)$$

where, in this case, $P_s = A_c^2 \overline{s^2(t)}/4$ and $B_T \simeq W$.

The result for VSB is very similar to Eq. (10.11-17). That this is so can be seen by considering the equation of a VSB wave

$$x(t) = \frac{A_c}{2}[s(t) \cos \omega_c t - q(t) \sin \omega_c t], \qquad (10.11-18)$$

where $q(t)$ depends on the transmittance of the VSB filter [Eq. (6.9-7)]. The term proportional to $q(t) \sin \omega_c t$ is removed by the synchronous detector in much the same way as is the term proportional to $\hat{s}(t) \sin \omega_c t$ in the SSB case. If the width of the vestigial band is small compared to W, then the signal power and noise power admitted by the synchronous detector are about the same as in SSB. Hence to a good approximation

$$\left(\frac{S}{\mathfrak{N}}\right)_0 \simeq \frac{P_s}{2WN_0} \qquad \text{(VSB)}. \qquad (10.11-19)$$

The input signal-to-noise ratio for SSB is computed at point (b) in Fig. 10.11-1. The mean-square value of $z(t)$ is

$$\overline{z^2(t)} = \frac{A_c^2}{4}[\overline{s^2(t)}] + 2WN_0, \qquad (10.11-20)$$

so that

$$\left(\frac{S}{\mathfrak{N}}\right)_i = \frac{P_s}{2WN_0} = \left(\frac{S}{\mathfrak{N}}\right)_0 \qquad \text{(SSB, VSB)}. \qquad (10.11-21)$$

Hence in SSB/VSB modulation, the output signal-to-noise ratio is identical to the input signal-to-noise ratio. Although less noise is admitted by the reduced bandwidths in SSB/VSB, the output signal-to-noise ratios are no greater than in DSBSC. The reason for this has, in effect, already been mentioned: The coherent addition of signal in DSBSC offsets the additional noise power admitted by the IF stage.

In deriving Eq. (10.11-20) use was made of the fact that $\overline{s(t)\hat{s}(t)} = 0$ and $\overline{\hat{s}^2(t)} = \overline{s^2(t)}$. To show the former is straightforward. We write

$$\overline{s(t)\hat{s}(t)} = \frac{1}{\pi} \int_{-\infty}^{\infty} \frac{\overline{s(t)s(\lambda)}}{t - \lambda} \, d\lambda$$

$$= \frac{1}{\pi} \int_{-\infty}^{\infty} \frac{R(t - \lambda)}{t - \lambda} \, d\lambda$$

$$= \frac{1}{\pi} \int_{-\infty}^{\infty} \frac{R(u)}{u} \, du = 0 \qquad \text{(i.e., integrand is odd)}. \quad (10.11-22)$$

To show that $\overline{\hat{s}^2(t)} = \overline{s^2(t)}$ we need only take the expectation of the square of the integral in Eq. (10.11-11). Thus

$$\overline{\hat{s}^2(t)} = \frac{1}{\pi^2} \int_{-\infty}^{\infty} \int_{-\infty}^{\infty} \frac{\overline{s(\lambda)s(\xi)}}{(t - \lambda)(t - \xi)} \, d\xi \, d\lambda$$

$$= \int_{-\infty}^{\infty} d\alpha \, R_s(\alpha) \left[\frac{1}{\pi^2} \int_{-\infty}^{\infty} \frac{d\zeta}{\zeta(\zeta + \alpha)}\right], \qquad (10.11-23)$$

where we let $\alpha \equiv \lambda - \xi$ and $\zeta = t - \lambda$. It is not difficult to show that the integral in brackets is $\delta(\alpha)$. Hence

$$\overline{\hat{s}^2(t)} = \int_{-\infty}^{\infty} R_s(\alpha) \, \delta(\alpha) \, d\alpha = R_s(0)$$

$$= \overline{s^2(t)}. \qquad (10.11\text{-}24)$$

A somewhat less esoteric way of arriving at the same result is to recall that $\hat{s}(t)$ is formed from $s(t)$ by shifting all Fourier components in $s(t)$ by 90°. The mere shifting of the spectrum by a constant phase does not alter the net power in the waveform. Since the power depends on the square magnitude of the spectrum, and since only the phase is affected in constructing the Hilbert transform, Eq. (10.11-24) is established.

The synchronous detection of AM is left as an exercise (Prob. 10-24). AM, however, is usually detected with an envelope detector, which we shall consider next.

Detection of AM with an Envelope Detector

The AM envelope detector is shown in Fig. 10.11-3.

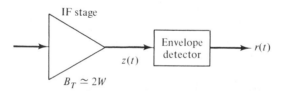

Figure 10.11-3. Envelope detector.

Signal Dominance

The simplest situation to deal with is when the input signal-to-noise ratio is high and the envelope detector is ideal. We shall assume this situation. After emerging from the IF stage, the signal is given by

$$z(t) = A_c[1 + ms(t)] \cos \omega_c t + n_c(t) \cos \omega_c t - n_s(t) \sin \omega_c t. \quad (10.11\text{-}25)$$

We can analyze the effect of the noise by considering a phasor representation for $z(t)$. This is furnished by

$$z(t) = \text{Re}\,[Z(t)]$$

$$Z(t) = Y(t)e^{j\omega_c t} \qquad (10.11\text{-}26)$$

$$Y(t) = A_c[1 + ms(t)] + n_c(t) + jn_s(t).$$

The phasor diagram is shown in Fig. 10.11-4. Only the complex amplitude $Y(t)$ is shown since $e^{j\omega_c t}$ conveys no information other than the fact that there is a superimposed rotation on $Y(t)$ of ω_c radians/sec. From Fig. 10.11-4, it

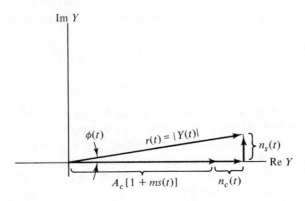

Figure 10.11-4. Phasor diagram for AM when $(\mathcal{S}/\mathfrak{N})_i \gg 1$.

is easily seen that $z(t)$ can be written as

$$z(t) = r(t) \cos [\omega_c t + \phi(t)], \tag{10.11-27}$$

where

$$r(t) = (\{A_c[1 + ms(t)] + n_c(t)\}^2 + n_s^2(t))^{1/2} \tag{10.11-28}$$

and

$$\phi(t) = \tan^{-1} \frac{n_s(t)}{A_c[1 + ms(t)] + n_c(t)}. \tag{10.11-29}$$

For large signal-to-noise ratios (also called signal dominance) we have $\overline{A_c^2[1 + ms(t)]^2} \gg \overline{n^2(t)}$. Under this condition we can write

$$r(t) = \{A_c[1 + ms(t)] + n_c(t)\}$$

$$\cdot \left(1 + \frac{n_s^2(t)}{\{A_c[1 + ms(t)] + n_c(t)\}^2}\right)^{1/2} \tag{10.11-30}$$

$$\simeq A_c[1 + ms(t)] + n_c(t), \qquad \left(\frac{\mathcal{S}}{\mathfrak{N}}\right)_i \gg 1. \tag{10.11-31}$$

The dc term in Eq. (10.11-31) does not convey information and is usually removed. Hence the detected signal is $A_c ms(t)$, the detected noise is the *in-phase* component of the Gaussian narrowband noise, and the output signal-to-noise ratio is

$$\left(\frac{\mathcal{S}}{\mathfrak{N}}\right)_0 = \frac{A_c^2 m^2 \overline{s^2(t)}}{\overline{n_c^2(t)}} \simeq \frac{A_c^2 m^2 \overline{s^2(t)}}{4WN_0}, \tag{10.11-32}$$

where we assumed, as in the DSB case, that the IF bandwidth $B_T \simeq 2W$.

The input signal-to-noise ratio is computed before the envelope detector. The mean-square value of $z(t)$ is

$$\overline{z^2(t)} = \tfrac{1}{2} A_c^2 [1 + m^2 \overline{s^2(t)}] + 4WN_0. \tag{10.11-33}$$

The input signal-to-noise ratio is therefore

$$\left(\frac{S}{\mathfrak{N}}\right)_i \simeq \frac{\frac{1}{2}A_c^2[1 + m^2\overline{s^2(t)}]}{4WN_0}$$

$$= \frac{P_c + 2P_{SB}}{4WN_0}, \qquad (10.11\text{-}34)$$

where P_c is the carrier power $= A_c^2/2$ and P_{SB} is the power in a sideband $= A_c^2 m^2 \overline{s^2(t)}/4$. From Eqs. (10.11-32) and (10.11-34) we can write

$$\left(\frac{S}{\mathfrak{N}}\right)_0 = \left(\frac{4P_{SB}}{P_c + 2P_{SB}}\right)\left(\frac{S}{\mathfrak{N}}\right)_i = \frac{2m^2\overline{s^2(t)}}{1 + m^2\overline{s^2(t)}}\left(\frac{S}{\mathfrak{N}}\right)_i. \qquad (10.11\text{-}35)$$

Clearly $(S/\mathfrak{N})_0$ cannot be larger than $(S/\mathfrak{N})_i$. The maximum sideband power for a full-load modulating tone, i.e., $ms(t) = \cos \omega_m t$, is $A_c^2/8$. In this case

$$\left(\frac{S}{\mathfrak{N}}\right)_0 = \frac{2}{3}\left(\frac{S}{\mathfrak{N}}\right)_i. \qquad (10.11\text{-}36)$$

It is more common, however, that $(S/\mathfrak{N})_0 \ll (S/\mathfrak{N})_i$. Comparing these results with the results for DSB, we see that because at least 50% of the radiated power goes into the carrier, more than twice as much AM power must be radiated to achieve the same $(S/\mathfrak{N})_0$ as in DSB. Hence AM is inferior in this respect to DSB.

Noise Dominance

When $(S/\mathfrak{N})_i \ll 1$, the envelope of the total wave is primarily determined by the envelope of the noise signal alone (Fig. 10.11-5). From the diagram, it can be seen that the envelope of the resultant is approximately given by

$$r(t) \simeq r_n(t) + A_c[1 + ms(t)] \cos \phi_n(t), \qquad (10.11\text{-}37)$$

where

$$\phi_n(t) = \tan^{-1}\frac{n_s(t)}{n_c(t)}. \qquad (10.11\text{-}38)$$

Figure 10.11-5. Phasor diagram for AM when $(S/\mathfrak{N})_i \ll 1$.

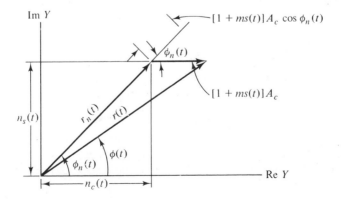

Interestingly, in this case we do not have a signal plus noise situation, and a conventional signal-to-noise calculation is meaningless. Since $\cos \phi_n(t)$ is a function of the noise angle $\phi_n(t)$ and is therefore a form of noise itself, the signal is corrupted by *multiplicative* noise and undergoes significant distortion. For example, when $\cos \phi_n(t) = 0$, the signal is not present at all. The conversion from masking by additive noise (which keeps the signal separate and intact) to corruption by multiplicative noise when $(\mathcal{S}/\mathcal{N})_i \ll 1$ is sometimes referred to as the *threshold effect* in AM envelope detection. It is possible to gain greater understanding of the threshold effect by considering what happens when an unmodulated carrier signal plus noise is detected by a square-law detector (Fig. 10.11-6). The noise is assumed to be NGP. The total input is written as

$$x(t) = [n_c(t) + A_c] \cos \omega_c t - n_s(t) \sin \omega_c t. \qquad (10.11\text{-}39)$$

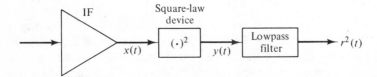

Figure 10.11-6. Block diagram of a square-law detector.

The output of the square-law detector is proportional to the squared envelope, which is

$$r^2(t) = [n_c(t) + A_c]^2 + n_s^2(t), \qquad (10.11\text{-}40)$$

and the mean-square value of the squared envelope is

$$\overline{r^4(t)} = \overline{\{[n_c(t) + A_c]^2 + n_s^2(t)\}^2}. \qquad (10.11\text{-}41)$$

How do we evaluate Eq. (10.11-41)? We already know from Sec. 10.8 that $\overline{n_c^2(t)} = \overline{n_s^2(t)} = \overline{n^2(t)} = \sigma^2$, that $n_c(t)$ and $n_s(t)$ are uncorrelated, and that $n_c(t)$ and $n_s(t)$ are Gaussian. But these facts imply that $n_c(t)$ and $n_s(t)$ are independent, and therefore so are $n_c^2(t)$ and $n_s^2(t)$ (Prob. 10-27). Hence $\overline{n_c^2(t)n_s^2(t)} = \overline{n_c^2(t)} \cdot \overline{n_s^2(t)} = \sigma^4$. Also from Eq. (9.10-46) with $n = 4$, we obtain $\overline{n_c^4(t)} = \overline{n_s^4(t)} = 3\sigma^4$. Putting all these results together and because the odd moments are zero, we can write

$$\overline{r^4(t)} = 8\sigma^4 + 8\sigma^2 A_c^2 + A_c^4. \qquad (10.11\text{-}42)$$

A reasonable definition of signal power for the square-law detector is the value of $\overline{r^4(t)}$ when noise is absent. The noise power would then be

$$\overline{r^4(t)} - \overline{r_s^4(t)} = 8\sigma^4 + 8\sigma^2 A_c^2, \qquad (10.11\text{-}43)$$

where $\overline{r_s^4(t)} = A_c^4$. The noise power thus consists of two terms: $8\sigma^4$ is the result of noise beating with noise, while $8\sigma^2 A_c^2$ is the result of noise beating with signal. Note that in envelope detection the presence of signal causes an increase in the noise. With the above definitions of signal and noise powers, we obtain

$$\left(\frac{S}{\mathfrak{N}}\right)_0 = \frac{A_c^4}{8\sigma^4 + 8A_c^2\sigma^2}.$$

However,

$$\frac{A_c^2}{2\sigma^2} = \frac{A_c^2/2}{4WN_0} = \left(\frac{S}{\mathfrak{N}}\right)_i,$$

from which it follows that

$$\left(\frac{S}{\mathfrak{N}}\right)_0 = \frac{1}{2} \cdot \frac{(S/\mathfrak{N})_i^2}{1 + 2(S/\mathfrak{N})_i}. \qquad (10.11\text{-}44)$$

Equation (10.11-44) sheds a good deal of light on the AM threshold effect. For $(S/\mathfrak{N})_i \ll 1$, we have

$$\left(\frac{S}{\mathfrak{N}}\right)_0 \simeq \frac{1}{2}\left(\frac{S}{\mathfrak{N}}\right)_i^2, \qquad (10.11\text{-}45)$$

i.e., a quadratic dependence on the input signal-to-noise ratio. For $(S/\mathfrak{N})_i \gg 1$, we have

$$\left(\frac{S}{\mathfrak{N}}\right)_0 \simeq \frac{1}{4}\left(\frac{S}{\mathfrak{N}}\right)_i, \qquad (10.11\text{-}46)$$

i.e., a linear dependence. The factor of $\frac{1}{4}$ results from considering a square-law device. A factor of $\frac{1}{2}$ would result from a piecewise linear detector. The threshold effect can thus be interpreted as a transition in $(S/\mathfrak{N})_0$ from a linear to a quadratic dependence $(S/\mathfrak{N})_i$. In other words, things get suddenly much worse when the input signal-to-noise ratio falls off below the threshold. This phenomenon is discussed at greater length in [10-2], p. 265, and [10-12], p. 103. Carlson ([10-13], p. 274) shows that a reasonable value of the threshold input signal-to-noise ratio is around 10. For other definitions of output signal-to-noise ratios including a discussion of the linear amplitude detector, see Panter ([10-14], p. 228).

The quadratic dependence of $(S/\mathfrak{N})_0$ on $(S/\mathfrak{N})_i$ when the input signal-to-noise ratio is small is known as the *small-signal suppression* of envelope detectors. Ordinary AM radios will sound very bad when $(S/\mathfrak{N})_i$ is anywhere near threshold; hence the large $(S/\mathfrak{N})_i$ situation is most common during normal listening.

Performance of AM and Derived Systems

We shall end this section by comparing and pointing out some of the salient features of the systems that have been discussed. In DSBSC, we found that synchronous detection furnished a factor of 2 (3 dB) in signal-to-noise improvement over the input signal-to-noise ratio, while no such improvement was manifest in SSB or VSB. Other things being equal, then,

DSBSC would seem to be slightly superior to SSB, assuming the power in the SSB signal is not raised. However, other things are not always equal, and SSB furnishes a significant bandwidth saving over DSB as well as greater ease of detection (see Secs. 6.7 and 6.8).

In ordinary AM there is actually a reduction in the output signal-to-noise ratio. In fact, the latter, even under conditions of a full-load modulating signal, is only two thirds as large as the input signal-to-noise ratio. More commonly, the modulation index satisfies $m \ll 1$, and the signal-to-noise performance of AM is much worse than that of either SSB or DSBSC, especially when a peak power constraint is applied. AM also suffers from a threshold effect which, although interesting, is not important under ordinary listening conditions. On the other hand, AM is easily detected with an envelope detector, a fact that accounts in great part for its leading role in commercial broadcasting. Table 10.11-1 summarizes some of the results of this section.

TABLE 10.11-1 COMPARISON OF AM-TYPE SYSTEMS

System	P_s	$r_i \equiv (\text{S}/\mathfrak{N})_i$	$r_0 \equiv (\text{S}/\mathfrak{N})_0$	r_0/r_i
DSBSC (synchronous det.)	$A_c^2 \overline{s^2}/2$	$P_s/4N_0W$	$P_s/2N_0W$	2
SSB, VSB (synchronous det.)	$A_c^2 \overline{s^2}/4$	$P_s/2N_0W$	$P_s/2N_0W$	1
AM (envelope det.)	$A_c^2(1 + m^2\overline{s^2})/2$	$P_s/4N_0W$	$m^2 \dfrac{\overline{s^2}A_c^2}{4N_0W}$	0.67 (max)

10.12 Noise in Angle-Modulated Systems

A general treatment of the effect of noise on angle-modulation systems is an exceedingly tedious affair, especially when input signal-to-noise ratios are not too large or too small. However, at the two extremes, i.e., very large or very small signal-to-noise ratios, a number of qualitative and quantitative results can be stated. We shall begin our analysis by recalling that an angle-modulated wave can be written as

$$x(t) = A_c \cos [\omega_c t + \phi(t)] \qquad \text{(FM or PM)}, \qquad (10.12\text{-}1)$$

where

$$\phi(t) = K's(t) \qquad (10.12\text{-}2)$$

in the case of PM and

$$\phi(t) = 2\pi K \int^t s(\lambda) \, d\lambda \qquad (10.12\text{-}3)$$

in the case of FM.

The important circuits in an FM demodulator are shown in Fig. 10.12-1. The signal $y(t)$ is angle-modulated carrier plus noise; i.e.,

$$y(t) = A_c \cos[\omega_c t + \phi(t)] + n(t). \tag{10.12-4}$$

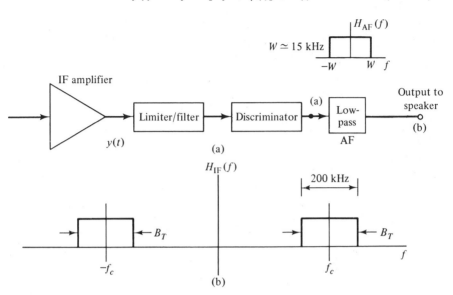

Figure 10.12-1. (a) Demodulator for FM; (b) assumed rectangular characteristic of IF filter.

The noise $n(t)$ is still an NGP since $B_T/f_c \ll 1$. The input signal-to-noise ratio is therefore

$$\left(\frac{S}{\mathfrak{N}}\right)_i = \frac{A_c^2/2}{2B_T N_0}, \tag{10.12-5}$$

which is independent of the signal $s(t)$. The computation of the output signal-to-noise ratio is more difficult. It is considered below.

Because $n(t)$ is narrowband, it is useful to write

$$n(t) = r_n(t) \cos[\omega_c t + \phi_n(t)], \tag{10.12-6}$$

where, as usual, $r_n(t)$ is Rayleigh distributed and $\phi_n(t)$ is uniformly distributed in $[-\pi, \pi]$. For $y(t)$ we write

$$y(t) = r(t) \cos[\omega_c t + \psi(t)], \tag{10.12-7}$$

where

$$r(t) = \{[A_c \cos \phi(t) + r_n(t) \cos \phi_n(t)]^2 + [A_c \sin \phi(t) + r_n(t) \sin \phi_n(t)]^2\}^{1/2} \tag{10.12-8}$$

and

$$\psi(t) = \tan^{-1} \frac{A_c \sin \phi(t) + r_n(t) \sin \phi_n(t)}{A_c \cos \phi(t) + r_n(t) \cos \phi_n(t)}. \tag{10.12-9}$$

As a check, we see that if there is no noise, i.e., $r_n(t) = 0$, then $\psi(t) = \phi(t)$. An ideal limiter would remove the envelope fluctuations in the input so that $r(t)$ is of no consequence. Hence in angle modulation, signal-to-noise ratios are derived from consideration of $\psi(t)$ only. Unfortunately the expression for $\psi(t)$ is too unwieldy for analysis. For this reason we shall do what we did in the envelope detector analysis—consider the two separate cases of signal dominance and noise dominance individually.

Signal Dominance, $A_c^2/2 \gg \overline{r_n^2}$

A phasor diagram (Fig. 10.12-2) for this situation is obtained from

$$y(t) = \text{Re} \left[Y(t)e^{j\omega_c t} \right], \tag{10.12-10}$$

where $Y(t)$ is given by

$$Y(t) = A_c e^{j\phi(t)} + r_n(t)e^{j\phi_n(t)}.$$

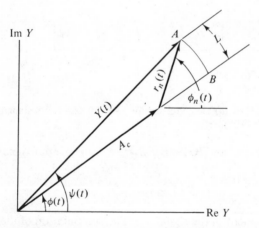

Figure 10.12-2. Phasor diagram for angle-modulated wave plus noise. The length of the vectors reflect the signal-dominance case.

From Fig. 10.12-2, it is clear that the length, L, of arc AB is

$$L = Y(t)[\psi(t) - \phi(t)]. \tag{10.12-11}$$

But $Y(t) \simeq A_c + r_n(t) \cos [\phi_n(t) - \phi(t)] \simeq A_c$, and $L \simeq r_n(t) \sin [\phi_n(t) - \phi(t)]$. Hence from Eq. (10.12-11) we obtain

$$\psi(t) = \phi(t) + \frac{r_n(t)}{A_c} \sin [\phi_n(t) - \phi(t)]. \tag{10.12-12}$$

Since for fixed t, $\phi_n(t)$ is an r.v. uniformly distributed in $[-\pi, \pi]$, we argue that $\phi_n(t) - \phi(t)$ is also uniformly distributed over $[-\phi(t) - \pi, -\phi(t) + \pi]$. For the purpose of computing an output signal-to-noise ratio, replacing

$\phi_n(t) - \phi(t)$ by $\phi_n(t)$ does not affect the computation. Therefore we permit ourselves to write

$$\psi(t) = \phi(t) + \frac{r_n(t)}{A_c} \sin \phi_n(t)$$

$$= \phi(t) + \frac{n_s(t)}{A_c}, \tag{10.12-13}$$

where $n_s(t) \equiv r_n(t) \sin \phi_n(t)$ is the quadrature component of Gaussian narrowband noise. The detected signal is

$$v(t) = \psi(t) = K's(t) + \frac{n_s(t)}{A_c} \qquad \text{(PM),} \qquad \text{(10.12-14a)}$$

and

$$v(t) = \frac{1}{2\pi} \dot{\psi}(t) = Ks(t) + \frac{1}{2\pi A_c} \dot{n}_s(t) \qquad \text{(FM).} \qquad \text{(10.12-14b)}$$

Power Spectrum of $n_s(t)$ and $\dot{n}_s(t)$

To proceed further it is helpful to consider the power spectrum of $n_s(t)$. Starting with Eqs. (10.8-3) and identifying $n_c(t), n_s(t)$ with $x_c(t)$ and $x_s(t)$, respectively, it is a simple matter to show that the autocorrelation of $n_s(t)$ is (Prob. 10-26)

$$R_{n_s}(\tau) = 2 \int_0^\infty W_n(\xi) \cos 2\pi(\xi - f_c)\tau \, d\xi$$

$$= R_{n_c}(\tau), \tag{10.12-15}$$

where $W_n(\xi)$ is the power spectrum of $n(t)$ and $R_{n_c}(\tau)$ is the autocorrelation function of $n_c(t)$. If we multiply both sides of Eq. (10.12-15) by $e^{-j2\pi f\tau} \, d\tau$ and integrate, we obtain [$u(f)$ is the unit-step function]

$$W_{n_s}(f) = W_n(f + f_c)u(f + f_c) + W_n(f_c - f)u(f_c - f)$$

$$= W_{n_c}(f). \tag{10.12-16}$$

Equation (10.12-16) can be used to confirm that $n_s(t)$ is a low-pass process. Thus, representing the power spectrum of $n(t)$ by

$$W_n(f) = N_0 \operatorname{rect}\left(\frac{f - f_c}{B_T}\right) + N_0 \operatorname{rect}\left(\frac{f + f_c}{B_T}\right), \tag{10.12-17}$$

we obtain, by direct substitution into Eq. (10.12-16),

$$W_{n_s}(f) = 2N_0 \operatorname{rect}\left(\frac{f}{B_T}\right), \tag{10.12-18}$$

so that

$$\overline{n_s^2(t)} = 2N_0 B_T,$$

as expected.

The power spectrum of $\dot{n}_s(t)$ can be determined from an elementary property of Fourier transforms. Recall that if an arbitrary function $x(t)$ has Fourier transform $X(f)$, then at points where $x(t)$ is continuous,

$$\dot{x}(t) \longleftrightarrow +j2\pi f X(f).$$

Hence $\dot{x}(t)$ can be obtained by passing $x(t)$ through a network with transfer function $H(f) = +j2\pi f$. From Table 10.6-1 we obtain, with $y \equiv \dot{x}$ and $H(f) = j2\pi f$,

$$W_{\dot{x}}(f) = |H(f)|^2 W_x(f)$$
$$= (2\pi f)^2 W_x(f). \tag{10.12-19}$$

Equation (10.12-19) enables us to write

$$W_{\dot{n}_s}(f) = 4\pi^2 f^2 W_{n_s}(f),$$

which, in the case of a rectangular band-pass characteristic, gives

$$W_{\dot{n}_s} = 8\pi^2 f^2 N_0 \operatorname{rect}\left(\frac{f}{B_T}\right). \tag{10.12-20}$$

Hence $\overline{\dot{n}_s^2}(t) = \frac{2}{3}\pi^2 N_0 B_T^3$.

Output Signal-to-Noise Ratio $(S/\mathfrak{N})_0$

With $v_1(t) \equiv n_s(t)/A_c$ and $v_2(t) \equiv \dot{n}_s(t)/2\pi A_c$, we can write Eqs. (10.12-14a) and (10.12-14b) as

$$v(t) = K's(t) + v_1(t) \quad \text{(PM)}, \tag{10.12-21a}$$

$$v(t) = Ks(t) + v_2(t) \quad \text{(FM)}, \tag{10.12-21b}$$

where the power spectra of $v_1(t)$, $v_2(t)$ are, respectively,

$$W_1(f) = 2\frac{N_0}{A_c^2} \operatorname{rect}\left(\frac{f}{B_T}\right) \quad \text{(PM)} \tag{10.12-22}$$

and

$$W_2(f) = 2\left(\frac{f}{A_c}\right)^2 N_0 \operatorname{rect}\left(\frac{f}{B_T}\right) \quad \text{(FM)}. \tag{10.12-23}$$

These are the spectra that would be observed at point (a) in Fig. 10.12-1. The spectra are shown in Fig. 10.12-3.

Since $B_T/2$ is generally much greater than the highest frequency, W, in $s(t)$, additional filtering helps to increase $(S/\mathfrak{N})_0$. Thus the final output of the demodulator is obtained at point (b) in Fig. 10.12-1(a) after passing the signal through a low-pass filter whose bandwidth is essentially W. Hence the noise power at the speaker input is

$$\mathfrak{N}_0 = \int_{-W}^{W} W_1(f)\,df = \frac{4N_0 W}{A_c^2} \quad \text{(PM)}, \tag{10.12-24}$$

$$\mathfrak{N}_0 = \int_{-W}^{W} W_2(f)\,df = \frac{4N_0 W^3}{3A_c^2} \quad \text{(FM)}. \tag{10.12-25}$$

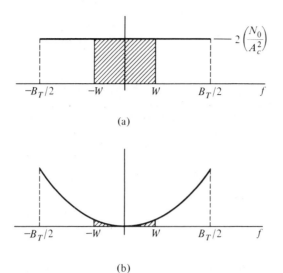

Figure 10.12-3. Detected noise spectra at output of discriminator: (a) PM noise spectrum; (b) FM noise spectrum. W is the bandwidth of the final audio low-pass amplifier.

With $P_c \equiv A_c^2/2$ representing the power in the input carrier signal, the final signal-to-noise ratios are

$$\left(\frac{S}{\mathfrak{N}}\right)_0 = P_c \frac{K'^2 \overline{s^2(t)}}{2N_0 W} \qquad \text{(PM)}, \qquad (10.12\text{-}26)$$

$$\left(\frac{S}{\mathfrak{N}}\right)_0 = 3P_c \left(\frac{K}{W}\right)^2 \frac{\overline{s^2(t)}}{2N_0 W} \qquad \text{(FM)}. \qquad (10.12\text{-}27)$$

We already know that because $|s(t)| \leq 1$, K represents the maximum frequency deviation, in hertz, of the FM signal. The ratio K/W is the parameter β^* of Sec. 7.5, i.e., the modulation index that is associated with maximum frequency deviation and highest signal frequency, W. Hence Eq. (10.12-27) can be written as

$$\left(\frac{S}{\mathfrak{N}}\right)_0 = 3P_c \beta^{*2} \frac{\overline{s^2(t)}}{2N_0 W}. \qquad (10.12\text{-}28)$$

The transmission bandwidth, B_T, is given by

$$B_T = 2W(2 + \beta^*),$$

using a conservative measure (Sec. 7.5). For wideband FM, $\beta^* \gg 1$ and $B_T \simeq 2W\beta^*$. Hence Eq. (10.12-27) can also be written in the following revealing form:

$$\left(\frac{S}{\mathfrak{N}}\right)_0 \simeq \frac{3}{4} \left(\frac{B_T}{W}\right)^2 \frac{P_c \overline{s^2(t)}}{2N_0 W} \qquad \text{(WBFM)}. \qquad (10.12\text{-}29)$$

Equation (10.12-29) states that, provided $(S/\mathfrak{N})_i \gg 1$, the output signal-to-noise ratio is a quadratic function of the *bandwidth expansion ratio* B_T/W. Increasing B_T furnishes a corresponding *quadratic* increase in $(S/\mathfrak{N})_0$. This is why wideband FM is so superior to narrowband FM. In commercial broadcast FM, $\beta^* \simeq 5$, so the results for WBFM are basically applicable to commercial broadcasting.

Last, it is instructive to compare the performance of FM with AM schemes. In particular we found that in the case of DSB,

$$\left(\frac{S}{\mathfrak{N}}\right)_{0,\,\text{DSB}} = \frac{(A_c^2/2)\overline{s^2(t)}}{2N_0 W} = P_c \frac{\overline{s^2(t)}}{2N_0 W}.$$

Accordingly, from Eq. (10.12-28),

$$\left(\frac{S}{\mathfrak{N}}\right)_{0,\,\text{FM}} = 3\beta^{*2}\left(\frac{S}{\mathfrak{N}}\right)_{0,\,\text{DSB}}. \qquad (10.12\text{-}30)$$

Hence WBFM can furnish a very significant increase in performance over AM systems under the constraint of equal power. The penalty is, of course, that FM requires much greater bandwidth.

Further Signal-to-Noise Improvement in FM by Preemphasis/Deemphasis (PDE) Filtering

In Sec. 7.11 we discussed the use of preemphasis (PE) and deemphasis (DE) filtering to reduce the effects of high-frequency interference and noise. We are now in a position to be somewhat more quantitative about the effects of PDE on noise in FM.

The preemphasis filtering distorts the signal somewhat but doesn't significantly change the signal bandwidth from W. The deemphasis filter restores the signal by high-frequency attenuation. The reader will recall that the transfer function of the deemphasis filter is proportional to

$$H_{\text{DE}}(f) = \frac{1}{1 + jf/f_1},$$

where $f_1 = (2\pi R_1 C)^{-1}$ and $R_1 C$ is the time constant of the filter shown in Fig. 7.11-2. In the presence of a DE filter, Eq. (10.12-25) must be modified to

$$[\mathfrak{N}_0]_{\text{DE}} = \int_{-W}^{W} W_2(f)\,|H_{\text{DE}}(f)|^2\,df$$

$$= \frac{N_0 f_1^3}{P_c} \int_{-W/f_1}^{W/f_1} \frac{f^2}{1 + f^2}\,df$$

$$= \frac{2N_0 f_1^3}{P_c}\left(\frac{W}{f_1} - \tan^{-1}\frac{W}{f_1}\right). \qquad (10.12\text{-}31)$$

Standard values for f_1 and W are 2.1 Hz and 15 kHz, respectively. Therefore $W/f_1 \simeq 7.2$ and $\tan^{-1} W/f_1 \simeq \pi/2$. Hence, to a first approximation, we can assume that $W/f_1 \gg \tan^{-1} W/f_1$. The output noise power with deemphasis filtering is then given by

$$(\mathfrak{N}_0)_{\mathrm{DE}} = \frac{2N_0 f_1^2 W}{P_c}. \qquad (10.12\text{-}32)$$

The ratio of signal-to-noise power with PDE to signal-to-noise power without PDE is

$$\frac{(\mathcal{S}/\mathfrak{N})_{0,\,\mathrm{PDE}}}{(\mathcal{S}/\mathfrak{N})_0} = \frac{1}{3}\left(\frac{W}{f_1}\right)^2. \qquad (10.12\text{-}33)$$

For the numbers given above, this amounts to $\simeq 12$ dB, an impressive gain in signal-to-noise ratio by any standard.

Noise Dominance, $A_c^2/2 \ll \overline{r_n^2}$

Returning now to Eq. (10.12-10), we find that when $A_c \ll [2\overline{r_n^2(t)}]^{1/2}$, the resulting phasor diagram is essentially dominated by the term $r_n(t)e^{j\phi_n(t)}$ (Fig. 10.12-4). Following the same line of reasoning that led up to Eq. (10.12-12), we obtain, for this case,

$$\psi(t) = \phi_n(t) + \frac{A_c}{r_n(t)}\sin[\phi(t) - \phi_n(t)]. \qquad (10.12\text{-}34)$$

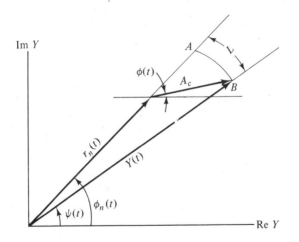

Figure 10.12-4. Phasor diagram for angle-modulated wave plus noise when noise dominates.

The dominant term is $\phi_n(t)$, the uniformly distributed phase of the narrowband noise. The second term is a highly nonlinear function of signal and noise, and since the signal doesn't stand alone, no meaningful signal-to-noise ratio can be computed. This case corresponds then to signal *obliteration*. The transition from signal obliteration for low $(\mathcal{S}/\mathfrak{N})_i \ll 1$ to a linear dependence of $(\mathcal{S}/\mathfrak{N})_0$ on P_c [Eq. (10.12-29)] when $(\mathcal{S}/\mathfrak{N})_i \gg 1$ is referred to as the

FM threshold effect. It is easy to gain the impression from Eq. (10.12-29) that indefinite improvement in $(S/\mathfrak{N})_0$ is possible by simply increasing B_T [and hence K since $B_T \simeq 2(K + 2W)$] without bound. However, increasing B_T also increases $\mathfrak{N}_i = 2B_T N_0$, and eventually $(S/\mathfrak{N})_i$ will fall below the threshold, in which case signal obliteration occurs.

The precise threshold signal-to-noise value depends on the criterion that is used. For example, suppose that the threshold criterion is determined by that level of carrier amplitude that exceeds the noise amplitude 99 % of the time. Then

$$P[r_n(t) \leq A_c] = 1 - P[r_n(t) > A_c] = 0.99$$

or

$$P[r_n(t) > A_c] = 0.01 = e^{-(S/\mathfrak{N})_i}. \qquad (10.12\text{-}35)$$

Equation (10.12-35) follows from the fact that $r_n(t)$ has a Rayleigh distribution, that

$$P[r_n(t) < A_c] = 1 - (1 - e^{-A_c^2/2\sigma^2}),$$

and that

$$\frac{A_c^2}{2\sigma^2} = \frac{A_c^2/2}{2B_T N_0} = \left(\frac{S}{\mathfrak{N}}\right)_i. \qquad (10.12\text{-}36)$$

The solution to Eq. (10.12-35) gives a 7-dB threshold level for the input carrier-to-noise power. Experimental results show that the threshold level is in the vicinity of 12–15 dB (Fig. 10.12-5).

Figure 10.12-5. Output signal-to-noise ratio as a function of input carrier-to-noise ratio. *Adapted from M.G. Crosby [10-15], with permission.*

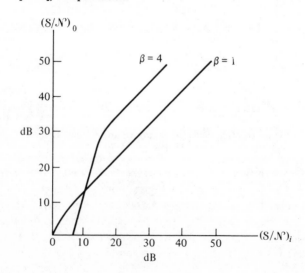

Techniques for lowering the threshold and thereby increasing the performance of FM receivers have been proposed. These techniques, called FM threshold extensions (FMTE), use frequency feedback or phase-locked loop detectors to extend the threshold. A thorough discussion of FMTE techniques is furnished by Panter [10-14], p. 478.

10.13 Summary

In this chapter we have studied random signals and considered the response of linear, time-invariant (LTI) systems to such signals. We introduced two extremely important functions—the correlation function and the power spectrum—and showed that they form a Fourier pair. We derived the input-output relations for correlation functions and power spectra for LTI systems.

The Gaussian process and the narrowband Gaussian process (NGP) were carefully studied. The latter figures prominently in the signal-to-noise analysis of communication systems. The signal-to-noise ratios for various types of AM and angle-modulated systems were considered. The quadratic noise characteristic at the output of FM discriminators was derived.

We ended the chapter by showing how FM preemphasis/deemphasis filtering, introduced in Chapter 7, can be used to improve the signal-to-noise ratio. Finally, the FM threshold effect was considered. FM systems are discussed in a number of books; among them are the very readable books by Schwartz [10-9] and Carlson [10-13]. Panter's book [10-14] has a very complete discussion of FM interference and other important problems.

PROBLEMS

10-1. In the experiment of throwing a fair die, define a random process $x(t)$ by

$$x(t) = kt^2,$$

where k is the outcome of a throw, $k = 1, \ldots, 6$. Compute the probability distribution function for $x(1)$ and $x(2)$. Is $x(t)$ a stationary process?

10-2. With reference to the die-throwing experiment, define a process by

$$x(t) = \cos\left(\frac{2\pi k}{6}\right)t,$$

where $k = 1, \ldots, 6$. Compute the probability distribution function for $x(1)$ and $x(2)$. Is $x(t)$ a stationary process?

10-3. Let $x(t)$ denote the "telegraph" signal of Sec. 10.4. Let α be a random variable that is independent of $x(t)$. Compute the autocorrelation function of $y(t) = \alpha x(t)$.

10-4. Let $x(t)$ be a random process given by

$$x(t) = A \cos(\omega t + \Theta),$$

where A is a constant and Θ is a uniformly distributed random variable in $(0, \pi/2)$. Compute $\overline{x(t)}$ and $R_{xx}(t, t + \tau) = \overline{x(t)x(t + \tau)}$. Is $x(t)$ a wss process?

10-5. A random process has sample functions $x(t) = X$, where X is a random variable not indexed by time. Let X be defined by

$$X = \begin{cases} \alpha, & \text{with probability } P(\alpha) = p \\ \beta, & \text{with probability } P(\beta) = 1 - p. \end{cases}$$

a. With $\hat{\mu}$ and μ denoting the time and statistical averages of $x(t)$, respectively, show that $\hat{\mu} \neq \mu$ and that therefore $x(t)$ is not ergodic. Are the time averages for different sample functions of $x(t)$ the same?
b. Compute the correlation function of $x(t)$.

10-6. A warehouse contains a very large number of oscillators that fall into one of two classes: those whose mean frequency is 1.0 kHz and those whose mean frequency is 1.0 MHz. Let $x(t)$ denote the output of an oscillator. Is $x(t)$ a stationary process? (Assume oscillators have been on for a very long, i.e., infinite time.) Is $x(t)$ an ergodic process?

10-7. Let $x(t)$ be a (real) stationary random process with correlation function $R_{xx}(\tau)$. Assuming that the various derivatives exist, show that

$$R_{x\dot{x}}(\tau) \equiv \overline{x(t)\dot{x}(t + \tau)}$$

is given by

$$R_{x\dot{x}}(\tau) = \frac{dR_{xx}(\tau)}{d\tau}.$$

10-8. Show that the correlation function of the process $\dot{x}(t)$, obtained by differentiating $x(t)$ in Prob. 10-7, is given by

$$R_{\dot{x}\dot{x}}(\tau) \equiv \overline{\dot{x}(t)\dot{x}(t + \tau)}$$

$$= -\frac{d^2 R_{xx}(\tau)}{d\tau^2}.$$

10-9. Generalize the result in Prob. 10-8 to the process

$$x^{(n)}(t) = \frac{d^n x(t)}{dt^n}.$$

10-10. Generalize the result in Prob. 10-8 to include nonstationary processes by writing

$$R_{\dot{x}\dot{x}}(t_1, t_2) = \overline{\dot{x}(t_1)\dot{x}(t_2)}$$

and showing that

$$R_{\dot{x}\dot{x}}(t_1, t_2) = \frac{\partial^2 R_{xx}(t_1, t_2)}{\partial t_1 \partial t_2}.$$

10-11. The truncated time average of a process $x(t)$ is

$$\hat{\mu}_T \equiv \frac{1}{2T} \int_{-T}^{T} x(t) \, dt.$$

a. Show that the variance of the r.v. $\hat{\mu}_T$ is given by

$$\sigma_{\hat{\mu}_T}^2 = \frac{1}{T} \int_0^{2T} \left(1 - \frac{\tau}{2T}\right)[R(\tau) - \mu^2]\, d\tau,$$

where μ is the expected value of $x(t)$. If $\sigma_{\hat{\mu}_T}^2 \rightarrow 0$ as $T \rightarrow \infty$, we say that $x(t)$ *is ergodic in the mean*.

b. Let $x(t) = \cos(\omega t + \Theta)$, where Θ is uniformly distributed in $[0, \pi]$. Is $x(t)$ ergodic?

10-12. Show that proving the ergodicity of the "correlation function"

$$\hat{R}_T(\tau) = \frac{1}{2T} \int_{-T}^{T} x(t)x(t + \tau)\, dt$$

requires knowledge of the fourth-order moments of $x(t)$.

10-13. Use the Wiener–Khintchine theorem to compute the power spectra of the random telegraph process of Sec. 10.4.

10-14. Consider a random process with autocorrelation function $R(\tau) = \exp(-a|\tau|)$ with $a > 0$.

a. Describe the asymptotic behavior of $W(f)$.

b. Repeat part a when $R(\tau)$ is given by

$$R(\tau) = \frac{2a}{a^2 + (2\pi\tau)^2}.$$

*10-15. A random process $x(t)$ is observed over $[-T, T]$. By directly computing the expected value of

$$\mathcal{W}_T(f) = \frac{1}{2T} \left| \int_{-T}^{T} x(t)e^{-j2\pi ft}\, dt \right|^2,$$

show that

$$W_T(f) \equiv \overline{\mathcal{W}}_T(f)$$

$$= 2T \int_{-\infty}^{\infty} W(\zeta)\, \text{sinc}^2\, 2T(f - \zeta)\, d\zeta.$$

Hence the series truncation due to the finite observation time produces an inherent "smoothing."

10-16. Explain why the sample correlation function is given by

$$\mathcal{R}_T(\tau) = \frac{1}{2T - |\tau|} \int_{-T+|\tau|/2}^{T-|\tau|/2} x\left(t - \frac{|\tau|}{2}\right)x\left(t + \frac{|\tau|}{2}\right) dt, \qquad \tau < 2T,$$

instead of

$$\mathcal{R}_T(\tau) \overset{(?)}{=} \frac{1}{2T} \int_{-T}^{T} x\left(t - \frac{|\tau|}{2}\right)x\left(t + \frac{|\tau|}{2}\right) dt, \qquad \tau < 2T,$$

when $x(t)$ is observed over $[-T, T]$.

*10-17. Let $\phi(t)$ be a zero-mean Gaussian random process with correlation function $R_\phi(\tau)$. Compute the autocorrelation function of $x(t) = e^{j\phi(t)}$. The process $e^{j\phi(t)}$ occurs in angle modulation and coherent optics.

10-18. Let X_c and X_s be two independent zero-mean Gaussian random variables with equal variance σ^2. Show that the autocorrelation function of

$$x(t) = X_c \cos \omega_c t - X_s \sin \omega_c t$$

is given by

$$R_x(\tau) = \sigma^2 \cos \omega_c \tau.$$

*10-19. (Sec. 10.8—the NGP) Justify that $x_c(t)$ and $x_s(t)$ are given, to a very good approximation, by Eqs. (10.8-3a) and (10.8-3b), respectively.

*10-20. Use Eqs. (10.8-3a) and (10.8-3b) to show that $\overline{x_c^2(t)} = \overline{x_s^2(t)} = \overline{x^2(t)}$ and that $\overline{x_s(t)x_c(t)} = 0$. *Hint*: Several of the resulting integrals can be simplified by using the appropriate Fourier transform pairs.

10-21. Starting out with Eq. (10.8-2), which describes the NGP, i.e.,

$$x(t) = x_c(t) \cos \omega_c t - x_s(t) \sin \omega_c t,$$

show that

$$R_x(\tau) = R_c(\tau) \cos \omega_c \tau - R_{cs}(\tau) \sin \omega_c \tau,$$

where $R_c(\tau) = \overline{x_c(t)x_c(t + \tau)}$ and $R_{cs}(\tau) = \overline{x_c(t)x_s(t + \tau)}$.

10-22. In Sec. 10.10, the bilateral clipper was used. Consider a variation of such a clipper defined by

$$y(t) = \begin{cases} 1, & x(t) \le x \\ 0, & x(t) > x. \end{cases}$$

Show that $\overline{y(t)} = F_X(x, t)$, where the latter is the probability distribution function (PDF) of $X = x(t)$.

10-23. For Prob. 10-22 show that $R_y(\tau)$ defined by $R_y(\tau) \equiv \overline{y(t + \tau)y(t)}$ is given by

$$R_y(\tau) = F_{X_1 X_2}(x, x; t, t + \tau),$$

where $F_{X_1 X_2}$ is the joint PDF of the r.v.'s $X_1 \equiv x(t)$ and $X_2 \equiv x(t + \tau)$.

10-24. Compute the output signal-to-noise ratio in the case of synchronous detection of an AM wave.

*10-25. An AM wave is detected by the square-law detector shown in Fig. P10-25.

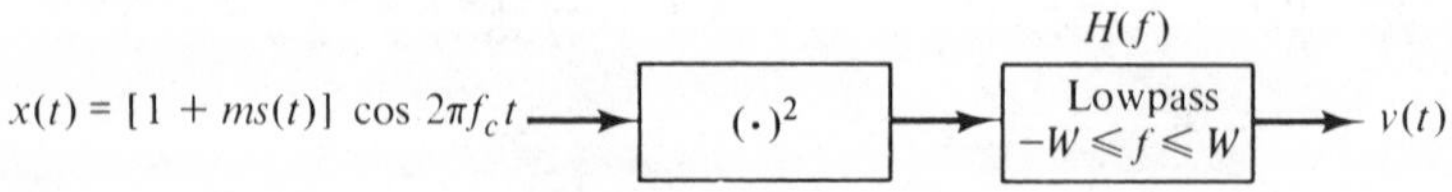

$$x(t) = [1 + ms(t)]\ \cos 2\pi f_c t \longrightarrow \boxed{(\cdot)^2} \longrightarrow \overset{H(f)}{\boxed{\begin{array}{c} \text{Lowpass} \\ -W \le f \le W \end{array}}} \longrightarrow v(t)$$

Figure P10-25

Let $s(t)$ be a stationary Gaussian process with power spectrum

$$W(f) = \begin{cases} N_0, & |f| < W \\ 0, & |f| > W. \end{cases}$$

With $W < f_c/2$ and $H(f)$ given by

$$H(f) = \begin{cases} 1, & |f| < W \\ 0, & |f| > W, \end{cases}$$

compute the output signal-to-noise ratio if the only "noise" is the nonlinear terms generated by the square-law device. Why should $m^2 \ll 1$?

10-26. Let $n(t) = n_c(t) \cos 2\pi f_c t - n_s(t) \sin 2\pi f_c t$ represent an NGP where $n_c(t)$ is $x_c(t)$ and $n_s(t)$ is $x_s(t)$ of Eq. (10.8-2). Show that

$$\overline{n_c(t)n_c(t + \tau)} = 2 \int_0^\infty W_n(\xi) \cos 2\pi(\xi - f_c)\tau \, d\xi$$

$$= \overline{n_s(t)n_s(t + \tau)},$$

where $W_n(\xi)$ is the power spectrum of $n(t)$.

10-27. Use the method of moment generating or characteristic functions discussed in Sec. 9.10 to prove that the r.v.'s $X_1 \equiv n_c^2(t)$ and $X_2 \equiv n_s^2(t)$ are independent; $n_c(t)$ and $n_s(t)$ are the in-phase and quadrature components of NGP noise. Recall that this result is required to obtain Eq. (10.11-42).

10-28. Work out the details leading to the mean-square value of the envelope square as given by Eq. (10.11-42).

10-29. Show that narrowband FM offers no signal-to-noise improvement over AM.

10-30. What signal-to-noise improvement is obtained with a PDE system characterized by a DE filter with transfer function

$$|H_{\mathrm{DE}}(f)| = e^{-|f|/f_1}.$$

Hint: Follow the technique leading to Eqs. (10.12-31)–(10.12-33).

REFERENCES

[10-1] A. PAPOULIS, *Probability, Random Variables and Stochastic Processes*, McGraw-Hill, New York, 1965.

[10-2] W. B. DAVENPORT, JR. and W. L. ROOT, *Introduction to Random Signals and Noise*, McGraw-Hill, New York, 1957.

[10-3] M. ROSENBLATT, "Some Comments on Narrow Band-Pass Filters," *Q. Appl. Math.*, **18**, No. 4, p. 387, 1961.

[10-4] H. STARK, "Smoothing of Irradiance Spectra with Finite-Bandwidth Windows, with Application to Particle-Size Analysis," *J. Opt. Soc. Am.*, **65**, pp. 1436–1442, Dec. 1975.

[10-5] A. PAPOULIS, "Minimum-Bias Windows for High-Resolution Spectral Estimates," *IEEE Trans. Information Theory*, IT-19, pp. 9–12, Jan. 1973.

[10-6] M. G. JENKINS and D. G. WATTS, *Spectral Analysis and Its Applications*, Holden-Day, San Francisco, 1968.

[10-7] R. B. BLACKMAN and J. W. TUKEY, *The Measurement of Power Spectra*, Dover, New York, 1958.

[10-8] H. NYQUIST, "Thermal Agitation of Electric Charge in Conductors," *Phys. Rev.*, **32**, pp. 110–113, July 1928. Also J. B. JOHNSON, "Thermal Agitation of Electricity in Conductors," *Phys. Rev.*, **32**, pp. 97–109, July 1928.

[10-9] M. SCHWARTZ, *Information Transmission, Modulation and Noise*, 2nd ed., McGraw-Hill, New York, 1970.

[10-10] J. H. VAN VLECK and D. MIDDLETON, "The Spectrum of Clipped Noise," *Proc. IEEE*, **54**, pp. 2–19, Jan. 1966.

[10-11] P. I. RICHARDS, "Computing Reliable Power Spectra," *IEEE Spectrum*, **4**, pp. 83–90, Jan. 1967.

[10-12] M. SCHWARTZ, W. R. BENNETT, and S. STEIN, *Communication Systems and Techniques*, McGraw-Hill, New York, 1966.

[10-13] B. A. CARLSON, *Communication Systems: An Introduction to Signals and Noise in Electrical Communications*, 2nd ed., McGraw-Hill, New York, 1975.

[10-14] P. F. PANTER, *Modulation, Noise, and Spectral Analysis*, McGraw-Hill, New York, 1965.

[10-15] M. G. CROSBY, "Frequency Modulation Noise Characteristics," *Proc. IRE*, **25**, pp. 472–514, Fig. 10, April 1937.

11

Signal Processing

11.1 Introduction

In this chapter we shall deal with the problem of extracting information from probabilistic signals, i.e., signals characterized by an inherent uncertainty. We might be uncertain whether a signal is there or not. We might be uncertain about a parameter of the signal or even about which signal out of a finite set is present. We might have a sine wave corrupted by noise and wish to be more certain about its amplitude, phase, or frequency. Given a probabilistic signal, we want to "operate" on it to reduce the uncertainty. The device that does the operation is frequently called a receiver (sometimes a filter), and the particular device that does the best job is called an *optimum receiver*. What kind of receiver is best? The question is too general to be properly answered. Clearly the best receiver depends on how we define "best." The definition of "best" depends on the criterion. Generally the criterion is a compromise between mathematical convenience and real-world considerations. Unfortunately, the criteria considered in this chapter will not include such real-world criteria as cost, i.e., system expense, reliability, ease of operation, etc. We shall limit ourselves to discussing certain basic mathematical principles in signal processing. The examples given below are situations involving probabilistic signals.

1. A search radar sends out a signal to determine the presence of aircraft. Is there an echo at time t in the noisy waveform observed at the receiver?

2. A sine wave $s(t)$ is transmitted and corrupted by additive noise. The frequency and phase are known, but the amplitude is to be estimated. What is the best estimate of the amplitude?

3. We wish to make measurements on a weak probabilistic signal with a strong source nearby. Is there a way to design a receiver that will adapt to the weak signal and ignore the source?

4. There are M different signals, one for each symbol in an alphabet. A signal appears at the receiver corrupted by additive Gaussian noise. Which of the M signals was sent?

5. EKG waveforms are monitored by a computer. Suddenly a waveform undergoes a change in shape and frequency. Can the computer recognize the onset of a stroke?

6. A photograph taken through an astronomical telescope may or may not contain an image of a faint star. The confusion arises because of the *grain noise* of the photographic emulsion. What processing scheme will increase the probability of detection of the weak image?

As can be seen, some of these problems involve estimation, and others detection. We shall give examples of both in what follows. Primarily for mathematical reasons but also because of a certain underlying reasonableness, the evaluation of the estimation procedure will be done frequently with the mean-square-error criterion.†

11.2 Estimation with an *RC* Filter

Before attacking the problem of optimum mean-square (m.s.) estimation, it will be illustrative to consider the problem of estimating a random sine wave of known amplitude and frequency, corrupted by noise. The noise $n(t)$ will be taken as signal-independent, additive, white noise with two-sided spectral density, N_0. For simplicity the "receiver" will be taken as the simple *RC* filter shown in Fig. 11.2-1. Although this particular problem may not have much practical significance, its solution will illustrate several important points associated with signal processing. The very important question of how the solution changes with different criteria will also be illustrated. The goodness of our estimate $y(t)$ will be measured by the *mean square-error* (m.s.e.) criterion. Mathematically, the m.s. estimation erro is described by

$$\overline{\epsilon^2(t)} = \overline{[y(t) - s(t)]^2}, \qquad (11.2-1)$$

---

†There are, however, many situations where the mean-square error (m.s.e.) is not the quantity of interest. For example, in filtering (enhancing) a picture the m.s.e. is not so important as sharpness of detail, at least from the viewer's point of view.

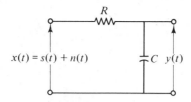

Figure 11.2-1. Estimating a signal $s(t)$ in the presence of noise, with an RC filter.

which can be written as

$$\overline{\epsilon^2(t)} = \overline{y^2(t)} + \overline{s^2(t)} - \overline{2y(t)s(t)}. \tag{11.2-2}$$

The random sine wave $s(t)$ is given by

$$s(t) = A \cos(\omega_0 t + \Theta), \tag{11.2-3}$$

where Θ is a random variable uniformly distributed in $[0, 2\pi]$. The output $y(t)$ can be decomposed into a signal-dependent output $y_s(t)$ and a noise-dependent output $y_n(t)$. Since signal and noise are uncorrelated, we obtain

$$\overline{y^2(t)} = \overline{y_s^2(t)} + \overline{y_n^2(t)}. \tag{11.2-4}$$

The m.s. value of the signal-dependent output is computed from

$$\overline{y_s^2(t)} = \frac{1}{2\pi} \int_{-\infty}^{\infty} |H(\omega)|^2 W_s(\omega)\, d\omega, \tag{11.2-5}$$

where the transfer function for the RC filter is given by

$$H(\omega) = \frac{1}{1 + j\omega\alpha}, \qquad \alpha \equiv RC, \tag{11.2-6}$$

and

$$W_s(\omega) = \mathfrak{F}\left[\frac{1}{2} A^2 \cos \omega_0 \tau\right]$$

$$= \frac{A^2}{2} \pi[\delta(\omega - \omega_0) + \delta(\omega + \omega_0)] \tag{11.2-7}$$

$$= \text{signal power spectrum}.$$

A direct substitution of Eqs. (11.2-7) and (11.2-6) into Eq. (11.2-5) gives

$$\overline{y_s^2(t)} = \frac{A^2}{2}\left(\frac{1}{1 + \alpha^2\omega_0^2}\right). \tag{11.2-8}$$

For $\overline{y_n^2(t)}$ we obtain

$$\overline{y_n^2(t)} = \frac{1}{2\pi} \int_{-\infty}^{\infty} \left(\frac{1}{1 + \alpha^2\omega^2}\right) N_0\, d\omega$$

$$= \frac{N_0}{2\alpha}. \tag{11.2-9}$$

Finally, $\overline{2y(t)s(t)}$ is computed to be

$$\overline{2y(t)s(t)} = \overline{2y_s(t)s(t)}$$

$$= 2\int_{-\infty}^{\infty} h(t-\tau)\overline{s(t)s(\tau)}\,d\tau$$

$$= A^2\left(\frac{1}{1+\alpha^2\omega_0^2}\right). \tag{11.2-10}$$

Use of Eqs. (11.2-4)–(11.2-10) in Eq. (11.2-2) enables us to write

$$\overline{\epsilon^2(t)} = \frac{N_0}{2\alpha} + \frac{A^2}{2}\left(\frac{\alpha^2\omega_0^2}{1+\alpha^2\omega_0^2}\right). \tag{11.2-11}$$

We now ask, What choice of α will minimize the m.s.e.? The minimum value of $\overline{\epsilon^2(t)}$ can be obtained by differentiating with respect to α and setting the derivative equal to zero.

This gives

$$\frac{\partial\overline{\epsilon^2}}{\partial\alpha} = \frac{A^2\omega_0^2\alpha}{(1+\alpha^2\omega_0^2)^2} - \frac{N_0}{2\alpha^2} = 0$$

or, equivalently,

$$\frac{(1+\alpha^2\omega_0^2)^2}{\omega_0^2\alpha^2} = \frac{2A^2\alpha}{N_0}. \tag{11.2-12}$$

The optimum value of α, say α_0, satisfies Eq. (11.2-12). We shall not attempt to compute the roots of Eq. (11.2-12) for the general case; we leave it as an exercise to the reader (Prob. 11-2) to show that when $2A^2/N_0\omega_0 \gg 1$ then an approximate value of α_0 is furnished by

$$\alpha_0 \simeq \frac{2A^2}{N_0\omega_0^2}. \tag{11.2-13}$$

We mention in passing that α_0 depends on ω_0 (as expected) and on the input signal-to-noise factor, $2A^2/N_0$.†

A different criterion might have produced a different result. For example suppose that the intent was not to minimize the m.s.e. as given in Eq. (11.2-1) but rather the suppression of the output noise. Do the two criteria furnish the same value of α? The suppression of the output noise might be measured by the signal-to-noise ratio at the output of the *RC* filter ([11-1], p. 329). Thus our criterion in this case would be

$$\max_{\alpha}\frac{\overline{y_s^2(t)}}{\overline{y_n^2(t)}}. \tag{11.2-14}$$

From Eqs. (11.2-8) and (11.2-9) we obtain

$$\frac{\overline{y_s^2(t)}}{\overline{y_n^2(t)}} = \frac{A^2}{N_0}\frac{\alpha}{1+\omega_0^2\alpha^2}, \tag{11.2-15}$$

†Note that this ratio is not equivalent to the signal-to-noise ratio used in Chapter 10.

which obtains a maximum at $\alpha = \alpha^* = \omega_0^{-1}$. In this case the optimum α does not depend on A or N_0. We are not suggesting that a simple RC filter should be used to filter out broadband noise in the presence of a sine-wave signal (a phase-lock loop or a tuned circuit would be much more effective). However, it should be clear to the reader that the "best" receiver is generally best only for a specific criterion and that the best receiver in one setting may in fact furnish poor results in another.

11.3 Mean-Square Estimation of Discrete Signals

The ideas presented in the previous section carry over directly to discrete signals. We shall, however, not restrict ourselves to a particular system such as the RC configuration in the previous section but shall treat the more general case instead.

In Chapter 5 we saw that a discrete linear filter is described by

$$y(i) = \sum_{j=-\infty}^{\infty} h(i, j)x(j). \tag{11.3-1}$$

The optimum linear filter is then the collection of constants or weights $h(i, j)$ for which the m.s.e.

$$\overline{\epsilon^2(i)} = \overline{|s(i) - y(i)|^2} \tag{11.3-2}$$

is a minimum.

Following the usual procedure, one finds the minimum by differentiation. The algebra is simplified if we use the fact that the operations of expectation and differentiation can be interchanged. Thus the optimum values of $h(i, j)$ are obtained from

$$\frac{\partial \overline{\epsilon^2(i)}}{\partial h(i, j)}\bigg|_{j=k} = 2\overline{\left[\epsilon(i)\frac{\partial \epsilon(i)}{\partial h(i, j)}\right]}\bigg|_{j=k} = 0, \qquad k = \ldots, -2, -1, 0, 1, \ldots. \tag{11.3-3}$$

Since $\epsilon(i) = s(i) - \displaystyle\sum_{j=-\infty}^{\infty} h(i, j)x(j)$, we have

$$\frac{\partial \epsilon(i)}{\partial h(i, j)}\bigg|_{j=k} = -x(k). \tag{11.3-4}$$

The index k is a particular value of j, and the computation is performed for all k. Then Eq. (11.3-3) can be put into the form

$$R_{sx}(i, k) = \sum_{j=-\infty}^{\infty} h(i, j)R_x(j, k), \tag{11.3-5}$$

where $R_{sx}(i, k) = \overline{s(i)x(k)}$ and $R_x(j, k) = \overline{x(j)x(k)}$. If the filter is causal, then $h(i, j) = 0$ for $j > i$. Then the optimum weights satisfy

$$R_{sx}(i, k) = \sum_{j=-\infty}^{i} h(i, j)R_x(j, k). \tag{11.3-6}$$

Frequently the input sequence is zero prior to, say, $j = 1$; then Eq. (11.3-6) takes the form

$$R_{sx}(i, k) = \sum_{j=1}^{i} h(i, j)R_x(j, k). \tag{11.3-7}$$

For a fixed value of i, Eq. (11.3-7) is a set of i simultaneous linear equations in the i unknown filter weights $h(i, j)$, $j = 1, \ldots, i$. With matrix notation, Eq. (11.3-7) can be written as

$$\mathbf{R}_{sx}(i) = \mathbf{R}_x\mathbf{h}(i), \tag{11.3-8}$$

where $\mathbf{R}_{sx}(i)$ consists of the elements $R_{sx}(i, k)$, $k = 1, \ldots, i$; the vector $\mathbf{h}(i)$ contains the unknown filter weights $h(i, j)$, $j = 1, \ldots, i$; and $\mathbf{R}_x$ is the covariance matrix of the received signal. For example, Eq. (11.3-7) for $i = 2$ takes the form

$$\begin{bmatrix} R_{sx}(2, 1) \\ R_{sx}(2, 2) \end{bmatrix} = \begin{bmatrix} R_x(1, 1) & R_x(2, 1) \\ R_x(1, 2) & R_x(2, 2) \end{bmatrix} \begin{bmatrix} h(2, 1) \\ h(2, 2) \end{bmatrix}.$$

The solution of Eq. (11.3-8) can be formally written in the form

$$\mathbf{h}_0(i) = \mathbf{R}_x^{-1}\mathbf{R}_{sx}(i). \tag{11.3-9}$$

The filter described by Eq. (11.3-9) is called the discrete Wiener filter after N. Wiener, who derived analog forms of the optimal smoothing and prediction filter [11-2]. Note that as the output time index i increases, the size of the covariance matrix that needs to be inverted grows, as do the vectors $\mathbf{R}_{sx}(i)$ and $\mathbf{h}_0(i)$. The solution of these equations as i becomes large is a formidable task even with a digital computer.

If the signal and noise are stationary and the filter is time-invariant, then Eq. (11.3-6) can be simplified by writing $n \equiv i - k$ and $l \equiv i - j$. In this case Eq. (11.3-6) becomes

$$R_{sx}(n) = \sum_{l=0}^{\infty} h(l)R_x(n - l). \tag{11.3-10}$$

We leave the details as an exercise (Prob. 11-3).

Equations (11.3-8) and (11.3-9) are completely general and require no assumptions of stationarity, additivity of signal and noise, or independence of signal and noise. However, by introducing some restrictive assumptions one can gain more insight into their meaning. Thus, suppose signal and noise are additive and independent and that the noise is zero-mean. Then

$$x(i) = s(i) + n(i), \qquad i = 1, 2, \ldots, \tag{11.3-11}$$

$$R_{sx}(i, k) = R_s(i, k), \tag{11.3-12a}$$

and

$$R_x(j, k) = R_s(j, k) + R_n(j, k), \tag{11.3-12b}$$

where $R_s(i, k) = \overline{s(i)s(k)}$ and $R_n(j, k) = \overline{n(j)n(k)}$ are elements of the signal and noise covariance matrices $\mathbf{R}_s$ and $\mathbf{R}_n$, respectively. Using Eqs. (11.3-12a) and (11.3-12b) in Eq. (11.3-9) results in†

$$\mathbf{h}_0(i) = (\mathbf{R}_s + \mathbf{R}_n)^{-1}\mathbf{R}_s(i), \qquad (11.3\text{-}13)$$

and the optimum estimate of the signal $s(i)$ is then furnished by

$$\hat{s}(i) = \mathbf{h}_0^T(i)\mathbf{x}$$

$$= \mathbf{R}_s^T(i)(\mathbf{R}_s + \mathbf{R}_n)^{-1}\mathbf{x}. \qquad (11.3\text{-}14)$$

In many practical problems causality is not a problem because the processing is not done in real time or the underlying indexing is not time but some other quantity such as position. For example, if there are m pieces of data stored on tape as a sequence $x(1), \ldots, x(m)$, then since m doesn't refer to actual time, we do not require that the output index i be constrained to $m \leq i$. Also in image processing, the constraint $m \leq i$ is meaningless if these indices refer to spatial position.

For problems of this type we can write

$$y(i) = \sum_{j=1}^{m} h(i, j)x(j), \qquad i = 1, \ldots, m, \qquad (11.3\text{-}15)$$

and the optimum weights satisfy

$$R_{sx}(i, k) = \sum_{j=1}^{m} h(i, j)R_x(j, k). \qquad (11.3\text{-}16)$$

For each value of i this equation specifies m filter weights. If we allow i to take on any integer value between 1 and m, there are m^2 weights altogether. The complete filter transmittance matrix has the form

$$\mathbf{H}_0 = \begin{bmatrix} h(1, 1) & \cdots & h(1, m) \\ \cdot & & \cdot \\ \cdot & & \cdot \\ \cdot & & \cdot \\ h(m, 1) & \cdots & h(m, m) \end{bmatrix}. \qquad (11.3\text{-}17)$$

A disadvantage of the optimum filter formulation given in Eq. (11.3-9) is that it involves the inversion of a large matrix, which is often a numerically tricky task. Also if the observations $x(j)$ represent a continuous stream of data, the requirement that the filter should operate on only m samples at a time is inconvenient. It is much better to compute the output recursively using each piece of data as it arrives and properly combining it with all the previous observations. A recursive filter that operates in this fashion is the

---

†It is interesting to note that in the continuous-time estimation problem the (non-causal) Wiener filter takes a similar form in the frequency domain: $H_0(\omega) = [W_s(\omega) + W_n(\omega)]^{-1}W_s(\omega)$, where $W_s(\omega)$ and $W_n(\omega)$ are the power spectra of signal and noise and $H_0(\omega)$ is the optimum transfer function. With sufficient delay, this filter can be made causal ([11-1], p. 338).

Kalman filter ([11-3] and [11-4], Chap. 7). The Kalman filter is identical to the Wiener filter for stationary inputs. We shall also present an *adaptive* filter in Sec. 11.5 which generates the Wiener solution recursively.

Before looking at some simple examples to illustrate some of the ideas presented in this section we shall digress briefly to consider techniques for modeling a discrete correlated noise process $\{n_j\}$. We shall switch to subscript notation† for convenience; i.e., $x(i)$ becomes x_i, $h(i, j)$ becomes h_{ij}, etc.

11.4 Discrete Noise Models; Parameter Estimation

Discrete white noise has a correlation function given by

$$R_z(i, j) = \overline{z_i z_j} = \sigma_z^2 \delta_{ij}, \tag{11.4-1}$$

where δ_{ij} is the Kronecker delta and has value

$$\delta_{ij} = \begin{cases} 1, & i = j \\ 0, & i \neq j. \end{cases} \tag{11.4-2}$$

We temporarily reserve the signal z_j for white noise and use the symbol n_j, $j = 1, 2, \ldots$, for any kind of noise, correlated or not. Correlated noise can be modeled in various ways. For example, we might expect n_j to depend on a linear aggregate of some previous values plus a white noise *shock*, z_j, which gives n_j some randomness on its own. In mathematical terms, n_j is described by

$$n_j = b_1 n_{j-1} + b_2 n_{j-2} + \cdots + b_p n_{j-p} + z_j. \tag{11.4-3}$$

Equation (11.4-3) is called an *autoregressive* (AR) model of order p. It shows that n_j depends on $n_{j-1}, \ldots, n_{j-p}$ plus a white-noise term z_j. Comparison with Eq. (5.10-4) shows that autoregressive noise is white noise passed through a recursive (IIR) filter. As shown in Chapter 5, recursive filters can be realized with feedback loops involving multipliers and delay elements. A first-order autoregressive noise process can be generated using a feedback loop containing a single delay operator and an amplifier of gain $|b| < 1$ (Fig. 11.4-1). Clearly n_j is described by

$$n_j = b n_{j-1} + z_j, \qquad |b| < 1. \tag{11.4-4}$$

The correlation function, $R_n(k) \equiv \overline{n_j n_{j+k}}$ of this process is easily computed. For example,

$$R_n(0) = \overline{n_j^2} = \overline{(b n_{j-1} + z_j)^2}$$
$$= \overline{b^2 n_{j-1}^2} + \overline{z_j^2}$$
$$= b^2 R_n(0) + \sigma_z^2.$$

---

†In Chapter 5, we switched from subscripts to functional notation to emphasize the time dependence of one set of variables with respect to another set of variables. In the present chapter, we switch back to subscript notation for reasons of compactness.

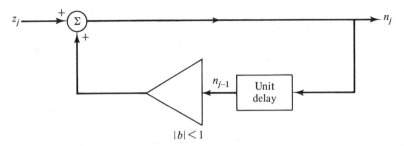

Figure 11.4-1. Generation of first-order autoregressive noise with a first-order recursive filter.

Hence

$$R_n(0) = \frac{\sigma_z^2}{1 - b^2},$$ (11.4-5)

where $\sigma_z^2 = \overline{z_j^2}$. Similarly, we find that

$$R_n(1) = \overline{n_j n_{j+1}} = \overline{n_j(bn_j + z_{j+1})}$$
$$= b\overline{n_j^2}$$
$$= bR_n(0).$$

In general, then, for k either positive or negative, we obtain

$$R_n(k) = \frac{b^{|k|}}{1 - b^2}\sigma_z^2.$$ (11.4-6)

The correlation function for $b = \frac{1}{2}$ and $b = -\frac{1}{2}$ is shown in Fig. 11.4-2. If b is close to unity, adjacent noise samples correlate strongly, and the correla-

Figure 11.4-2. Correlation function for a first-order autoregressive process with parameter b: (a) $b = \frac{1}{2}$; (b) $b = -\frac{1}{2}$.

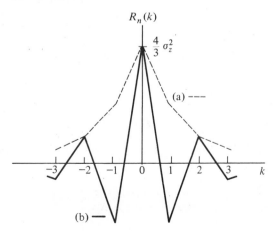

tion function will tend to retain significant amplitude for relatively long lags. If b is small but finite, the correlation function decreases rapidly in value but never goes completely to zero; this only happens when $b = 0$, which corresponds to white noise. It is worth noting that for positive b this correlation function is quite similar in form to $\sigma^2 e^{-2\lambda|\tau|}$, the correlation function of the telegraph signal. [See Eq. (10.4-6).]

Another widely used model for discrete noise and other discrete random processes is the so-called *moving-average* (MA) model [11-5]. In the MA model the noise is given by a weighted average of present and earlier white-noise shocks, z_j, z_{j-1}, ... Thus a moving-average process of order q is described by

$$n_j = z_j - a_1 z_{j-1} - a_2 z_{j-2} - \cdots - a_q z_{j-q}. \tag{11.4-7}$$

We again note that this equation is similar to Eq. (5.11-2), which describes the nonrecursive (FIR) digital filter. The term *moving average*, despite widespread use, is not completely accurate because we do not require the sum of the weights, i.e., $1 - a_1 - \cdots - a_q$, to be unity. Consider a first-order MA process

$$n_j = z_j - a z_{j-1}, \qquad |a| < 1.$$

Its autocorrelation function is easily computed to be (Prob. 11-4)

$$R_n(l) = R_z(l) - a R_z(l-1) - a R_z(l+1) + a^2 R_z(l), \tag{11.4-8}$$

where $R_z(l) = \overline{z_j z_{j+l}}$. This correlation function is shown in Fig. 11.4-3. Because this is a first-order MA process, nonzero correlation extends up only to the first lag interval (compare with the first-order AR process).

EXAMPLE

Estimating a constant signal in noise ([11-4], p. 279). Our observation is described by

$$x_j = s + n_j,$$

where the signal s is fixed but unknown. We model s as an r.v. with $\overline{s^2} = P_s$. For simplicity we assume that our estimate, $\hat{s}$, is based on two observations; i.e.,

$$\hat{s} \equiv y_2 = h_1 x_1 + h_2 x_2.$$

Signal and noise are assumed independent, with $\overline{n_j} = 0$ and $\overline{n_j^2} \equiv \sigma_n^2$. We use Eq. (11.3-16) with $m = 2$ and with $h(i, j) \equiv h_j$, independent of i. Since the signal is constant and independent from the noise, $R_{sx}(i, k) = P_s$ for all i and k. Also, to simplify the notation we use R_{jk} for $R_x(j, k)$. We then obtain

$$P_s = h_1 R_{11} + h_2 R_{21}$$
$$P_s = h_1 R_{12} + h_2 R_{22}. \tag{11.4-9}$$

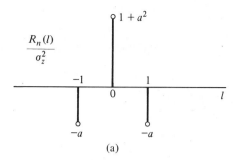

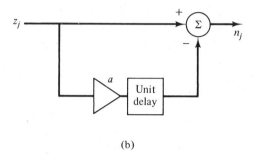

Figure 11.4-3. (a) Autocorrelation of first-order MA process; (b) method of generating a first-order MA process.

The coefficients R_{ij} are

$$R_{11} = \overline{x_1^2} = \overline{(s + n_1)(s + n_1)} = P_s + \sigma_n^2$$

$$R_{22} = \overline{x_2^2} = \overline{(s + n_2)(s + n_2)} = P_s + \sigma_n^2 \qquad (11.4\text{-}10)$$

$$R_{12} = R_{21} = \overline{x_1 x_2} = \overline{(s + n_1)(s + n_2)} = P_s + \overline{n_1 n_2}.$$

We assume that the noise is a first-order MA process. Its correlation function is therefore given by Eq. (11.4-8) and

$$\overline{n_1 n_2} = R_n(1) = \frac{-a}{1 + a^2}\sigma_z^2, \qquad [\sigma_n^2 = (1 + a^2)\sigma_z^2 \text{ and } |a| < 1].$$

$$(11.4\text{-}11)$$

For simplicity let $\tilde{b} \equiv -a/(1 + a^2)$. Then the optimum filter coefficients h_1 and h_2 satisfy the pair of equations

$$(P_s + \sigma_n^2)h_1 + (P_s + \tilde{b}\sigma_n^2)h_2 = P_s$$

$$(P_s + \tilde{b}\sigma_n^2)h_1 + (P_s + \sigma_n^2)h_2 = P_s,$$

for which the solution is

$$h_1 = h_2 \equiv h_0 = \frac{1}{2 + (P_s/\sigma_n^2)^{-1}(1 + \tilde{b})}. \qquad (11.4\text{-}12)$$

The best m.s. estimate of s is then furnished by

$$\hat{s} = \frac{1}{2 + (P_s/\sigma_n^2)^{-1}(1 + \tilde{b})}(x_1 + x_2).$$ (11.4-13)

If the noise were uncorrelated, $\tilde{b}$ would be zero, and the best estimate would be given by

$$\hat{s} = \frac{1}{2 + (P_s/\sigma_n^2)^{-1}}(x_1 + x_2) \qquad \text{(uncorrelated noise)}.$$ (11.4-14)

The estimate furnished by the commonly used *sample mean*, for $m = 2$, is

$$\hat{s}_{SM} = \tfrac{1}{2}(x_1 + x_2).$$ (11.4-15)

Observe that this is the same as Eq. (11.4-14) if $P_s/\sigma_n^2 \longrightarrow \infty$, in other words, in the absence of noise. When noise is present the Wiener filter gives a better estimate. To see this, we compute the m.s.e. for arbitrary h_1, h_2. This is given by

$$\overline{\epsilon^2} = \overline{(\hat{s} - s)^2} = h_1^2\sigma_n^2 + h_2^2\sigma_n^2 + 2h_1h_2\tilde{b}\sigma_n^2 + P_s(h_1 + h_2 - 1)^2.$$ (11.4-16)

For the sample-mean estimate, $h_1 = h_2 = \tfrac{1}{2}$. Hence, from Eq. (11.4-16),

$$\frac{\overline{\epsilon_{SM}^2}}{\sigma_n^2} = \frac{1}{2}(1 + \tilde{b}) \qquad \text{(sample-mean estimate)}.$$ (11.4-17)

For the Wiener filter, $h_1 = h_2 = h_0$ of Eq. (11.4-12). Substitution of these results into Eq. (11.4-16) furnishes

$$\frac{\overline{\epsilon_0^2}}{\sigma_n^2} = \frac{1}{2}K(1 + \tilde{b}) \qquad \text{(optimum estimate)},$$ (11.4-18)

where K is given by

$$K = \frac{1}{1 + (P_s/\sigma_n^2)^{-1}[(1 + \tilde{b})/2]} < 1, \qquad \text{since } |\tilde{b}| < \frac{1}{2}.$$ (11.4-19)

Hence $\overline{\epsilon_0^2} < \overline{\epsilon_{SM}^2}$, as expected (Probs. 11-5 to 11-7).

In the example just considered, s might be the constant depth of a patch of ocean and x_j the sounding at t_j, or s might be the average value of transmittance of uniformly exposed photographic film and x_j the microdensitometer trace at position r_j. There are in fact numerous situations that can be modeled in this way. They all have in common that a single number is to be estimated. This may be true even if the signal is not a constant. Thus, consider the estimation of a sine wave of known frequency but unknown amplitude. Then the signal s_i might be given by

$$s_i = A \cos \omega t_i.$$

It is much better to estimate A than to try to estimate the s_i individually. A more general model of this same sort is furnished by observations of the form

$$x_i = \alpha_i s + n_i,$$ (11.4-20)

where the α_i's are a sequence of known coefficients and s is to be estimated. In all of these examples s is referred to as a *parameter*, and the estimation procedure is called *parameter estimation*. There may be several parameters, for instance, A and B in the sequence

$$s_i = A + Bt_i,$$

which represents a straight line passing through A for $t_i = 0$ and having a slope B. A more sophisticated example is the range from an earthbound radar to an orbiting satellite. The orbit is determined by six parameters which one would want to estimate from the radar observations in order to predict future positions of the satellite. These more general problems are modeled by

$$x_i = \sum_{j=1}^{L} \alpha_{ij} s_j + n_i \qquad (11.4\text{-}21)$$

for a signal determined by L parameters. L is generally a relatively small number, but the model given in Eq. (11.3-11) is a special case of Eq. (11.4-21) in which $\alpha_{ii} = 1$ and $\alpha_{ij} = 0$ for $i \neq j$. If s is a parameter vector of a lower dimensionality than the observation vector x, the equation for the Wiener estimate [Eq. (11.3-14)] is changed to (Prob. 11-8)

$$\hat{\mathbf{s}} = \mathbf{R}_s^T \mathbf{A}^T (\mathbf{A} \mathbf{R}_s \mathbf{A}^T + \mathbf{R}_n)^{-1} \mathbf{x}, \qquad (11.4\text{-}22)$$

where **A** is the matrix of the α_{ij}'s in Eq. (11.4-21). We see that Eq. (11.3-14) is a special case of Eq. (11.4-22) for $\mathbf{A} = \mathbf{I}$, the unit matrix.

For our second example, consider the parameter estimation problem modeled by Eq. (11.4-20). The single parameter s is to be estimated; the α_i are known. For simplicity, assume that n_j is stationary white noise with correlation function $\overline{n_i n_j} = \sigma_n^2 \delta_{ij}$. As in the earlier example, we assume the $\overline{n_j} = 0$, that $\overline{n_j s} = 0$, and that $\overline{s^2} = P_s$. R_{ij} is given by

$$R_{ij} = \overline{(\alpha_i s + n_i)(\alpha_j s + n_j)}$$
$$= \alpha_i \alpha_j P_s + \sigma_n^2 \delta_{ij}, \qquad i, j = 1, \ldots, m, \qquad (11.4\text{-}23)$$

and $R_{sx}(i)$ is given by

$$R_{sx}(i) = \overline{x_i s} = \alpha_i P_s, \qquad i = 1, \ldots, m. \qquad (11.4\text{-}24)$$

If these results are substituted into Eq. (11.3-16), we obtain the system of equations

$$\sigma_n^2 h_i + P_s \alpha_i \sum_{j=1}^{m} \alpha_j h_j = \alpha_i P_s, \qquad i = 1, \ldots, m. \qquad (11.4\text{-}25)$$

With K defined by $K \equiv \sum_{j=1}^{m} \alpha_j h_j$, Eq. (11.4-25) can be solved for h_i as

$$h_i = \frac{\alpha_i P_s}{\sigma_n^2}(1 - K), \qquad i = 1, \ldots, m. \qquad (11.4\text{-}26)$$

Now if in Eq. (11.4-26) we multiply both sides by α_i and sum over the i's, we obtain

$$K\left(1 + \frac{P_s}{\sigma_n^2} \cdot \sum_{i=1}^{m} \alpha_i^2\right) = \frac{P_s}{\sigma_n^2} \sum_{i=1}^{m} \alpha_i^2, \qquad (11.4\text{-}27)$$

which gives the solutions for K and $1 - K$ as

$$K = \frac{P_s \sum_{i=1}^{m} \alpha_i^2}{\sigma_n^2 + P_s \sum_{i=1}^{m} \alpha_i^2} \qquad (11.4\text{-}28)$$

and

$$1 - K = \frac{\sigma_n^2}{\sigma_n^2 + P_s \sum_{i=1}^{m} \alpha_i^2}. \qquad (11.4\text{-}29)$$

Hence, the optimum filter has weights given by

$$h_i = \left[\frac{P_s}{\sigma_n^2 + P_s \sum_{i=1}^{m} \alpha_i} \right] \alpha_i, \qquad i = 1, \ldots, m, \qquad (11.4\text{-}30)$$

and the minimum mean-square estimate is

$$\hat{s} = \sum_{i=1}^{m} h_i x_i. \qquad (11.4\text{-}31)$$

It is interesting to observe that the h_i are directly proportional to the signal coefficients α_i. We say that h_i is *matched* to the signal $\alpha_i s$. This filter is a particular example of the *matched* filter to be discussed in Sec. 11.6.

The example just discussed is a special case of the general parameter estimation problem given in Eq. (11.4-21). We leave it as an exercise for the reader to show that Eq. (11.4-22) leads to the result given here if the conditions of the example are substituted.

11.5   Mean-Square Estimation Using Adaptive Filters

We found that the Wiener filter is the best filter, in the m.s. sense, for cancelling the noise that corrupts a signal. However, the design of a fixed Wiener filter requires *a priori* knowledge of signal and noise statistics which may not be readily available. Moreover, if the parameters of the signal and noise change, the original Wiener filter may not be optimum anymore, resulting in a decrease in the output signal-to-noise ratio. For this reason, adaptive filters, which hold many advantages over fixed filters, have become widely used in recent times. Adaptive filters are used in antenna array processing [11-6], in fetal electrocardiography, in speech processing, in cancelling periodic interference signals, in self-tuning filters [11-7], and in high-speed digital communication systems between computers [11-8]. In the last application, which is in commercial use, the adaptive filters are used to reduce intersymbol interference (Sec. 4.4).

Adaptive filters are frequently constructed as transversal, or tapped delay-line, filters (see Sec. 5-11). Such a filter is shown in Fig. 11.5-1. The weights

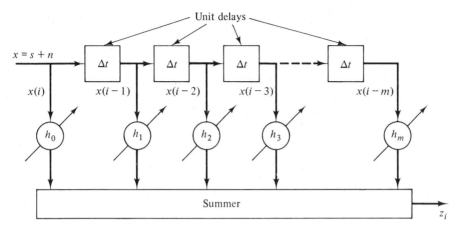

Figure 11.5-1. Transversal filter; the arrows are meant to indicate that the $\{h_j\}$ are time-dependent.

h_j are adjustable and multiply delayed values of the input. Therefore the output at time $t_i = i\,\Delta t$ is given by

$$z_i = \sum_{j=0}^{m} h_{ij} x_{i-j}. \tag{11.5-1}$$

Note that $x_{i-j} \equiv x[i-j)\,\Delta t] \equiv x(i-j)$. In Eq. (11.5-1) the inclusion of the subscript i in h_{ij} is to show that the weights generally depend on time.

The operation of the adaptive filter depends on the fact that the mean-square error between z and s, or at least its gradient with respect to the h's, can be determined from measurements made in the filter. Very little *a priori* knowledge is generally required. In the following discussion we assume that the noise covariance is known, but adaptive filters can also be designed without this knowledge. Generally speaking, however, the system must have some information permitting it to distinguish the signal from the noise.

In the following explanation of the adaptive system we assume that the system is digital; i.e., the delays shown in Fig. 11.5-1 are produced by a shift register, and the signals s, n, x, and z are digital sequences, as in Sec. 11.4. We suppose that at some reference time t_i, the h's have some arbitrary value given by the vector $\mathbf{h}_i = (h_{i0}, h_{i1}, h_{i2}, \ldots, h_{im})^T$. (The superscript T means transpose.) The mean-square error at this reference time is given by

$$\overline{\epsilon_i^2} = \overline{(s_i - z_i)^2} = \overline{\left(s_i - \sum_{j=0}^{m} h_{ij} x_{i-j}\right)^2} \tag{11.5-2}$$

and is seen to be a quadratic function of the h_{ij}'s. In fact, if we plot $\overline{\epsilon_i^2}$ as a function of any one of the h's, say h_{ik}, we get a parabola as shown in Fig. 11.5-2. The minimum point of the parabola occurs at $h_{ik} = h^*$. As a function of all the h's, the m.s. error will be a *hyperparaboloid* in $m+1$ dimensions.

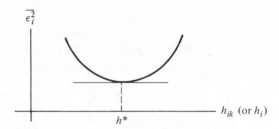

Figure 11.5-2. Variation of m.s. error with any one filter weight.

The "bottom" of the hyperparaboloid is the point where $\overline{\epsilon_i^2}$ is smallest and the corresponding values of h_{ij} are the optimum values.

To find this point we use what is referred to as a gradient method. This is best illustrated for the one-dimensional case shown in Fig. 11.5-2. Suppose that h_i has some value $h_i \neq h^*$. We find the slope of the $\overline{\epsilon^2}$ curve at this point; this is $d\overline{\epsilon^2}/dh|_{h=h_i}$. If the slope is negative, we are to the left of the minimum, and if it is positive, we are to the right. Therefore, if at the next time increment we make

$$h_{i+1} = h_i - \mu_i \frac{d\overline{\epsilon^2}}{dh}\bigg|_{h=h_i}, \qquad (11.5\text{-}3)$$

where μ_i is a small positive number, then h_{i+1} is shifted in the right direction to reduce $\overline{\epsilon^2}$. It is important not to make μ_i too large since otherwise h_{i+1} may overshoot the optimum value, and we may end up with a larger value of $\overline{\epsilon^2}$ than we started with. However, if μ_i is small enough, h_{i+1} will be better than h_i. We can then repeat the process. When the slope becomes very small we know that we are close to the optimum and can stop.

For the n-dimensional case the procedure is the same except that we compute the gradient $\nabla\overline{\epsilon_i^2}$ rather than the slope. The gradient is the vector given by

$$\nabla\overline{\epsilon_i^2} = \left[\frac{\partial\overline{\epsilon_i^2}}{\partial h_{i0}}, \frac{\partial\overline{\epsilon_i^2}}{\partial h_{i1}}, \ldots, \frac{\partial\overline{\epsilon_i^2}}{\partial h_{im}}\right]^T, \qquad (11.5\text{-}4)$$

and the updating algorithm becomes

$$\mathbf{h}_{i+1} = \mathbf{h}_i - \mu_i \nabla\overline{\epsilon_i^2}. \qquad (11.5\text{-}5)$$

Note that since $\mathbf{h}_i$ is a vector, Eq. (11.5-5) implies the simultaneous adjustment of $m + 1$ filter weights h_{ij}. The adjustment of each of the filter weights is proportional to its component of the gradient vector.

There remains the question of how to compute the gradient. The m.s. error at the time t_i is

$$\begin{aligned}
\overline{\epsilon_i^2} &= \overline{(s_i - z_i)^2} = \overline{(x_i - n_i - z_i)^2} \\
&= \overline{(x_i - z_i)^2} + \overline{n_i^2} - \overline{2n_i(x_i - z_i)} \\
&= \overline{(x_i - z_i)^2} + \overline{n_i^2} - \overline{2n_i(s_i + n_i - z_i)} \\
&= \overline{(x_i - z_i)^2} - \overline{n_i^2} + 2\sum_{j=0}^{m} h_{ij}R_n(j). \qquad (11.5\text{-}6)
\end{aligned}$$

In going from the third line to the fourth we use the fact that $\overline{s_i n_i} = 0$. Also $R_n(j)$ is the noise correlation function, which we assume to be known.

The desired gradient can now in principle be obtained by differentiating Eq. (11.5-6) with respect to the h_{ij} as in Eq. (11.3-3). Since z_i is a function of the h_{ij}, as shown in Eq. (11.5-1), this differentiation gives

$$\frac{\partial \overline{\epsilon_i^2}}{\partial h_{ij}} = -2\overline{(x_i - z_i)x_{i-j}} + 2R_n(j). \tag{11.5-7}$$

The averaging that still remains to be done requires knowledge of the correlation of the x's, which we do not have. Therefore the computation for the gradient cannot be completed. A way out of this difficulty is furnished by the *stochastic approximation* technique of Robbins and Monro [11-9], according to which the h_i will converge to the optimum if the averaging indicated in Eq. (11.5-5) is simply omitted. Thus the algorithm becomes

$$\begin{aligned} \mathbf{h}_{i+1} &= \mathbf{h}_i - \mu_i \nabla \epsilon_i^2 \\ &= \mathbf{h}_i + 2\mu_i \{(x_i - z_i)\mathbf{X}_i - \mathbf{R}_n\}, \end{aligned} \tag{11.5-8}$$

where the vectors $\mathbf{X}_i$ and $\mathbf{R}_n$ are given by

$$\mathbf{X}_i = (x_i, x_{i-1}, x_{i-2}, \ldots, x_{i-m})^T$$
$$\mathbf{R}_n = (R_n(0), R_n(1), \ldots, R_n(m))^T,$$

respectively. Since we assume the noise correlation $R_n(j)$ to be known, the algorithm of Eq. (11.5-8) can be implemented. In fact, since x_{i-j} is the signal at the jth point on the shift register, we derive the correction for each filter weight by simply multiplying the appropriate x_{i-j} by the difference between the input and output of the filter. A complete block diagram of the filter is shown in Fig. 11.5-3.

It was shown by Robbins and Monro [11-9] that for the $\mathbf{h}_i$ to converge to the optimum $\mathbf{h}^*$ the μ_i must decrease with i. The precise conditions are

(i) $\mu_i > 0$

(ii) $\lim_{i \to \infty} \mu_i = 0$

(iii) $\sum_{i=1}^{\infty} \mu_i = \infty$

(iv) $\sum_{i=1}^{\infty} \mu_i^2 < \infty.$

A particular choice of μ_i that satisfies these conditions is $\mu_i = 1/i$. As the μ_i decrease, the effectiveness of the adaptation is reduced and eventually ceases completely. This kind of behavior is all right if signal and noise inputs are truly stationary but is unsatisfactory for a filter operating in a slowly varying environment.

Widrow's LMS algorithm [11-6] gets around this difficulty by using a constant value of $\mu_i = \mu$ satisfying the inequality

$$0 < \mu < \lambda_{\text{max}}^{-1},$$

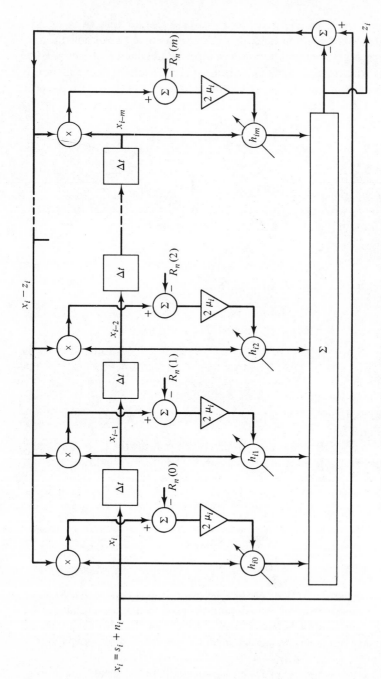

Figure 11.5-3. Adaptive filter for known noise correlation.

where λ_{max} is the largest eigenvalue of the correlation matrix of the x's. Even though we have assumed this matrix to be unknown, one could presumably know enough about it to set some bound on μ. Since the Widrow algorithm does not satisfy the Robbins-Monro conditions, it does not, strictly speaking, converge. As a matter of fact the h's tend to jitter around the optimum value. By adjusting μ the jitter can be held to an acceptable level, yet the filter is capable of continuous adaptation.

It is left as an exercise to show that the filter, operating under conditions (i)–(iv), converges to the "unrealizable" Wiener form, Eq. (11.3-13). Since the filter is clearly being realized, the designation "unrealizable" appears to be a misnomer. The explanation lies in the way the delay-line filter operates. When a signal x is first applied to the filter there will be no signal at later stages of the delay line, and therefore most of the terms in the sum of Eq. (11.5-1) are zero. It is only after m time increments that the filter reaches a steady state. Thus there is built into the system a time delay of m time units, and therefore the output is a delayed version of the input. Thus the "unrealizable" filter can be realized if a sufficient amount of delay is introduced. This is done automatically by the delay-line filter.

The adaptive algorithm can be modified to accommodate *a priori* knowledge about signal and noise other than the noise correlation function. For instance, it is possible to operate the filter if only the signal correlation function is known. This modification in the algorithm is left as an exercise (Prob. 11-10). Also, if a noise source, say n_1, that is correlated with the noise, say n_0, in the received signal is available, an adaptive algorithm can be based on it. We shall not go into great details vis-à-vis this scheme (see Widrow et al. [11-7], [11-10]) except to point out that output feedback to the adaptive filter can be used to minimize the mean-square error. The concept is easily demonstrated with the aid of Fig. 11.5-4. The input to the system at A is called the primary input and consists of signal s plus noise n_0. The input at B is the noise reference input and consists of noise n_1 correlated in some unknown way

Figure 11.5-4. Noise cancelling with an adaptive filter.

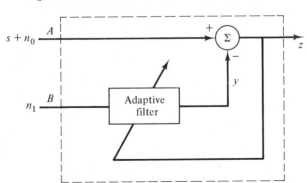

with n_0 but uncorrelated with s. The object of the adaptive filter is to generate an output y from n_1 which looks like n_0. If $n_0 = n_1$, $z = s$ is the desired output, and the output signal-to-noise ratio becomes infinite.

Surprisingly, it turns out that little or no *a priori* knowledge of s, n_0, and n_1 is required to make the filter effective. To understand why, we need only refer to Fig. 11.5-4. The m.s.e. between z and the desired output s is

$$\overline{\epsilon^2} = \overline{(z - s)^2} = \overline{(s + n_0 - y - s)^2} = \overline{(n_0 - y)^2}. \qquad (11.5\text{-}9)$$

Clearly $\overline{\epsilon^2}$ can be made zero by having $y = n_0$. The m.s. value of the output is

$$\overline{z^2} = \overline{s^2} + \overline{(n_0 - y)^2} + \overline{2s(n_0 - y)}. \qquad (11.5\text{-}10)$$

For a given set of filter weights, the output y depends essentially only on n_1; therefore if n_1 is uncorrelated with s, then y and $n_0 - y$ will also be uncorrelated with s.[†] The expected value of the last term in Eq. (11.5-10) is therefore zero, and the m.s. value of z is

$$\overline{z^2} = \overline{s^2} + \overline{(n_0 - y)^2}. \qquad (11.5\text{-}11)$$

Comparing Eq. (11.5-11) with Eq. (11.5-9), we see that minimizing the m.s.e. is completely equivalent to minimizing the average output power, since $\overline{s^2}$ is unaffected by the filter. Hence the adaptive filter *minimizes the m.s.e. by adjusting its impulse response so as to drive the system output z to a minimum.*

11.6   Linear Detection of Signals: The Matched Filter

In the previous sections we have dealt with the problem of estimating the parameters of a signal corrupted by noise. We shall now consider another important problem, widely studied in communication engineering, namely that of detecting a signal in the presence of noise or other signals. Examples of situations where signal detection is a problem are (1) the detection of a radar echo in receiver noise and/or ground clutter, (2) the detection of sonar signals in ocean waters, (3) the detection of faint stars in telescope images, and (4) the detection of a symbol, message, or code word from a known set. If the input noise is white, the best filter will turn out to be the so-called *matched* filter, already encountered in Eq. (11.4-30). For reasons that will shortly be understood, the matched filter is also called a correlator. The correlation product of a signal $s(t)$ is the integral of the product of s with its shifted self. The matched filter is a fairly ubiquitous object and appears in image enhancement, pattern recognition, lensless imaging of X-ray and radioactive sources, and other modern research endeavors.

[†]There is a slight dependence of y on the signal s through the filter weights $\{h\}$ since these are adjusted by z, which in turn depends on s. However, this dependence is generally very slight.

The matched filter is a special case of the optimum linear filter for maximizing the output signal-to-noise ratio $(S/\mathfrak{N})_0$. Using $(S/\mathfrak{N})_0$ as the quantity to be maximized is a reasonable criterion in view of the fact that our ability to make a correct decision regarding the presence or absence of a signal, at a particular instant of time, will strongly depend on the extent to which the signal has been extracted from the noise. The latter is a measure of the signal-to-noise improvement furnished by the receiver.

Mathematical Formulation

The input signal $x(t)$ consists of signal $s(t)$ and additive noise $n(t)$ with spectral density $W_n(\omega)$. Let t_0 denote the instant of time at which we want to maximize the output signal-to-noise ratio. The output signal at t_0, $y(t_0)$, is decomposed into a signal portion, $y_s(t_0)$, and a noise portion, $y_n(t_0)$, respectively given by

$$y_s(t_0) = \int_{-\infty}^{\infty} s(\tau)h(t_0 - \tau)\, d\tau, \tag{11.6-1}$$

$$y_n(t_0) = \int_{-\infty}^{\infty} n(\tau)h(t_0 - \tau)\, d\tau. \tag{11.6-2}$$

Our problem is to maximize the ratio

$$\left(\frac{S}{\mathfrak{N}}\right)_0 = \frac{y_s^2(t_0)}{y_n^2(t_0)} \tag{11.6-3}$$

by considering all linear, time-invariant impulse responses $\{h(\tau)\}$. (See Fig. 11.6-1.) The solution to this problem is easily obtained using the Schwarz

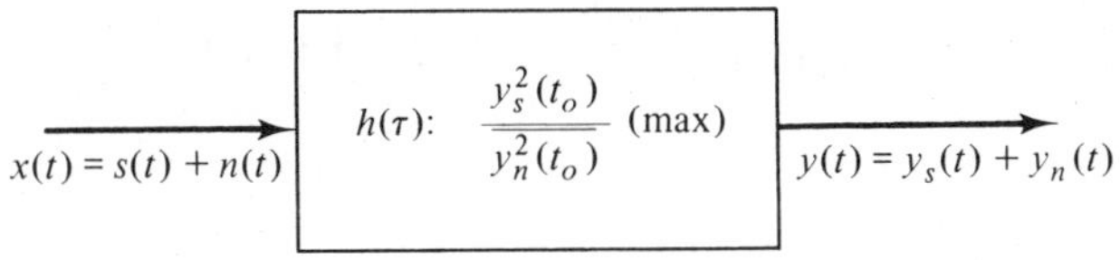

Figure 11.6-1. Matched filter objective.

inequality ([11-11], p. 63). One form of the Schwarz inequality states if $A_1(\omega)$ and $A_2(\omega)$ are any complex signals, then

$$\left| \int_{-\infty}^{\infty} A_1(\omega)A_2(\omega)\, d\omega \right|^2 \le \int_{-\infty}^{\infty} |A_1(\omega)|^2\, d\omega \int_{-\infty}^{\infty} |A_2(\omega)|^2\, d\omega \tag{11.6-4}$$

with equality if and only if

$$A_1(\omega) = kA_2^*(\omega), \tag{11.6-5}$$

where k is a constant. To use the Schwarz inequality we assume that $A_1(\omega)$ and $A_2(\omega)$ are square-integrable functions.

Returning now to the problem at hand, we can rewrite Eq. (11.6-3) as

$$\left(\frac{S}{\mathfrak{N}}\right)_0 = \frac{\left|\int_{-\infty}^{\infty} H(\omega)S(\omega)e^{j\omega t_0}\, d\omega\right|^2}{2\pi \int_{-\infty}^{\infty} |H(\omega)|^2 W_n(\omega)\, d\omega}, \tag{11.6-6}$$

where we have used the fact that $y_s(t_0)$ and $\overline{y_n^2(t_0)}$ can be, respectively, written as

$$y_s(t_0) = \frac{1}{2\pi} \int_{-\infty}^{\infty} H(\omega)S(\omega)e^{j\omega t_0}\, d\omega, \tag{11.6-7}$$

$$\overline{y_n^2(t_0)} = \frac{1}{2\pi} \int_{-\infty}^{\infty} |H(\omega)|^2 W_n(\omega)\, d\omega. \tag{11.6-8}$$

$S(\omega)$ and $H(\omega)$ are the Fourier transforms of $s(t)$ and $h(t)$, respectively. In Eq. (11.6-4), let

$$A_1(\omega) \equiv H(\omega)\sqrt{W_n(\omega)}, \tag{11.6-9a}$$

$$A_2(\omega) \equiv \frac{S(\omega)}{\sqrt{W_n(\omega)}} e^{j\omega t_0}. \tag{11.6-9b}$$

It therefore follows that

$$\left|\int_{-\infty}^{\infty} H(\omega)S(\omega)e^{j\omega t_0}\, d\omega\right|^2 \leq \int_{-\infty}^{\infty} |H(\omega)|^2 W_n(\omega)\, d\omega \int_{-\infty}^{\infty} \left\{\frac{|S(\omega)|^2}{W_n(\omega)}\right\} d\omega \tag{11.6-10}$$

or, since both factors on the right are positive,

$$\frac{\left|\int_{-\infty}^{\infty} H(\omega)S(\omega)e^{j\omega t_0}\, d\omega\right|^2}{2\pi \int_{-\infty}^{\infty} |H(\omega)|^2 W_n(\omega)\, d\omega} \leq \frac{1}{2\pi} \int_{-\infty}^{\infty} \left\{\frac{|S(\omega)|^2}{W_n(\omega)}\right\} d\omega. \tag{11.6-11}$$

But the term on the left is precisely the output signal-to-noise ratio. Hence

$$\left(\frac{S}{\mathfrak{N}}\right)_0 \leq \frac{1}{2\pi} \int_{-\infty}^{\infty} \frac{|S(\omega)|^2}{W_n(\omega)}\, d\omega, \tag{11.6-12}$$

with equality if and only if

$$H(\omega)\sqrt{W_n(\omega)} = k\frac{S^*(\omega)}{\sqrt{W_n(\omega)}} e^{-j\omega t_0}. \tag{11.6-13}$$

The optimum filter, therefore, has the transfer function

$$H(\omega) = H_0(\omega) \equiv k\frac{S^*(\omega)}{W_n(\omega)} e^{-j\omega t_0}. \tag{11.6-14}$$

In contrast to the Wiener filter, discussed in Sec. 11.3, which required knowledge only of correlation functions or power spectra, $H_0(\omega)$ requires knowledge of the amplitude and *phase* of the signal, in effect knowledge of the complete signal shape. Because the phase of $S(\omega)$ must be known, this type of filter is sometimes known as a *coherent* filter as opposed to the Wiener filter, which is phase-independent and therefore "incoherent." Furthermore, if the signal spectrum is not zero at frequencies for which the noise spectrum is zero, the transmittance of the filter becomes infinite, as does the signal-to-noise ratio. The physical interpretation of this is that if we observe nonzero spectral components in spectral regions where we know *the noise to be zero*, the probability of signal detection can be made unity. In practice, however, the noise spectrum is never zero, although it could be small in relation to the signal.

The matched filter results when the noise spectrum is white, i.e., when

$$W_n(\omega) = N_0.$$

Equation (11.6-14), with $H_M(\omega)$ denoting the transmittance of the matched filter, now becomes

$$H_M(\omega) = KS^*(\omega)e^{-j\omega t_0}, \tag{11.6-15}$$

where $K \equiv k/N_0$. K represents a gain factor and can be conveniently set to unity without loss of generality. Hence the transmittance of the matched filter will be taken as

$$H_M(\omega) = S^*(\omega)e^{-j\omega t_0}, \tag{11.6-16}$$

and the impulse response is

$$
\begin{aligned}
h_M(t) &= \frac{1}{2\pi} \int_{-\infty}^{\infty} H_M(\omega)e^{j\omega t} \, d\omega \\
&= s^*(t_0 - t) \\
&= s(t_0 - t) \qquad \text{if } s(t) \text{ is real.}
\end{aligned} \tag{11.6-17}
$$

The impulse response is seen to be a reversed and delayed image of the input. Hence the optimum filter must be *matched* to the particular signal of interest. Note that if $s(t)$ is of finite duration, $h_M(t)$ can be made causal by proper choice of t_0.

The signal portion of the response of the matched filter at $t = t_0$ is

$$y_s(t_0) = \int_{-\infty}^{\infty} s(\tau)s^*\{\tau - (t - t_0)\} \, d\tau \bigg|_{t=t_0} \tag{11.6-18}$$

$$= \int_{-\infty}^{\infty} |s(\tau)|^2 \, d\tau \tag{11.6-19}$$

$$= E_s,$$

where E_s denotes the energy in the signal. Integrals of the form of Eq. (11.6-18) are denoted correlation products, which accounts for the designation *correlator* for the matched filter. The matched filter can therefore be realized by a system in which a signal is multiplied by its shifted self and integrated.

The mean-square noise power is given by

$$\overline{y_n^2(t)} = \int_{-\infty}^{\infty} N_0 |H_M(\omega)|^2 \frac{d\omega}{2\pi}$$

$$= N_0 E_s. \tag{11.6-20}$$

The peak signal-to-noise ratio occurs at $t = t_0$ [to prove this for himself, the reader should apply the Schwarz inequality to Eq. (11.6-18)] and is given by

$$\left(\frac{S}{\mathfrak{N}}\right)_0 = \frac{E_s^2}{N_0 E_s} = \frac{E_s}{N_0}. \tag{11.6-21}$$

This interesting result shows that the maximum signal-to-noise ratio for the matched filter is *independent* of signal shape and depends only on the ratio of *signal energy* to noise spectral density. It has been found that the exact shape of the matched filter is not nearly so important as the proper choice of system bandwidth. Thus for simple rectangular pulses in white noise, other filters such as the *RC* filter, ideal low-pass filter, and Gaussian filter furnish signal-to-noise results only $\simeq 1$ dB lower than the matched filter, provided that the system bandwidth in each case is optimized ([11-12], p. 67). The use of more complicated signals, as is frequently the practice in radar, sonar, and optics, imposes stricter constraints on deviations from the matched filter transmittance. Matched filters are often used with coded signals, i.e., signals that have significant amplitude over relatively large intervals in both the time and frequency domain. Such signals are characterized by a large time-bandwidth (*TW*) product, the latter being essentially the product of the signal duration *T* by the signal bandwidth *W*. For a simple rectangular pulse, $TW \simeq 1$, while, for example, a $TW \simeq 1000$ is possible with linear ramp-modulated FM pulses† [11-13]. The reason for using large *TW* signals is that the selectivity of the matched filter is greatly enhanced. In other words, the output of the matched filter when it is matched to a large *TW* signal is much greater than the output when the "wrong," i.e., nonmatched, input appears. This is important in pattern recognition and other applications where discrimination between classes is the object.

11.7   Optimum Receivers

We shall now consider the problem of how best to detect the signal $s(t)$ from a more fundamental point of view than maximization of signal-to-noise ratio at the output of a linear filter. Instead we shall consider the state of mind of

---

†These are pulses whose frequency increases linearly with time during the duration *T* of the pulse.

an observer confronted with some kind of a noisy received signal and asked to decide whether it contains something of interest, and if so what. Typical examples are in radar, where the question would be whether the signal on the screen contains some target echo, and if so, where it is. Or the received signal might be a chest X-ray, and the question is whether one or more of the vague shadows in it are from a cancer. Several other examples have already been mentioned.

The general problem is best illustrated with the help of a block diagram, as shown in Fig. 11.7-1. In this model we assume that there is a message

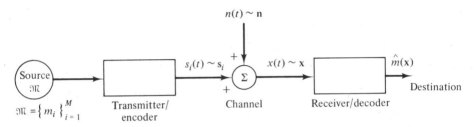

Figure 11.7-1. Block diagram of a communication system. The symbol $\sim$ means "is represented by."

source $\mathfrak{M}$ containing a finite number of messages $\{m_i\}$. In the transmitter/ encoder the messages $\{m_i\}$ are converted into signals $\{s_i(t)\}$ which are sent out over the channel. The mapping is presumably one to one. The channel is noisy, and therefore a noisy and possibly distorted version of the signal is received. In much of the subsequent discussion we assume that the noise is added to the signal so that the received signal is $x(t) = s_i(t) + n(t)$, as shown in Fig. 11.7-1. However, the general formulation is capable of handling other ways in which the noise and signal interact (for instance, fading,† which is multiplicative noise). The received signal is processed in the receiver, and a best estimate of the original message $\hat{m}_i$ is delivered to the destination.

We find it convenient to regard all signals as vectors. As pointed out in Chapters 4 and 5, continuous waveforms can in practice always be converted to vectors by sampling. An analytically more useful vector representation of signals is by resolution into orthonormal functions, as discussed in Chapter 2. However, regardless of the method actually used, $s_i(t)$, $n(t)$, $x(t)$ will be represented by the vectors $\mathbf{s}_i$, $\mathbf{n}$, $\mathbf{x}$. This is indicated in Fig. 11.7-1. In the case of sampling, the dimensions of these vectors are generally on the order of $2TW$ for observation times T and bandwidths W. (See Chapter 4.)

---

†In fading, an incident wave is typically broken up by the medium into several waves, each with its own (random) phase. At the receiver, the sum of the waves is observed and if the incident beam is an unmodulated carrier, the received signal is a randomly modulated AM-type wave.

Note that the different messages m_i may also represent continuous quantities. For instance, in radar one is interested in the distance to the target. Although this is clearly a continuous quantity, there is little loss in discretizing it and estimating it, say, to the nearest kilometer. Thus the parameter estimation problem already mentioned in Sec. 11.4 is also included in the formulation of this section.

A rational basis for making the decision is the cost of making it. We define a *cost matrix*

$$\mathbf{C} = \begin{bmatrix} C_{11}, & C_{12}, & \ldots, & C_{1M} \\ \cdot & \cdot & & \cdot \\ \cdot & \cdot & & \cdot \\ \cdot & \cdot & & \cdot \\ C_{M1}, & C_{M2}, & \ldots, & C_{MM} \end{bmatrix}, \tag{11.7-1}$$

where C_{ij} is the cost, i.e., penalty of deciding that the transmitted message was m_i when in fact it was m_j, and where elements C_{ii}, $i = 1, \ldots, M$, represent the costs of correct decisions and are often made zero. However, since some correct decisions may be better than others, one sometimes assigns nonzero values to some of the C_{ii} as well. Obviously the cost of making an error should be larger than the cost of a correct decision if C_{ij} is to have the meaning normally associated with the word *cost*.

Because of the randomness of the noise, any single decision is generally random, and the associated cost is a random variable. Therefore a reasonable basis for making decisions is the average cost, or *risk*, of the decision, given by

$$\mathcal{R} = \sum_{i=1}^{M} \sum_{j=1}^{M} P(\hat{m}_i, m_j) C_{ij}, \tag{11.7-2}$$

where the joint probability $P(\hat{m}_i, m_j)$ in a somewhat compressed notation† is the joint probability of m_j being the true message and m_i having been decided. In practice it may be difficult to assign reasonable values to the cost matrix $\mathbf{C}$. However, supposing this to have been done, then we presumably would wish to design our decision maker so as to minimize $\mathcal{R}$. This is the so-called Bayes criterion, and the resulting receiver is a Bayes receiver.

In the model considered here it is assumed that the selection of the actual message at the source is random; i.e., we can think of an urn holding all messages, and the message m_j is picked from the urn with probability P_j. The number P_j is sometimes called the *a priori* probability of m_j. This having been done, there is (according to the model in Fig. 11.7-1) a probabilistic mapping from the message m_j to the received signal $\mathbf{x}$; we denote the conditional probability density of the resulting $\mathbf{x}$ by $f(\mathbf{x} \mid m_j)$. An interpretation of

---

†Specifically, $P(\hat{m}_i, m_j) \equiv P[\hat{m}(\mathbf{x}) = m_i, m_j]$, which should be read as the joint probability that the receiver decision function $\hat{m}(\mathbf{x})$ assumes the value m_i and that m_j was transmitted. This joint probability depends on m_j, on $\mathbf{x}$, and *on the decision function* $\hat{m}(\cdot)$.

$f(\mathbf{x}|m_j)$ is furnished by

$$P[\mathbf{x} < \mathbf{x}' < \mathbf{x} + d\mathbf{x}, \text{ given that } m_j \text{ is true}] = f(\mathbf{x}|m_j)\, d\mathbf{x}$$

$$\text{for } j = 1, \ldots, M. \tag{11.7-3}$$

The symbol $d\mathbf{x}$ is shorthand for the N-dimensional hypervolume $dx_1 \ldots dx_N$. (For convenience we drop the prime on $\mathbf{x}'$ and use $\mathbf{x}$ to denote both the random vector and the values that it can take on.)

Decisions are made in accordance with a decision rule. Such a rule is essentially a partition of the space of all possible received signals $\mathbf{x}$ into disjoint decision regions $\mathfrak{X}_1, \mathfrak{X}_2, \ldots, \mathfrak{X}_M$ such that if $\mathbf{x}$ falls into the region $\mathfrak{X}_i$, the decision is that the message m_i was sent. For instance, if the signals m_1 and m_2 are, respectively, a positive and a negative dc voltage, a reasonable partition might be to choose m_1 when the average of the received $\mathbf{x}$'s is positive and to choose m_2 otherwise. The regions $\mathfrak{X}_1$ and $\mathfrak{X}_2$ would contain all vectors $\mathbf{x}$ having positive and negative mean values, respectively. Note that to find out in which region $\mathbf{x}$ belongs may involve some processing, e.g., taking the average, or something more complicated. There are generally very many decision rules, some reasonable and others not. For instance, we could decide always to choose m_1, independent of what is observed. Or we might decide to pick m_1 whenever any one element of the vector $\mathbf{x}$ exceeds some positive number. We can represent the decision process as shown in Fig. 11.7-2.

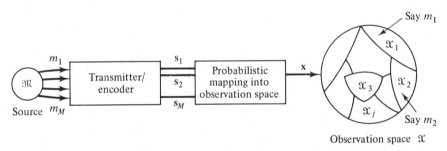

Figure 11.7-2. Optimum receiver problem.

We should mention that the decision process is sometimes generalized to include random decision rules. Such rules are defined in terms of the probability of deciding in favor of m_i given that $\mathbf{x}$ is observed. This generalization has some mathematical advantages, but we need not consider it here.

In terms of the additional functions defined, we may rewrite Eq. (11.7-2) in the form

$$\mathfrak{R} = \sum_{i=1}^{M} \sum_{j=1}^{M} P(\hat{m}_i|m_j) P_j C_{ij}$$

$$= \sum_{i=1}^{M} \sum_{j=1}^{M} P_j C_{ij} \int_{\mathfrak{X}_i} f(\mathbf{x}|m_j)\, d\mathbf{x}. \tag{11.7-4}$$

The integral is N-fold, and the fact that it is over the region $\mathscr{X}_i$ implies that the decision is in favor of m_i. The probability density $f(\mathbf{x}|m_j)$ indicates that the observations are actually due to m_j.

We now seek to adjust the partitioning $\{\mathscr{X}_i\}$ so as to minimize $\mathfrak{R}$. The optimum partitioning can be found with the help of Bayes' rule:

$$f(\mathbf{x}|m_j) = \frac{f(m_j|\mathbf{x})f(\mathbf{x})}{P_j}$$

or (11.7-5)

$$P_j f(\mathbf{x}|m_j) = f(\mathbf{x})f(m_j|\mathbf{x}),$$

where $f(\mathbf{x})$ is the unconstrained pdf of $\mathbf{x}$. Use of Eq. (11.7-5) in Eq. (11.7-4) furnishes

$$\mathfrak{R} = \sum_{i=1}^{M} \sum_{j=1}^{M} C_{ij} \int_{\mathscr{X}_i} f(m_j|\mathbf{x})f(\mathbf{x})\,d\mathbf{x} \qquad (11.7\text{-}6)$$

$$= \sum_{i=1}^{M} \int_{\mathscr{X}_i} \beta_i(\mathbf{x})f(\mathbf{x})\,d\mathbf{x}, \qquad (11.7\text{-}7)$$

where

$$\beta_i(\mathbf{x}) \equiv \sum_{j=1}^{M} C_{ij} f(m_j|\mathbf{x}), \qquad i = 1, \ldots, M. \qquad (11.7\text{-}8)$$

We are now in a position to find the optimum regions $\mathscr{X}_1^*, \ldots, \mathscr{X}_M^*$ which partition $\mathscr{X}$ so as to minimize the risk. For simplicity, let $i = 1, 2$, and refer to Fig. 11-7.3. In this case $\mathfrak{R}$ is given by

$$\mathfrak{R} = \int_{\mathscr{X}_1} \beta_1(\mathbf{x})f(\mathbf{x})\,d\mathbf{x} + \int_{\mathscr{X}_2} \beta_2(\mathbf{x})f(\mathbf{x})\,d\mathbf{x}. \qquad (11.7\text{-}9)$$

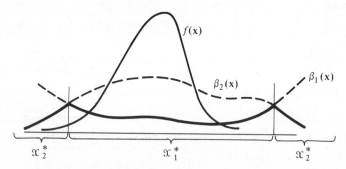

Figure 11.7-3. Optimum partitioning of $\mathscr{X}$ into two decisions regions.

Since $f(\mathbf{x})$ is everywhere nonnegative, $\mathfrak{R}$ is minimized by choosing the partitions $\mathscr{X}_1^*$ and $\mathscr{X}_2^*$ such that $\mathbf{x}$ is in $\mathscr{X}_1^*$ whenever $\beta_1(\mathbf{x}) < \beta_2(\mathbf{x})$ and in $\mathscr{X}_2^*$ whenever $\beta_2(\mathbf{x}) < \beta_1(\mathbf{x})$. Any other choice of $\mathscr{X}_1$ and $\mathscr{X}_2$ would increase $\mathfrak{R}$ (for example, try the obvious $\mathscr{X}_1 = \mathscr{X}_2^*$ and $\mathscr{X}_2 = \mathscr{X}_1^*$).

The result is easily generalized to M decision regions (Fig. 11.7-4). We find that the optimum partitioning is equivalent to the following (Bayes) decision rule: *For every observed* $\mathbf{x}$, *compute*

$$\beta_i(\mathbf{x}) = \sum_{j=1}^{M} C_{ij} f(m_j \mid \mathbf{x}), \qquad i = 1, \ldots, M, \qquad (11.7\text{-}10)$$

and choose that m_i *for which* β_i *is smallest.*

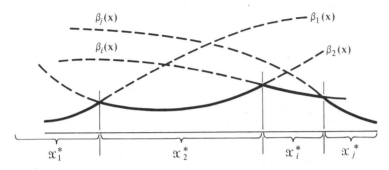

Figure 11.7-4. Optimum partitioning in the general case.

The function $f(m_j \mid \mathbf{x})$ is the probability that m_j is the message if $\mathbf{x}$ was received. It is referred to as the *a posteriori* probability of m_j given $\mathbf{x}$ and it contains all the information available to the receiver after reception of $\mathbf{x}$. Therefore it forms a natural basis for making the decision. The function β_i is the *a posteriori* average cost, or *a posteriori* risk, associated with the message m_i and the particular received signal $\mathbf{x}$. Hence the Bayes decision rule is to decide in such a way as to always minimize the *a posteriori* risk.

Equation (11.7-5) can be written in the equivalent form

$$
\begin{aligned}
f(m_j \mid \mathbf{x}) &= \frac{P_j f(\mathbf{x} \mid m_j)}{f(\mathbf{x})} \\
&= \frac{P_j f(\mathbf{x} \mid m_j)}{\displaystyle\sum_{j=1}^{M} P_j f(\mathbf{x} \mid m_j)}.
\end{aligned}
\qquad (11.7\text{-}11)
$$

The second line is the form in which Bayes' theorem is usually given [cf. Eq. (8.4-4)]. We see that the computation of the *a posteriori* risk involves the *a priori* probability P_j, the probabilistic mapping $f(\mathbf{x} \mid m_j)$, and the cost matrix $\mathbf{C}$. Not all of these are always known. In these cases the Bayes receiver cannot be implemented, and the receiver design must be based on other criteria ([11-14], Chap. 2).

The Zero-One Cost Function

An important special case occurs when the cost function is

$$C_{ij} = 1 - \delta_{ij}, \qquad (11.7\text{-}12)$$

where δ_{ij} is the Kronecker delta. This gives zero cost for correct decisions and unity cost for any incorrect decision; i.e., any error is as bad as any other. Then Eq. (11.7-8) becomes

$$\beta_i(\mathbf{x}) = \sum_{j \neq i}^{M} f(m_j|\mathbf{x}) = 1 - f(m_i|\mathbf{x}).$$

Hence the optimum decision rule, i.e.,

$$\min_i \beta_i(\mathbf{x}) \qquad \text{for each } \mathbf{x},$$

in this case is equivalent to

$$\max_i f(m_i|\mathbf{x}) \qquad \text{for each } \mathbf{x}. \qquad (11.7\text{-}13)$$

Since the optimum receiver maximizes the *a posteriori* probability, it is sometimes called a maximum *a posteriori* probability receiver or MAP receiver for short. The risk function for the zero-one loss is given by

$$\mathfrak{R} = \sum_{i=1}^{M} \sum_{j \neq i}^{M} P_j \int_{\mathfrak{X}_i} f(\mathbf{x}|m_j)\,d\mathbf{x}$$

$$= \sum_{i=1}^{M} \sum_{j \neq i}^{M} \int_{\mathfrak{X}_i} f(m_j|\mathbf{x})f(\mathbf{x})\,d\mathbf{x}$$

$$= 1 - \sum_{i=1}^{M} \int_{\mathfrak{X}_i} f(m_i|\mathbf{x})f(\mathbf{x})\,d\mathbf{x}$$

$$= 1 - \text{probability of a correct decision}. \qquad (11.7\text{-}14)$$

Hence minimizing the risk in this case is equivalent to maximizing the probability of the receiver making a correct decision. Since $f(\mathbf{x}) \geq 0$, the probability of making a correct decision is maximized by using the MAP decision rule.

Another way of describing the MAP decision rule is as follows: $\hat{m}(\mathbf{x}) = m_i$ if

$$f(m_i|\mathbf{x}) > f(m_k|\mathbf{x}) \qquad \text{for } k = 1, \ldots, M, \quad k \neq i. \qquad (11.7\text{-}15)$$

Equivalently, the Bayes rule enables us to write the following: $\hat{m}(\mathbf{x}) = m_i$ if

$$\frac{f(\mathbf{x}|m_i)P_i}{f(\mathbf{x})} > \frac{f(\mathbf{x}|m_k)P_k}{f(\mathbf{x})} \qquad \text{for every } k \neq i \qquad (11.7\text{-}16)$$

or, since $f(\mathbf{x})$ is irrelevant, $\hat{m}(\mathbf{x}) = m_i$ if

$$f(\mathbf{x}|m_i)P_i > f(\mathbf{x}|m_k)P_k \qquad \text{for every } k \neq i. \qquad (11.7\text{-}17)$$

If the $\{P_i\}$ are not known or are assumed equal, i.e.,

$$P_1 = P_2 = \cdots = P_M = \frac{1}{M},$$

the optimum receiver chooses $\hat{m}(\mathbf{x}) = m_i$ if $f(\mathbf{x}\,|\,m_i)$ is the maximum conditional pdf from the set $\{f(\mathbf{x}\,|\,m_j)\}_{j=1}^{M}$. This kind of a receiver is known as a *maximum likelihood* (ML) receiver. It minimizes the probability of error if all *a priori* probabilities $\{P_j\}$ are equal.

Note that if the *a priori* probabilities are not equal, there generally exists a MAP receiver having a smaller error probability than the ML receiver. However, if this same MAP receiver is faced with a different set of *a priori* probabilities, its error probability could be larger than that of the ML receiver; in fact for certain unfavorable *a priori* distributions its performance might be much worse. Since *a priori* probabilities are frequently unavailable, one would want a receiver that guards against the possibility of a particularly unfavorable *a priori* distribution resulting in very large errors. The receiver whose *worst-case performance is better than* that of any other receiver is called a *minimax receiver.*† For a discussion of such receivers, see [11-15], p. 264. It turns out that under conditions frequently encountered in practice the ML receiver is, in fact, minimax. For this reason and also because its design is independent of *a priori* information, the ML receiver is often preferred over other types.

The Likelihood Ratio

The computation that implements the optimum decision rule can be done serially. For example, Eq. (11.7-17) enables us to write the following: If

$$\frac{f(\mathbf{x}\,|\,m_i)}{f(\mathbf{x}\,|\,m_k)} > \frac{P_k}{P_i}, \tag{11.7-18}$$

reject m_k for consideration and test $f(\mathbf{x}\,|\,m_{k+1})$. Otherwise, i.e., if Eq. (11.7-18) is not true, reject m_i, replace $f(\mathbf{x}\,|\,m_i)$ with $f(\mathbf{x}\,|\,m_k)$, and test the next pdf. The ratio

$$LR_{ik}(\mathbf{x}) \equiv \frac{f(\mathbf{x}\,|\,m_i)}{f(\mathbf{x}\,|\,m_k)} \tag{11.7-19}$$

is known as the *likelihood ratio*, and a test of the form of Eq. (11.7-18) is known as a *likelihood-ratio test*.

EXAMPLE

Let m_i, $i = 1, 2, 3$, represent three source messages, and let C_{ij} be the zero-one cost function; i.e.,

$$C_{ij} = \begin{cases} 1, & i \neq j \\ 0, & i = j. \end{cases}$$

†Minimax is a contraction of *minimizing the maximum* risk or probability of error, whichever the case may be.

The *a priori* probability of m_i is P_i, $i = 1, 2, 3$. Define

$$LR_{21}(\mathbf{x}) = \frac{f(\mathbf{x}\,|\,m_2)}{f(\mathbf{x}\,|\,m_1)}$$

$$LR_{31}(\mathbf{x}) = \frac{f(\mathbf{x}\,|\,m_3)}{f(\mathbf{x}\,|\,m_1)}$$

$$LR_{23}(\mathbf{x}) = \frac{LR_{21}(\mathbf{x})}{LR_{31}(\mathbf{x})} = \frac{f(\mathbf{x}\,|\,m_2)}{f(\mathbf{x}\,|\,m_3)}.$$

The decision boundaries are established by

$$LR_{21}(\mathbf{x}) = \frac{P_1}{P_2}; \quad \text{i.e., if } LR_{21}(\mathbf{x}) \begin{cases} > P_1/P_2, & \text{don't choose } m_1 \\ < P_1/P_2, & \text{don't choose } m_2 \end{cases}$$

$$LR_{31}(\mathbf{x}) = \frac{P_1}{P_3}; \quad \text{i.e., if } LR_{31}(\mathbf{x}) \begin{cases} > P_1/P_3, & \text{don't choose } m_1 \\ < P_1/P_3, & \text{don't choose } m_3 \end{cases}$$

$$LR_{23}(\mathbf{x}) = \frac{P_3}{P_2}; \quad \text{i.e., if } LR_{23}(\mathbf{x}) \begin{cases} > P_3/P_2, & \text{don't choose } m_3 \\ < P_3/P_2, & \text{don't choose } m_2. \end{cases}$$

The decision space is shown in Fig. 11.7-5. Note that the optimum partitioning is unambiguous and can be represented in a two-dimensional coordinate system even though three messages were involved. This is true more generally; in the case of M decision possibilities, the minimum dimension of the decision space is $M - 1$. The boundaries of the decision regions are hyperplanes.

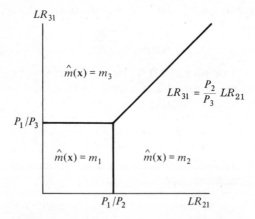

Figure 11.7-5. Optimum decision regions in likelihood-ratio space for the case of three messages.

11.8 Optimum Reception of Known Signals in Gaussian White Noise

The problem we shall consider now is the detection of a signal corrupted by stationary, white, additive Gaussian noise. The noise is assumed to be independent of the signal and to have a zero-mean value. The basic communication system is that shown in Fig. 11.7-1, and we assume that the receiver knows the waveforms of all the possible signals $\{s_i(t)\}$. What it does not know is which one was transmitted.

As indicated in the previous section, all signals are regarded as vectors. The vector representation of the transmitted signal $s_i(t)$ is

$$\mathbf{s}_i = (s_{i1}, \ldots, s_{iN})^T, \qquad i = 1, \ldots, M, \tag{11.8-1}$$

while the noise and received vectors are, respectively, given by

$$\mathbf{n} = (n_1, \ldots, n_N)^T, \tag{11.8-2}$$

$$\mathbf{x} = (x_1, \ldots, x_N)^T. \tag{11.8-3}$$

As in Sec. 11.7, the components of these vectors are either samples of the continuous-time waveforms or the values along the coordinates $\phi_1, \ldots, \phi_N$ of a suitable set $\{\phi_i(t)\}$ which spans an N-dimensional subspace of $L_2(T)$. The $\{\phi_i(t)\}$ could, for example, be obtained by the Gram-Schmidt procedure. In that case the components would simply be the coefficients of the truncated, generalized Fourier series discussed in Chapter 2. The assumption that the noise is white, is stationary, and has zero mean implies that

$$\bar{n}_1 = \bar{n}_2 = \cdots = \bar{n}_N = 0 \qquad \text{(zero-mean property)} \tag{11.8-4}$$

$$\overline{n_i n_j} = 0 \quad \text{for } i \neq j \qquad \text{(white-noise property)} \tag{11.8-5}$$

$$\overline{n_1^2} = \overline{n_2^2} = \cdots = \overline{n_N^2} = \sigma_n^2 \qquad \text{(isotropy property)}. \tag{11.8-6}$$

Since the noise is also Gaussian, Eq. (11-8.5) implies independence of the elements of the vector $\mathbf{n}$ so that the joint probability density is simply the product of the one-dimensional densities. Finally, since $\mathbf{x} = \mathbf{s}_i + \mathbf{n}$, $\mathbf{x}$ is Gaussian with the mean value $\mathbf{s}_i$. Hence, the pdf of $\mathbf{x}$ for a particular $\mathbf{s}_i$ is

$$f(\mathbf{x} \mid \mathbf{s}_i) = \frac{1}{(\sqrt{2\pi\sigma_n^2})^N} \exp\left[-\frac{1}{2\sigma_n^2} \sum_{k=1}^{N} (x_k - s_{ik})^2 \right]. \tag{11.8-7}$$

As mentioned in previous sections, there are a number of ways in which a receiver can be optimized. We first consider the MAP receiver; it chooses $\mathbf{s}_i$ so as to maximize the product $P_i f(\mathbf{x} \mid \mathbf{s}_i)$. Maximization of this product is equivalent to maximizing its logarithm (or any monotonically increasing function of it). Hence the receiver chooses the value of i which *maximizes*

$$\ln P_i - \frac{N}{2} \ln (2\pi\sigma_n^2) - \frac{1}{2\sigma_n^2} \sum_{k=1}^{N} (x_k - s_{ik})^2. \tag{11.8-8}$$

Since the term $N/2 \ln(2\pi\sigma_n^2)$ is irrelevant to the maximization process, the receiver chooses the value of i which *minimizes* the decision function

$$d_i(\mathbf{x}) = \| \mathbf{x} - \mathbf{s}_i \|^2 - 2\sigma_n^2 \ln P_i, \tag{11.8-9}$$

where $\| \mathbf{x} - \mathbf{s}_i \|^2 = \sum_{k=1}^{N} (x_k - s_{ik})^2$ is the Euclidean distance between the vectors $\mathbf{x}$ and $\mathbf{s}_i$. To illustrate how optimum decision regions are determined, consider the simple case of M equiprobable signals; i.e.,

$$P_1 = P_2 = \cdots = P_M = \frac{1}{M}. \tag{11.8-10}$$

Since all the P_i are equal, $d_i(\mathbf{x})$ does not depend on P_i in this case. Under these conditions the decision rule is reduced to finding the value of i which minimizes

$$\| \mathbf{x} - \mathbf{s}_i \|,$$

i.e., the distance from $\mathbf{x}$ to $\mathbf{s}_i$. The receiver action is $\hat{m}(\mathbf{x}) = m_k$ if $\mathbf{s}_k$ is closest, in the sense of Euclidean distance, to the received vector $\mathbf{x}$. Given any two signals $\mathbf{s}_1$ and $\mathbf{s}_2$, we say that the vector $\mathbf{x}$ is *closer* to $\mathbf{s}_1$ than to $\mathbf{s}_2$ if it lies on the $\mathbf{s}_1$ side of the perpendicular bisector of the line joining $\mathbf{s}_1$ to $\mathbf{s}_2$.

The decision regions for three equienergetic, equiprobable signals is shown in Fig. 11-8.1. A receiver that seeks to minimize $\| \mathbf{x} - \mathbf{s}_i \|$ is sometimes called a *minimum-distance* receiver. The decision boundaries in Fig. 11.8-1

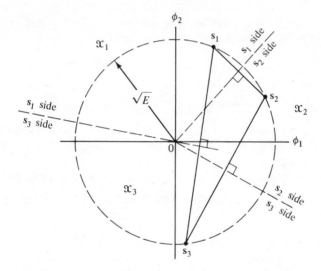

Figure 11.8-1. Optimum decision regions for equiprobable signals of equal energy in the presence of isotropic Gaussian noise.

are easily established by repeated application of the minimum-distance rule. In this example the decision regions are partitions of the finite dimensional observation space. In contrast, the decision regions established in Sec. 11.7 are partitions of the space of *likelihood ratios*.

If the *a priori* probabilities of the $\{m_i\}$ are not the same, the optimum receiver must modify the minimum-distance rule to include a bias that depends on $\{P_i\}$. Consider two signals $\mathbf{s}_1$ and $\mathbf{s}_2$. The decision boundary is given by

$$\|\mathbf{x} - \mathbf{s}_2\|^2 - 2\sigma_n^2 \ln P_2 = \|\mathbf{x} - \mathbf{s}_1\|^2 - 2\sigma_n^2 \ln P_1$$

or

$$\|\mathbf{x} - \mathbf{s}_2\|^2 = \|\mathbf{x} - \mathbf{s}_1\|^2 + K, \qquad (11.8\text{-}11)$$

where

$$K \equiv 2\sigma_n^2 \ln \frac{P_2}{P_1}. \qquad (11.8\text{-}12)$$

Suppose that $P_2 > P_1$. Then although the received signal $\mathbf{x}$ might be closer to $\mathbf{s}_1$, the optimum receiver might still decide in favor of $\mathbf{s}_2$ (i.e., m_2) because m_2 was the more *likely* message. We leave it as an exercise for the reader (Prob. 11-17) to modify the decision boundaries in Fig. 11.8-1 when the $\{P_i\}$ are not all equal.

The Optimum Receiver as a Correlator

The decision function $d_i(\mathbf{x})$, Eq. (11.8-9), can be rewritten as

$$d_i(\mathbf{x}) = \|\mathbf{x}\|^2 - 2\sigma_n^2 \ln P_i + E_i - 2 \sum_{k=1}^N x_k s_{ik}, \qquad (11.8\text{-}13)$$

where $E_i = \|\mathbf{s}_i\|^2$ is the energy of the ith signal. In minimizing this function with respect to the $\mathbf{s}_i$, $\|\mathbf{x}\|$ is irrelevant. The quantity

$$2\xi_i = 2\sigma_n^2 \ln P_i - E_i \qquad (11.8\text{-}14)$$

is a known constant for each i. Hence the basic operation of the receiver is to process the observations $\mathbf{x}$ to generate the *inner product*

$$\mathbf{x} \cdot \mathbf{s}_i = \sum_{k=1}^N x_k s_{ik} \qquad (11.8\text{-}15)$$

for each $\mathbf{s}_i$ and then to choose that value of i for which the modified decision function

$$d_i(\mathbf{x}) \equiv \mathbf{x} \cdot \mathbf{s}_i + \xi_i \qquad (11.8\text{-}16)$$

is a maximum. For the case where the x_k and s_{ik} are coefficients of an orthonormal expansion, then, as shown in Chapter 2, $\mathbf{x} \cdot \mathbf{s}_i$ can be written as

$$y_i(T) \equiv \mathbf{x} \cdot \mathbf{s}_i = \int_0^T x(t) s_i(t)\, dt. \qquad (11.8\text{-}17)$$

For the case where the x_k and s_{ik} are regarded as samples of continuous, band-limited waveforms $x(t)$ and $s_i(t)$ we can still write the inner product as an integral†:

$$\sum_{k=1}^N x_k s_{ik} \simeq 2W \int_{-T/2}^{T/2} x(t) s_i(t)\, dt,$$

†The shift of the interval of integration from $[0, T]$ to $[-T/2, T/2]$ is done for mathematical convenience.

where we have used the fact that

$$x(t) = \sum_{k=-\infty}^{\infty} x_k \, \text{sinc} \, (2Wt - k)$$

$$s_i(t) = \sum_{k=-\infty}^{\infty} s_{ik} \, \text{sinc} \, (2Wt - k)$$

and

$$\int_{-TW}^{TW} \text{sinc} \, (t - m) \, \text{sinc} \, (t - n) \, dt \simeq \delta_{mn}$$

if $T \gg W^{-1}$, W being the bandwidth of the channel. Recall from Sec. 10.4 that the correlation function of an ergodic process $x(t)$ can be estimated from

$$\mathcal{R}_{xx}(\tau) = \frac{1}{2T} \int_{-T}^{T} x(t)x(t + \tau) \, dt$$

and, by extension, that the cross correlation can be estimated from

$$\mathcal{R}_{xy}(\tau) = \frac{1}{2T} \int_{-T}^{T} x(t)y(t + \tau) \, dt$$

$$= \frac{1}{2T} \int_{-T}^{T} x(t)y(t) \, dt \qquad \text{at } \tau = 0.$$

Thus we see that the optimum MAP receiver is basically a correlator which correlates the received signal with the M possible transmitted signals s_i. In the special but important case of equiprobable and equienergetic signals the constant ξ_i, defined in Eq. (11.8-14), is independent of i, and the estimated signal $\hat{s}_i$ is then the one that correlates best with the received signal. If the ξ_i are not equal, they enter as a bias after the correlation operation. This is shown in Fig. 11.8-2.

Figure 11.8-2. Optimum (MAP) receiver.

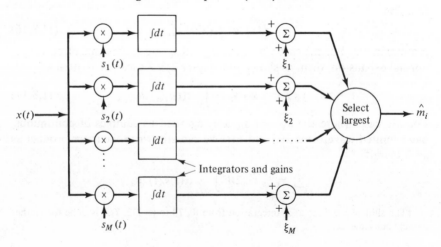

As pointed out in Sec. 11.6, the correlator and matched filter are closely related; in fact either can be used to implement the other. Thus an alternative scheme for implementing the optimum receiver utilizes a bank of matched filters instead of the multipliers and integrators shown in Fig. 11.8-2. The impulse responses of the M filters are $h_i(t) = s_i(T - t)$, $i = 1, \ldots, M$, and it is assumed that these are realizable. All the matched filter outputs are sampled at time $t = T$. This scheme is illustrated in Fig. 11.8-3.

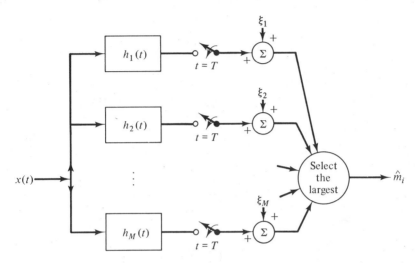

Figure 11.8-3. Realization of the optimum receiver with M matched filters.

11.9 Design of Optimum Receivers Using Orthogonal Signals

The optimum filter circuits shown in Figs. 11.8-2 and 11.8-3 can be simplified if the signals $\{s_i(t), i = 1, \ldots, M\}$ are linearly dependent. As an example, consider a binary system using signals $s_1 = s(t)$ and $s_2 = -s(t)$. Our discussion thus far implies that this calls for a receiver containing two correlators or matched filters. However, the same optimum result can clearly be obtained by using a single filter matched to $s(t)$ and to decide in favor of s_1 or s_2 depending on whether the output exceeds some positive or negative threshold, respectively.

To show how the design of optimum receivers can be streamlined with the help of orthogonal signals, we recapitulate for the reader's benefit some results from Chapter 2. Specifically, the signals $s_i(t)$ can be expanded in terms of a set of orthogonal functions:

$$s_i(t) = \sum_{j=1}^{N} s_{ij}\phi_j(t), \qquad (11.9\text{-}1)$$

where the $\phi_j(t)$ are functions satisfying

$$\int_0^T \phi_i(t)\phi_j^*(t)\,dt = 0, \qquad i \neq j. \tag{11.9-2}$$

It is usually convenient to also scale the functions ϕ_j in order to make them *orthonormal*, i.e., such that they satisfy

$$\int_0^T |\phi_j(t)|^2\,dt = 1. \tag{11.9-3}$$

Then

$$s_{ij} = \int_0^T s_i(t)\phi_j^*(t)\,dt, \tag{11.9-4}$$

which follows directly from Eqs. (11.9-1) and (11.9-2). The asterisk in Eq. (11.9-2), indicating complex conjugation, implies that the $\phi_j(t)$ could be complex functions.

We saw in the previous section that the critical operation performed by the optimum receiver designed to detect one of M known signals in Gaussian white noise is the formation of the inner product [Eq. (11.8-17)]:

$$y_i(T) = \int_0^T x(t)s_i(t)\,dt, \qquad i = 1, \ldots, M. \tag{11.9-5}$$

Figure 11.9-1. Optimum detector using orthogonal signal expansion.

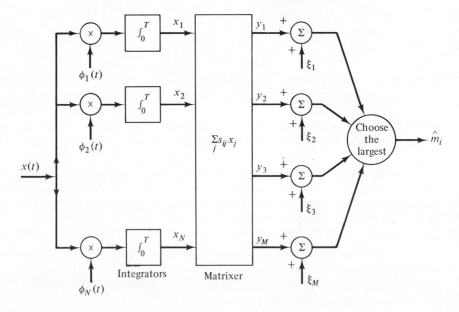

Substituting Eq. (11.9-1) into Eq. (11.9-5) gives

$$y_i(T) = \int_0^T x(t) \sum_{j=1}^N s_{ij}\phi_j(t)\, dt$$

$$= \sum_{j=1}^N s_{ij} \int_0^T x(t)\phi_j(t)\, dt = \sum_{j=1}^N s_{ij}x_j, \qquad (11.9\text{-}6a)$$

where

$$x_j \equiv \int_0^T x(t)\phi_j(t)\, dt. \qquad (11.9\text{-}6b)$$

This implies that the receiver can be implemented using N correlators or filters matched to the functions $\phi_j(t)$. The N correlator outputs x_j are then combined in a matrix circuit to implement the summation over N. For the binary example above such a *matrixer* would be very simple, and in general for $M \gg N$ the net effort could be a considerable simplification of the receiver circuitry. The circuit is shown in Fig. 11.9-1.

11.10   Binary Signals in Gaussian Noise; Signal Classes

In the last three sections we considered the problem of optimizing a receiver for a given set of signals. However, it is clear that certain kinds of signals are easier to separate than others. Therefore there exists a complementary problem of designing signals (subject to some power or energy constraint) such that the overall system consisting of transmitter and optimum receiver is optimized. The general signal design problem is quite difficult and not completely solved. However, there are a number of standard signal classes that, if not optimum under every criterion, are generally very good. A study of the properties of such signal classes together with a comparative analysis of their performance, as measured by the probability of error, is furnished in [11-15]. We shall undertake a brief discussion of signal classes by considering the special but important case of the detection of binary signals in Gaussian noise.

Consider the problem of the optimum detection of two equiprobable signals s_i, $i = 1, 2$, of equal energy. From Eq. (11.8-16) we find that the optimum receiver computes $d_i(x) = x \cdot s_i + \xi_i$ for each i and chooses the largest. Since $E_1 = E_2 = E$ and the *a priori* probabilities are $P_1 = P_2 = \frac{1}{2}$, the decision function is independent of ξ_i and the actual decision is based on $d_i(x) \equiv x \cdot s_i$. The signal s_i that maximizes $d_i(x)$ is the presumed transmitted signal. Hence the receiver makes an error whenever s_1 is sent and $d_2(x) - d_1(x) > 0$ and whenever s_2 is sent and $d_2(x) - d_1(x) < 0$.

The total probability of error is then

$$P(\epsilon) = P(\epsilon, s_1) + P(\epsilon, s_2)$$
$$= P_1 P(\epsilon | s_1) + P_2(\epsilon | s_2) \tag{11.10-1}$$
$$= \tfrac{1}{2} P(d_2 - d_1 > 0 | s_1) + \tfrac{1}{2} P(d_2 - d_1 < 0 | s_2), \tag{11.10-2}$$

where $d_i \equiv d_i(\mathbf{x})$, $i = 1, 2$. Define Z by

$$Z \equiv d_2 - d_1 = \mathbf{x} \cdot (\mathbf{s}_2 - \mathbf{s}_1)$$
$$= \int_0^T x(t)[s_2(t) - s_1(t)] \, dt. \tag{11.10-3}$$

Also assume that the noise $n(t)$ is white (an approximation) so that

$$R_n(\tau) = N_0 \delta(\tau)$$

and

$$\overline{n(t + \tau)s(t)} = 0 \qquad \text{for all } \tau.$$

Since Z is the result of a linear operation on Gaussian r.v.'s, it is itself Gaussian. We leave it as an exercise for the reader (Prob. 11-18) to show that the expected value of Z given s_i is

$$E[Z | s_1] \equiv \mu_1 = -E(1 - r) \tag{11.10-4a}$$
$$E[Z | s_2] \equiv \mu_2 = E(1 - r), \tag{11.10-4b}$$

where

$$r \equiv \frac{1}{E} \int_0^T s_1(t)s_2(t) \, dt. \tag{11.10-4c}$$

The variance of Z is the same whether $s_1(t)$ or $s_2(t)$ is transmitted. It is given by

$$\sigma_Z^2 = \overline{\left| \int_{-\infty}^{\infty} n(t)[s_2(t) - s_1(t)] \, dt \right|^2}$$
$$= \int_{-\infty}^{\infty} d\lambda \int_{-\infty}^{\infty} d\tau \, R_n(\tau - \lambda)[s_2(\tau) - s_1(\tau)][s_2(\lambda) - s_1(\lambda)]$$
$$= N_0 \int_{-\infty}^{\infty} [s_2(\tau) - s_1(\tau)]^2 \, d\tau$$
$$= 2N_0 E(1 - r). \tag{11.10-5}$$

Hence the probability of error can be written explicitly as

$$P(\epsilon) = \frac{1}{2} \frac{1}{\sqrt{2\pi\sigma_Z^2}} \int_0^{\infty} \exp\left\{ -\frac{1}{2}\left[\frac{x + E(1 - r)}{\sigma_Z} \right]^2 \right\} dx$$
$$+ \frac{1}{2} \frac{1}{\sqrt{2\pi\sigma_Z^2}} \int_{-\infty}^0 \exp\left\{ -\frac{1}{2}\left[\frac{x - E(1 - r)}{\sigma_Z} \right]^2 \right\} dx. \tag{11.10-6}$$

By a symmetry argument, the two integrals can be shown to have equal value.

A reduction to standard form finally results in†

$$P(\epsilon) = \Phi_c \left[\sqrt{\frac{E(1-r)}{2N_0}} \right]. \tag{11.10-7}$$

By varying the parameter r, we are in a position to compare different classes of signals. If the signals are *orthogonal*, $r = 0$, and the result is

$$P(\epsilon) = \Phi_c \left(\sqrt{\frac{E}{2N_0}} \right). \tag{11.10-8}$$

If $r = -1$, the signals are said to be *antipodal*. The probability of error is then

$$P(\epsilon) = \Phi_c \left(\sqrt{\frac{E}{N_0}} \right). \tag{11.10-9}$$

Note that this is a *lower* error probability than for orthogonal signals even though the energy outlay is the same. Conversely the same error probability can be achieved by antipodal signals with only one-half the energy required for orthogonal signals. It may be said therefore that antipodal signals are superior to orthogonal signals within the context of this discussion. Figure 11.10-1 shows the vector representation of antipodal and orthogonal signals in two-dimensional space.

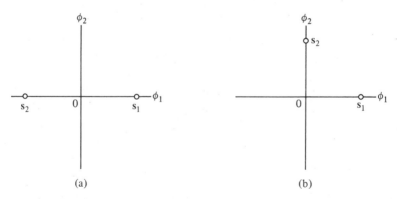

(a)   (b)

Figure 11.10-1. Binary signal classes: (a) antipodal signals; (b) orthogonal signals.

In practice the use of antipodal signals involves synchronization problems that are absent when the signals are orthogonal. For instance, $s_1(t) = A \cos \omega_0 t$ and $s_2(t) = -A \cos \omega_0 t$ are antipodal, but to tell which is which requires a precise knowledge of the phase. On the other hand, $A \cos \omega_1 t$ and $A \cos \omega_2 t$ are essentially orthogonal over T if the two frequencies ω_1

†Recall that $\Phi_c(\alpha) \equiv [2\pi]^{-1/2} \int_\alpha^\infty \exp\left[-\frac{1}{2}u^2\right] du = \frac{1}{2} - \Phi(\alpha)$, where $\Phi(\alpha)$ was defined in Eq. (9.3-8) in Chapter 9.

and ω_2 differ by a sufficiently large amount (i.e., $T\omega_1 \gg 1$, $T\omega_2 \gg 1$, $T|\omega_2 - \omega_1| \gg 1$), and they are easily separated by narrow filters tuned to the appropriate frequencies.

The computations of error probabilities when more than two signals are involved are essentially direct extensions of the calculations associated with two signals. Frequently the computations can be simplified by considering a signal set that is equivalent to the original with respect to the probability of error but, by virtue of rotation or translation, occupies different regions in signal space. This idea can be used to generate so-called minimum-energy signal classes as well as other signal classes. We briefly describe some of these below.

Some Well-Known Signal Classes

Minimum-Energy Signals

The average energy associated with a signal set s_i', $i = 1, \ldots, M$, is

$$\bar{E}' = \sum_{i=1}^{M} E_i' P_i = \sum_{i=1}^{M} \|s_i'\|^2 P_i, \tag{11.10-10}$$

where E_i', P_i are the ith signal energy and *a priori* probability, respectively. If each vector s_i' is translated by the same vector $\boldsymbol{\alpha}$, the probability of error $P(\epsilon)$ is unaffected. However, the average energy of the new set is

$$\bar{E}(\boldsymbol{\alpha}) = \sum_{i=1}^{M} \|s_i' - \boldsymbol{\alpha}\|^2 P_i. \tag{11.10-11}$$

When the translation vector $\boldsymbol{\alpha}$ satisfies (Prob. 11-19)

$$\boldsymbol{\alpha} = \boldsymbol{\alpha}_0 \equiv \sum_{i=1}^{M} s_i' P_i, \tag{11.10-12}$$

then the average energy of the translated cluster is a minimum; i.e.,

$$\bar{E}(\boldsymbol{\alpha}_0) = \bar{E}_{\min} = \sum_{i=1}^{M} \|s_i' - \boldsymbol{\alpha}_0\|^2 P_i. \tag{11.10-13}$$

The translated cluster represents a minimum-energy configuration, and the resulting signal set $\{s_i = s_i' - \boldsymbol{\alpha}_0\}$ is *minimum energy*. Such signals are often desirable since they represent an efficient allocation of energy for a given probability of error. An example is shown in Fig. 11.10-2.

Rectangular Signals: Vertices of a Hypercube

A signal frequently used in communication systems is a string of N binary pulses where each pulse in the string can take on the value $\pm C$. Such signals are, for instance, generated in a standard antipodal PCM system (cf. Sec. 4.6). If the energy of the string of N pulses is E, then the pulse height is $\pm\sqrt{E/N}$. For $N = 2$ there are four possibilities: $\sqrt{E/2}\,(1, 1)$, $\sqrt{E/2}\,(1, -1)$,

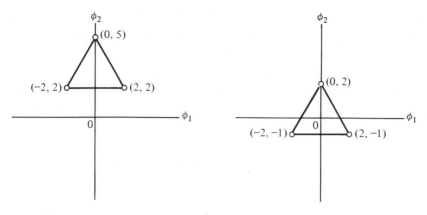

Figure 11.10-2. (a) Three equiprobable signals; (b) the minimum-energy configuration.

$\sqrt{E/2}\,(-1,\,1)$, and $\sqrt{E/2}\,(-1,\,-1)$. These four possibilities can be represented as the corners of a square, as shown in Fig. 11.10-3(a). For $N = 3$ there are eight possibilities, representable as the vertices of a cube. This is shown in Fig. 11.10-3(b). In general, for any positive integer N there are 2^N signals representable in terms of the vertices of an N-dimensional hypercube. The set $\phi_i(t)$, $i = 1, \ldots, N$, of orthogonal functions can take on many forms; a simple example is the set of single pulses of unit height occurring at the ith time interval. These are indicated in the figure. Then the coefficients s_{ij} of $s(t)$ of Eq. (11.9-1) take on values of $\pm\sqrt{E/N}$.

A receiver operating on this signal set makes an error whenever it locates the received signal on the wrong vertex of the hypercube. This is equivalent to deciding the wrong sign for any one of the coefficients s_{ij} in Eq. (11.9-1).

Simplex Signals

Orthogonal signals do not generally form a minimum-energy configuration. A set of minimum-energy signals derived from an orthogonal set, for example, through use of Eq. (11.10-13) is called a *simplex* set. The simplex signals furnish the minimum probability of error under conditions of (1) white Gaussian noise, (2) mean-energy constraint, and (3) equal *a priori* probabilities $\{P_i\}$; i.e., $P_i = 1/M$, $i = 1, \ldots, M$.

The construction of a simplex set can be illustrated by considering two equally likely orthogonal signals, shown in Fig. 11.10-4(a), where $\mathbf{s}_1 = \sqrt{E}\,\boldsymbol{\phi}_1$, $\mathbf{s}_2 = \sqrt{E}\,\boldsymbol{\phi}_2$. The optimum translation vector $\boldsymbol{\alpha}_0$ is given by

$$\boldsymbol{\alpha}_0 = \sum_{i=1}^{2} \mathbf{s}_i P_i \qquad (11.10\text{-}14)$$

$$= \left(\frac{\sqrt{E}}{2}, \frac{\sqrt{E}}{2}\right)^T. \qquad (11.10\text{-}15)$$

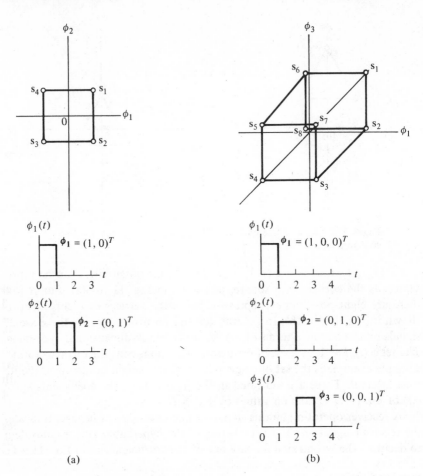

Figure 11.10-3. Rectangular signals on the vertices of a hypercube: (a) $N = 2$, four signals; (b) $N = 3$, eight signals.

The corresponding simplex signals are given by

$$\tilde{\mathbf{s}}_1 = \mathbf{s}_1 - \boldsymbol{\alpha}_0 = \left(\frac{\sqrt{E}}{2}, -\frac{\sqrt{E}}{2}\right)^T$$

$$\tilde{\mathbf{s}}_2 = \mathbf{s}_2 - \boldsymbol{\alpha}_0 = \left(-\frac{\sqrt{E}}{2}, \frac{\sqrt{E}}{2}\right)^T$$

and are shown in Fig. 11.10-4(c). The simplex signals in this case are seen to be antipodal.

Biorthogonal Signals

A biorthogonal set is a mixture of orthogonal and antipodal signals that is constructed from an orthogonal set by adding the mirror image of the origi-

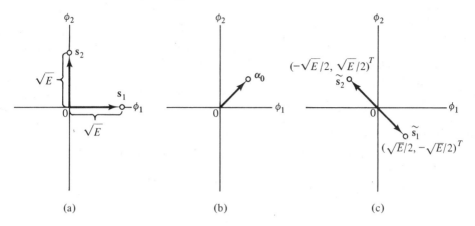

Figure 11.10-4. Generation of simplex signals from orthogonal signals: (a) orthogonal signals; (b) optimum translation vector; (c) derived simplex set $\{\tilde{\mathbf{s}}_i = \mathbf{s}_i - \boldsymbol{\alpha}_0\}$.

nal signals. Thus if $\mathbf{s}_1, \ldots, \mathbf{s}_N$ is the original orthogonal set, the biorthogonal set consists of $\{\mathbf{s}_i\}$ and $\{-\mathbf{s}_i\}$, leading to $M = 2N$ signals.

For large M, biorthogonal signals have almost as low an error probability as orthogonal signals. Biorthogonal signals are used primarily where bandwidth is at a premium since they require only half the bandwidth of the same number of orthogonal signals.

In situations where both energy and bandwidth are at a great premium, as in deep-space communications, antipodal binary signals are used. These signals furnish the smallest error probability for a given energy-to-noise ratio. For this reason the extra complication introduced by the need for phase coherence is regarded as worthwhile.

11.11 Conclusion

The material in this chapter serves as an introduction to signal processing, a subject of considerable research interest since World War II. A large number of existing books and articles discuss and elaborate on the various topics touched on here. Much of modern communication theory is concerned with the science of extracting signals from noise, signal design, coding, digital signal processing, image processing, adaptive filtering, etc. Fortunately, the interested reader is lucky to have a plethora of good textbooks available for further study in these areas. The list is long, but our own favorites include the books by Schwartz and Shaw [11-4], Wozencraft and Jacobs [11-15], and Van Trees [11-14]. The book by Wainstein and Zubakov [11-16] is an excellent translation from the Russian and should be of particular interest to readers with an interest in radar. A long list of references is given on p. 706 of [11-15].

PROBLEMS

11-1. (a) For the *RLC* circuit shown in Fig. P11-1, compute the signal-to-noise ratio

$$\left(\frac{S}{\mathfrak{N}}\right)_0 = \frac{\overline{y_s^2(t)}}{\overline{y_n^2(t)}},$$

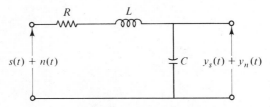

Figure P11-1

where $y_s(t)$ and $y_n(t)$ are the outputs due to signal $s(t)$ and uncorrelated noise $n(t)$, respectively. Let the input signal be $s(t) = A\cos(\omega_0 t + \Theta)$, where Θ is a uniformly distributed r.v. and $n(t)$ is white noise with spectral density N_0. (b) Investigate $(S/\mathfrak{N})_0$ as a function of the parameter $\beta = R\sqrt{C/L}$ if $\omega_r \equiv [LC]^{-1/2}$ is held fixed. At what value of β is $(S/\mathfrak{N})_0$ maximized? (c) Remove the restriction on holding ω_r fixed, and find a set of parameters that will maximize $(S/\mathfrak{N})_0$. Make any reasonable assumptions.

11-2. For the *RC* filter shown in Fig. 11.2-1, show that the minimum m.s.e. in estimating $s(t)$ from $y(t)$ is obtained when the parameter $\alpha \equiv RC$ is approximately

$$\alpha = \alpha_0 = \frac{2A^2}{N_0\omega_0^2}$$

when $2A^2/N_0\omega_0 \gg 1$.

11-3. Show that for a discrete, causal, time-invariant system excited by a stationary signal, x, the optimum [i.e., as defined by Eqs. (11.3-1) and (11.3-2)] filter weights $h(l)$ satisfy

$$R_{sx}(n) = \sum_{l=0}^{\infty} h(l)R_x(n-l),$$

where $R_{sx}(n)$ and $R_x(n)$ are the cross-correlation of s and x and autocorrelation of x respectively.

11-4. a. Show that the autocorrelation function $R_n(l)$ for the first-order MA process

$$n_j = z_j - az_{j-1}, \qquad |a| < 1,$$

where n_j is the MA process and z_j is a white-noise shock, is given by

$$R_n(l) = (1 + a^2)R_z(l) - aR_z(l-1) - aR_z(l+1).$$

b. Show that the MA process of order q, i.e., Eq. (11.4-7), has nonzero correlation terms only up to the qth lag interval.

11-5. (Discrete Wiener filter). Consider the problem of estimating the r.v. s from two measurements as was done in Sec. 11.4. Let the observations be

$$x_j = s + n_j, \qquad j = 1, 2.$$

The Wiener filter has the form

$$\hat{s} = h_1 x_1 + h_2 x_2.$$

Assume a first-order AR process. Compute the optimum filter weights h_1 and h_2.

11-6. In Prob. 11-5, show that the Wiener estimate furnishes a better estimate than the sample mean. "Better" here means a lower m.s.e.

11-7. In Prob. 11-6, investigate the behavior of the Wiener estimate when (a) the signal-to-noise ratio $\overline{s^2}/\sigma_n^2$ becomes very large and (b) the noise becomes uncorrelated. Explain why, even when the noise is uncorrelated, the Wiener estimate $\hat{s}$ and the sample estimate $\hat{s}_{SM}$ are not the same.

11-8. Assume that we want to estimate the parameter vector $s = (s_1, \ldots, s_L)^T$ whose components s_j, $j = 1, \ldots, L$, are related to the observed x_i at $t = t_i$ according to

$$x_i = \sum_{j=1}^{L} \alpha_{ij} s_j + n_i,$$

where n_i is signal-independent noise. Show that for this case Eq. (11.3-14) is generalized to

$$\hat{s} = R_s^T A^T (A R_s A^T + R_n)^{-1} x,$$

where $\hat{s}$ is the Wiener filter estimate of s, A is the matrix of the α_{ij}'s, and R_s and R_n are the autocorrelation matrices of signal and noise, respectively.

11-9. Assume that we are interested in estimating the scalar parameter s from L samples of the observed signal x_i; i.e.,

$$x_i = is + n_i, \qquad i = 1, \ldots, L.$$

The estimate of s is $\hat{s}$, given by

$$\hat{s} = \sum_{i=1}^{L} h_i x_i,$$

where the h_i, $i = 1, \ldots, L$, are chosen to minimize the m.s.e. between $\hat{s}$ and s. Show that if $L = 3$, the filter weights are ([11-4], p. 285)

$$h_1 = \frac{1}{14 + (\sigma_n^2/P_s)}, \qquad h_i = ih_1, \quad i = 2, 3,$$

where σ_n^2 and P_s are as defined in Sec. 11.4. Assume zero-mean white noise; i.e., $\overline{n_i n_j} = 0$ for $i \neq j$, $\overline{n_i^2} = \sigma_n^2$, $\overline{n_i} = 0$.

11-10. Derive a recursive algorithm for an adaptive filter as in Eq. (11.5-8), but assume that the signal covariance rather than the noise covariance is known.

11-11. In antenna or sonar array processing, the sampled signals appear simultaneously at all L receiving elements. A block diagram of the adaptive filter for minimizing the mean-square error ϵ_j at the jth time is shown in Fig. P11-11. The error ϵ_j is the difference between the desired output r_j and the actual output y_j. The weights $\{h_i\}$ are adjustable.

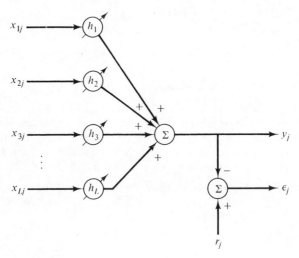

Figure P11-11

a. If the input vector is $\mathbf{x} = (x_{1j}, x_{2j}, \ldots, x_{Lj})^T$ and the weight vector is $\mathbf{h}_j = (h_{1j}, h_{2j}, \ldots, h_{Lj})$, show that

$$y_j = \mathbf{h}_j^T \mathbf{x}_j = \mathbf{x}_j^T \mathbf{h}_j.$$

b. Let it be understood that all quantities from here on refer to the jth time. The subscript j will therefore be dropped, and other subscripts with different meanings will be introduced. Show that the square error is

$$\epsilon^2 = r^2 - 2r\mathbf{x}^T\mathbf{h} + \mathbf{h}^T\mathbf{x}\mathbf{x}^T\mathbf{h}.$$

c. Show that the mean-square error is

$$\overline{\epsilon^2} = \overline{r^2} - 2\mathbf{R}_{rx}^T\mathbf{h} + \mathbf{h}^T\mathbf{R}_{xx}\mathbf{h},$$

where $\mathbf{R}_{rx} = (\overline{rx_1}, \ldots, \overline{rx_L})^T$ and

$$\mathbf{R}_{xx} = \begin{bmatrix} \overline{x_1 x_1} & \cdots & \overline{x_1 x_L} \\ \vdots & & \vdots \\ \overline{x_L x_1} & \cdots & \overline{x_L x_L} \end{bmatrix}.$$

11-12. This is a continuation of Prob. 11-11.
a. Compute the gradient vector $\nabla \overline{\epsilon^2}$, and set it equal to zero to show that

the optimum weights are given by

$$\mathbf{h}_0 = \mathbf{R}_{xx}^{-1}\mathbf{R}_{rx}.$$

b. Define $\mathbf{g} \equiv \mathbf{h} - \mathbf{h}_0$ to be the difference between an arbitrary weight vector and the optimum. Show that the mean-square error can be written as

$$\overline{\epsilon^2} = [\overline{\epsilon^2}]_{\min} + \mathbf{g}^T\mathbf{R}_{xx}\mathbf{g}$$

and that

$$\nabla\overline{\epsilon^2} = 2\mathbf{R}_{xx}\mathbf{g}.$$

11-13. In this problem we continue the discussion in Probs. 11-12 and 11-11. In Prob. 11-11, $\overline{\epsilon^2}$ was seen to be a quadratic function of the weights. The surface described by this quadratic function is a hyperparaboloid. The principal axes of the surface are the set of coordinates in which the quadratic function contains no cross terms. To find these coordinates, proceed to write

$$\mathbf{R}_{xx} = \mathbf{U}\mathbf{\Lambda}\mathbf{U}^T,$$

where $\mathbf{U}$ is the modal matrix of $\mathbf{R}_{xx}$, i.e., the matrix of eigenvectors, and $\mathbf{\Lambda}$ is the diagonal matrix of eigenvalues, i.e., $\mathbf{\Lambda} = \text{diag}[\lambda_1, \ldots, \lambda_L]$. Show that $\overline{\epsilon^2}$ can be written as

$$\overline{\epsilon^2} = [\overline{\epsilon^2}]_{\min} + \mathbf{g}'^T\mathbf{\Lambda}\mathbf{g}'.$$

Relate $\mathbf{g}'$ to $\mathbf{g}$ in Prob. 11-12. Show that $\overline{\epsilon^2}$ does not contain any cross terms.

11-14. A matched filter has impulse response

$$h_M(t) = \cos \omega_c(t_0 - t) \cdot \text{rect}\frac{t_0 - t}{T}, \qquad T \gg \frac{2\pi}{\omega_c}, \quad t_0 > \frac{T}{2}.$$

The input is given by $x(t) = A \cos \omega_c(t - t_1) \cdot \text{rect}[(t - t_1)/T] + n(t)$, where $n(t)$ has uniform spectral density N_0.
a. Compute the maximum signal-to-noise ratio at the output.
b. At what time does this occur?

11-15. Justify the arguments in Eq. (11.6-18).

11-16. Let m_i, $i = 1, 2, 3$, be three messages with *a priori* probabilities $P_1 = \frac{1}{6}$, $P_2 = \frac{2}{6}$, $P_3 = \frac{3}{6}$. The conditional probabilities are $f(x|m_1) = 2 \exp(-2x)$, $f(x|m_2) = 3 \exp(-3x)$, and $f(x|m_3) = 4 \exp(-4x)$, $x \geq 0$. Assume that the likelihood-ratio test is applied.
a. Find the optimum decision regions in likelihood-ratio space as in Fig. 11.7-5.
b. Find the optimum partitioning of the observation space $\mathcal{X}$.

11-17. Consider the detection of equienergetic signals in isotropic Gaussian noise as in Sec. 11-8. Use Eqs. (11.8-11) and (11.8-12) to modify the decision boundaries when the *a priori* probabilities are not equal. Consider two signals in the ϕ_1-ϕ_2 coordinate system given by $\mathbf{s}_1 = (5, 0)$, $\mathbf{s}_2 = (0, 5)$. Let

$P_1 = \frac{1}{5}$, $P_2 = \frac{4}{5}$. Take $\sigma_n^2 = 1$. Draw the decision regions in the ϕ_1-ϕ_2 coordinate system. Under what conditions will the decision boundaries be straight lines (or planes, in higher dimensions)?

11-18. In the problem of detecting binary signals in Gaussian noise, show that $Z \equiv \mathbf{x} \cdot (\mathbf{s}_2 - \mathbf{s}_1)$ satisfies

$$E(Z\,|\,\mathbf{s}_1) = -E(1 - r)$$
$$E(Z\,|\,\mathbf{s}_2) = E(1 - r),$$

where r is defined as in Eq. (11.10-4c).

11-19. Consider the average energy $\bar{E}(\boldsymbol{\alpha})$ of a set of signals $\{\mathbf{s}'_i\}$ translated by a vector $\boldsymbol{\alpha}$; i.e.,

$$\bar{E}(\boldsymbol{\alpha}) = \sum_{i=1}^{M} \|\mathbf{s}'_i - \boldsymbol{\alpha}\|^2 P_i.$$

Show that $\bar{E}(\boldsymbol{\alpha})$ is minimized when $\boldsymbol{\alpha} = \boldsymbol{\alpha}_0 = \sum_{i=1}^{M} \mathbf{s}'_i P_i$.

11-20. Work out the details of the example illustrated in Fig. 11.10-2.

11-21. Consider the two signals

$$\phi_1(t) = \sqrt{\frac{2}{T}} \cos 2\pi f_1 t$$

$$\phi_2(t) = \sqrt{\frac{2}{T}} \cos 2\pi f_2 t, \qquad f_2 < f_1.$$

a. Show that if $Tf_1 \gg 1$, $Tf_2 \gg 1$, and $T(f_2 - f_1) \gg 1$, then $(\phi_1, \phi_2) \simeq 0$, and hence $\phi_1(t)$ and $\phi_2(t)$ are orthogonal over T.
b. Let $s_1(t) = \sqrt{E}\,\phi_1(t)$ and $s_2(t) = 2\sqrt{E}\,\phi_2(t)$. Assume that $P_1 = P_2$. Compute the simplex signals.
c. In part b, assume that $P_1 = 4P_2$; compute the simplex signals.

11-22. Describe the decision regions for the following signal sets:
a. Rectangular, with $N = 2$, $M = 4$ as in Fig. 11.10-3.
b. Orthogonal, with (1) $N = 2$, (2) $N = 3$.
c. Simplex, with $N = M = 2$.
d. Biorthogonal, with $N = 2$, $M = 4$.
Equiprobable signals are assumed.

REFERENCES

[11-1] R. J. Schwarz and B. Friedland, *Linear Systems*, McGraw-Hill, New York, 1965.

[11-2] N. Wiener, *The Extrapolation, Interpolating and Smoothing of Stationary Time Series*, Wiley, New York, 1949.

[11-3] R. E. Kalman, "A New Approach to Linear Filtering and Prediction," *Trans. ASME D*, **82**, pp. 35–45, March 1960.

[11-4]   M. Schwartz and L. Shaw, *Signal Processing*, McGraw-Hill, New York, 1970.

[11-5]   E. P. G. Box and G. M. Jenkins, *Time Series Analysis: Forecasting and Control*, Holden-Day, San Francisco, 1970.

[11-6]   B. Widrow, P. Mantey, L. Griffiths, and B. Goode, "Adaptive Antenna Systems," *Proc. IEEE*, **55**, pp. 2143–2159, Dec. 1967.

[11-7]   B. Widrow et al., "Adaptive Noise Cancelling: Principles and Applications," *Proc. IEEE*, **63**, pp. 1692–1716, Dec. 1975.

[11-8]   R. Lucky, "Automatic Equalization for Digital Communication," *Bell System Tech. J.*, **44**, pp. 547–588, April 1965.

[11-9]   H. Robbins and S. Monro, "A Stochastic Approximation Method," *Ann. Math. Statistics*, **22**, No. 1, 400–407, 1951.

[11-10] B. Widrow et al., "Stationary and Nonstationary Learning Characteristics of the LMS Adaptive Filter," *Proc. IEEE*, **64**, pp. 1151–1161, Aug. 1976.

[11-11] A. Papoulis, *The Fourier Integral and Its Applications*, McGraw-Hill, New York, 1962.

[11-12] M. Schwartz, W. R. Bennett and S. Stein, *Communication Systems and Techniques*, McGraw-Hill, New York, 1966.

[11-13] R. S. Berkowitz, ed., *Modern Radar*, Wiley, New York, 1965.

[11-14] H. L. van Trees, *Detection, Estimation and Modulation Theory*, Wiley, New York, 1968.

[11-15] J. M. Wozencraft and I. W. Jacobs, *Principles of Communication Engineering*, Wiley, New York, 1965.

[11-16] L. A. Wainstein and V. D. Zubakov, *Extraction of Signals from Noise*, Prentice-Hall, Englewood Cliffs, N.J., 1962.

Radar and Sonar

12.1 Introduction

Radar and sonar are used to determine the range, velocity, and direction of objects that are remote from an observer. Radar employs electromagnetic radiation, usually at microwave frequencies, and is generally used to observe airborne objects and objects in outer space.† Sonar uses sound waves and can therefore be used under water where the electromagnetic waves of radar do not penetrate. Sonar systems are used by certain animals, notably bats and dolphins, to orient themselves in space and to locate prey. Simple forms of sonar have been used at least since World War I to detect submarines. The fact that radio waves could be reflected from objects—the basis for radar—was already known to Heinrich Hertz in the late nineteenth century. One of the earliest pulsed radars was constructed by L. C. Young and R. M. Page at the U.S. Naval Research Laboratory in 1934. The major development of radar took place in the years prior to and during World War II. Important components such as the cavity magnetron, required to generate the necessary high microwave powers, as well as the PPI display (described in Sec. 12-2) were invented by British scientists and engineers working under the direction of Sir William Watson-Watt. The major center for radar development in this country was the Radiation Laboratory at the Massachusetts Institute of Technology. The work of this laboratory during the years from 1940 to 1945 has been summarized in the 28-volume Radiation Laboratory

---

†A notable exception is "side-looking" radar which furnishes high-resolution terrain imagery, even through clouds from an airplane.

Series [12-1], which at the time of its publication was one of the most complete compendia of the state of the art in electronics, microwaves, etc., available.

Radar and sonar systems can be active or passive. An active system employs a powerful transmitter that sends signals toward the target and analyzes the resulting echo. A passive system uses only a receiver and obtains target information from signals radiated by the target itself. Most radars are active,† whereas many sonars are passive. In an active system range is determined in terms of the round-trip time between the emission of a signal and the reception of the echo. This process is referred to as *echo location*. By analyzing the frequency content of the echo, an active system can also determine the Doppler shift from which target velocity can be estimated. Radars generally use highly directional, very narrow-beam antennas that are rotated mechanically or electronically to sweep out various regions in space. The narrow beam permits estimation of target direction from the fact that echoes are received only from the direction in which the antenna is pointed. Antenna beam patterns can be shaped in a variety of ways, each of which determines a different kind of radar. For instance, the beam can be very narrow in azimuth and rather broad in elevation, and such a system (termed a search radar) will show all targets at a given azimuth, independent of elevation. Other systems, for instance, tracking radars, may have a beam that is as narrow as possible in all directions.

Since passive systems emit no signals, they cannot employ echo location. They are therefore limited to detection (i.e., to decide whether a certain kind of target is present or not) and estimation of direction. Directional information can be used indirectly, i.e., via some sort of triangulation, to determine range, but the range information obtainable in passive systems is usually rather imprecise. The main advantage of passive systems is that they do not reveal their presence to the target or to anyone else who might be listening. For this reason passive sonars are commonly used on submarines.

In the following discussion we consider mainly the simple active radar. Except for certain details that relate specifically to electromagnetic radiation, this discussion applies also to active sonars. Some of the special problems posed by passive operation are considered at the end of the chapter.

12.2 Pulsed Radar

In the most common form of radar, the transmitter sends out a narrow beam of very short-duration microwave pulses. Echoes from these pulses are received and amplified in a receiver and displayed on a cathode-ray tube. An accurate linear sweep voltage is applied to the cathode-ray tube in synchro-

---

†Again a notable exception is the class of *radio telescopes*, which receive electromagnetic radiation from self-luminous sources such as radio stars.

nism with the transmitted pulse. The received echo deflects or, more commonly, brightens the cathode-ray trace, and therefore the range to the target can be read off directly from the cathode-ray display. A block diagram of a typical radar system is shown in Fig. 12.2-1.

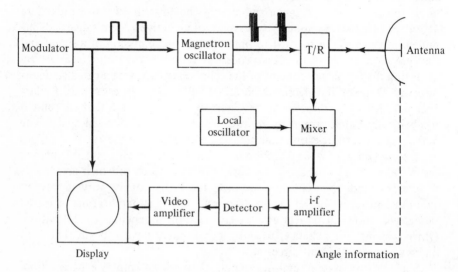

Figure 12.2-1. Block diagram of a radar system.

The magnetron is the source of the transmitted power, which even in a small radar may be in excess of 100 kW. The carrier frequency is determined by the construction of the magnetron and may be anywhere between 200 MHz to 30 GHz. A common value for small radars is 10 GHz (3 cm or X band).†

The magnetron is pulsed by power pulses derived from the modulator. Typical pulse lengths are on the order of milliseconds, resulting in duty ratios on the order of 0.001. Hence the 100-kW peak magnetron power may correspond to about 100 W of average power. A single antenna is typically used for both transmitter and receiver. An electronic switch, the T/R box (for transmit/receive), disconnects the receiver from the antenna during the time that the transmitter is on so as to protect delicate receiver input circuits from the large transmitted power. The receiver is a typical superheterodyne, as described in Chapter 6; it consists of a local oscillator, mixer, IF amplifier, detector, and video amplifier.

The most commonly used antenna in small radars operating at wave-

---

†The term *X band* was originally used for military secrecy. It denotes the band from 0.52×10^{10} Hz to 1.09×10^{10} Hz. Other bands, in 10^{10} Hz, are the *K* band, 1.09–3.6; *Q* band, 3.6–4.6; L band, 0.039–0.155, and S band, 0.155–0.52.

lengths shorter than 10 cm is the parabolic dish. The dish is frequently illuminated by an electromagnetic horn placed at the end of the waveguide connecting the antenna to the radar set proper. The horn is located at the focus of the paraboloid, and the effect is to produce a highly collimated beam similar to that from a search light. A beam having a circular cross section is produced by a circular dish. A vertical fan-shaped beam is produced by an antenna whose width is larger than its height, because the beam width is inversely proportional to the antenna width.

In most small radars the antenna is rotated mechanically. A typical example is the radar used on ships to detect other ships or obstacles in the neighborhood. In these radars the antenna radiates a fan-shaped beam that is rotated through a 360° angle every few seconds. The angular position of the antenna as well as the receiver output is coupled to the display device. A popular form of display in azimuthal search radars is the PPI scope (for plan position indicator). In this scope the spot starts each sweep from the center of the screen and sweeps radially outward at a constant rate. The direction of the sweep corresponds to the antenna position; e.g., if the antenna points north, the sweep is vertically up. The received signal is applied as z-axis modulation so that any echo in the beam results in a bright spot on the screen. Since the spot moves out from the center at the beginning of the sweep, the farther away the target is, the farther out toward the edge of the screen it will appear. A long-persistence phosphor is used so that the image remains on the screen for the time taken to complete one revolution of the antenna. Thus the PPI display sweeps out a map of all the targets surrounding the radar. A timing diagram for a typical radar is shown in Fig. 12.2-2. A typical PPI display is shown in Fig. 12.2-3.

Tracking radars use a rapidly rotating conical scan as shown in Fig. 12.2-4. Such a scan may be produced by slightly offsetting the feed from the focus of the dish and rotating the feed rather than the entire dish. In this way very high scanning speeds can be obtained. Tracking is accomplished by coupling a synchronous switch (also called a commutator) to the rotating beam so that signals reflected from the target during each quarter-revolution of the scan can be separated. For example, if the target is above the axis of the conical scan, the echo in the upper quarter is larger than in the lower. This difference can be used to rotate the entire antenna upward until the signals in the upper and lower quarters are the same. Sideways tracking is accomplished in a similar way.

A major problem associated with radars used in ground control of aircraft and similar applications is the fact that stationary objects near the radar result in strong echoes that mask those from the desired targets. These large undesired echoes are referred to as ground clutter. (There is also sea clutter, which plagues radars operating on ships.) Ground clutter can be combatted by employing MTI (moving-target indicator) radars. MTI radars detect not

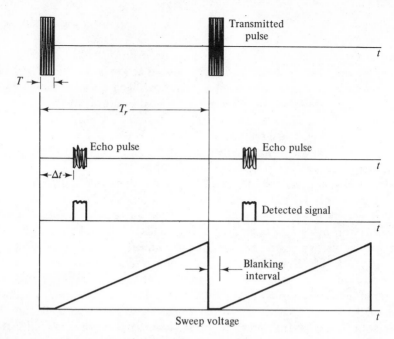

Figure 12.2-2. Radar timing diagram.

Figure 12.2-3. Typical PPI display from an airborne radar (Hazeltine E-2C) on an aircraft flying up the northeastern coast of the United States. Cape Cod, Nantucket Island, Martha's Vineyard and Long Island appear clearly. The short lines in the ocean are radar targets. *Source: Grumman Aerospace Corporation, Bethpage, NY.*

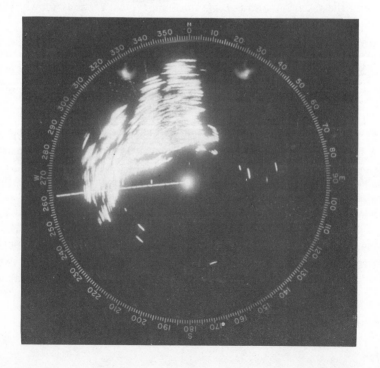

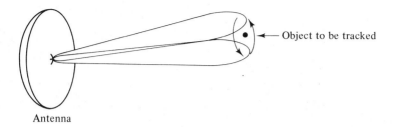

Object to be tracked

Antenna

Figure 12.2-4. Conical scan for a tracking radar.

only the range of targets but also their velocity by analyzing the Doppler shift in the echo return. MTI radars can be designed to suppress all displays except those of targets moving within a narrow range of velocities.

12.3 The Radar Equation

If a target is at a range R from the radar, the round-trip time for the signal is given by

$$\tau = \frac{2R}{c}, \tag{12.3-1}$$

where c is the speed of light (or of sound in sonar). Two point reflectors whose range differs by a distance ΔR are said to be resolvable if

$$\Delta R > \frac{cT}{2}, \tag{12.3-2}$$

where T is the pulse length.† ΔR is also the smallest range that the radar can observe. Thus the pulse length T is determined by the minimum desired range and by the desired target resolution.

For a target at maximum range, the round-trip time for the pulse has to be less than the pulse repetition period T_r:

$$R_{\max} < \frac{c}{2}T_r. \tag{12.3-3}$$

Since the transmitted signal is periodic, the fact that an echo from a target at greater than maximum range has a round-trip time of more than the pulse repetition period is not apparent in the display; in fact the indicated range is always $c\tau/2 \bmod (R_{\max})$. Thus targets at greater than maximum range appear at the fractional part of $R/R_{\max}$ and result in a range ambiguity. In practical radars this ambiguity is largely eliminated by making sure that echoes received from very distant targets are too weak to yield a noticeable

---

†Without explicitly stating so, we always assume that noise is present. Otherwise the resolution could be unlimited.

signal. This is done by matching the radiated power to the maximum range for which the equipment is designed.

The maximum range yielding detectable echoes can be easily computed subject to the following assumptions:

1. No large obstacles between antenna and target.
2. Single straight-line transmission path (i.e., no reflections or refractions).
3. Homogeneous transport medium.

Suppose that the transmitted power is P_0 watts. If the transmitter were omnidirectional, i.e., if the power were radiated uniformly in all directions, then the power density at a distance R from the transmitter would be $P_0/4\pi R^2$ watts per unit area. Actually the power is radiated from a highly directional antenna having a maximum gain $G_0 \gg 1$ at the center of the beam. Hence the power density at the center of the beam and at the distance R is $P_0 G_0/4\pi R^2$ watts per unit area.

The amount of power intercepted by the target depends on the radar cross section α of the target. This is in units of area and depends on the physical dimensions of the target and on its electrical characteristics (good conductors have a larger cross section than poor conductors). Targets having complicated shapes, such as airplanes, have cross sections that vary widely and randomly with small changes of target aspect. This is referred to as target scintillation. However, a crude estimate for α, which ignores this variability, is simply the area of the projection of the target on a plane perpendicular to a ray from the radar to the target.

A target with cross section α constitutes a radiator sending $P_0 G_0 \alpha/4\pi R^2$ watts back to the radar. The target is regarded as a uniform radiator (by the definition of α), and therefore the power density at the radar receiver is the radiated power again divided by $4\pi R^2$ or $P_0 G_0 \alpha/(4\pi R^2)^2$ watts per unit area. Finally the amount of power, P_r, intercepted by the receiving antenna is

$$P_r = \frac{P_0 G_0 \alpha A}{(4\pi R^2)^2}, \tag{12.3-4}$$

where A is the effective area or aperture of the receiving antenna.

The antenna gain G_0 is related to the aperture by the relation [12-2]

$$G_0 = \frac{4\pi A \kappa}{\lambda^2}, \tag{12.3-5}$$

where λ is the wavelength of the signal and κ is an empirical factor that takes into account the fact that the aperture may not be uniformly illuminated. For uniform illumination κ would be 1; in practice a value of 0.6 may be used.

Substituting this into Eq. (12.3-4) gives

$$P_r = \frac{P_0 A^2 \alpha \kappa}{4\pi R^4 \lambda^2}. \tag{12.3-6}$$

The maximum range is the value R for which P_r is a barely detectable amount. As shown in the previous chapter, this depends on the ratio of *signal energy* to *noise spectral level* [cf. Eq. (11.6-21)]. The energy in a single pulse of received echo is $P_r T$, where T is the pulse length. The target is generally illuminated by a number of pulses; the echoes returned from these pulses usually add *incoherently* in the receiver. This actually happens in the cathode-ray display tube where the individual echo pulses all add to the brightness of a given spot. The incoherent addition means that if n pulses contribute to the intensity of a given spot, the received signal energy-to-noise ratio is $nP_r T/N_0$, where N_0 is the spectral level of the noise (assumed to be white in the narrow signal band of interest). For unambiguous detection of a radar target this number must exceed some threshold level; for example, to take into account the uncertainty in range and direction one might require that $nP_r T/N_0 \geq 100$ for clear detection. Using this in Eq. (12.3-6) and solving for R, we obtain

$$R_{\max} \leq \left(\frac{nP_0 T A^2 \alpha \kappa}{400\pi\lambda^2 N_0} \right)^{1/4}. \tag{12.3-7}$$

An example may be instructive. Suppose that $P_0 T = 0.1$ joule, corresponding to a 1-μs pulse with 100 kW of peak power. Let $A = 1$ m^2, $\alpha = 10$ m^2 (small airplane), $\kappa = 0.6$ and $\lambda = 3.0$ cm (X band). The noise is caused by random molecular motion (mostly in the receiver input circuits) and is proportional to the temperature. The minimum noise level is kT, where $k = 1.38 \times 10^{-23}$ joule/°C is the Boltzmann constant and T is the temperature in degrees Kelvin (about 300° on a warm day). Thus the noise spectral density is at least 400×10^{-21} joule. A more conservative value that accounts for all sorts of imperfections in the electronics might be 10 times more, or 400×10^{-20}. Finally, assume that the number of pulses that illuminate a given spot is 10. Then

$$R_{\max} = \left(\frac{10 \times 0.1 \times 1 \times 10 \times 0.6}{400\pi \times (0.03)^2 \times 400 \times 10^{-20}} \right)^{1/4} = 33.9 \text{ km}.$$

Note that because of the fourth-power relation between received energy and range the result is not very sensitive to factors of 2 or 3 in some of the assumed values. It is clear that significant increases in range would require substantial increase in transmitted power, much larger antennas, and much less noise. The noise in particular can be reduced by operating the system at very low temperature, and therefore maser amplifiers, cooled to liquid helium temperatures, are used in the most critical applications, such as in radar astronomy [12-3]. By these means radar ranges on the order of hundreds of millions of kilometers, i.e., to the nearer planets, have been achieved.

12.4 Radar Signal Processors

Radar echoes are generally very weak and heavily masked by noise. For this reason fairly sophisticated processors are needed in the receiver; in fact, much of the development of signal processing discussed in the last chapter was originally inspired by the requirements of radar.

One purpose of the signal processor is to generate an estimate of the range to targets illuminated by the radar beam. The range is proportional to the time that it takes a radar signal pulse to make the round trip from the radar to the target. Hence range estimation is equivalent to estimation of the time delay between the time of departure of a given transmitted pulse and the time of arrival of the resulting echo.

Another function performed by high-performance radars such as the ones used in air traffic control is to estimate the radial velocity of the target. This permits the radar to discriminate between targets moving at different velocities. Signals reflected from a moving target experience a Doppler shift, and therefore the carrier frequency of the echo returned from a moving target differs from the transmitter frequency. Velocity estimation is therefore equivalent to estimation of the center frequency of the echo signal.

In this section we shall consider only the range estimation problem in detail. For a thorough analysis of the combined range and velocity estimation problem, the reader is referred to a number of excellent books ([12-4], [12-5], and [12-6]) that have been written on radar signal processing. Some of the principal results are summarized at the end of Sec. 12.5.

The Target Model

Suppose that the transmitted signal is $s(t)$. Typically this consists of a sequence of narrow pulses modulating a microwave carrier. The echo returned from a stationary target arrives back at the radar after a round-trip time

$$\tau_0 = \frac{2R}{c}, \qquad (12.4\text{-}1)$$

where, as before, R is the range and c is the speed of light. Hence the echo component of the received signal is given by

$$x(t) = as(t - \tau_0), \qquad (12.4\text{-}2)$$

where a is an amplitude factor involving signal power, target cross section, etc., as described in the previous section.

Equation (12.4-2) is strictly accurate only for a point target. More complicated targets can be modeled as a collection of reflectors at slightly different distances from the antenna. Hence the return signal would consist of a superposition of a number of terms of the form given in Eq. (12.4-2), all with somewhat different values of a and τ_0. Scintillation may cause a and τ_0 to be random time functions. We shall disregard all of these complications here.

We assume that the received signal is corrupted by additive noise $n(t)$; hence if $z(t)$ is the received signal,

$$z(t) = s(t - \tau_0) + n(t), \tag{12.4-3}$$

where the amplitude factor a has been absorbed into $s(t - \tau_0)$ for simplicity. In practice $n(t)$ may be considerably larger than $s(t - \tau_0)$.

The Optimum Receiver

The receiver is basically a device that operates on the signal $z(t)$ to produce an estimate of the time delay τ_0. Because $z(t)$ contains a large amount of random noise, the estimate of τ_0 will usually be in error. An optimum receiver is designed to minimize this error in some specified sense; for instance, it might be designed to minimize the mean-square error.

A convenient point of view is to regard the optimum receiver as a computer of the *a posteriori* probability density of the time delay, given that a sample of received signal $z(t)$ has been observed for an interval of duration T. In a somewhat compressed notation we use $p(\tau \mid z)$ to signify this *a posteriori* pdf. The argument of the pdf is τ, which may be regarded as a hypothetical time delay. Hopefully $p(\tau \mid z)$ will peak for values of τ near the true value τ_0. In fact, the receiver may only deliver the peak value of τ as its estimate of τ_0; this would be a maximum *a posteriori* probability (MAP) type of processor (cf. Sec 11.7).

By Bayes' rule the desired *a posteriori* probability density is

$$p(\tau \mid z) = \frac{p(\tau)p(z \mid \tau)}{p(z)}. \tag{12.4-4}$$

Since z is known, $p(z)$ is a normalizing constant which ensures that the integral of the right-hand side over all τ is unity. The function $p(\tau)$ is the *a priori* pdf and contains information known in advance of any processing of the received signal. For instance, we may know that the target range is somewhere between 10 and 50 km and that all of these possibilities are equiprobable. Then $p(\tau)$ is a constant over some *a priori* range interval $\mathfrak{I}$ and zero outside. We may, on the other hand, have information tending to give more weight to some ranges than to others. In any case the *a priori* probability is known or assumed independently from any observations and therefore does not affect the signal processor directly.

We are left with the likelihood function, $p(z|\tau)$, as the fundamental output of the signal processor. We see from Eq. (12.4-3) that $z(t)$ is the sum of signal and noise. As in Sec. 11.8 we assume that the noise is Gaussian and has zero-mean value. Furthermore, the noise is assumed to be white; i.e., it has a flat spectral density over a frequency band $-W \leq f \leq W$ large enough to include the bandwidth of the echo signal. Then by use of the sampling theorem, $z(t)$ can be represented by samples $z(t_i)$ taken at the rate of 2W samples/sec. If the echo is assumed to have the hypothetical delay τ, the ith sample of received signal is

$$z(t_i) = s(t_i - \tau) + n(t_i). \tag{12.4-5}$$

If sampling is at the Nyquist rate, these samples are all statistically independent Gaussian random variables with identical variance† σ^2. Also, since the $n(t_i)$ have zero mean, the $z(t_i)$ have the mean value $s(t_i - \tau)$. The number of samples needed to represent the observed signal $z(t)$ over the time interval T is (approximately) $2TW$. Thus the desired likelihood function—the joint pdf of the $z(t_i)$ given τ—can be written in the form

$$p(\mathbf{z}|\tau) = \frac{1}{(\sqrt{2\pi\sigma^2})^{2WT}} \exp\left\{-\frac{1}{2\sigma^2} \sum_{i=1}^{2TW} [z(t_i) - s(t_i - \tau)]^2\right\}, \tag{12.4-6}$$

where $\mathbf{z}$ is used to denote the vector of sample values of $z(t)$; i.e.,

$$\mathbf{z} = [z(t_1), z(t_2), \ldots, z(t_{2TW})]^T. \tag{12.4-7}$$

(The superscript T stands for transposition.) As in several other places in this book we use the same symbol for a random variable and for the values it can assume.

The summation in the exponent of Eq. (12.4-6) can be replaced by an integral using the following argument: For simplicity we write $y_i \equiv z(t_i) - s(t_i - \tau)$. By the sampling theorem [Eq. (4.2-5)], the samples y_i represent the function $y(t)$, given by

$$y(t) = \sum_{i=-\infty}^{\infty} y_i \operatorname{sinc}(2Wt - i). \tag{12.4-8}$$

If we multiply the left-hand side by $y(t)$ and integrate from $-\infty$ to ∞ and do the same on the right except that we use the sampling theorem representation for $y(t)$, we get

$$\int_{-\infty}^{\infty} y^2(t)\, dt = \sum_{i=-\infty}^{\infty} \sum_{j=-\infty}^{\infty} y_i y_j \int_{-\infty}^{\infty} \operatorname{sinc}(2Wt - i) \operatorname{sinc}(2Wt - j)\, dt. \tag{12.4-9}$$

†Since the noise is actually band-limited, samples are generally *not* independent. However, if sampling is precisely at the Nyquist rate, the samples are uncorrelated and hence independent. The uncorrelatedness of the samples follows from the fact that the correlation function of white noise of spectral density N_0 and band-limited to W Hz is $R_n(\tau) = 2WN_0 \operatorname{sinc} 2W\tau$. Hence noise samples every $\tau = (2W)^{-1}$ sec apart are uncorrelated.

However, the sinc functions are orthogonal, and

$$\int_{-\infty}^{\infty} \text{sinc}^2 (2Wt - i) \, dt = \frac{1}{2W}. \tag{12.4-10}$$

Hence, Eq. (12.4-9) becomes

$$\int_{-\infty}^{\infty} y^2(t) \, dt = \frac{1}{2W} \sum_{i=-\infty}^{\infty} y_i^2. \tag{12.4-11}$$

This relation is exact if the number of samples is infinite. If there are only $2TW$ samples but TW is large, then to a very good approximation

$$\sum_{i=1}^{2TW} y_i^2 = 2W \int_0^T y^2(t) \, dt. \tag{12.4-12}$$

We also observe that if the spectral density of the noise is given by

$$W_n(f) = \begin{cases} N_0, & -W \le f \le W \\ 0, & \text{otherwise,} \end{cases} \tag{12.4-13}$$

then

$$\sigma^2 = 2WN_0. \tag{12.4-14}$$

This follows since the variance is the same as the noise power, which, in turn, is the area under the spectral density curve.

If we use Eq. (12.4-14) in Eq. (12.4-12) and substitute for $y(t)$, we finally find that the exponent in Eq. (12.4-6) can be written in the form

$$\frac{1}{2\sigma^2} \sum_{i=1}^{2TW} [z(t_i) - s(t_i - \tau)]^2 = \frac{1}{2N_0} \int_0^T [z(t) - s(t - \tau)]^2 \, dt. \tag{12.4-15}$$

We shall now examine this expression more closely. Expanding the integrand, we find that

$$\int_0^T [z(t) - s(t - \tau)]^2 \, dt$$

$$= \int_0^T z^2(t) \, dt + \int_0^T s^2(t - \tau) \, dt - 2 \int_0^T z(t)s(t - \tau) \, dt. \tag{12.4-16}$$

The first term is, for a given $z(t)$, not a function of τ. In this respect it is similar to $p(z)$ in Eq. (12.4-4) and can be absorbed into the normalizing constant. Also, if the interval $[0, T]$ encompasses the duration of all possible echoes, the second term is also independent of τ; moreover,

$$\int_0^T s^2(t - \tau) \, dt = E, \tag{12.4-17}$$

where E is the energy in the return echo. There remains the cross-product term, and Eq. (12.4-6) can be written in the form†

$$p(\mathbf{z} \mid \tau) = k' \exp \left[\frac{1}{N_0} \int_0^T z(t)s(t - \tau) \, dt \right]. \tag{12.4-18}$$

†Even though the vector $\mathbf{z}$ does not appear explicitly on the right side of Eq. (12.4-18), the notation reflects the fact that we can represent $z(t)$ by its sampled values over $[0, T]$.

Finally, the desired *a posteriori* probability of τ is given by substituting Eq. (12.4-18) into Eq. (12.4-4):

$$p(\tau \mid \mathbf{z}) = k'' p(\tau) \exp \left[\frac{1}{N_0} \int_0^T z(t) s(t - \tau) \, dt \right], \qquad (12.4\text{-}19)$$

where k'' is another constant, independent of τ. Equation (12.4-19) says that the ideal receiver operates on the received signal $z(t)$ to generate the function

$$g(\tau) = \frac{1}{N_0} \int_0^T z(t) s(t - \tau) \, dt \qquad (12.4\text{-}20)$$

and then passes the result through a nonlinear device with input-output relation $y = \exp(x)$. Finally it weights the output with the supposedly known *a priori* pdf of τ. For an MAP estimator this procedure can be simplified, especially if the *a priori* pdf is constant for τ in some range $\mathfrak{I}$. In this case the receiver computes $g(\tau)$ for all τ in the range $\mathfrak{I}$ and finds the point (or points) where this is a maximum.

A simple approximate way of performing the computation is to use a matched filter (Sec. 11-6). To see how this works here we consider the expression

$$y(\tau) = \int_{-\infty}^{\infty} z(t) h(\tau - t) \, dt. \qquad (12.4\text{-}21)$$

This is the output at time τ of a filter with the impulse response $h(t)$ when the received signal $z(t)$ is applied at its input. A filter matched to $s(t)$ has the impulse response $h(t) = (1/N_0) s(T - t)$, where T is at least as large as the signal duration [Eq. (11.6-17)]. The multiplication by the constant $1/N_0$ does not materially affect anything but results in an equation that is more similar to Eq. (12.4-20). Letting $h(t)$ be the matched filter impulse response gives

$$y(\tau) = \frac{1}{N_0} \int_{-\infty}^{\infty} z(t) s(T + t - \tau) \, dt$$

$$= \frac{1}{N_0} \int_{-\infty}^{\infty} z(t - T) s(t - \tau) \, dt. \qquad (12.4\text{-}22)$$

The second line is obtained by a simple change of the variable of integration.

The limits of integration in Eq. (12.4-20) could have been extended to $\pm\infty$ since the interval $[0, T]$ is assumed to contain all significant values of $s(t)$. Therefore Eq. (12.4-22) is quite similar to Eq. (12.4-20) except that the correlation is between $s(t - \tau)$ and the input signal delayed by a time T. If $z(t)$ contains the component $s(t - \tau_0)$, $y(\tau)$ will peak at $\tau = \tau_0 + T$, but since T is presumably known, τ_0 is easily obtained from this. A MAP receiver based on this idea is shown in Fig. 12.4-1. The estimate $\hat{\tau}$ is the value of delay where output is largest, less the known constant delay T.

In the interest of brevity we have omitted certain details from our discussion. One of these is that the radar echo is actually a modulated micro-

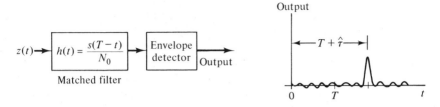

Figure 12.4-1. Matched filter to determine range. The estimated delay is $\hat{\tau}$; the filter output peaks at $T + \hat{\tau}$.

wave signal, i.e., a narrowband signal centered at a very high frequency. The modulation may be AM, FM, or PM or a combination of these. The matched filter is therefore a band-pass filter with a structure that matches the modulation of the echo. For simplicity, suppose that the echo is only amplitude modulated. Then in Eqs. (12.4-20) and (12.4-22) we replace $z(t)$ by $\zeta(t) \cos(\omega_c t + \theta')$ and $s(t)$ by $\sigma(t) \cos(\omega_c t + \theta'')$, where ω_c is the carrier frequency (typically 10 GHz), $\zeta(t)$ and $\sigma(t)$ are relatively low-frequency modulating functions, and θ' and θ'' are arbitrary, possibly time-varying phase angles. If one makes these substitutions in Eq. (12.4-20) and uses the fact that ω_c is much larger than the bandwidth of $\sigma(t)$ and $\zeta(t)$, one finds (see Prob. 12-4) that $g(\tau)$ has the form $G(\tau) \cos(\omega_c \tau + \theta)$, where $\theta \equiv \theta' - \theta''$ and

$$G(\tau) = \frac{1}{2N_0} \int_{-\infty}^{\infty} \zeta(t)\sigma(t - \tau)\, dt. \qquad (12.4\text{-}23)$$

The general form of $g(\tau)$ is therefore as shown in Fig. 12.4-2. By Eq. (12.4-19) the *a posteriori* probability of τ has the same high-frequency oscillation as $g(\tau)$; values of τ where $g(\tau)$ is large are highly probable, and closely adjacent values of τ where $g(\tau)$ is small are highly improbable. Intuitively this oscil-

Figure 12.4-2. Fine structure of $g(\tau)$.

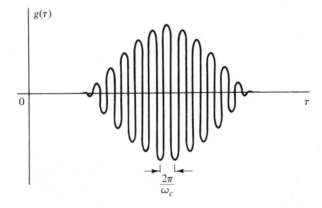

lation in the *a posteriori* pdf indicates that most echo delays do not satisfy $\tau_0 - \tau = 2\pi m/\omega_c$; i.e., the phase of the carrier in the echo does not match the phase of the carrier in the transmitted signal.

In practice the detailed delay information conveyed by the high-frequency oscillation is quite useless. For a 10-GHz carrier frequency, the range increment corresponding to one carrier cycle is 1.5 cm and is therefore much smaller than the expected range estimation error. In effect the system tries to tell us that the target is exactly at one of a number of points that are 1.5 cm apart, but it doesn't tell us which one.

The fine structure in Fig. 12.4-2 moreover implies precise knowledge of the phase of the transmitted signal, but actually the phase angle is unknown. A convenient assumption is in fact that the phase angle is a random variable, uniformly distributed over $[0, 2\pi]$ and independent of τ. The *a posteriori* probability including θ then takes the form

$$p(\tau, \theta \,|\, \mathbf{z}) = k'' p(\tau, \theta) p(\mathbf{z} \,|\, \tau, \theta)$$

$$= \begin{cases} \dfrac{k''}{2\pi} p(\tau) \exp\left[G(\tau) \cos\left(\omega_c \tau + \theta\right)\right], & 0 \le \theta \le 2\pi \\ 0, & \text{otherwise.} \end{cases} \quad (12.4\text{-}24)$$

Note that this shows the fine structure discussed above. The *a posteriori* pdf of interest is obtained by averaging over θ. This gives

$$p(\tau \,|\, \mathbf{z}) = \int_0^{2\pi} p(\tau, \theta \,|\, \mathbf{z}) \, d\theta$$

$$= k'' p(\tau) \cdot \frac{1}{2\pi} \int_0^{2\pi} \exp\left[G(\tau) \cos\left(\omega_c \tau + \theta\right)\right] d\theta$$

$$= k'' p(\tau) I_0[G(\tau)], \quad (12.4\text{-}25)$$

where $I_0(\cdot)$ is the zero-order modified Bessel function of the first kind. One of the defining equations for $I_0(\cdot)$ is

$$I_0(x) = \frac{1}{2\pi} \int_0^{2\pi} e^{x \cos \theta} \, d\theta. \quad (12.4\text{-}26)$$

The Bessel function is monotonically increasing just like e^x, and the nonlinear operation involved in generation $p(\tau \,|\, \mathbf{z})$ from the matched filter output $g(\tau)$ is therefore not materially altered. The important result is that the quantity of interest is $G(\tau)$, the low-frequency envelope of $g(\tau)$. This explains the presence of the envelope detector in Fig. 12.4-1.

12.5 Estimation Error

Radar systems suffer from several different kinds of error. One class of errors is the *ambiguity* errors. These are typically very large errors; for instance, a range of 5 km might be estimated as being 15 km. One of the ambiguity errors, called *noise ambiguity*, arises from the fact that if the signal-to-noise ratio is small the function $G(\tau)$, and therefore the *a posteriori* pdf, will have many peaks caused by the noise. This is illustrated in Fig. 12.5-1. We have shown the peak near the time delay τ_0 to be largest, but it

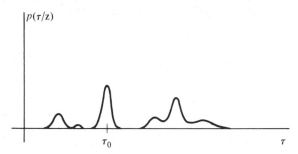

Figure 12.5-1. *A posteriori* pdf showing numerous noise-induced peaks.

is quite possible for some other peak to be larger than the true peak. When this happens the result is a completely erroneous estimate of τ. The noise ambiguity problem has been studied in [12-7] and [12-8], and the effect is qualitatively similar to the FM threshold considered in Chapter 10. If the SNR is low enough so that noise ambiguity becomes a problem, the radar is essentially useless. We therefore assume a sufficiently large SNR so that we can ignore this kind of error.

Another form of ambiguity, called signal ambiguity, results from the fact that certain kinds of transmitted signal correlate strongly for values of τ that are different from the true value. For example, most radars employ a periodic pulse sequence. This has a periodic correlation product. Delays τ that differ from the true value τ_0 by an integral number of pulse repetition periods all result in strong correlation. As noted earlier, this type of ambiguity can sometimes be eliminated by limiting the *a priori* range of τ values to be considered.

If ambiguity can be eliminated as a factor causing radar errors, there still remains the possibility that the peak in $p(\tau \,|\, z)$ near the true value τ_0 does not coincide precisely with τ_0. Although compared to ambiguity errors this is generally a relatively small error, it determines the basic radar accuracy. This error can be related to the width of the peak of $p(\tau \,|\, z)$ around τ_0. This is directly related to the width of $G(\tau)$ near τ_0, and we shall therefore examine this function in more detail.

Suppose that there is a target at the time range delay τ_0. Then the received signal is $z(t) = s(t - \tau_0) + n(t)$. To convert to envelope functions, assume for convenience that $s(t - \tau_0) = \sigma(t - \tau_0) \cos \omega_c t$; i.e., take the arbitrary carrier phase $\theta'' - \omega_c \tau_0$ to be zero. Also, recall from Chapter 10 [Eq.

(10.8-2)] that narrowband noise can be written in the form

$$n(t) = n_c(t) \cos \omega_c t - n_s(t) \sin \omega_c t,$$

where $n_c(t)$ and $n_s(t)$ are low-frequency noise envelopes. For large signal-to-noise ratio the effect of the quadrature noise term $n_s(t)$ is negligible (as in Fig. 10.11-4), and we therefore have approximately

$$\zeta(t) \approx \sigma(t - \tau_0) + n_c(t). \qquad (12.5\text{-}1)$$

If we substitute this into the expression for $G(\tau)$ [Eq. (12.4-23)], we get

$$G(\tau) = G_s(\tau) + G_n(\tau), \qquad (12.5\text{-}2)$$

where

$$G_s(\tau) = \frac{1}{2N_0} \int_{-\infty}^{\infty} \sigma(t - \tau_0)\sigma(t - \tau) \, dt \qquad (12.5\text{-}3)$$

and

$$G_n(\tau) = \frac{1}{2N_0} \int_{-\infty}^{\infty} \sigma(t - \tau)n_c(t) \, dt. \qquad (12.5\text{-}4)$$

The function $G_s(\tau)$ and $G_n(\tau)$ are, respectively, the signal and noise components of $G(\tau)$. $G_s(\tau)$ has a peak value for $\tau = \tau_0$, and if there is no signal ambiguity, this is the only peak. $G_n(\tau)$ is a random process. It may have peaks that are larger than the peak in $G_s(\tau)$; this is the cause for the noise ambiguity referred to earlier. If the SNR is large enough so that noise ambiguity is not a factor, then $G_n(\tau)$ causes only a small perturbation in the output (see Prob. 12-5). The amount by which these perturbations affect the delay estimate depends mostly on the shape of $G_s(\tau)$ near the peak. If the peak is needle-like, a small noise perturbation can displace it only a small amount, while a broad peak would be fairly sensitive to noise effects. We are led, therefore, to consider only the function $G_s(\tau)$.

If $G_s(\tau)$ is a twice differentiable function in the neighborhood of its peak, its behavior near the peak can be determined by performing a Taylor expansion. It turns out, unfortunately, that for some practically important radar signals $G_s(\tau)$ is not twice differentiable. In particular if $\sigma(t)$ is a square pulse, $G_s(\tau)$ is triangular (see Prob. 12-6). However, since square pulses are an idealization that never exactly occurs in practice, it is probably safe to assume that a Taylor expansion is usually possible. Assuming this to be so, we have

$$G_s(\tau) = G_s(\tau_0) + \frac{1}{2}(\tau - \tau_0)^2 \left.\frac{d^2 G_s(\tau)}{d\tau^2}\right|_{\tau = \tau_0} + \cdots, \qquad (12.5\text{-}5)$$

where the second derivative is evaluated for $\tau = \tau_0$. The linear term of the Taylor expansion doesn't appear because $G_s(\tau)$ has a maximum at τ_0, and therefore the first derivative vanishes. We use the definition of $G_s(\tau)$ given in Eq. (12.5-3) to calculate the Taylor coefficients as follows:

$$G_s(\tau_0) = \frac{1}{2N_0} \int_{-\infty}^{\infty} \sigma^2(t - \tau_0)\, dt = \frac{E}{N_0}, \tag{12.5-6}$$

where E is the signal energy.

Also,

$$\frac{d^2 G_s(\tau)}{d\tau^2}\bigg|_{\tau=\tau_0} = \frac{1}{2N_0} \int_{-\infty}^{\infty} \sigma(t - \tau_0)\ddot{\sigma}(t - \tau_0)\, dt$$

$$= -\frac{2\pi^2}{N_0} \int_{-\infty}^{\infty} f^2 |S(f)|^2\, df \equiv -\frac{E\beta^2}{N_0}, \tag{12.5-7}$$

where $S(f)$ is the Fourier transform of $\sigma(t)$.† The second line follows from Parseval's theorem and the fact that the Fourier transform of $\ddot{\sigma}(t)$ is $-4\pi^2 f^2 S(f)$. Equivalently, Eq. (12.5-7) can be written as

$$\beta \equiv 2\pi \left[\frac{\int_{-\infty}^{\infty} f^2 |S(f)|^2\, df}{2E} \right]^{1/2}. \tag{12.5-8}$$

The quantity β has dimensions of frequency and is referred to as the *Gabor bandwidth* of the transmitted signal [12-10]. Note that it is the square root of the normalized second moment of the signal spectrum about its center and is therefore a reasonable measure of bandwidth.‡ It is, of course, necessary for $|S(f)|$ to decrease sufficiently rapidly as $f \rightarrow \infty$ so that the integral in Eq. (12.5-7) exists, and we assume this to be the case. Here again, however, there are functions, notably the square pulse, for which this assumption fails.

We now use the expansion for $G(\tau)$ in the expression for the *a posteriori* probability, Eq. (12.4-25). We assume that the *a priori* probability of τ is constant—this will generally be true for small τ neighborhoods. Also, the argument of the Bessel function in Eq. (12.4-25) is large; otherwise the assumption that the noise component of $G(\tau)$ is negligible would not be valid. But for large argument $I_0(z)$ behaves very approximately like e^z ([12-11], p. 377), and therefore Eq. (12.4-25) becomes

$$p(\tau\,|\,\mathbf{z}) \simeq k_0 \exp[G(\tau)]$$

$$= k_0 \left(\exp\left\{ D^2\left[1 - \frac{\beta^2}{2}(\tau - \tau_0)^2 \right] \right\} \right), \tag{12.5-9}$$

---

†Equation (12.5-7) presupposes that the radar signal has no frequency or phase modulation. The expression for the second derivative is more complicated otherwise ([12-9], pp. 4–6).

‡β may differ considerably from what is usually regarded as the bandwidth. See [12-9], pp. 4–12.

where k_0 is a new normalizing constant and where $D^2 \equiv E/N_0$ is a constant often called the *detection index*. We see that τ is distributed approximately according to a Gaussian distribution with mean value τ_0. The standard deviation of the delay measurement is given by

$$\sigma_\tau \equiv (\tau - \tau_0)_{\text{rms}} = \frac{1}{\beta D}, \qquad (12.5\text{-}10)$$

and therefore the standard deviation of the range measurement is given by

$$\sigma_R = \frac{c\sigma_\tau}{2} = \frac{c}{2\beta D}. \qquad (12.5\text{-}11)$$

Thus the standard deviation of the range measurement is inversely proportional to the Gabor signal bandwidth and detection index D. Equation (12.5-11) is one of the key results in radar theory.

There is a contradiction implied in Eq. (12.5-9) since it indicates the peak of $p(\tau \mid \mathbf{z})$ to be exactly at τ_0. Thus the MAP estimate would be error-free. The contradiction is, of course, due to the neglect of the noise term $G_n(\tau)$. A more careful analysis, which considers this term, shows that the peak is itself random with mean τ_0 and variances as obtained above [12-8].

For the sake of keeping the discussion of radar signal processing as simple as possible we have considered only the range estimation problem. The estimation of radial velocity is a very similar problem and yields similar results. We shall present these results here without any derivations; for details, see, for instance, [12-5], pp. 41–75.†

The accuracy with which the Doppler shift in the echo can be estimated depends on a quantity α called the *Gabor signal duration* [12-10]. This is the time analog of the Gabor bandwidth encountered earlier and defined by

$$\alpha = 2\pi \left\{ \frac{\int_{-\infty}^{\infty} t^2 \sigma^2(t)\, dt}{2E} - \left[\frac{\int_{-\infty}^{\infty} t\sigma^2(t)\, dt}{2E} \right]^2 \right\}^{1/2}. \qquad (12.5\text{-}12)$$

If the radial velocity is the only parameter to be estimated, the standard deviation of the Doppler frequency estimate is

$$\sigma_\eta = \frac{1}{\alpha D}. \qquad (12.5\text{-}13)$$

where, as before, $D^2 \equiv E/N_0$. Also, since the Doppler shift is $2f_c\dot{R}/c$, where $\dot{R}$ is the radial velocity, f_c the carrier frequency in hertz, and c the speed of light, the rms velocity error is given by

$$\sigma_{\dot{R}} = \frac{c}{2f_c}\sigma_\eta = \frac{c}{2\alpha f_c D}. \qquad (12.5\text{-}14)$$

A somewhat more complicated problem is presented by the joint estimation of range and velocity. This is an example of a joint parameter estimation

†See also Probs. 12-7 and 12-11.

problem, and it can be shown ([12-4] and [12-5]) that the combined rms
range and velocity errors are given by

$$\sigma_R = \frac{\alpha c}{2D\sqrt{\alpha^2 \beta^2 - \rho^2}}, \qquad (12.5\text{-}15)$$

$$\sigma_{\dot{R}} = \frac{\beta c}{2Df_c\sqrt{\alpha^2 \beta^2 - \rho^2}}. \qquad (12.5\text{-}16)$$

The parameter ρ is a measure of the correlation between time and frequency
in the signal pulse. An expression for ρ is given in [12-9], pp. 4–6. If the time
origin is chosen so that

$$\int_{-\infty}^{\infty} t\sigma^2(t)\, dt = 0,$$

then

$$\rho = 2\pi \frac{\int t\dot{\Omega}(t)\sigma^2(t)\, dt}{2E}, \qquad (12.5\text{-}17)$$

where $\Omega(t)$ is the phase modulation of the signal. For signals that are not
frequency modulated ρ is zero, and in this case Eqs. (12.5-15) and (12.5-16)
are seen to reduce to Eqs. (12.5-11) and (12.5-14). If $\rho \neq 0$, lack of knowledge of one of the parameters increases the error in estimating the other.
This is a general property of multiple parameter estimators ([12-12], p. 81).

We see that better range accuracies are achieved by larger signal bandwidth. Since short pulses have large bandwidth, this implies that short pulses
enhance range accuracy, in accordance with intuition. Similarly, long-duration signals are required for good Doppler accuracy. This is again in accordance with intuition since one cannot expect a good frequency estimate if
one has only a few carrier cycles to look at.

To achieve *both* good range and Doppler accuracy requires signals that
combine large bandwidth and long duration. This cannot be accomplished
with simple pulse signals, because for such signals the product of bandwidth
and duration is generally on the order of unity [12-13] (see also Probs. 12-8
and 12-9). Hence large bandwidth goes with short time duration and vice
versa. To obtain large duration-bandwidth products one resorts to more
complicated signals. A commonly used approach is to use both frequency and
amplitude modulation. For instance, the frequency of the pulse might increase linearly from the beginning to the end of the pulse. Such a pulse is
called a *chirp* pulse because the rising pitch sounds a little like a bird chirping.
It is easily shown (see Prob. 12-9) that the product of time duration and
bandwidth for a single chirp pulse can be made arbitrarily large.

Chirp pulses are used by some animals that employ sonar. This is particularly true of certain bats. The most advanced of these in the scale of evolu-
tion use pulses in which brief periods of no FM alternate with periods of FM
[12-14].

The design of sophisticated signals for use in high-performance radars is not simple, and much effort has gone into this problem since the earliest days of radar. Many of the commonly used techniques are discussed in the books by Rihaczek [12-5] and Cook and Bernfeld [12-6], already referred to earlier.

We point out finally that the use of chirp signals is an example of a technique referred to as *pulse compression*. A chirp pulse may be quite long, but since each instant is associated with a unique frequency, the radar processor can, in effect, combine the different parts of the pulse in such a way as to put out a very narrow spike which accurately defines the target delay time. Qualitatively this time compression is performed in the matched filter, which delays the different frequencies occurring in the pulse by different amounts so that the output is a very short pulse. A simple example illustrating this principle is given in Prob. 12-10. The phenomenon of obtaining different delays for different frequencies is called *dispersion*. Space itself is usually a *nondispersive* medium.

12.6 Passive Systems

As implied by their name, passive systems do not employ a transmitter; they only have a receiver which listens (in the case of sonar) to signals of possible interest. Submarine sonars are frequently passive since active "pinging"† would give away their presence. Passive systems are also used in radio astronomy and in the pinpointing of seismic disturbances. In these applications, echo location is either impractical or inappropriate. However, it is possible to obtain directional versus intensity information with passive systems.

The received signal in most of these applications is random and noise-like. For instance, a ship moving through water causes a certain amount of turbulence, which generates a characteristic flow noise. Propeller and machinery noises have characteristic spectra but are otherwise also essentially random. The passive sonar system must be able to detect these "signals" in a general noise background. It can do this if there are features that distinguish the "signal" from the "noise."

One of the most important distinguishing features of the signal is that it comes from a particular direction. This is in contrast to ambient background noise, which is frequently isotropic; i.e., it comes uniformly from all directions. The directionality of the signal can be used in a directional receiver.

Since the signals of interest in sonar are often very broadband, with important frequency components extending down to a few hundred hertz, it

---

†Readers who are familiar with war movies involving submarines will recall the "pings" that characterize the submarine sonar.

is generally impractical to use parabolic reflectors; they would tend to be too bulky. Instead sonars commonly employ receiving arrays. By inserting variable delays in the lines coming from the array elements, such an array can be steered so that signals coming from any desired direction add in phase. The delays are frequently produced by shift-register delay lines as described in Chapter 5, and they can be changed very rapidly under computer control. Hence the control of the direction in which an array is "pointed" is much more flexible than the mechanical rotation used with parabolic dish antennas. For this reason such electronically steered array antennas, also called *phased arrays*, are used quite frequently with active radars. Phased arrays make it possible, for instance, to construct search radars that will examine a particular spot in the sky for a time long enough to be certain that no target is present there and then move on to another spot. The length of time need not be the same for all spots, and the search need not be sequential.

A simple array processor is shown in Fig. 12.6-1(b). The array elements are shown located at equally spaced points on a straight line, although other spacings are possible and are in fact used. Assume that a plane wave, as shown in Fig. 12.6-1(a), arrives at the array. The wavefront makes an angle θ with a line passing through the array. Suppose that at time t the signal component on the first element is $s(t)$. This signal component will impinge on

Figure 12.6-1. Array processor: (a) a plane wave arrives at the array, making an angle θ with a line drawn through the array elements. (b) The delayed outputs from the array elements are summed, squared, and passed through a smoothing filter.

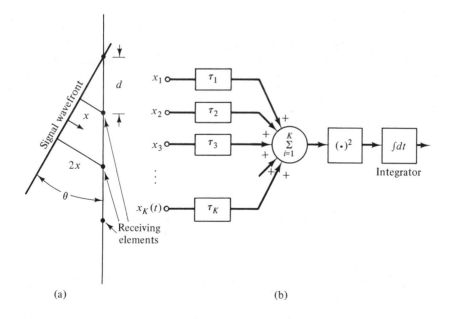

(a)   (b)

the second element of the array after a delay $\hat{\tau}_1 = x/c$, where x is the perpendicular distance from the wavefront to the element. Clearly, $x = d \sin \theta$. Hence $s(t)$ arrives at the second element at time $t + (d \sin \theta/c)$. It will arrive at the third element at the time $t + (2d \sin \theta/c)$, etc. Thus if we consider the vector

$$\mathbf{s}(t) = [s_1(t), s_2(t), \ldots, s_K(t)]^T \tag{12.6-1}$$

of signal components at the K array elements, we see that

$$s_i(t) = s(t - \hat{\tau}_i), \tag{12.6-2}$$

where

$$\hat{\tau}_i = \frac{(i - 1)d \sin \theta}{c}. \tag{12.6-3}$$

As shown in Fig. 12.6-1(b), the signal from the ith element is delayed by an amount τ_i, and then all the delayed signals are summed. The array is steered in the direction θ by adjusting the processing delays so that

$$\hat{\tau}_i + \tau_i = \tau, \tag{12.6-4}$$

where τ is a constant delay. Thus the signal from array element 1 is delayed by a larger amount than the signal from element 2, which is delayed by a larger amount than the signal from element 3, etc. Hence the signal components emerging at each of the lines going into the summer is $s(t - \hat{\tau}_i - \tau_i)$ $= s(t - \tau)$, and therefore the signal component out of the summer is $Ks(t - \tau)$; i.e., the signal components are added coherently. Note that this happens even though the signals may be random and noise-like. However, it is required that the field have a well-defined wavefront. This means that there exists a surface where the relative phase between different points is zero. A plane wave (the zero-phase surface is a plane) is characteristic of such fields, which are said to have a high degree of *spatial coherence*. Also, it should be clear that by using different kinds of processing delays the same effect can be obtained with arbitrary array geometries. Although line arrays are quite common, spherical arrays are also often used.†

The complete signal picked up at each array element is

$$x_i(t) = s_i(t) + n_i(t). \tag{12.6-5}$$

Hence the output of the summer is

$$y(t) = \sum_{i=1}^{K} [s(t - \tau) + n_i(t - \tau_i)]$$

$$= Ks(t - \tau) + \sum_{i=1}^{K} n_i(t - \tau_i). \tag{12.6-6}$$

---

†Another important type of array which uses nonequal spacings between elements is the so-called *nonredundant* array. In this type of array no spacing between any two elements is ever repeated.

After the squaring and integrating operations, the output of the processor is approximately $\langle y^2(t) \rangle$, where the $\langle \cdot \rangle$ indicates time average. For ergodic processes this is the same as the ensemble average; also the fixed delay τ is irrelevant. Hence $\langle \; \rangle$ is replaced by the ensemble average (the overbar), and there results

$$\overline{y^2(t)} \equiv \overline{y^2} = K^2 \overline{s^2(t)} + 2K \sum_{i=1}^{K} \overline{s(t)n_i(t - \tau_i)} + \sum_{i=1}^{K} \sum_{j=1}^{K} \overline{n_i(t - \tau_i)n_j(t - \tau_j)}.$$

(12.6-7)

We assume that signal and noise are independent and zero-mean. Therefore the cross-product term vanishes. Also, we assume for simplicity that the noises at the different array elements are independent. This is not necessarily true in practice, especially if the array elements are physically close to each other. However, with this assumption Eq. (12.6-7) becomes

$$\begin{aligned} \overline{y^2} &= K^2 P_s + K P_n \\ &= K(K P_s + P_n), \end{aligned}$$

(12.6-8)

where P_s and P_n are, respectively, the signal and noise powers at each array element. Equation (12.6-8) is nothing more than a demonstration of the fact that the signal components add coherently while the noise components add incoherently. This is due to the spatial *coherence* of the signal and the spatial *incoherence* of the noise.

Observe that the signal-to-noise ratio P_s/P_n at each array element is multiplied by K in the output. The number K is therefore called the processing gain or, more commonly, the *array gain*. With the simplifying assumptions made here, the array gain is identical to the number of array elements. This is not generally true. The array gain decreases if the noise at different array elements is correlated. This can be understood qualitatively by considering two adjacent array elements in which the noise correlation is unity. In this case both signal and noise are correlated exactly the same way, and the two elements are equivalent to a single one. For negatively correlated noise the array gain may also exceed the number of array elements.

The result obtained thus far is for an array properly steered "on target." If the array is steered slightly "off target," the delayed signal components will not be exactly identical. Then after squaring and averaging, the signal component has the form

$$\sum_{i=1}^{K} \sum_{j=1}^{K} \overline{s(t - \tau_i)s(t - \tau_j)} = \sum_{i=1}^{K} \sum_{j=1}^{K} R_s(\tau_i - \tau_j),$$

where the τ_i and τ_j are now the small errors between the on-target and off-target delays and $R_s(\tau)$ is the correlation function of the signal component. Since $|R_s(\tau_i - \tau_j)| \leq R_s(0)$, this shows that the signal component is reduced. If the amount of off-steering is sufficient to make $R_s(\tau_i - \tau_j) \simeq 0$ for $i \neq j$, then the signal component will be KP_s rather than $K^2 P_s$. Thus the output of the array, plotted as a function of azimuth angle θ, will appear as in Fig. 12.6-2. It is assumed that $P_s/P_n \ll 1$ but that $K \gg 1$. It can be shown [12-15] that the width of the peak around $\theta = \theta_0$ is given by

$$\Delta\theta = \frac{kc}{L\beta},$$

where L is the total array length, c is the wave velocity, β is the Gabor bandwidth [cf. Eq. (12.5-8)] of the signal, and k is a constant on the order of unity which depends on the array geometry.

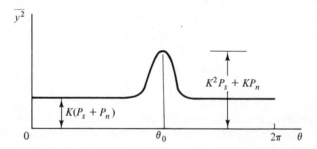

Figure 12.6-2. Output from an array detector for a single target located at θ_0.

The array processor shown in Fig. 12.6-1 is very elementary and makes no use of differences that might exist between the spectral properties of the signal and noise. If substantial differences exist, the processor can be improved by inserting a Wiener filter between the summer and the squarer. It can be shown ([12-16] and [12-17]) that this is an optimum processor for Gaussian signals in Gaussian noise, where the signal is from a point source and the noise between different array elements is uncorrelated. For other situations the optimum processor is more complicated.

12.7 Summary

In this chapter we have dealt with some elementary and with some not-so-elementary properties of radar and sonar. We considered first the general operation of simple active radars and discussed block diagrams, typical radar display devices, etc. We then derived the radar range equation and showed that the maximum range is proportional, among other things, to the fourth root of the transmitted power. We discussed the form of the typical radar-echo signal and showed how target range is translated into time delay. We found the optimum processor for this kind of signal to consist of a filter matched to the shape of the transmitted pulse.

To study the actual performance of a radar system we computed the *a posteriori* probability density function of the delay. We found this pdf to be approximately Gaussian, and we were able to show that the range error is inversely proportional to the bandwidth of the transmitted signal. We concluded by analogy that the radial velocity error is inversely proportional to the signal duration.

In the final section we considered passive systems and array processors which depend on spatial coherence of the signal as compared to the spatial incoherence of the noise to detect the signal and estimate its direction.

PROBLEMS

12-1. a. At its closest approach the planet Venus is about 42 million km from earth. What is the power in a radar echo if the peak power in the transmitted signal is 1 MW and if the antenna has a diameter of 60 m. Assume that the radar cross section of Venus is equal to the area of its equatorial circle and that the diameter of Venus is about the same as that of the earth (about 12,500 km). The radar frequency is 10 GHz.

      b. The spectral density of thermal noise is given by $2kT_e$, where $k = 1.38 \times 10^{-23}$ is the Boltzmann constant and T_e is the effective noise temperature. Suppose that the minimum obtainable temperature is 40°K. What integration time would be needed for the radar echo from Venus to be detectable? (Assume that the energy-to-noise ratio $nP_rT/N_0 \geq 100$ for clear detection.)

12-2. What should be the pulse length and minimum pulse repetition frequency for a radar capable of resolving terrain features larger than 10 m at a distance of up to 10 km?

12-3. A simplified block diagram of a coherent MTI radar is shown in Fig. P12-3. The radar carrier frequency is $f_c + f_i$ and is derived from a stable local oscillator (STALO) and a coherent oscillator (COHO) whose outputs are combined in a mixer. The sum frequency component is passed on to the modulator which pulses the carrier on and off with a pulse repetition period T_r and pulse duration T. The carrier pulses are amplified in the klystron amplifier and sent to the antenna. The return echo is heterodyned down to an intermediate frequency by mixing with the STALO output, then amplified, and finally coherently demodulated by further mixing with the COHO output signal.

      a. Show that the demodulator output consists of a train of pulses that are suppressed-carrier-modulated (DSBSC) by the Doppler frequency.

      b. The canceller is a special filter which is supposed to reject signals having zero or small Doppler frequency and pass signals with large Doppler frequency. What is the transfer function of the canceller? Sketch the magnitude of its transfer function. Verify that the system does discriminate properly against stationary targets.

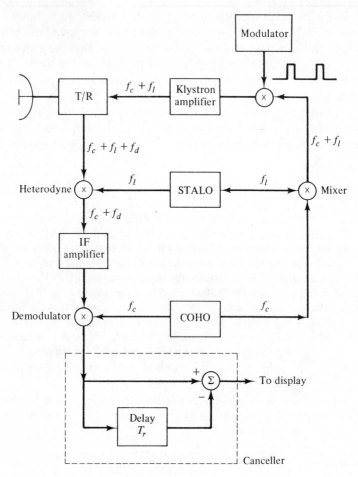

Figure P12-3

12-4.   Consider the two narrowband signals

$$x(t) = a(t) \cos (2\pi f_0 t + \theta_1)$$
$$y(t) = b(t) \cos (2\pi f_0 t + \theta_2),$$

where $a(t)$ and $b(t)$ are low-frequency envelope functions and f_0 is a high carrier frequency. Assume that the bandwidth of $a(t)$ and $b(t)$ is W and that $f_0 \gg W$. Show that if $g(\tau) = \int_{-\infty}^{\infty} x(t)y(t-\tau)\,dt$, then

$$g(\tau) \simeq G(\tau) \cos (2\pi f_0 \tau + \theta_3),$$

where

$$G(\tau) = \tfrac{1}{2} \int_{-\infty}^{\infty} a(t)b(t-\tau)\,dt.$$

12-5. This problem concerns the functions $G_s(\tau)$ and $G_n(\tau)$ defined in Eqs. (12.5-3) and (12.5-4).

    a. Show that $G_s(\tau) \le G_s(\tau_0) = E/N_0$.

    b. Show that $\overline{[G_n(\tau)]^2} = E/N_0$ and that therefore

$$\frac{G_s^2(\tau)_{\max}}{[G_n(\tau)]^2} = E/N_0.$$

Hence

$$G_s(\tau_0)_{\max} \gg [G_n(\tau)]_{\mathrm{rms}} \quad \text{if } E/N_0 \gg 1.$$

12-6. Find $G_s(\tau)$ if

    a. $\sigma(t) = \mathrm{rect}\, t/T$.

    b. $\sigma(t) = \exp[-(t/T)^2]$.

12-7. The transmitted pulse in a radar is given by

$$s(t) = e^{-t^2/2T^2} \cos 2\pi f_c t,$$

where f_c is the microwave carrier frequency. Because of target motion the echo has a Doppler shift; therefore, in the absence of noise the received signal is given by

$$s_r(t) = a e^{-(t-\tau_0)^2/2T^2} \cos[2\pi(f_c + \eta_0)t + \theta],$$

where τ_0 is the time delay, η_0 the true Doppler shift, θ is a random phase angle, and a is an attenuation factor.

    a. Show that if the delay is known the *a posteriori* pdf $p(\eta \,|\, z)$ is a function of

$$g(\eta) = \frac{a}{N_0} \int_{-\infty}^{\infty} z(t) e^{-(t-\tau_0)^2/2T^2} \cos[2\pi(f_c + \eta)t + \theta']\, dt,$$

where $z(t)$ is the received signal which includes noise, and θ' is another random phase angle.

    b. For simplicity, assume $\tau_0 = 0$. (If τ_0 is known, this is just a shift of the time axis.) Separate $g(\eta)$ into a signal component $g_s(\eta)$ and a noise component $g_n(\eta)$. Show that the signal component can be written in the form

$$g_s(\eta) = G_s(\eta) \cos \theta'',$$

where

$$G_s(\eta) = \frac{a^2}{2N_0} \int_{-\infty}^{\infty} e^{-t^2/T^2} \cos 2\pi(\eta - \eta_0)t\, dt$$

and where θ'' is another random phase angle. By the argument used in deriving Eq. (12.4-25) the function $G_s(\eta)$ determines the Doppler estimate in error.

    c. Expand $G_s(\eta)$ in a Taylor series about $\eta = \eta_0$. Evaluate the coefficient of $(\eta - \eta_0)^2$, and show that the rms error is given by Eq. (12.5-13). What is α? (Note that the shift in time origin used here causes the second term in the definition of α to vanish.)

12-8.   a. Find α and β for the Gaussian pulse

$$\sigma(t) = e^{-t^2/2T^2}.$$

Find $\alpha\beta$, and observe that it is a constant independent of T.

        b. Repeat for the triangular pulse

$$\sigma(t) = \begin{cases} 0, & |t| \geq \dfrac{a}{2} \\[2mm] 1 - \dfrac{2|t|}{a}, & |t| \leq \dfrac{a}{2}. \end{cases}$$

12-9.   A Gaussian chirp pulse is given by

$$s(t) = e^{-t^2/2T^2} \cos(2\pi f_c t + \gamma t^2),$$

where f_c is the carrier frequency. Find α, β, and show that $\alpha\beta$ can be made as large as desired by choice of γ. *Note*: For phase-modulated signals, it is convenient to use a complex notation; i.e., let

$$s(t) = \text{Re } u(t)e^{j\omega_c t},$$

where $u(t) = e^{-t^2/2T^2 + j\gamma t^2}$. Then

$$\beta = 2\pi \left[\frac{\int f^2 |U(f)|^2 \, df}{2E} \right]^{1/2},$$

where $U(f)$ is the Fourier transform of $u(t)$.

12-10.   Figure P12-10 shows a system that can be used to generate a chirp pulse. The system employs a tapped delay line, with a tap spacing of Δt. A narrow-band filter is connected to each tap, and the summed outputs of the filters are assumed to produce the output pulse. The input to the delay line is a delta-function pulse. Assume that the transfer function of the ith narrow-band filter is given by

$$|H_i(f)| = \text{rect}\left(\frac{f - f_i}{\Delta f}\right);$$

Figure P12-10

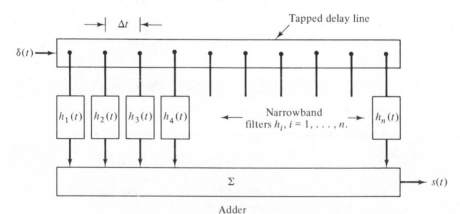

i.e., the passband for all the filters is the same, and the center frequencies are at f_i, $i = 1, \ldots, n$. To generate a chirp pulse whose frequency increases with time we can let $f_i = i\,\Delta f$, $i = 1, \ldots, n$.

   a. Find an expression for the output signal $s(t)$ if $f_i = i\,\Delta f$.

   b. Show that the matched filter for the signal generated by the circuit shown in Fig. P12-10 is *the same as* Fig. P12-10 except that the input is applied at the right-hand end of the delay line.

   c. Find the output of this matched filter. Observe that although the duration of $s(t)$ is essentially equal to $n\,\Delta t$, which is the total delay furnished by the delay line, the matched filter output will be a narrow pulse.

   d. Show qualitatively that the midfrequencies of the band-pass filters can have arbitrary values (i.e., not necessarily $i\,\Delta f$) without affecting the operation of the system in an essential way.

12-11. The bivariate, zero-mean, Gaussian probability density function for two random variables X and Y has the form

$$p(x, y) = k\,\exp\left[-\tfrac{1}{2}(ax^2 + 2bxy + cy^2)\right],$$

where k is a normalizing constant determined by the constraint

$$\int_{-\infty}^{\infty}\int_{-\infty}^{\infty} p(x, y)\,dx\,dy = 1.$$

   a. Verify that

$$k = \frac{\sqrt{ac - b^2}}{2\pi}.$$

   b. Find the mean and variance of X if Y is known. Show in particular that the variance of X is $1/a$.

   c. If Y is unknown, show that the variance of X is $c/(ac - b^2)$. [Note that if $\bar{X}$ is the mean value of X, then

$$\text{var}\,(X) = \int\!\!\int_{-\infty}^{\infty} dx\,dy(x - \bar{X})^2 p(x, y).\Bigg]$$

   d. If both delay and Doppler shift are to be simultaneously estimated in a radar receiver, then a function $Q(\tau, \eta)$ can be defined to take the place of $G(\tau)$ defined in Eq. (12.4-23), with τ and η being the hypothetical values of time delay and Doppler shift. Under suitable conditions $Q(\tau, \eta)$ will have a single peak near the true values of τ and η, and the values at the peak constitute an MAP estimate. Using the approach employed to obtain Eq. (12.5-9), show that $p(\tau, \eta\,|\,z)$ has the approximate form

$$p(\tau, \eta\,|\,z) \simeq k\,\exp\left\{D^2\left[1 - \frac{a}{2}(\tau - \tau_0)^2 - \frac{b}{2}(\eta - \eta_0)^2\right.\right.$$
$$\left.\left. -\,c(\tau - \tau_0)(\eta - \eta_0)\right]\right\},$$

and by use of this result and that of part c, verify that Eqs. (12.5-15) and (12.5-16) have the proper form.

REFERENCES

[12-1] L. N. RIDENOUR, ed.-in-chief, Radiation Laboratory Series, Vols. 1–28, McGraw-Hill, New York, 1948.

[12-2] *Reference Data for Radio Engineers*, 5th ed., Howard W. Sams & Co., Inc., Publishers, Indianapolis, Ind., 1968, pp. 25–42.

[12-3] W. G. MELBOURNE, "Navigation Between the Planets," *Sci. Am.*, **234**, No. 6, pp. 58–74, June 1976.

[12-4] M. I. SKOLNIK, *Introduction to Radar Systems*, McGraw-Hill, New York, 1962, p. 117.

[12-5] A. W. RIHACZEK, *Principles of High Resolution Radar*, McGraw-Hill, New York, 1969.

[12-6] C. E. COOK and M. BERNFELD, *Radar Signals*, Academic Press, New York, 1967.

[12-7] P. M. WOODWARD, *Probability and Information Theory with Applications to Radar*, McGraw-Hill, New York, 1953.

[12-8] P. M. WOODWARD and I. L. DAVIES, "A Theory of Radar Information," *Phil. Mag.*, **41**, No. 7, pp. 1001–1017, 1950.

[12-9] M. I. SKOLNIK, ed., *Radar Handbook*, McGraw-Hill, New York, 1970.

[12-10] D. GABOR, "Theory of Communication," *J. Inst. Elec. Engr.*, **93**, part III, p. 429, 1946.

[12-11] M. ABRAMOWITZ and I. A. STEGUN, *Handbook of Mathematical Functions*, Applied Mathematics Series No. 55, U.S. Department of Commerce, National Bureau of Standards, Washington, D.C., 1965.

[12-12] H. L. VAN TREES, *Detection, Estimation, and Modulation Theory*, Part I, Wiley, New York, 1968.

[12-13] A. PAPOULIS, *The Fourier Integral and Its Application*, McGraw-Hill, New York, 1962, pp. 62–64.

[12-14] A. NOVICK, "Echolocation in Bats—Some Aspects of Pulse Design," *Am. Sci.*, **59**, No. 2, pp. 198–209, March 1971.

[12-15] F. BRYN, "Optimum Signal Processing of Three-Dimensional Arrays Operating on Gaussian Signals and Noise," *J. Acoust. Soc. Am.*, **34**, pp. 289–297, March 1962.

[12-16] P. M. SCHULTHEISS and F. B. TUTEUR, "Optimum and Suboptimum Detection of Directional Gaussian Signals in an Isotropic Gaussian Noise Field, Part I. Likelihood Ratio and Power Detectors," *IEEE Trans. Military Electron.*, **MIL-9**, Nos. 3 and 4, pp. 197–208, July–Oct. 1965.

[12-17] D. J. EDELBLUTE, J. M. FISK, and G. L. KINNISON, "Criteria for Optimum Signal Detection Theory for Arrays," *J. Acoust. Soc. Am.*, **41**, pp. 199–205, Jan. 1967.

13

Image-Processing Systems and Two-Dimensional Transforms

13.1 Introduction

It is a fact that well over 90% of the informational inputs to a normal human being is through his vision. Vision is the perception of images, and images are the most important form of two-dimensional signals. Images are intimately connected with optics, and optical systems can frequently be modeled as linear two-dimensional systems. Since the early 1960's, there has been a tremendous amount of research into image processing, medical imaging, pattern recognition of images, ultrasonic and microwave imaging, image enhancement, image coding, image transmission, and lensless imaging using holographic techniques. The analysis and design of such systems make copious use of the mathematics of communication theory. Indeed modern optics is often called Fourier optics, and its study falls well within the realm of communication engineering.

A number of textbooks have been written in the previous decade dealing with the rapidly expanding fields of optical information processing and holography. For the most part these books discuss optics from the viewpoint of the systems and communications theory disciplines of electrical engineering. Optical systems inherently handle two-dimensional signals, and to analyze them the reader must become acquainted with the basic aspects of two-dimensional systems theory.

In this chapter we discuss some elementary properties of two-dimensional systems. Some of the results do not appear in any other textbook, although they can be found in the periodical technical literature. The authors believe that 2-D systems and transforms will play an ever-increasing role in modern communication systems and that it is the communication-theory specialist who will play a central role in modern optical systems research for the following reasons:

1. The mathematics of communication theory is ideally suited for the analysis of optical systems.
2. More and more signal processing is done with optical devices.
3. For nearly every optical phenomenon, there is a corresponding analogous electrical phenomenon.

13.2 Two-Dimensional Fourier Transforms

The 2-D Fourier transform of a 2-D function $f(x, y)$ is defined by

$$F(u, v) = \int_{-\infty}^{\infty} \int_{-\infty}^{\infty} f(x, y) e^{-j2\pi(ux+vy)} \, dx \, dy, \qquad (13.2\text{-}1)$$

where $f(x, y)$ may be real or complex, u and v are so-called spatial frequencies, and $F(u, v)$ is the Fourier transform of $f(x,y)$. The *inverse* Fourier transform of $F(u, v)$ recovers $f(x, y)$ and is given by

$$f(x, y) = \int_{-\infty}^{\infty} \int_{-\infty}^{\infty} F(u, v) e^{j2\pi(ux+vy)} \, du \, dv. \qquad (13.2\text{-}2)$$

The integral in Eq. (13.2-1) doesn't always converge. A set of *sufficient* conditions for convergence of the 2-D Fourier transform is

1. f must be absolutely integrable; i.e.,

$$\int_{-\infty}^{\infty} \int_{-\infty}^{\infty} |f(x, y)| \, dx \, dy = M < \infty. \qquad (13.2\text{-}3)$$

2. f must be a reasonably well-behaved function; i.e., (a) f must have only a finite set of discontinuities and maxima and minima in any finite rectangle, and (b) f must have no infinite discontinuities.

We emphasize that these are sufficient conditions. For example, the 2-D Dirac function $\delta(x, y)$ fails to satisfy condition 2(b), yet its Fourier transform is

$$F(u, v) = \int_{-\infty}^{\infty} \int_{-\infty}^{\infty} \delta(x, y) e^{-j2\pi(ux+vy)} \, dx \, dy = 1. \qquad (13.2\text{-}4)$$

Also for $f(x, y) = \cos 2\pi u_0 x$, we have

$$\int_{-\infty}^{\infty} \int_{-\infty}^{\infty} |\cos 2\pi u_0 x| \, dx \, dy = \infty; \qquad (13.2\text{-}5)$$

thus condition 1 is not satisfied, yet its Fourier transform is, at least formally,

$$F(u, v) = \int_{-\infty}^{\infty} \int_{-\infty}^{\infty} \left[\frac{e^{-j2\pi(u-u_0)x} + e^{j2\pi(u+u_0)x}}{2} \right] e^{j2\pi vy} \, dx \, dy$$

$$= \tfrac{1}{2}[\delta(u - u_0, v) + \delta(u + u_0, v)], \tag{13.2-6}$$

where $\delta(u, v)$ is defined by $\delta(u, v) = \delta(u)\delta(v)$. Equation (13.2-6) is a very useful engineering result even though δ is not an ordinary function. (See Sec. 2-11.) An example for which Eq. (13.2-1) doesn't converge is

$$f(x, y) = e^{\alpha x + \beta y} u(x)u(y), \qquad \text{Re } \alpha > 0, \text{Re } \beta > 0. \tag{13.2-7}$$

We can view the 2-D Fourier transform (2-DFT) in several ways. One way is to view it as a succession of one-dimensional transforms. Thus with $Q(u, y)$ denoting the one-dimensional Fourier transform of $f(x, y)$, we have

$$Q(u, y) = \int_{-\infty}^{\infty} f(x, y)e^{-j2\pi ux} \, dx$$

and

$$F(u, v) = \int_{-\infty}^{\infty} Q(u, y)e^{-j2\pi vy} \, dy.$$

Equation (13.2-2) suggests that the 2-DFT can be viewed as a decomposition of $f(x, y)$ into elementary complex sinusoids of infinitesimal amplitude $F(u,v) \, du \, dv$. In other words, $f(x,y)$ can be built up by a coherent (i.e., phase-preserving) addition of a very large number of complex sinusoids of different amplitudes and spatial frequencies. What physical meaning do the functions $\{e^{j2\pi(ux+vy)}\}$ have? In optics, it can be shown that $e^{j2\pi(ux+vy)}$ can be used to represent a unit plane wave of illumination in the plane $z = 0$ with direction cosines

$$\alpha = u\lambda, \qquad \beta = v\lambda, \qquad \gamma = \sqrt{1 - \alpha^2 - \beta^2},$$

where α is associated with the angle between the wave normal and the x axis, β is associated with the wave normal and the y axis, γ is associated with the wave normal and the z axis, and λ is the wavelength of the illumination ([13-1], p. 48).

We shall frequently denote a Fourier transform pair by the symbology $F(u, v) \longleftrightarrow f(x, y)$.

13.3 Two-Dimensional Fourier Transform Theorems

Many of the results given below are direct extensions of corresponding one-dimensional results. Most of the proofs therefore will either be omitted or left as exercises. The reader should refer to Chapter 2 of this book or to Goodman [13-1] for more details.

Linearity

The 2-DFT is a linear operator. This means that if α and β are any complex constants and $f_1(\cdot)$ and $f_2(\cdot)$ are any two functions satisfying the sufficiency conditions given in Sec. 13.2, then

$$\mathcal{F}_2(\alpha f_1 + \beta f_2) = \alpha F_1(u, v) + \beta F_2(u, v), \qquad (13.3\text{-}1)$$

where $\mathcal{F}_2$ denotes the 2-D Fourier operator.

Scaling

If $f(x, y) \longleftrightarrow F(u, v)$, then

$$f(ax, by) = \frac{1}{|ab|} F\left(\frac{u}{a}, \frac{v}{b}\right). \qquad (13.3\text{-}2)$$

This is a direct extension of Eq. (2.9-6).

Parseval's Theorem

Parseval's theorem is a special case of the two dimensional convolution theorem. It states that

$$\int_{-\infty}^{\infty} \int_{-\infty}^{\infty} |f(x, y)|^2 \, dx \, dy = \int_{-\infty}^{\infty} \int_{-\infty}^{\infty} |F(u, v)|^2 \, du \, dv, \qquad (13.3\text{-}3)$$

which, in the notation of Chapter 2 [e.g., Eq. (2.7-4)], can be succinctly written as

$$\| f \|^2 = \| F \|^2 \quad \text{or} \quad \| f \| = \| F \|. \qquad (13.3\text{-}4)$$

In optics, this is quite an important result.[†] It states that in a lossless, passive optical system, the total illumination power on one side of a lens must equal the total illumination power in the focal plane on its other side.

Shift Theorem

If $f(x, y) \longleftrightarrow F(u, v)$, then

$$f(x - a, y - b) \longleftrightarrow F(u, v) e^{-j2\pi(ua+vb)}.$$

This result is a direct extension of Eq. (2.9-5). Note that

$$|\mathcal{F}_2[f(x - a, y - b)]| = |\mathcal{F}_2[f(x, y)]|, \qquad (13.3\text{-}5)$$

which is a result of considerable utility in optics, where it is somewhat erroneously called the "shift-invariance property of the Fourier transform."

[†]A result which, to be fully understood, requires knowledge of the Fourier property of lenses. See Sec. 13.12.

Symmetry Properties

Let $f(x, y)$ be a real function of (x, y). Then

$$F(u, v) = F^*(-u, -v)$$
$$F(u, -v) = F^*(-u, v)$$
$$F(-u, v) = F^*(u, -v). \tag{13.3-6}$$

This is the 2-D extension of the Hermitian property of Eq. (2.9-23). From the above, we easily determine with $F(u, v) = A(u, v) + jB(u, v)$ and $A(u, v)$ and $B(u, v)$ real that

$$A(u, v) = A(-u, -v)$$
$$A(-u, v) = A(u, -v)$$
$$B(u, v) = -B(-u, -v) \tag{13.3-7}$$
$$B(-u, v) = -B(u, -v).$$

Further, with $R(u, v) = \{[A(u, v)]^2 + [B(u, v)]^2\}^{1/2}$ we obtain

$$R(u, v) = R(-u, -v)$$
$$R(-u, v) = R(u, -v) \tag{13.3-8}$$
$$R(u, -v) = R(-u, v).$$

Also, with $\theta(u, v) = \tan^{-1}[B(u, v)/A(u, v)]$, we obtain

$$\theta(u, v) = -\theta(-u, -v)$$
$$\theta(u, -v) = -\theta(-u, v) \tag{13.3-9}$$
$$\theta(-u, v) = -\theta(u, -v).$$

If $f(x, y)$ is real and equal to $f(-x, -y)$, then it is easy to show, either from duality or from the definition of the inverse transform, that $F(u, v)$ is real and $F(u, v) = F(-u, -v)$. If $f(x, y)$ is pure imaginary, then certain skew Hermitian properties hold (see Prob. 13-3).

Convolution and Correlation

Let $f(x, y)$ and $h(x, y)$ be two functions with Fourier transforms $F(u, v)$ and $H(u, v)$, respectively. If $g(x, y)$ denotes their convolution† product, i.e.,

$$g(x, y) = \int_{-\infty}^{\infty} \int_{-\infty}^{\infty} f(\xi, \eta) h(x - \xi, y - \eta) \, d\xi \, d\eta$$
$$\equiv f(x, y) * h(x, y), \tag{13.3-10}$$

then $G(u, v) = F(u, v)H(u, v)$, where $G(u, v) = \mathcal{F}_2[g]$. This is a direct extension of the very important convolution theorem introduced in Sec. 2.9. If we use

---

†The centered asterisk denotes convolution, whereas the asterisk on a function denotes complex conjugation.

the same notation and define $q(x, y)$ as the correlation product of f and h, i.e.,

$$q(x, y) = \int_{-\infty}^{\infty} \int_{-\infty}^{\infty} f(\xi, \eta) h^*(\xi - x, \eta - y) \, d\xi \, d\eta, \qquad (13.3\text{-}11)$$

then

$$Q(u, v) = \mathfrak{F}_2[q] = F(u, v) H^*(u, v). \qquad (13.3\text{-}12)$$

A special case of the above is the self-correlation product theorem, which gives

$$\mathfrak{F}_2\left[\int_{-\infty}^{\infty} \int_{-\infty}^{\infty} f(\xi, \eta) f^*(\xi - x, \eta - y) \, d\xi \, d\eta\right] = |F(u, v)|^2. \quad (13.3\text{-}13)$$

Similarly, it is easy to show that

$$\int_{-\infty}^{\infty} \int_{-\infty}^{\infty} H(\xi, \eta) H^*(\xi - u, \eta - v) \, d\xi \, d\eta = \mathfrak{F}_2[|h|^2]. \qquad (13.3\text{-}14)$$

Equation (13.3-11) is sometimes denoted in symbolic form as

$$q(x, y) = f(x, y) \circledast h(x, y), \qquad (13.3\text{-}15)$$

where $\circledast$ denotes correlation product. Note that in general

$$(f_1 * f_2) * f_3 = f_1 * (f_2 * f_3)$$

and

$$(f_1 \circledast f_2) * f_3 = f_1 \circledast (f_2 \circledast f_3).$$

Fourier Identity

At points where the function $f(x, y)$ is continuous, it can be shown that

$$\mathfrak{F}_2^{-1} \mathfrak{F}_2[f(x, y)] = \mathfrak{F}_2 \mathfrak{F}_2^{-1}[f(x, y)] = f(x, y)$$

and

$$\mathfrak{F}_2 \mathfrak{F}_2[f(x, y)] = f(-x, -y). \qquad (13.3\text{-}16)$$

At points where $f(x, y)$ is discontinuous, the two successive transforms give an average of $f(x, y)$ in a small neighborhood of the point of discontinuity.

13.4 Separable and Circularly Symmetric Functions

A function f is said to be separable in rectangular coordinates if

$$f(x, y) = f_1(x) f_2(y). \qquad (13.4\text{-}1)$$

It is separable in polar coordinates (r, θ) if

$$f(r, \theta) = f_1(r) f_2(\theta). \qquad (13.4\text{-}2)$$

By applying the definition of the 2-DFT to Eq. (13.4-1), we obtain

$$F(u, v) = F_1(u) F_2(v), \qquad (13.4\text{-}3)$$

where

$$F_1(u) = \int_{-\infty}^{\infty} f_1(x)e^{-j2\pi ux}\,dx = \mathcal{F}[f_1]$$

and

$$F_2(v) = \int_{-\infty}^{\infty} f_2(y)e^{-j2\pi vy}\,dy = \mathcal{F}[f_2].$$

Thus the 2-DFT of a function separable in x and y is itself separable into a product of two transforms, one depending only on u and the other depending only on v.

A very important class of two-dimensional functions is comprised of functions that depend only on $r = (x^2 + y^2)^{1/2}$. Such functions are said to possess *circular symmetry*. The 2-DFT of such functions can be computed as follows. Let $f(x, y) = f(r)$; the 2-DFT is

$$F(u, v) = \int_{-\infty}^{\infty} \int_{-\infty}^{\infty} f(x, y)e^{-j2\pi(ux+vy)}\,dx\,dy. \qquad (13.4\text{-}4)$$

With

$$
\begin{aligned}
x &= r\cos\theta, & u &= \rho\cos\phi \\
y &= r\sin\theta, & v &= \rho\sin\phi \\
r^2 &= x^2 + y^2, & \rho^2 &= u^2 + v^2 \\
\theta &= \tan^{-1}\left(\frac{y}{x}\right), & \phi &= \tan^{-1}\left(\frac{v}{u}\right),
\end{aligned}
\qquad (13.4\text{-}5)
$$

we obtain

$$
\begin{aligned}
F(\rho\cos\phi, \rho\sin\phi) &= \int_0^{2\pi} d\theta \int_0^{\infty} f(r)e^{-j2\pi r\rho(\cos\theta\cos\phi+\sin\theta\sin\phi)}r\,dr. \\
&= \int_0^{\infty} dr\, rf(r) \int_0^{2\pi} e^{-j2\pi r\rho\cos(\theta-\phi)}\,d\theta. \qquad (13.4\text{-}6)
\end{aligned}
$$

If, in Eq. (13.4-6), we use the identity

$$J_0(x) = \frac{1}{2\pi}\int_0^{2\pi} e^{-jx\cos(\theta-\phi)}\,d\theta, \qquad (13.4\text{-}7)$$

where $J_0(x)$ is a Bessel function of the first kind, of order zero, we obtain

$$
\begin{aligned}
F(\rho\cos\phi, \rho\sin\phi) &= 2\pi \int_0^{\infty} rf(r)J_0(2\pi r\rho)\,dr \\
&\equiv \tilde{F}(\rho). \qquad (13.4\text{-}8)
\end{aligned}
$$

Equation (13.4-8) is an important result because it shows that the dependence on ϕ has vanished and that the 2-DFT of a circularly symmetric function is itself circularly symmetric with $\rho = (u^2 + v^2)^{1/2}$. The 2-DFT of a circularly symmetric function can be computed from the one-dimensional integral given in Eq. (13.4-8), which is known as a *Hankel transform*. We shall use the following notation to avoid confusion:

$$f(x, y) \longleftrightarrow F(u, v), \qquad \text{a 2-DFT pair}$$

$$f(x) \longleftrightarrow F(u), \qquad \text{an ordinary 1-DFT pair} \qquad (13.4\text{--}9)$$

$$f(x) \overset{h}{\longleftrightarrow} \tilde{F}(u), \qquad \text{a Hankel transform pair.}$$

It is a simple matter to show, starting with Eq. (13.2-2), that the inverse 2-DFT of a circularly symmetric function $F(u, v) \equiv \tilde{F}(\rho)$ is given by

$$f(x, y) = f(r) = 2\pi \int_0^\infty \rho \tilde{F}(\rho) J_0(2\pi r \rho) \, d\rho, \qquad (13.4\text{-}10)$$

which is again a Hankel transform. Hence the Hankel transform is self-reciprocal, and

$$f(r) = \mathcal{H}\mathcal{H}[f(r)].† \qquad (13.4\text{-}11)$$

13.5   Profile Functions; Relation Between Hankel and Fourier Transforms; Line-Spread Function

Let $f(x, y)$ be a function of circular symmetry. Then $f(x, y) = f[(x^2 + y^2)^{1/2}] = f(r)$. The function

$$p(x) = \int_{-\infty}^{\infty} f(x, y) \, dy \qquad (13.5\text{-}1)$$

is called the x profile of $f(x, y)$; a similar expression holds for the y profile of $f(x, y)$. In general f need not be circularly symmetric to have a profile. Let $p(x) \longleftrightarrow P(u)$ be ordinary Fourier transform pairs; then clearly

$$F(u, 0) = \int_{-\infty}^{\infty} dx \, e^{-j2\pi ux} \int_{-\infty}^{\infty} f(x, y) \, dy$$

$$= P(u). \qquad (13.5\text{-}2)$$

Suppose that we write ρ for u in $P(u)$; then

$$P(\rho) = \int_{-\infty}^{\infty} p(x) e^{-j2\pi \rho x} \, dx$$

$$= \int_{-\infty}^{\infty} dx \, e^{-j2\pi \rho x} \int_{-\infty}^{\infty} f[(x^2 + y^2)^{1/2}] \, dy$$

$$= 2\pi \int_0^\infty r f(r) J_0(2\pi r \rho) \, dr$$

$$= \tilde{F}(\rho). \qquad (13.5\text{-}3)$$

Equation (13.5-3) is a most important result in 2-D transform theory. It states that the Hankel transform of a circularly symmetric function is simply the ordinary Fourier transform of the profile function. This result can be used

†The symbol $\mathcal{H}$ is used for the Hankel transform operator.

to compute Hankel transform pairs from known Fourier transform pairs and vice versa.

EXAMPLE

See [13-2] p. 153. Let $f(r)$ be defined by

$$f(r) = \begin{cases} 1, & r < a \\ 0, & \text{otherwise.} \end{cases} \tag{13.5-4}$$

Then

$$p(x) = \begin{cases} \displaystyle\int_{-\sqrt{a^2-x^2}}^{\sqrt{a^2-x^2}} 1 \, dy = 2\sqrt{a^2 - x^2}, & |x| < a \\ 0, & |x| > a. \end{cases}$$

Hence

$$p(x) = 2\sqrt{a^2 - x^2}\, \text{rect}\left(\frac{x}{2a}\right). \tag{13.5-5}$$

With the aid of the identify $\displaystyle\int_0^x rJ_0(r) \, dr = xJ_1(x)$ the Hankel transform of $f(r)$ is easily computed to be

$$\tilde{F}(\rho) = 2\pi \int_0^a rJ_0(2\pi r \rho) \, dr = 2\pi a^2 \frac{J_1(2\pi\rho a)}{2\pi\rho a}. \tag{13.5-6}$$

Hence

$$\sqrt{a^2 - x^2}\, \text{rect}\left(\frac{x}{2a}\right) \longleftrightarrow \pi a^2 \frac{J_1(2\pi\rho a)}{2\pi\rho a}, \tag{13.5-7}$$

where $J_1(x)$ is the first-order Bessel function of the first kind.

13.6   Some Two-Dimensional Fourier Transform Pairs

In Chapter 2, we introduced some useful one-dimensional functions and their Fourier transforms. By direct extension, we obtain Table 13.6-1.

TABLE 13.6-1   FOURIER PAIRS

Function	Transform	
rect (x) rect (y)	sinc (u) sinc (v)	(13.6-1)
$\delta(x, y)$	1	(13.6-2)
$e^{j\pi(x+y)}$	$\delta(u - \frac{1}{2}, v - \frac{1}{2})$	(13.6-3)
sgn (x) sgn (y)	$\dfrac{1}{j\pi u}\dfrac{1}{j\pi v}$	(13.6-4)
comb (x) comb (y)	comb (u) comb (v)	(13.6-5)
circ $(\sqrt{x^2 + y^2})$	$2\pi\dfrac{J_1(2\pi\rho)}{2\pi\rho}$	(13.6-6)
tri (x) tri (y)	sinc^2 (u) sinc^2 (v)	(13.6-7)
$e^{-\pi(x^2+y^2)}$	$e^{-\pi(u^2+v^2)}$	(13.6-8)

For the reader's benefit, the definitions of the functions in Table 13.6-1 are repeated here:

$$\text{rect}(x) = \begin{cases} 1, & |x| < 1 \\ 0, & \text{otherwise,} \end{cases} \tag{13.6-9}$$

$$\text{comb}(x)\dagger = \sum_{n=-\infty}^{\infty} \delta(x - n), \tag{13.6-10}$$

$$\text{tri}(x) = [1 - |x|] \text{ rect}\left(\frac{x}{2}\right), \tag{13.6-11}$$

$$\text{circ}(\sqrt{x^2 + y^2}) = \begin{cases} 1, & \sqrt{x^2 + y^2} \le 1 \\ 0, & \text{otherwise,} \end{cases} \tag{13.6-12}$$

$$\delta(x, y) = 0, \qquad x, y \text{ not both zero,}$$

$$\int_{-\infty}^{\infty} \int_{-\infty}^{\infty} \delta(x, y) \, dx \, dy = 1, \tag{13.6-13}$$

$$\text{sgn}(xy) = \text{sgn}(x)\,\text{sgn}(y) = \begin{cases} 1, & xy > 0 \\ 0, & xy = 0 \\ -1, & xy < 0. \end{cases} \tag{13.6-14}$$

13.7 Some Two-Dimensional Hankel Transform Pairs; Convolution of Circularly Symmetric Functions

Table 13.7-1 is a very short table of Hankel transform pairs. More extensive tables can be found in the literature [13-3].

TABLE 13.7-1   HANKEL TRANSFORM PAIRS

Function	Transform	
$\dfrac{1}{r}$	$\dfrac{1}{\rho}$	(13.7-1)
$\delta(r - r_0)$	$2\pi r_0 J_0(2\pi r_0 \rho)$	(13.7-2)
e^{-ar^2} (Re $a > 0$)	$\dfrac{\pi}{a} e^{-\pi^2 \rho^2/a}$	(13.7-3)
e^{jar^2} (Im $a > 0$)	$\dfrac{j\pi}{a} e^{-j\pi^2 \rho^2/a}$	(13.7-4)

Proof of First Pair

$$\tilde{F}(\rho) = 2\pi \int_0^\infty \left(\frac{1}{r}\right) r J_0(2\pi r \rho) \, dr$$

$$= \frac{1}{\rho} \int_0^\infty J_0(\alpha) \, d\alpha.$$

†This function was referred to as comb (1) in Sec. 2.12. The notation used here is standard in optics.

However, $\int_0^\infty J_v(\alpha)\,d\alpha = 1$ for Re $v > -1$ [13-4], Eq. (11.4-17). Hence

$$\frac{1}{r} \longleftrightarrow \frac{1}{\rho}. \tag{13.7-5}$$

Proof of Second Pair

$$\tilde{F}(\rho) = 2\pi \int_0^\infty r\delta(r - r_0)J_0(2\pi r\rho)\,dr.$$

From the definition of the delta function, we have

$$\tilde{F}(\rho) = 2\pi r_0 J_0(2\pi r_0\rho).$$

Hence

$$\delta(r - r_0) \longleftrightarrow 2\pi r_0 J_0(2\pi r_0\rho). \tag{13.7-6}$$

We leave the proofs of the other pairs as an exercise for the reader.

A useful formula is the Hankel transform convolution theorem, which states that the Hankel transform of the convolution of two circularly symmetric functions is the product of their Hankel transform. Thus with

$$f(x,y) \equiv \int_{-\infty}^\infty \int_{-\infty}^\infty f_1(\sqrt{\xi^2 + \eta^2})f_2(\sqrt{(x - \xi)^2 + (y - \eta)^2})\,d\xi\,d\eta,$$

$$\tag{13.7-7a}$$

we have

$$\tilde{F}(\rho) = \tilde{F}_1(\rho)\tilde{F}_2(\rho), \tag{13.7-7b}$$

where as usual, the functions with the tilde denote Hankel transforms. The proof follows directly from the two-dimensional Fourier convolution theorem in (13.3-10) and the definition of the Hankel transform.

EXAMPLE

Convolution of two circ $(\cdot)$ *functions.* The convolution of circ $(\cdot)$ functions is a very important result in 2-D systems theory. It appears in the theory of imaging ([13-1], p. 119) and in two-dimensional random processes [13-5]. From Eq. (13.6-6) and the Fourier scaling law, we have that

$$\text{circ}\left(\frac{r}{a}\right) \overset{h}{\longleftrightarrow} 2\pi a^2 \left[\frac{J_1(2\pi a\rho)}{2\pi a\rho}\right], \tag{13.7-8}$$

where circ $(r/a) = 1$ for $0 < r < a$ and zero otherwise. Since the convolution product will clearly be symmetric, it suffices to compute Eq. (13.7-7a) for the circles centered on the ξ axis. For a separation x for which there is overlap (Fig. 13.7-1), the area of overlap which is the convolution product is four times the area of the region ⑥ whose vertices are at B, C, and E. The area of sector ABE is $(\theta/2\pi)\pi a^2$.

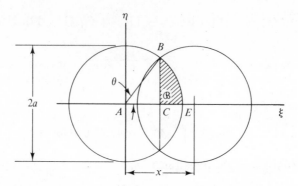

Figure 13.7-1. Convolution of two circ (·) functions.

The area of triangle ABC is $\frac{1}{2}(x/2)\sqrt{a^2 - (x/2)^2}$. Hence the area of $\mathcal{B}$ is

$$\text{area } (\mathcal{B}) = \left(\frac{\theta}{2\pi}\right)\pi a^2 - \frac{1}{2}\left(\frac{x}{2}\right)\sqrt{a^2 - \left(\frac{x}{2}\right)^2},$$

and the total area of overlap is

$$f(x) = \begin{cases} 2a^2\left[\cos^{-1}\left(\frac{x}{2a}\right) - \left(\frac{x}{2a}\right)\sqrt{1 - \left(\frac{x}{2a}\right)^2}\right], & x < 2a \\ 0, & \text{otherwise.} \end{cases}$$

More generally, the convolution for arbitrary displacement r is given by

$$f(r) = \begin{cases} 2a^2\left[\cos^{-1}\left(\frac{r}{2a}\right) - \left(\frac{r}{2a}\right)\sqrt{1 - \left(\frac{r}{2a}\right)^2}\right], & r \leq 2a \\ 0, & \text{otherwise.} \end{cases} \tag{13.7-9}$$

The Hankel transform of $f(r)$ is, from Eqs. (13.7-7) and (13.7-8),

$$F(\rho) = [2\pi a^2]^2 \left[\frac{J_1(2\pi a\rho)}{2\pi a\rho}\right]^2. \tag{13.7-10}$$

13.8 Hilbert Transforms

We denote the two-dimensional Hilbert transform operator by $\mathcal{L}_2$ (having already used $\mathcal{K}$ for the Hankel transform operator) and define the two-dimensional Hilbert transform by

$$\mathcal{L}_2[f(x,y)] = \mathring{f}(x,y) = \frac{1}{\pi^2} \int_{-\infty}^{\infty} \int_{-\infty}^{\infty} \frac{f(\xi,\eta)}{(x-\xi)(y-\eta)} \, d\xi \, d\eta. \tag{13.8-1}$$

The one-dimensional Hilbert transform with respect to a variable, say x, will be denoted by

$$\hat{f}_x(x,\eta) = \frac{1}{\pi} \int_{-\infty}^{\infty} \frac{f(\xi,\eta)}{x-\xi} \, d\xi. \tag{13.8-2}$$

Note that the order of the arguments is important. The above integrals are all *Cauchy principal value* integrals, and care must be used in an actual evaluation.† We note that Eq. (13.8-1) is a two-dimensional convolution of $f(x, y)$ and the function $l(x, y) = (\pi^2 xy)^{-1}$; this fact enables us to quickly determine the inverse 2-D Hilbert transform. To that end we make the following definitions:

$$L(u, v) = \mathcal{F}_2[l(x, y)], \tag{13.8-3}$$

$$F(u, v) = \mathcal{F}_2[f(x, y)], \tag{13.8-4}$$

$$Y(u, v) = \mathcal{F}_2[\hat{f}(x, y)], \tag{13.8-5}$$

and

$$G(u, v) = \mathcal{F}_2[g(x, y)],$$

where

$$g(x, y) = \frac{1}{\pi^2} \int_{-\infty}^{\infty} \int_{-\infty}^{\infty} \frac{\hat{f}(\xi, \eta)}{(x - \xi)(y - \eta)} d\xi \, d\eta. \tag{13.8-6}$$

Now $L(u, v)$ is the 2-DFT of a product of separable functions each of which is the impulse response of the quadrature filter, i.e., the kernel of the one-dimensional Hilbert transform. Hence

$$L(u, v) = \frac{1}{j\pi} \int_{-\infty}^{\infty} \frac{\sin 2\pi ux}{x} dx \cdot \frac{1}{j\pi} \int_{-\infty}^{\infty} \frac{\sin 2\pi vy}{y} dy$$

$$= \begin{cases} -1, & uv > 0 \\ +1, & uv < 0, \end{cases} \tag{13.8-7}$$

where we used the result that

$$\frac{1}{j\pi} \int_{-\infty}^{\infty} \frac{\sin tx}{x} dx = \begin{cases} -j, & t > 0 \\ +j, & t < 0. \end{cases} \tag{13.8-8}$$

The 2-DFT's of Eqs. (13.8-1) and (13.8-6) give, respectively,

$$Y(u, v) = F(u, v)L(u, v), \tag{13.8-9}$$

$$G(u, v) = Y(u, v)L(u, v), \tag{13.8-10}$$

so that, by direct substitution of Eq. (13.8-9) into Eq. (13.8-10), we obtain

$$G(u, v) = F(u, v)L^2(u, v) = F(u, v). \tag{13.8-11}$$

Hence at all points where $f(x, y)$ is continuous, $g(x, y) = f(x, y)$, and the inverse 2-D Hilbert transform is just another 2-D Hilbert transform; i.e.,

$$f(x, y) = \frac{1}{\pi^2} \int_{-\infty}^{\infty} \int_{-\infty}^{\infty} \frac{\hat{f}(\xi, \eta)}{(x - \xi)(y - \eta)} d\xi \, d\eta. \tag{13.8-12}$$

---

†CPV integrals are discussed in Secs. 2.10 and 2.13. An example of CPV integration is given in Sec. 2.13 with respect to the one-dimensional Hilbert transform.

We shall now give, without proof, an extension of the Hilbert transform multiplication theorem [see Eq. (2.13-16)]. First we observe that whereas the concepts of low-pass and high-pass are well defined in one-dimensional signal theory, it is frequently not so clear which is a high-pass or low-pass function in two dimensions (Fig. 13.8-1). To clear up these concepts in two-dimensions we shall make the following definitions.

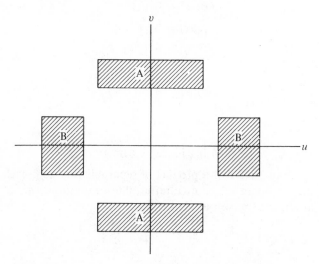

Figure 13.8-1. Region *A* represents the spectral basewidth of a signal which is low-pass along the *u* frequency but high-pass along the *v* frequency with respect to a signal whose basewidth is *B*.

Support

For any function $g(x, y)$, the support of g, written supp g, is defined to be the set of points (x, y) at which g is different from zero.

Low-Pass with Cutoff Vector (u_0, v_0)

A function $f(x, y)$ with $\mathfrak{F}_2[f] = F(u, v)$ is said to be low-pass with cutoff vector $\boldsymbol{\Omega}_0 = (u_0, v_0)$ if

$$\max_{\text{all } u \in \text{supp } F} |u| = u_0 \quad \text{and} \quad \max_{\text{all } v \in \text{supp } F} |v| = v_0. \tag{13.8-13}$$

We remind the reader that in Eq. (13.8-13) supp F is the set of points at which $F(u, v)$ is not zero.

High-Pass Function; Strong Separability from Low-Pass

A function $g(x, y)$ with $\mathfrak{F}_2[g] = G(u, v)$ is said to be high-pass and strongly separable from the low-pass function $f(x, y)$ with cutoff vector

$\Omega_0 = (u_0, v_0)$ if

$$\min_{\text{all } u \in \text{supp } G} |u| > u_0 \quad \text{and} \quad \min_{\text{all } v \in \text{supp } G} |v| > v_0. \qquad (13.8\text{-}14)$$

Spectral Disjointness

We say that $g(x, y)$ and $f(x, y)$ are spectrally disjoint if

$$[\text{supp } F] \cap [\text{supp } G] = \varnothing \qquad (\varnothing = \text{empty set}), \qquad (13.8\text{-}15)$$

where F and G are the 2-DFT's of f and g, respectively. In Fig. 13.8-2 A is low-pass with cutoff vector (u_0, v_0), and B is high-pass with respect to A.

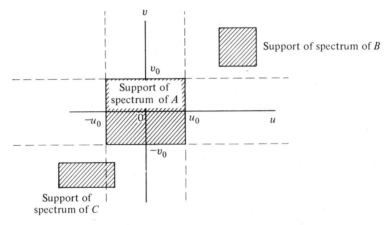

Figure 13.8-2. Illustrating low-pass, high-pass, and spectral disjointedness in two dimensions.

However, although A and C are spectrally disjoint, C is not high-pass strongly separable with respect to A.

The Hilbert transform multiplication theorem can now be stated without ambiguity.

Let $f(x, y)$ be low-pass with cutoff vector Ω_0, and let $g(x, y)$ be high-pass and strongly separable from $f(x, y)$. Then

$$\mathcal{L}_2[f(x, y)\, g(x, y)] = f(x, y)\mathcal{L}_2[g(x, y)]. \qquad (13.8\text{-}16)$$

To prove Eq. (13.8-16) it suffices to show that

$$I(\Omega) = \int_{-\infty}^{\infty} \int_{-\infty}^{\infty} [\text{sgn}(\xi)\, \text{sgn}(\eta) - \text{sgn}(u)\, \text{sgn}(v)]$$
$$\cdot\, G(\xi, \eta) F(u - \xi, v - \eta)\, d\xi\, d\eta = 0. \qquad (13.8\text{-}17)$$

We leave this as an exercise for the reader (Prob.13-11). The Hilbert transform of a two-dimensional signal can be realized by a method suggested by Eq.

(13.8-7); i.e., shift the phase of the signal spectrum by 180° whenever $uv > 0$. In optics this can be done in a straightforward manner. Just as in the one-dimensional case, only the high-frequency signal in the product of low- and high-frequency signals gets Hilbert-transformed. A proof of the multi-dimensional Hilbert transform is furnished by Stark [13-6] and extended by Bedrosian and Stark [13-7].

13.9 Two-Dimensional Sampling Theory

In Chapter 4 we discussed the Whittaker-Shannon sampling theorem ([13-8] and [13-9]). The baseband version of this theorem states that a function $h(t)$ band-limited to W Hz can be recovered from its uniformly spaced samples provided that the sampling period T_s satisfies the condition

$$T_s \leq [2W]^{-1}. \tag{13.9-1}$$

The definition of bandwidth here is the strict one, i.e., the number W such that $H(f) = 0$ for $|f| > W$, where $H(f)$ is the Fourier transform of $h(t)$. There are other versions of the sampling theorem that enable signal recovery from unequally spaced samples or from samples of higher derivatives of $h(t)$. In general, however, the interpolating functions in these cases are more complicated than the simple $\mathrm{sinc}(\cdot)$ functions that are used in the uniform sampling theorem. An important extension of the basic sampling theorem is its high-frequency variation that states that a modulated waveform of bandwidth W can be recovered from samples obtained at the rate furnished in Eq. (13.9-1) even though the highest frequency in the signal is the sum of carrier frequency f_c and W.

The uniform sampling theorem is easily extended to two dimensions. Let us consider a function $f(x, y)$ which is low-pass with cutoff vector $\mathbf{B}_0 = (B_x, B_y)$; then the region in the uv plane for which $F(u, v) = \mathfrak{F}_2[f(x,y)]$ is not zero (i.e., supp F) can be framed by a rectangle of length $2B_x$ along the u direction and length $2B_y$ along the v direction, as shown in Fig. 13.9-1. Now construct the function

$$G(u, v) = F(u, v) * \sum_{n=-\infty}^{\infty} \sum_{m=-\infty}^{\infty} \delta\left(u - \frac{m}{\Delta x}, v - \frac{n}{\Delta y}\right),$$

where $\Delta x \leq [2B_x]^{-1}$ and $\Delta y \leq [2B_y]^{-1}$. The region representing supp G is shown shaded in Fig. 13.9-2. Clearly we can recover the spectrum $F(u, v)$ from $G(u, v)$ by simply passing $G(u, v)$ through a linear, low-pass 2-D filter with two-dimensional transfer function $H(u, v)$ given by

$$H(u, v) = \mathrm{rect}\left(\frac{u}{2B_x}\right) \mathrm{rect}\left(\frac{v}{2B_y}\right). \tag{13.9-2}$$

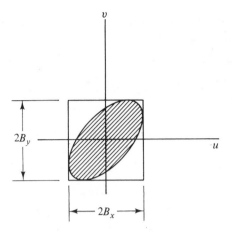

Figure 13.9-1. Support of a two-dimensional band-limited function (crosshatched region).

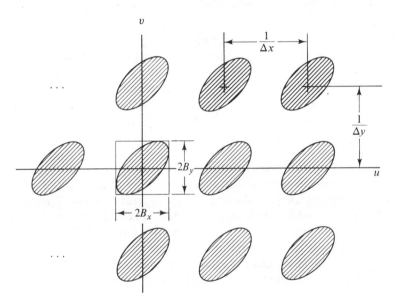

Figure 13.9-2. Support of $G(u, v)$.

Hence we can write the following identity:

$$F(u, v) = G(u, v)H(u, v)$$

$$= \left[F(u, v) * \sum_{n=-\infty}^{\infty} \sum_{m=-\infty}^{\infty} \delta\left(u - \frac{m}{\Delta x}, v - \frac{n}{\Delta y}\right) \right]$$

$$\cdot \text{rect}\left(\frac{u}{2B_x}\right) \text{rect}\left(\frac{v}{2B_y}\right). \tag{13.9-3}$$

We leave it as a simple exercise for the reader to show that the inverse Fourier transform of Eq. (13.9-3) is

$$f(x, y) = 4B_x B_y \, \Delta x \, \Delta y$$

$$\cdot \sum_{n=-\infty}^{\infty} \sum_{m=-\infty}^{\infty} f(m \, \Delta x, n \, \Delta y) \, \text{sinc} \, 2B_x(x - m \, \Delta x) \, \text{sinc} \, 2B_y(y - n \, \Delta y).$$

$$(13.9\text{-}4)$$

It is clear from Fig. 13.9-2 that the recovery of $F(u, v)$ requires that $[\Delta y]^{-1} \geq 2B_y$ and $[\Delta x]^{-1} \geq 2B_x$. The limiting case occurs when

$$[\Delta y]^{-1} = 2B_y$$
$$[\Delta x]^{-1} = 2B_x,$$

$$(13.9\text{-}5)$$

in which case the two-dimensional sampling theorem takes its simplest form; i.e.,

$$f(x, y) = \sum_{n=-\infty}^{\infty} \sum_{m=-\infty}^{\infty} f\left(\frac{m}{2B_x}, \frac{n}{2B_y}\right) \, \text{sinc} \, (2B_x x - m) \, \text{sinc} \, (2B_y y - n).$$

$$(13.9\text{-}6)$$

If the spacing between samples is greater than given in Eq. (13.9-5), that is, $\Delta x \geq [2B_x]^{-1}$ or $\Delta y \geq [2B_y]^{-1}$, then *aliasing* occurs, and recovery of $f(x,y)$ from the samples is not possible.

Equations (13.9-4) and (13.9-6) are fundamental results and are used in image processing by computer, radio astronomy, and two-dimensional spectral analysis. It is useful to recall that a band-limited function can never be limited in the xy domain. Similarly, a function limited in the xy domain can never be truly band-limited. Nevertheless, if $f(x, y)$ is band-limited, then the portion of $f(x, y)$ extending over the region $-x_M \leq x \leq x_M$ and $-y_M \leq y \leq y_M$ can approximately be specified by $(2x_M)(2y_M)(2B_x)(2B_y) = 16x_M y_M B_x B_y$ complex numbers or $32x_M y_M B_x B_y$ real numbers. This product is called the *space-bandwidth* product of the truncated form of $f(x, y)$. If most of the significant energy in $f(x, y)$ is within the region $|x| < x_M$ and $|y| < y_M$, then the contributions of the weighted sinc ($\cdot$) terms from outside this region will be small, and the approximate reconstruction of $f(x, y)$ from the $32x_M y_M B_x B_y$ real numbers will, at least away from the edges, be generally excellent.

Other Forms of the Two-Dimensional Sampling Theorem

Suppose we had framed the region containing supp F by the smallest possible circle of radius B, rather than a rectangle, as shown in Fig. 13.9-3. We can develop an alternative form of the sampling theorem by constructing, as before, the function

$$G(u, v) = F(u, v) * \sum_{n=-\infty}^{\infty} \sum_{m=-\infty}^{\infty} \delta(u - 2mB, v - 2nB). \quad (13.9\text{-}7)$$

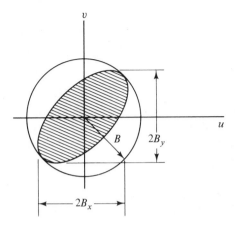

Figure 13.9-3. Relating to an alternative form
of the two-dimensional sampling theorem.

Recovery of $F(u, v)$ is possible if the signal $g(x,y)$ is passed through a linear
filter with transfer function given by

$$H(u, v) = \text{circ}\left(\frac{\sqrt{u^2 + v^2}}{B}\right),\tag{13.9-8}$$

so that we have the identity

$$F(u, v) = \left[F(u, v) * \sum_{n=-\infty}^{\infty}\sum_{m=-\infty}^{\infty} \delta(u - 2mB, v - 2nB)\right]\text{circ}\left(\frac{\sqrt{u^2 + v^2}}{B}\right).\tag{13.9-9}$$

The 2-DFT inverse of Eq. (13.9-9) gives

$$f(x, y) = \frac{\pi}{2}\sum_{n=-\infty}^{\infty}\sum_{m=-\infty}^{\infty} f\left(\frac{m}{2B}, \frac{n}{2B}\right)\frac{J_1\{2\pi B\sqrt{[x - (m/2B)]^2 + [y - (n/2B)]^2}\}}{2\pi B\sqrt{[x - (m/2B)]^2 + [y - (n/2B)]^2}}.\tag{13.9-10}$$

Equation (13.9-10) uses an interpolating function of the form $x^{-1}J_1(x)$.
Depending on how we choose to frame supp F, we can have various expansions for $f(x, y)$.

13.10   Two-Dimensional Random Variables and Processes

Two-dimensional scalar random variables are used to model situations
where random variations are associated with two arguments† such as position is the xy plane. Thus a two-dimensional r.v. is a scalar function of two

---

†In the terminology of Sec. 10.2, a two-dimensional random variable would have two
indexing sets.

variables. This nomenclature should not be confused with two-dimensional *vector* r.v.'s such as

$$\mathbf{X} = (X_1, X_2)^T, \qquad (13.10\text{-}1)$$

where X_1, X_2 are random variables and T denotes transpose. The following are examples of two-dimensional random variables:

1. The amplitude transmittance $t(x, y)$ of uniformly exposed film; this quantity is associated with the *grain noise* at (x, y).
2. The solar emission observed on Earth† $S_\lambda(x, y)$ at wavelength λ, at coordinates (x, y).
3. The gray-tone function (intensity transmittance) of a chest x-ray film.
4. The brightness function† $B(x, y)$ of a TV picture.
5. The acoustic signal† $S(x, y)$ received by a passive sonar at bearing x and elevation y.

In all these cases we see that the random variable, say $Z(\cdot)$, depends on the pair (x, y). Such a pair can be associated with a point in the xy plane, which is the terminus of the vector $\mathbf{s}$ given by

$$\mathbf{s} = x\mathbf{i}_x + y\mathbf{i}_y, \qquad (13.10\text{-}2)$$

where $\mathbf{i}_x$ and $\mathbf{i}_y$ are orthogonal unit vectors parallel to the coordinate axes. For any given vector $\mathbf{s}$, $Z(\mathbf{s})$ is the random variable describing the random phenomenon at (x, y). When $\mathbf{s}$ is used to index Z, a collection of random variables is formed, one for each value of $\mathbf{s}$. This collection of random variables constitutes a *two-dimensional random process*. In optics, $Z(\mathbf{s})$ is frequently a complex random variable, and the process $\{Z(\mathbf{s})\}$ is therefore a complex random process. A *wide-sense stationary* 2-D random process requires the following:

$$\text{(i)} \quad \overline{Z(\mathbf{s})} = Z_0 \qquad \text{independent of } \mathbf{s}; \qquad (13.10\text{-}3)$$

i.e., the expected value of Z does not depend on the argument.

$$\text{(ii)} \quad \overline{Z(\mathbf{s} + \mathbf{l})Z^*(\mathbf{s})} \equiv R_Z(\mathbf{s} + \mathbf{l}, \mathbf{s})$$
$$= R_Z(\mathbf{l}); \qquad (13.10\text{-}4)$$

i.e., the autocorrelation function of $\{Z(\mathbf{s})\}$ is only a function of the difference of the two vectors.

$$\text{(iii)} \quad \overline{|Z(\mathbf{s})|^2} < \infty. \qquad (13.10\text{-}5)$$

It is frequently useful to deal with the zero-mean random variable $z(\mathbf{s}) = Z(\mathbf{s}) - Z_0$. The covariance of $\{Z(\mathbf{s})\}$ is the autocorrelation function of $\{z(\mathbf{s})\}$ and is given by

$$R_z(\mathbf{l}) = R_Z(\mathbf{l}) - |Z_0|^2. \qquad (13.10\text{-}6)$$

†At a given instant of time.

From the definition of $R_Z(\mathbf{l})$, it follows that $R_Z(\mathbf{l}) = R_Z^*(-\mathbf{l})$. Hence in rectangular coordinates (where $\mathbf{l} = \alpha \mathbf{i}_\alpha + \beta \mathbf{i}_\beta$), there results

$$R_Z(\alpha, \beta) = R_Z^*(-\alpha, -\beta) \tag{13.10-7}$$

from which it follows that

$$R_Z(\alpha, -\beta) = R_Z^*(-\alpha, \beta)$$
$$R_Z(-\alpha, \beta) = R_Z^*(\alpha, -\beta). \tag{13.10-8}$$

In many situations, where additional information might be lacking, it is reasonable to model the covariance function $R_z(\mathbf{l})$ as a function only of the magnitude of $\mathbf{l}$; i.e., the covariance function is a function of a scalar. In this case the statistics are independent of the direction of the coordinate axes, and the covariance depends only on the distance from the origin of the α, β coordinate axes. Hence, with γ denoting the magnitude of $\mathbf{l}$, for this case we can write:

$$R_z(\alpha, \beta) = R_z(\gamma), \quad \text{where } \gamma = (\alpha^2 + \beta^2)^{1/2}, \tag{13.10-9}$$

and since

$$R_z(\alpha, \beta) = R_z(-\alpha, -\beta) = R_z^*(\alpha, \beta), \tag{13.10-10}$$

the covariance function $R_z(\alpha, \beta)$ is a real function. When the covariance may be written as a function of the single argument γ as in Eq. (13.10-9), the random process is said to be *isotropic*. The covariance of an isotropic process possesses circular symmetry in the sense that the loci of constant $R_z(\gamma)$ are circles about the origin.

Power Spectrum

The two-dimensional spectral density of a stationary random process is the Fourier transform of its autocorrelation $R(\alpha, \beta)$; i.e.,†

$$W(u, v) = \int_{-\infty}^{\infty} \int_{-\infty}^{\infty} R(\alpha, \beta) e^{-j2\pi(u\alpha + v\beta)} \, d\alpha \, d\beta. \tag{13.10-11}$$

The power spectrum is a real function of the arguments u, v. This is easily seen by considering the conjugate of Eq. (13.10-11) and using the fact that $R^*(\alpha, \beta) = R(-\alpha, -\beta)$. $W(-u, v)$ is given by

$$W(-u, v) = \int_{-\infty}^{\infty} \int_{-\infty}^{\infty} R(\alpha, \beta) e^{-j2\pi(-u\alpha + v\beta)} \, d\alpha \, d\beta$$

$$= \int_{-\infty}^{\infty} \int_{-\infty}^{\infty} R(-\alpha, \beta) e^{-j2\pi(u\alpha + v\beta)} \, d\alpha \, d\beta. \tag{13.10-12}$$

Since $R(-\alpha, \beta)$ is not generally equal to $R(\alpha, \beta)$, it may be concluded from the uniqueness property of the Fourier transform that $W(-u, v)$ is not equal

---

†We dispense with the subscript z since only a single r.v. is involved.

to $W(u, v)$. An identical argument can be applied to show that $W(u, v)$ is not equal to $W(u, -v)$.

An exception occurs in the isotropic case, for which $R(\alpha, \beta) = R(\sqrt{\alpha^2 + \beta^2}) \equiv R(\gamma)$. Following the procedure in Eq. (13.4-5), we write

$$
\begin{aligned}
\alpha &= \gamma \cos \theta \\
\beta &= \gamma \sin \theta \\
u &= \rho \cos \zeta \\
v &= \rho \sin \zeta.
\end{aligned}
\tag{13.10-13}
$$

Equation (13.10–11) may then be written as

$$W(\rho \cos \zeta, \rho \sin \zeta) = 2\pi \int_0^\infty \gamma R(\gamma) J_0(2\pi\rho\gamma) \, d\gamma \equiv W(\rho). \tag{13.10-14}$$

In the isotropic case, $W(u, v)$ is an even function with respect to both arguments.† The inverse relation is

$$R(\gamma) = 2\pi \int_0^\infty \rho W(\rho) J_0(2\pi\rho\gamma) \, d\rho \tag{13.10-15}$$

[see Eq. (13.4-10)].

Examples of Two-Dimensional Processes

The Random Checkerboard Process (RCP)

The RCP is defined as the limit, with $N \to \infty$, of the process

$$f_N(x, y) = \sum_{k=-Q}^{Q} \sum_{l=-Q}^{Q} c_{kl} \operatorname{rect}\left(\frac{x - k\Delta}{\Delta}\right) \operatorname{rect}\left(\frac{y - l\Delta}{\Delta}\right), \tag{13.10-16}$$

where $2Q + 1$ is the number of squares on a side, $N = (2Q + 1)^2$, Δ is the width of a square, and

$$c_{kl} = \begin{cases} 1, & \text{with probability } \mu \\ 0, & \text{with probability } 1 - \mu. \end{cases} \tag{13.10-17}$$

The Fourier transform of Eq. (13.10-16) gives

$$
\begin{aligned}
F_N(u, v) &= \Delta^2 \operatorname{sinc} u\Delta \operatorname{sinc} v\Delta \\
&\quad \cdot \sum_{k=-Q}^{Q} \sum_{l=-Q}^{Q} c_{kl} e^{-j2\pi(uk + vl)\Delta},
\end{aligned}
\tag{13.10-18}
$$

and the expected value of the square magnitude of Eq. (13.10-18) is

$$
\begin{aligned}
\overline{|F_N(u, v)|^2} &= (\Delta \operatorname{sinc} u\Delta)^2 (\Delta \operatorname{sinc} v\Delta)^2 \\
&\quad \cdot \sum_{k=-Q}^{Q} \sum_{l=-Q}^{Q} \sum_{m=-Q}^{Q} \sum_{n=-Q}^{Q} \overline{c_{kl} c_{mn}} e^{-j2\pi u(k - m)\Delta} \\
&\quad \cdot e^{-j2\pi v(l - n)\Delta},
\end{aligned}
\tag{13.10-19}
$$

†That is, $W(u, v) = W(-u, -v) = W(-u, v) = W(u, -v)$.

where

$$\overline{c_{kl}c_{mn}} = \begin{cases} \mu, & k = m, \, l = n \\ \mu^2, & \text{otherwise.} \end{cases}$$

We leave it to the reader to show that Eq. (13.10-19) can be written as

$$\overline{|F_N(u, v)|^2} = (\Delta \text{ sinc } u\Delta)^2(\Delta \text{ sinc } v\Delta)^2$$
$$\cdot \left\{ \mu(1 - \mu)N + \mu^2 N^2 \left[\frac{\text{sinc}^2 u(2Q + 1)\Delta}{\text{sinc}^2 u\Delta} \right]\left[\frac{\text{sinc}^2 v(2Q + 1)\Delta}{\text{sinc}^2 v\Delta} \right] \right\}.$$

$$(13.10\text{-}20)$$

The power spectrum, $W_{RC}(u, v)$, of the RC process is then

$$W_{RC}(u, v) = \lim_{N\to\infty} \frac{\overline{|F_N(u, v)|^2}}{N\Delta^2} = \Delta^2\mu(1 - \mu) \text{ sinc}^2 u\Delta \text{ sinc}^2 v\Delta$$
$$+ \mu^2\delta(u, v). \qquad (13.10\text{-}21)$$

A realization of the *RC* process is shown in Fig. 13.10-1(a). The sample power spectrum is shown in Fig. 13.10-1(b). For fixed Δ, μ is the parameter needed to characterize the process; μ is simply the expected value of the process.

The correlation function of this process is the inverse Fourier transform of Eq. (13.10-21). This turns out to be

$$R_{RC}(\alpha, \beta) = \mu(1 - \mu)\left(1 - \frac{|\alpha|}{\Delta}\right)\left(1 - \frac{|\beta|}{\Delta}\right) \text{rect}\left(\frac{\alpha}{2\Delta}\right)$$
$$\cdot \text{rect}\left(\frac{\beta}{2\Delta}\right) + \mu^2. \qquad (13.10\text{-}22)$$

The RC process is wide-sense stationary but *not isotropic*. The RC is an important 2-D process, and it is used in film-grain noise modeling [13-10] and studies in visual perception [13-11]. Some interesting experimental effects related to the RC model are discussed in [13–12].

The Overlapping Circular Grain (OCG) Process

A realization of the OCG process is shown in Fig. 13.10-2. It consists of circular grains of uniform transmittance (say unity) sprinkled on a black background. The number of grain centers per unit area is Poisson distributed.

The analysis of the OCG process is considerably more difficult than that of the RC process, and we shall not do this here. The derivation of the correlation function is somewhat tedious, and the reader can consult references for help, e.g., [13-13], p. 115. By using the fact that this process has isotropic statistics, it can be shown that the correlation function is given by

$$R_{OCG}(\gamma) = \begin{cases} \mu^2 \exp\left[\pi R^2 g(\gamma)d\right], & \gamma \le 2R \\ \mu^2, & \gamma > 2R, \end{cases} \qquad (13.10\text{-}23)$$

(a)

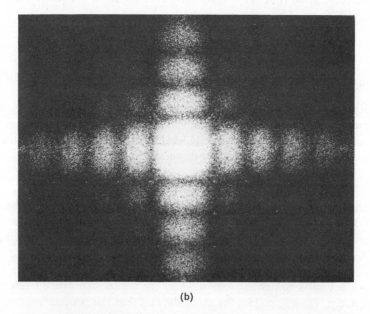

(b)

Figure 13.10-1. (a) Realization of an RC process; (b) sample power spectrum of the process in (a).

where μ is the mean value of the OCG process and is given by

$$\mu = e^{-\pi R^2 d} \tag{13.10-24}$$

Figure 13.10-2. Realization of the OCG model.

and where R is the radius of the cell, d is the Poisson parameter and is the average number of grain centers per unit area, and $g(\gamma)$ is proportional to the convolution of two circ ($\cdot$) functions of radius R each and has the value

$$g(\gamma) = \frac{2}{\pi}\left[\cos^{-1}\left(\frac{\gamma}{2R}\right) - \left(\frac{\gamma}{2R}\right)\sqrt{1 - \left(\frac{\gamma}{2R}\right)^2}\right]. \qquad (13.10\text{-}25)$$

This model is widely used for modeling film-grain noise ([13-5] and [13-14]). A sample power spectrum is shown in Fig. 13.10-3. The waveform on the

Figure 13.10-3. Sample power spectrum of OCG process.

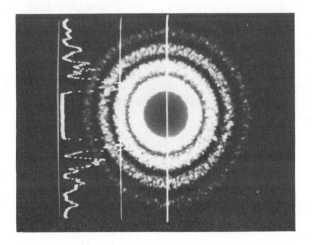

left shows the brightness level sensed by the scan line in the center. The center region is blocked out to protect the video circuitry against excessive light. Other lines are for video reference only.

13.11 Two-Dimensional Linear Systems (2-DLS's)

The basic input-output relations for 2-DLS's are, for the most part, direct extensions of the one-dimensional theory. It is customary to represent a system by an operator, say $\mathcal{C}\{\ \}$, that acts on an input signal f_1 to produce an output signal f_2. For two-dimensional systems we shall define the input signal coordinate system by the pair (x_1, y_1) and the output pair by (x_2, y_2). Given any f_1 from the domain of $\mathcal{C}$, there is an associated f_2 which is generated by the system. This is essentially the meaning of the statement

$$f_2(x_2, y_2) = \mathcal{C}\{f_1(x_1, y_1)\}, \tag{13.11-1}$$

where $f_1(x_1, y_1)$ is the input signal and $f_2(x_2, y_2)$ is the output signal.

In what follows we shall discuss only linear operators. A linear operator is defined by

$$\mathcal{C}\{\alpha_1 f_1 + \beta_1 g_1\} = \alpha_1 \mathcal{C}\{f_1\} + \beta_1 \mathcal{C}\{g_1\}, \tag{13.11-2}$$

where f_1 and g_1 are any inputs in the domain of $\mathcal{C}$ and α_1 and β_1 are any constants. When the system is linear, we can compute the response of the system to an arbitrary input if we know how the system responds to certain elementary functions into which the input can be decomposed. One of the most useful decompositions of a function is in terms of two-dimensional Dirac δ functions [13-15], which are not functions in the strict sense of the word.† This is a direct extension of the approach taken in Sec. 3.2. The decomposition of $f_1(x_1, y_1)$ in terms of δ functions is furnished by the sifting integral, which can also be used as the definition of the δ function. Thus we write

$$f_1(x_1, y_1) = \int_{-\infty}^{\infty} \int_{-\infty}^{\infty} f_1(\xi, \eta)\delta(x_1 - \xi, y_1 - \eta)\, d\xi\, d\eta, \tag{13.11-3}$$

which indicates that $f_1(x_1, y_1)$ is a linear combination of shifted δ functions with appropriate weights. Note that this is a direct extension of the familiar one-dimensional result

$$f(t) = \int_{-\infty}^{\infty} f(\xi)\delta(t - \xi)\, d\xi. \tag{13.11-4}$$

Use of Eq. (13.11-3) in Eq. (13.11-1), together with the linearity property

---

†See, for example, the appendix in [13-15].

expressed in Eq. (13.11-2), enables us to write

$$f_2(x_2, y_2) = \mathcal{Q}[f_1(x_1, y_1)]$$

$$= \mathcal{Q}\left[\int_{-\infty}^{\infty}\int_{-\infty}^{\infty} f_1(\xi, \eta)\delta(x_1 - \xi, y_1 - \eta)\, d\xi\, d\eta\right]$$

$$= \int_{-\infty}^{\infty}\int_{-\infty}^{\infty} f_1(\xi, \eta)\mathcal{Q}[\delta(x_1 - \xi, y_1 - \eta)]\, d\xi\, d\eta. \quad (13.11\text{-}5)$$

We assign the symbol $h(\cdot)$ to $\mathcal{Q}[\delta(\cdot)]$, i.e.,

$$h(x_2, y_2; \xi, \eta) \equiv \mathcal{Q}[\delta(x_1 - \xi, y_1 - \eta)], \quad\quad\quad (13.11\text{-}6)$$

and call $h(\cdot)$ the impulse response of the system. The term impulse response is quite appropriate since it is precisely the output of the system when the input is a δ function displaced to $x_1 = \xi, y_1 = \eta$. In optics $h(\cdot)$ is often called the point-spread function of the system.

The integral

$$f_2(x_2, y_2) = \int_{-\infty}^{\infty}\int_{-\infty}^{\infty} f_1(\xi, \eta)h(x_2, y_2; \xi, \eta)\, d\xi\, d\eta \quad\quad (13.11\text{-}7)$$

is conceivably the most important integral in 2-DLS theory: It is known as the *superposition integral* and enables us to compute any f_2 in the range of $\mathcal{Q}$ for any f_1 in the domain of $\mathcal{Q}$ provided that we know $h(x_2, y_2; \xi, \eta)$. But knowing $h(\cdot)$ is indeed a tall order because it requires that we place an impulse at $\xi = x_1, \eta = y_1$ and measure the response at every pair (x_2, y_2) in the output plane. We must then repeat this process for every (x_1, y_1) in the input plane in order to get a complete specification of $h(\cdot)$. The term *plane* is used in a general sense here simply to mean the appropriate set of points, i.e., input or output.

Because of the practical difficulty of completely specifying a linear system by the extremely large family $\{h(\cdot)\}$, we check first whether our system can be modeled as an *invariant* system [Eq. (13.11-8)]. Invariant systems constitute the most important subclass of linear systems because (1) the specification of such systems is much more concise and (2) these systems can be analyzed by transform techniques. So great are the advantages of dealing with invariant systems that even when the system is not invariant over the whole input plane we find it advantageous to divide the input plane into a number of smaller regions, say N, within which the system is invariant. This enables us to specify the system by a family of N functions $\{h_i(\cdot), (x_i, y_i) \in \mathcal{X}_i, i = 1, \ldots, N\}$, where $\mathcal{X}_i$ is the ith region within which the system is invariant.

An invariant system has the property that

$$h(x_2, y_2; \xi, \eta) = h(x_2 - \xi, y_2 - \eta), \quad\quad\quad (13.11\text{-}8)$$

which implies that if we shift the input δ function by a certain amount, the output shifts by the same amount; this property removes the need to put δ

functions all over the input plane and then measure the corresponding responses in the output plane in order to specify the system. If we use Eq. (13.11-8) in Eq. (13.11-7), we obtain the two-dimensional convolution integral

$$f_2(x_2, y_2) = \int_{-\infty}^{\infty} \int_{-\infty}^{\infty} f_1(\xi, \eta) h(x_2 - \xi, y_2 - \eta) \, d\xi \, d\eta. \qquad (13.11\text{-}9)$$

The 2-DFT of both sides of Eq. (13.11-9) gives

$$F_2(u, v) = F_1(u, v) H(u, v), \qquad (13.11\text{-}10)$$

where F_2 is the output spectrum, F_1 is the input spectrum, and H is the *transfer function* in the uv frequency plane.

2-DLS computations can be made using either Eq. (13.11-9) or Eq. (13.11-10) followed by an inverse 2-DFT. The considerations as to which to use in a particular situation are the same as in the one-dimensional case. When transforms and inverse transforms are easily computed or are tabulated, then Eq. (13.11-10) is preferable. Otherwise the direct approach should be considered [Eq. (13.11-9)].

Stability

Before turning to examples, we briefly discuss the stability of two-dimensional systems. An invariant 2-DLS is said to be stable if its response to any bounded input is bounded. Thus if $|f_1(x_1, y_1)| < M$, then $|f_2(x_2, y_2)| < MI$, where I does not depend on f_1. With I defined by

$$I = \int_{-\infty}^{\infty} \int_{-\infty}^{\infty} |h(\xi, \eta)| \, d\xi \, d\eta < \infty, \qquad (13.11\text{-}11)$$

we have

$$|f_2(x_2, y_2)| = \left| \int_{-\infty}^{\infty} \int_{-\infty}^{\infty} f_1(x_2 - \xi, y_2 - \eta) h(\xi, \eta) \, d\xi \, d\eta \right|$$

$$\leq M \int_{-\infty}^{\infty} \int_{-\infty}^{\infty} |h(\xi, \eta)| \, d\xi \, d\eta = MI. \qquad (13.11\text{-}12)$$

Hence stability in the above sense is ensured, provided Eq. (13.11-11) is valid.†

13.12   Image Processing

As far back as the early 1950's it became evident that many of the problems occurring in optical image formation bore a strong resemblance to the problems encountered in communication theory such as filtering, detection,

---

†Since space rather than time variables are usually involved in 2-D systems, stability is much less of a problem and therefore of less interest than in systems in which time is the independent variable.

and estimation ([13-16] and [13-17]). To a large extent modern optics and communication theory have become overlapping fields, an event that is reflected in the fact that much of the research in optical image processing is done by scientists and engineers with an electrical communication theory background.

Modern image processing is often categorized as optical or digital, although a third category—hybrid (optical-digital) processing—is now recognized. Digital image processing received much impetus from the discovery of the fast Fourier transform (FFT), which enables two-dimensional convolutions to be rapidly evaluated by frequency-domain techniques. Thus to carry out the operation

$$g = f * h \quad \text{[Eq. (13.3-10)]} \tag{13.12-1}$$

one uses the FFT to compute

$$\mathcal{F}_2[f] = F(u, v) \tag{13.12-2}$$

$$\mathcal{F}_2[h] = H(u, v), \tag{13.12-3}$$

so that

$$G(u, v) = F(u, v)H(u, v). \tag{13.12-4}$$

The FFT is then used to obtain the filtered image as

$$g(x, y) = \mathcal{F}_2^{-1}[FH]. \tag{13.12-5}$$

The operations given in Eqs. (13.12-1)–(13.12-5) are typical of the operations used in image coding, image enhancement, image-noise reduction, two-dimensional matched filtering, and pattern recognition. Nonlinear operations, such as are needed to correct nonlinear aberrations in images, are also easily implemented with a digital computer. Basically, the most compelling reasons for using a digital computer for image processing are (1) the computational and logical flexibility and (2) the accuracy that is provided by the computer.

Optical image processing is frequently done with coherent, i.e., laser-illuminated, systems. Such illumination is characterized by a high degree of monochromaticity with regions of highly correlated phase. Coherent optical systems furnish the flexibility of realizing a large number of complex-valued transfer functions and bipolar impulse responses. Optical processors have two significant advantages over digital computers: (1) They furnish a parallel processing capability, and (2) the input image need not be digitized for processing. The parallel processing capability is especially important if there is a great deal of image data to be processed. Once the system is assembled and the required filter is realized (which may not be a trivial task), the optical system processes the image at the speed of light. On the minus side, optical systems lack the computational flexibility of a digital computer, especially vis-à-vis logical and desired nonlinear operation. Also they are generally nowhere nearly so accurate.

The most widely used coherent optical configuration is shown in Fig. 13.12-1; it is discussed in [13-18], which, at the time of its publication,

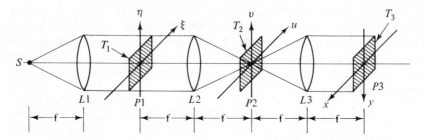

Figure 13.12-1. Coherent optical system. *S* is the source of light; *L*1 is the collimating lens; *P*1 is the input plane; *L*2 and *L*3 are two identical Fourier-transforming lenses; *P*2 is the Fourier plane; *P*3 is the output plane.

stirred considerable interest. The most important property of this system is that it uses frequency-domain techniques to realize convolutions and correlations. For simplicity we shall assume that the lenses are all identical and that their focal length is f. The source of coherent illumination, S, is typically a helium-neon laser of modest power, i.e., one milliwatt to a fraction of a watt. Lens $L1$ furnishes a parallel or *collimated* beam of light which is essentially a section of a uniform amplitude, monochromatic, plane wave. The image to be filtered (henceforth called the *object* T_1) is placed in plane $P1$, which corresponds to the *front-focal plane* of lens $L2$. We ask the reader to take it on faith that the optical field produced in the plane $P2$ is the two-dimensional Fourier transform of the transmittance $f(\xi, \eta)$ of the object. In other words, in the *spatial frequency uv* plane, there results a field proportional to

$$F(u, v) = \int_{-\infty}^{\infty} \int_{-\infty}^{\infty} f(\xi, \eta) e^{-j2\pi(u\xi + v\eta)} \, d\xi \, d\eta. \qquad (13.12\text{-}6)$$

This result is known as the *Fourier transform property* (FTP) of a lens and is derived in many places ([13-1], p. 83; [13-19], p. 115). Because of the FTP, the *back-focal plane* of $L2$ is also known as the Fourier plane of lens $L2$. The spatial frequencies *uv* are directly proportional to displacement and their units are cycles per millimeter.

The transparency T_2 is the filter. It contains a complex transmittance proportional to the transfer function $H(u, v)$, where $H(u, v)$ is the 2-DFT of the desired impulse response $h(\xi, \eta)$. A method of realizing $H(u, v)$ through the interference and recording of coherent light waves is credited to A. B. Vander Lugt [13-20]. The term *transmittance* here has the same meaning as in one-dimensional signal theory; i.e., it is the ratio of the output signal level to the input. Here, of course, the signals are light waves rather than voltage levels. Hence immediately after T_2, there is a field proportional to

$$G(u, v) \equiv F(u, v)H(u, v), \qquad (13.12\text{-}7)$$

which is identical to Eq. (13.12-4).

Finally there remains the job of computing the inverse Fourier transform of Eq. (13.12-7) in order to realize

$$g = f * h. \qquad (13.12\text{-}8)$$

The inverse transform computation is done by lens $L3$, which produces the Fourier transform of $G(u, v)$ in the plane $P3$. Without the reflection of the xy system in $P3$, the output field would be proportional to

$$g(-x, -y) = \mathcal{F}_2[G(u, v)]. \qquad (13.12\text{-}9)$$

For this reason the xy system is reflected as shown in Fig. 13.12-1. In this system, what is obtained is

$$g(x, y) = \mathcal{F}_2^{-1}[G(u, v)], \qquad (13.12\text{-}10)$$

which is the desired output. The filtered image is stored on T_3 for a hard-copy record, or directly viewed by a human observer, or scanned by a TV-type camera. In the last case, the TV video signal can be digitized and sent to a computer for additional processing. Systems using both optical and digital computers are sometimes called hybrid image processors [13-21].

We shall give some examples of the type of operations easily implemented with a coherent optical system of the type shown in Fig. 13.12-1.

Low-Pass Filtering; Image Smoothing

A mask with a rectangular hole centered about the origin represents an ideal two-dimensional low-pass filter (Fig. 13.12-2). It is described by a transfer function

$$H(u, v) = \text{rect}\left(\frac{u}{2u_0}\right) \text{rect}\left(\frac{v}{2v_0}\right). \qquad (13.12\text{-}11)$$

With $f(\xi, \eta)$ denoting the input, the smoothed image appearing in $P3$ is given by

$$g(x, y) = 4u_0 v_0 \int_{-\infty}^{\infty} \int_{-\infty}^{\infty} f(\xi, \eta) \, \text{sinc} \, 2u_0(x - \xi) \cdot \text{sinc} \, 2v_0(y - \eta) \, d\xi \, d\eta. \qquad (13.12\text{-}12)$$

Band-Pass Filtering; Modulation

An ideal band-pass filter is shown in Fig. 13.12-3. The output image in this case is also smoothed in both the x and y directions but has superimposed on it a one-dimensional sinusoidal pattern whose spatial frequency is u_c

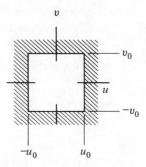

Figure 13.12-2. Ideal two-dimensional low-pass filter. The shaded area represents the opaque region, while the clear area has unity transmittance.

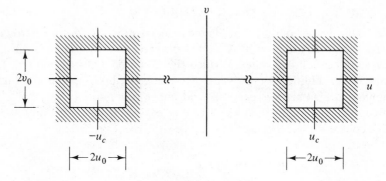

Figure 13.12-3. Filter which is band-pass in the *u* direction and low-pass in the *v* direction.

cycles per millimeter in the *x* direction. The band-pass filtered image is described by

$$g(x, y) = 8u_0v_0 \int_{-\infty}^{\infty} \int_{-\infty}^{\infty} f(\xi, \eta) \operatorname{sinc} 2u_0(x - \xi) \operatorname{sinc} 2v_0(y - \eta)$$

$$\cdot \cos 2\pi u_c(x - \xi) \, d\xi \, d\eta. \tag{13.12-13}$$

If a sinusoidal grating $t(\xi, \eta) = \frac{1}{2} + \frac{1}{2} \cos 2\pi u_c \xi$ is placed next to the object $f(\xi, \eta)$ and the frequency-plane filter of Fig. 13.12-3 is used, a two-dimensional spatial DSBSC signal is generated in $P3$.

Image Deblurring

Suppose that an object described by $f_0(\xi, \eta)$ is blurred in a manner such that the blurring can be described by a space-invariant blurring function

$b(\xi, \eta)$. The blurred image is then given by

$$f_b(\xi, \eta) = \int_{-\infty}^{\infty} \int_{-\infty}^{\infty} f_0(\alpha, \beta) b(\xi - \alpha, \eta - \beta) \, d\alpha \, d\beta. \qquad (13.12\text{-}14)$$

The field in the Fourier plane is

$$F_b(u, v) = F_0(u, v) B(u, v), \qquad (13.12\text{-}15)$$

where $F_b = \mathcal{F}_2[f_b]$, $F_0 = \mathcal{F}_2[f_0]$, and $B = \mathcal{F}_2[b]$. If the Fourier plane filter $H(u, v)$ is proportional to $B^{-1}(u, v)$, the filtered image is

$$g(x, y) = \mathcal{F}_2^{-1}[F_0(u, v) B(u, v) B^{-1}(u, v)]$$
$$= f_0(x, y). \qquad (13.12\text{-}16)$$

Thus the ideal image is reconstructed. This technique is also called *inverse filtering* and is widely used when the input scene is relatively noise-free. When noise is present, the processed image is often very noisy. An example of this type of image processing is given in [13-22].

Visualization of Phase Objects

An object with transmittance

$$f(\xi, \eta) = e^{j\phi(\xi, \eta)} \qquad (13.12\text{-}17)$$

is said to be a *phase object*. It is not visible to the eye because the eye or any other sensor responds only to the intensity $|f(\xi, \eta)|^2$, which in this case is unity. To see such objects, a spatial filtering technique known as central dark field is often used. Assume that $\phi(\xi, \eta) \ll 1$. Then

$$e^{j\phi(\xi, \eta)} \simeq 1 + j\phi(\xi, \eta), \qquad (13.12\text{-}18)$$

and its Fourier transform is

$$F(u, v) \simeq \delta(u, v) + j\Phi(u, v), \qquad (13.12\text{-}19)$$

where $\Phi(u, v) = \mathcal{F}_2[\phi(\xi, \eta)]$. Let the Fourier plane mask consist of a small opaque disk at $u = v = 0$. The transfer function associated with such a structure is

$$H(u, v) \simeq \begin{cases} 0, & u = v = 0 \\ 1, & \text{otherwise.} \end{cases} \qquad (13.12\text{-}20)$$

Then

$$G(u, v) = F(u, v) H(u, v) = j\Phi(u, v) \qquad (13.12\text{-}21)$$

and

$$g(x, y) = j\phi(x, y). \qquad (13.12\text{-}22)$$

The intensity in the recorded image is $|\phi(x, y)|^2$, which is easily discernible to the eye. Another technique is the so-called Schlieren method, described below.

Schlieren Method for Observing Phase Objects

Again let $f(\xi, \eta) = e^{j\phi(\xi, \eta)}$. The Fourier plane "mask" in this case consists of a knife edge that blocks out one half of the Fourier plane light (Fig. 13.12-4). The transmittance of the knife edge can be written as

$$H(u, v) = \tfrac{1}{2}[1 + \text{sgn}(u)]. \tag{13.12-23}$$

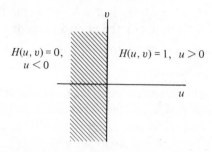

$H(u, v) = 0,$ $H(u, v) = 1, \ u > 0$
$u < 0$

Figure 13.12-4. Knife edge in the Fourier plane that blocks half of the Fourier plane light. $H(u, v)$ is the associated transfer function.

This mask will generate a Hilbert transform component in the reconstructed image. If $\phi(\xi, \eta) \ll 1$, the intensity of the reconstructed image is

$$|g(x, y)|^2 = \frac{1}{4}\left[1 - \frac{2}{\pi}\int_{-\infty}^{\infty} \frac{\phi(\xi, y)}{x - \xi}\, d\xi\right]. \tag{13.12-24}$$

Hence the image intensity is seen to depend on $\phi(\xi, \eta)$. We shall leave the details as an exercise for the reader (Prob. 13-16).

13.13 Summary

In this chapter we have investigated the mathematical properties of two-dimensional signals and transforms. We showed that many of the input-output systems relations are direct extensions of the one-dimensional results derived in Chapters 2 and 3. We also found that 2-D functions with circular symmetry are determined by their one-dimensional profile functions and that the Hankel transform of the 2-D functions is the ordinary 1-D Fourier transform of their profiles.

We saw that the 2-D sampling theorem can take several interesting forms and that the concepts of high-pass and low-pass in 2-D are more subtle than in the one-dimensional case and need to be carefully defined.

We considered several 2-D random processes and discussed the important subclass of *isotropic* processes. We showed that for the latter the correlation and power spectrum functions depend on only one variable.

Finally we considered several applications of 2-D image processing to low-pass and band-pass filtering, image deblurring, and phase-object visualization. Goodman [13-1] and O'Neill [13-13] discuss image-processing systems from a communication-theoretic viewpoint. Papoulis [13-2] furnishes a complete discussion of 2-D Fourier transforms and associated theory.

PROBLEMS

13-1. Prove Eqs. (13.3-1)–(13.3-3).

13-2. Prove Eqs. (13.3-6)–(13.3-9).

13-3. Show that if $f(x, y)$ is purely imaginary, then the real part, $A(u, v)$, and the imaginary part, $B(u, v)$, of its Fourier transform obey

$$A(u, v) = -A(-u, -v)$$
$$B(u, v) = B(-u, -v).$$

13-4. a. Write an expression for the transmittance, $t(x, y)$, of the doubly infinite object shown in Fig. P13-4. The clear areas have unity trans-

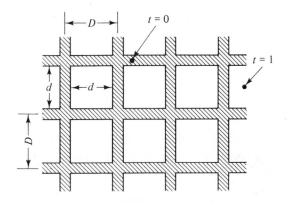

Figure P13-4

mittance, while the shaded areas have zero transmittance. This kind of object is an example of a two-dimensional *diffraction grating* of use in optics.

b. Compute the Fourier transform of $t(x, y)$, and sketch its magnitude.

13-5. Let $f(x, y)$ be a double periodic function in x and y with periods α and β, respectively; i.e.,

$$f(x, y) = f(x + m\alpha, y + n\beta)$$

for any m, n and any x, y. Let $f(x, y)$ be represented by the Fourier series

$$\sum_{m=-\infty}^{\infty} \sum_{n=-\infty}^{\infty} c_{mn} e^{j2\pi(mx/\alpha + ny/\beta)}.$$

Show that the coefficients c_{mn} are given by

$$c_{mn} = \frac{1}{\alpha\beta} \int_{-\alpha/2}^{\alpha/2} \int_{-\beta/2}^{\beta/2} f(x, y) e^{-j2\pi(mx/\alpha + ny/\beta)} \, dx \, dy.$$

13-6. Derive the expressions in Table 13.6-1.

13-7. Derive Eqs. (13.7-3) and (13.7-4).

13-8. Show that the 2-DFT of a function $f(x, y)$ that can be transformed into $f(r)e^{jm\theta}$ [for example, through use of the transformation of Eq. (13.4-5)] is given by

$$\tilde{F}_m(\rho)e^{jm(\phi - \pi/2)},$$

where $\tilde{F}_m(\rho)$ is a "generalized" Hankel transform of order m, i.e.,

$$\tilde{F}_m(\rho) = 2\pi \int_0^\infty r f(r) J_m(2\pi r \rho) \, dr,$$

and $J_m(2\pi r \rho)$ is the Bessel function of the first kind, of order m.

13-9. a. Explain why an arbitrary function $f(x, y)$ considered as a function of r and θ can be expanded as a Fourier series as follows:

$$f(r\cos\theta, r\sin\theta) = \sum_{m=-\infty}^{\infty} f_m(r)e^{jm\theta}.$$

b. Using the results from Prob. 13-8, show that the Fourier transform of an arbitrary function $f(x, y)$ can be written as an infinite sum of Hankel transforms:

$$F(\rho\cos\phi, \rho\sin\phi) = \sum_{m=-\infty}^{\infty} \tilde{F}_{mm}(\rho)e^{jm(\phi - \pi/2)},$$

where $\tilde{F}_{mm}(\rho)$ is the mth-order Hankel transform of $f_m(r)$.

13-10. Using the results of Prob. 13-9, part b, compute the response of a two-dimensional circularly symmetric system to an arbitrary input

$$f(r\cos\theta, r\sin\theta) = \sum_{m=-\infty}^{\infty} f_m(r)e^{jm\theta}.$$

Write the result as

$$g(r\cos\theta, r\sin\theta) = \sum_{m=-\infty}^{\infty} g_{mm}(r)e^{jm\theta},$$

and compute the coefficients g_{mm}.

*13-11. Establish Eq. (13.8-17), and use this result to show that

$$\mathcal{L}_2[f(x, y)g(x, y)] = f(x, y)\mathcal{L}_2[g(x, y)],$$

where $f(x, y)$ is low-pass and $g(x, y)$ is high-pass and strongly separable from $f(x, y)$.

13-12. Use the results from Sec. 13.5 to establish the result that

$$\int_0^\infty \sin ay J_0(ry)\, dy = \begin{cases} \dfrac{1}{\sqrt{a^2 - r^2}}, & r < a \\ 0, & r < a. \end{cases}$$

13-13. Let $f_i(x, y)$, $i = 1, 2$, be two arbitrary functions, real or complex, space-limited to a region R. Establish the 2-D version of the Schwarz inequality, which states that

$$\left| \iint_R f_1(x, y) f_2(x, y)\, dx\, dy \right|^2 \le \int_R |f_1(x, y)|^2\, dx\, dy \int_R |f_2(x, y)|^2\, dx\, dy.$$

Hint: Write

$$\iint_R f_1(x, y) f_2(x, y)\, dx\, dy = \left| \iint_R f_1(x, y) f_2(x, y)\, dx\, dy \right| e^{j\theta},$$

where θ is the phase angle of the integral. Then expand the expression

$$\iint_R |f_1(x, y) - \lambda e^{j\theta} f_2^*(x, y)|^2\, dx\, dy > 0,$$

where λ is a real constant, and consider the roots of the quadratic in λ.

13-14. The impulse response of a two-dimensional system is given by

$$h(x, y) = u_0 v_0 \operatorname{sinc}(x u_0) \operatorname{sinc}(y v_0).$$

a. Compute the transfer function $H(u, v) = \mathcal{F}_2[h(x, y)]$. In optics $H(u, v)$ is known as the *coherent* optical transfer function of the system, and $h(x, y)$ is known as the *point-spread function*.

b. Compute $G(u, v) = H(u, v) \circledast H(u, v)$, where $\circledast$ is the correlation product. In optics $G(u, v)$ is known as the (unnormalized) *incoherent* optical transfer function. Plot $G(u, v)$, and compare the cutoff frequencies for the coherent vs. incoherent case. Which system furnishes a higher cutoff frequency?

13-15. Holography is a method of reconstructing an object from its energy spectrum. Let $t(x, y)$ represent the object, and consider the quantity

$$g(x, y) = t(x - x_0, y - y_0) + a\delta(x, y).$$

Assume that $t(x, y)$ is space limited; i.e.,

$$t(x, y) = t(x, y) \operatorname{rect}\left(\frac{x}{D}\right) \operatorname{rect}\left(\frac{y}{D}\right).$$

Explain how, if x_0 and y_0 are large enough, $t(x, y)$ can be recovered from the energy spectrum

$$S(u, v) = |\mathcal{F}_2[g(x, y)]|^2 \equiv |G(u, v)|^2.$$

Note that the energy spectrum can be recorded on photographic film. *Hint*: Consider the inverse transform of $S(u, v)$.

13-16.   Derive Eq. (13.12-24), which is valid for Schlieren imaging when the object $f(\xi, \eta) = \exp[j\phi(\xi, \eta)]$ satisfies $\phi(\xi, \eta) \ll 1$.

REFERENCES

[13-1]   J. W. GOODMAN, *Introduction to Fourier Optics*, McGraw-Hill, New York, 1968.

[13-2]   A. PAPOULIS, *Systems and Transforms with Application in Optics*, McGraw-Hill, New York, 1968.

[13-3]   H. BATEMAN, *Tables of Integral Transforms*, Vol. II, McGraw-Hill, New York, 1954.

[13-4]   M. ABRAMOWITZ and I. A. STEGUN, eds., *Handbook of Mathematical Functions*, Dover, New York, 1965.

[13-5]   B. PICINBONO, "Modele Statistique Suggèré par la Distribution de Grains d'Argent dans les Film Photographique," *Compt. Ren.*, **240**, pp. 2206–2208, 1955.

[13-6]   H. STARK, "An Extension of the Hilbert Transform Product Theorem," *Proc. IEEE*, **59**, No. 9, pp. 1359–1360, Sept. 1971.

[13-7]   P. BEDROSIAN and H. STARK, "Comment on 'An Extension of the Hilbert Transform Product Theorem'," *Proc. IEEE*, **60**, pp. 228–229, No. 2, Feb. 1972.

[13-8]   E. T. WHITTAKER, "On the Functions Which Are Represented by the Expansion of the Interpolation-Theory," *Proc. Roy. Soc. Edinburgh*, **35**, pp. 181–194, 1915.

[13-9]   C. E. SHANNON, "Communication in the Presence of Noise," *Proc. IRE*, **37**, pp. 10–21, Jan. 1949.

[13-10]   J. W. GOODMAN, "Film Grain Noise in Wavefront-Reconstruction Imaging," *J. Opt. Soc. Am.*, **57**, pp. 493–502, April 1967.

[13-11]   B. JULESZ, *Foundations of Cyclopean Perception*, University of Chicago Press, Chicago, 1971, Chap. 8.

[13-12]   H. STARK and F. B. TUTEUR, "Higher Orders in the Diffraction Pattern of Random Scenes in Coherent Optical Systems," *J. Opt. Soc. Am.*, **63**, pp. 675–685, June 1973.

[13-13]   E. L. O'NEILL, *Introduction to Statistical Optics*, Addison-Wesley, Reading, Mass., 1963.

[13-14]   M. SAVELLI, "Resultats Pratiques de l'Etude d'un Modele à Trois Paramètres pour la Representation des Propriétés Statistiques de la Granularité des Film Photographiques et Notamment de Propriétés Spectrales," *Compt. Ren.*, **246**, pp. 3605–3608, 1958.

[13-15] A. PAPOULIS, *The Fourier Integral and Its Applications*, McGraw-Hill, New York, 1965.

[13-16] P. ELIAS, "Optics and Communication Theory," *J. Opt. Soc. Am.*, **43**, pp. 229–232, 1953.

[13-17] E. L. O'NEILL, "Spatial Filtering in Optics," *IRE Trans. Information Theory*, **IT-2**, pp. 56–65, 1956.

[13-18] L. J. CUTRONA, E. N. LEITH, C. J. PALERMO, and L. J. PORCELLO, "Optical Data Processing and Filtering Systems," *IRE Trans. Information Theory*, **IT-6**, pp. 386–400, 1960.

[13-19] R. J. COLLIER, C. B. BURCKHARDT, and L. H. LIN, *Optical Holography*, Academic Press, New York, 1971.

[13-20] A. B. VANDER LUGT, "Signal Detection by Complex Spatial Filtering," *IEEE Trans. Information Theory*, **IT-10**, pp. 139–145, 1964.

[13-21] H. STARK, "An Optical-Digital Approach to the Pattern Recognition of Coal-Workers' Pneumoconiosis," *IEEE Trans. Systems, Man, Cybernetics*, **SMC-7**, pp. 788–793, Nov. 1976.

[13-22] B. A. KRUSOS, "Restoration of Radiographic Images by Optical Spatial Filtering," *Opt. Eng.*, **13**, pp. 208–218, May–June 1974.

Appendix

An Example of a Communication System: Viking I

Late in the summer of 1975, the National Aeronautics and Space Administration (NASA) launched two unmanned spacecrafts to carry out a series of experiments designed to find traces of biological life as it exists, or may have existed, on Mars. The spacecrafts, christened Viking I and II, both consist of two connected sections—an Orbiter and Lander—which separate in Martian orbital flight. The Orbiter remains in a synchronous Martian orbit, which amounts to one revolution every 24.6 hours. The Lander continued on to a soft landing on Mars and remained in radio contact with the Orbiter for approximately 20 minutes each day. During the period of radio contact, the Lander transmitter—a 30-watt, 400-MHz device—relayed information to the Orbiter at variable bit rates up to 16,000 bits sec. The term bit, short for binary digit, has a slightly different meaning here from that in information theory (see Chapter 1). A pulse that has one of two levels (i.e., ASK), one of two frequencies (FSK), or one of two phases (PSK) represents *one bit*.

The information received at the Orbiter is recorded and relayed to Earth, either immediately or after some delay, at 4000 bits/sec. The reduced information rate in the Orbiter-earth link is due to the great distance involved. An FSK modulation scheme is used for the Lander-Orbiter channel. The reason for choosing FSK over PSK even though the latter offers a 3-dB advantage† has to do with existing multipath problems on Mars. FSK, being incoherent, is much less susceptible to multipath distortion than PSK, where loss of coherence is a major problem.

---

†FSK signals form an orthogonal signal set. PSK signals form an *antipodal* signal set. For the same probability of error, PSK signals have a 3-dB advantage over FSK. The theory is developed in Chapter 11.

The FSK system is derived from two, mutually incoherent, crystal oscillators. Crystal oscillators are used because of their extreme stability. Signaling consists of switching between two frequencies. However, since the carrier signals are incoherent with each other, there is no phase continuity at the instant of a carrier switch; therefore, the spectrum of the incoherent FSK signal drops off slowly ($\sim 1/f$), and excessive bandwidth is required. To avoid this problem, a makeshift coherence is established by the following scheme. The two oscillator outputs, say $x_1(t) = \cos(\omega_1 t + \theta_1)$ and $x_2(t) = \cos(\omega_2 t + \theta_2)$, are mixed together to produce a down-converted beat signal at frequency $\omega_2 - \omega_1$. The down-converted beat signal is proportional to $x(t) = \cos[(\omega_2 - \omega_1)t + \theta_2 - \theta_1]$. At the maximum value of $x(t)$, $\arg x_1(t) = \arg x_2(t)$, and at that instant the two signals have the same phase and are said to be *phase coherent*.[†] The instants $\{t_n\}$ at which phase coherence occurs are related to the instants $\{\tilde{t}_n\}$ at which positive-slope zero crossings are detected, according to

$$t_n = \tilde{t}_n + \frac{\pi}{2(\omega_2 - \omega_1)}.$$

It is more convenient to determine $\tilde{t}_n$ than t_n. If signaling is confined during the $\{t_n\}$, the spectral spread of the FSK is reduced and roll-off tends to be like $1/f^2$ rather than $1/f$. At the Orbiter, the FSK is demodulated with a discriminator.

The Orbiter-Earth channel is a one-way *down-link*, i.e., Mars to Earth, and the modulation is PSK. The nominal carrier frequency is 2.2 GHz. The spectral band 1.55–5.2 GHz is called S band, and the Orbiter-Earth link is called the S-band relay.

In addition to the one-way Lander-Orbiter-Earth channel, there is a *two-way* direct link between the Lander and Earth. Called the S-band *direct* link, it provides for direct transmission of high-volume scientific and imaging data and up-link (Earth-to-Mars) command signals. The S-band direct link uses a 63-lb, 113-W transmitter/receiver operating at a nominal carrier frequency of 2.2 GHz. Earth commands are PSK pulses at 4 bits/sec on a 380-Hz subcarrier. Phase coherence is derived from the up-link signals with the help of a 20-Hz bandwidth phase-lock loop. Loss of coherence for the Viking is expected to be commonplace—at least once per Martian day. Reestablishment of coherence is done by a maneuver at the Earth station. A transmitted signal is swept slowly through the receiver frequency aperture. When the transmitted input frequency coincides with the idler frequency of the Lander receiver VCO, phase lock occurs. Since the Earth station doesn't know when this happens, the transmitted signal is swept through the whole receiver range and returned to center.

[†]A loose term actually. True phase coherence implies a phase-lock condition between two signals.

Viking Lander Communications

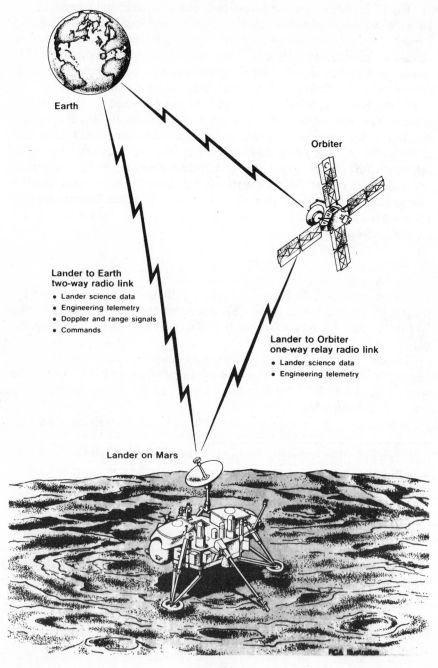

Earth

Orbiter

**Lander to Earth
two-way radio link**

- Lander science data
- Engineering telemetry
- Doppler and range signals
- Commands

**Lander to Orbiter
one-way relay radio link**

- Lander science data
- Engineering telemetry

Lander on Mars

RCA Illustration

Figure A-1

Scientific information is sent on the S-band direct link at rates of 250 to 1000 bits/sec. Approximately 2 hr of transmission to earth are required each day to return the scientific and engineering data. PSK modulation is used here also. Because of the 36-min round-trip transmission time and the great length of the channel, command signals are kept to a minimum, and the Lander's functions have been programmed to occur automatically.

The S-band direct link uses two antennas to communicate with Earth. An S-band low-gain antenna with a broad directivity pattern is used to receive signals from earth. This antenna weighs only a few ounces. It does not have to be pointed directly at Earth to receive signals.

Another antenna—an S-band high-gain antenna with a narrow directivity pattern—is also in use. Consisting of a 30-in. parabolic dish of traditional appearance, the antenna is under servo control to aim it in the direction of Earth for maximum effective transmission.

Additional antennas are located on the Lander and Orbiter for transmission on the Lander-Orbiter and Orbiter-Earth links. A sketch of the Viking communication links is shown in Fig. A-1.

We wish to thank Dr. Max Feryzska [A-1] for supplying much of the information in this Appendix.

REFERENCES

[A-1]   Personal communication with Dr. Max Feryzska of RCA Astro-Electronics, Princeton, N.J., 1976.

Index

Index